AF348978

Springer Geophysics

For further volumes:
http://www.springer.com/series/10173

Vadim Surkov • Masashi Hayakawa

Ultra and Extremely Low Frequency Electromagnetic Fields

 Springer

Vadim Surkov
National Research Nuclear
 University MEPhI
Moscow, Russia

Masashi Hayakawa
Advanced Wireless Communications
 research Center
The University of Electro-Communications
Tokyo, Japan

ISBN 978-4-431-54366-4   ISBN 978-4-431-54367-1 (eBook)
DOI 10.1007/978-4-431-54367-1
Springer Tokyo Heidelberg New York Dordrecht London

Library of Congress Control Number: 2014938022

Preface

This book deals with the electromagnetic fields observed on the ground and in space in the ULF (ultra low frequency, $f < 3$ Hz) and ELF (extremely low frequency, $3 < f < 3 \times 10^3$ Hz) bands. It recently has been agreed that the observation of these ULF/ELF electromagnetic fields can provide us with important information not only on the upper atmosphere (magnetosphere/ionosphere), but also on the lower atmosphere and the Earth's crust (lithosphere).

The main emphasis of ULF/ELF electromagnetic fields in early studies was on large-scale geomagnetic variations, and their sources are thought to be due to the interaction of solar wind with the Earth's magnetosphere. This interaction results in the generation of magnetohydrodynamical (MHD) waves, perturbation of the ionospheric plasma due to the solar radiation, precipitation of energetic particles, and so on. As a parallel development, the Earth–ionosphere resonance cavity and the class of geomagnetic field-line resonances became an important branch of ULF/ELF field studies. The properties of those ULF geomagnetic variations or geomagnetic pulsations have been well understood and were used as a tool of "Space Weather", in which the parameters of the ionosphere/magnetosphere have been continuously monitored in real time in order to forecast coming disasters affecting ground-based technological human systems (pipelines, satellite communications, GPS techniques, and so on) and also satellite operation. With the recent discovery of the ionospheric Alfvén resonator (IAR) in the topside ionosphere, an entirely new realm of ULF oscillations became an interesting topic for geophysicists.

There has been a lot of progress in the last few decades in the field of local and seismogenic effects, which can be associated with large-scale rock deformations and tectonic activity deep in the Earth's crust. These effects have been extensively examined both in the laboratory and in the field. Especially among many seismogenic phenomena, ULF/ELF effects are the most promising candidates for short-term predictions of natural disasters such as earthquakes, volcano eruptions, tsunami waves, and so on.

In this book we place a major emphasis on physical mechanisms and sources of these natural ULF/ELF electromagnetic fields. During the course of the book, some of these mechanisms of magnetospheric origin are discussed in detail because they are fundamental. The global electromagnetic resonances excited by lightning activity and recently discovered sprites, blue jets, and other gigantic mesospheric discharges are our priority. Moreover, much attention will be paid to the entirely new studies of electromagnetic phenomena caused by rock deformation/fracture including the seismoelectric and geomagnetic perturbations and other ULF/ELF effects possibly associated with natural disasters.

This book is written for the broad geophysical community, scientists, lecturers, post- and undergraduates as well as for physicists interested in geophysical prospects of ULF/ELF electromagnetic fields. In order to create no difficulties in a book intended for general readers, we try to offer a simple and clear presentation of the physical picture by omitting the complicated calculations, which will be found in the Appendices. To understand the main ideas and contents of this book, the reader is required to have a basic knowledge of electrodynamics (as from a standard education course) and a general understanding of the Earth and its environment.

Moscow, Russia   Vadim Surkov
Tokyo, Japan   Masashi Hayakawa

Contents

Introduction

Special credit has been given in past decades to the study of ultra and extremely low frequency (ULF and ELF) electromagnetic fields of natural and man-made origins. In early studies the main emphasis was on largescale geomagnetic variations with periods from tenths of seconds to several minutes. The sources of these variations were thought to be the interaction of solar wind with the Earth's magnetosphere resulting in the generation of magnetohydrodynamic (MHD) waves, perturbations of the ionospheric plasma due to solar radiation, precipitation of energetic particles and so on. In a parallel development, the Earth–ionosphere resonance cavity and the class of geomagnetic field-line resonances became an important branch of ULF/ELF field studies due to their role in spectra of global electromagnetic resonances. With the discovery of the ionospheric Alfvén resonator (IAR) located in the topside ionosphere, a new realm of ULF oscillations became amenable to scientific study.

Considerable recent attention has been focused on local electromagnetic effects, which can be associated with largescale rock deformations and tectonic activity deep in the Earth's crust. These effects have been extensively examined both in laboratory and in situ measurements for several decades. An investigation of these phenomena can be of great importance for nonseismic prediction of impending natural disasters such as earthquakes, volcano eruptions, and tsunamis.

In this book we put the major emphasis on physical mechanisms and sources of these ULF/ELF natural electromagnetic noises. Some of these mechanisms of magnetospheric origin will be treated in detail and others in a more sketchy fashion. The interested reader is referred to the books cited in the text for details about the ULF/ELF fields of the magnetospheric origin. The global electromagnetic resonances excited by lightning activity and recently discovered sprites, blue jets and other gigantic mesospheric discharges are our priority. Much emphasis is put on studies of electromagnetic phenomena caused by rock deformation/fracture including the ULF/ELF effects possibly associated with tectonic activity, earthquakes and other natural disasters. One of the challenges of this research is to know enough about electric structure and physical processes in the rocks deep in the Earth's crust.

Such disciplines as electrodynamics, plasma physics, gas dynamics, magnetohydrodynamics, and rock and fluid mechanics among others, are required for adequate modeling of a variety of ULF/ELF effects. We cannot give a detailed review of all of these disciplines, but we try to give readers some insights into how these different sources and physical mechanisms may operate and how they may be coupled.

Part I deals with the basics of the Earth's electromagnetic field. After some introductory material, the study is begun in earnest with the origin of the Earth's electromagnetic field. Here the hydromagnetic dynamo is considered as a main source of the Earth's magnetic field. We study the interaction between solar wind and the Earth's magnetosphere including the key role played by magnetic storms in space weather and the Earth's environment. Basics of MHD waves in the magnetosphere/ionosphere are discussed. We cannot come close to exploring these topics in any detail, but we need a framework within which to organize the Earth's resonance cavity effects. Then we focus our attention on atmospheric electricity including conventional mechanisms for lightning discharge. Global lightning activity is treated as a driving force for the global electric circuit operating in the atmosphere. In the remainder of this part we deal with recently discovered gigantic electric discharges, also known as transient luminous events (TLEs), which occur above a large thunderstorm at stratospheric and mesospheric altitudes. Here the main emphasis is on the low frequency effects.

In Part II we study the global electromagnetic resonances in the ULF/ELF frequency range. We first deal with the structure and models of the Earth–ionosphere cavity resonator. Schumann resonances excited by global lightning activity in the Earth–ionosphere resonance cavity are studied. Then we deal with the IAR which is localized at altitudes below the one to two the Earth's radius. Models and possible physical mechanisms for IAR excitation are studied. The IAR resonance spectra observed at night and during the day are compared with those derived from the dispersion relation of the IAR. The nature of magnetospheric MHD resonances and ULF pulsations in the frequency range below 1 Hz are also discussed to some extent. Finally, we examine source mechanisms of natural electromagnetic ULF noises.

Part III covers a variety of electromagnetic phenomena associated with deformation and fracture of rocks. Local geomagnetic perturbations that originate from seismic wave propagation in conductive rocks are examined. Prominence is given to the electromagnetic noise resulting from crack formation in rocks. This part deals with laboratory studies of strong electric fields in cracks and pores, including electromagnetic and particle emissions from fracturing rocks and shock polarization effects. Electrokinetic and seismoelectric phenomena in water-saturated rocks are examined as well. In the remainder of this part we discuss electromagnetic phenomena associated with largescale natural and man-made disasters such as earthquakes, volcano eruptions, tsunamis, hurricanes and underground explosions. The lithosphere–atmosphere–ionosphere coupling, which is highlighted through these catastrophic phenomena, is emphasized.

Part I
Electromagnetic Field of the Earth

Chapter 1
The Earth's Magnetic Field

Abstract This chapter deals with the basics of the Earth's magnetic field. Hydromagnetic dynamo operating in the Earth's fluid outer core is treated as a main source of the Earth's magnetic field. Here we discuss the interaction between solar wind and the Earth's magnetosphere that forms the global magnetospheric configuration as well as the impacts of the magnetic storms on the Earth environment and space weather. Basics of Magnetohydrodynamic (MHD) waves in the magnetosphere/ionosphere are studied.

Keywords Hydromagnetic dynamo • Magnetic storm • Magnetosphere • Magnetohydrodynamic (MHD) waves • Solar wind

1.1 Origin of the Earth's Magnetic Field

1.1.1 The Earth's Interior Structure

In this section we deal first with an overview of the origin of the Earth's magnetic field. We start with a brief study of the internal construction of the Earth since a source of the geomagnetic field is contained (hidden) inside the Earth's core. According to modern seismic data, the Earth's interior can be split into three basic classes: solid crust, mantle and core which is in turn divided into outer core and inner core, as illustrated in Fig. 1.1 (Jeffreys 1970; Bott 1982; Zharkov 1983; Brown and Mussett 1993). The Earth's crust forms continents and oceanic basement. The effective depth of the crust is about 35 km. The crust is divided from the rocks of dense mantle with a sharp seismic boundary, which has been termed Mohorovičic boundary (e.g., see the book by Nikolaevskiy 1996). The velocities of longitudinal and transverse seismic waves enhance sharply at this boundary. The silicate cover or the Earth mantle lies in the range of depths 35–2885 km.

V. Surkov and M. Hayakawa, *Ultra and Extremely Low Frequency Electromagnetic Fields*, Springer Geophysics, DOI 10.1007/978-4-431-54367-1__1,
© Springer Japan 2014

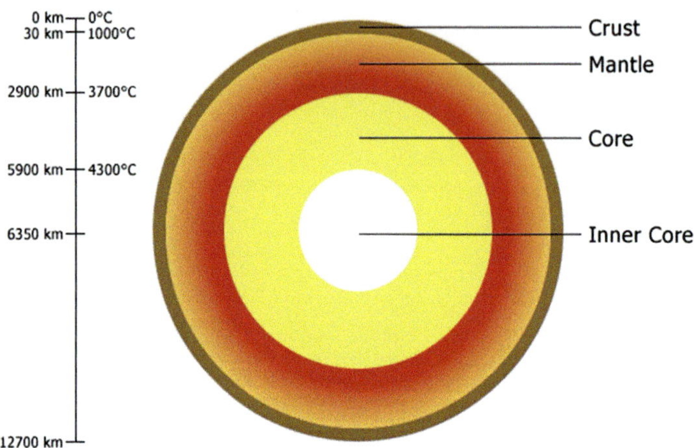

Fig. 1.1 Sketch of the Earth interior. Taken from the site http://blog.world-mysteries.com/science/earth-crust-displacement-and-the-british-establishment/

The presence of the molten core inside the Earth has been discovered by Gutenberg (1914). At the mantle–core interface the velocity of longitudinal seismic wave (P-wave) decreases abruptly from 13.6 km/s in the mantle down to 8.1 km/s in the core, while the velocity of transverse wave (S-wave) falls off stepwise with depth from 7.3 km/s up to zero. This means that the shear modulus inside the outer core equals zero and hence the core is liquid. The outer core is about 30 % of the Earth mass and the core radius is evaluated as 3,483 km. The outer core consists of the mixture of molted sulfur (12 %) and iron (88 %). The volume of molten material inside the core is five times as much as the Moon.

There is a solid inner core inside the outer core. The inner core is about 1.7 % of the Earth mass and the core radius is evaluated as $1,240 \pm 10$ km. The inner core consists of ferro-nickel-alloy (80 % Fe and 20 % Ni). Present-day estimation of the parameters in the Earth center is as follows: the pressure is 3.6×10^{11} Pa, mass density is 12.5×10^3 kg/m^3, and temperature is about 5.3×10^3 K, that is close to the temperature of the Sun surface. The core conductivity is assumed to be $(7 \pm 2) \times 10^5$ S/m.

According to contemporary data, the heat flow from the Earth's interior is not only due to thermal conductivity but also due to convection of the melted material. The heated material of the core goes up due to its decreased density. At the boundary between the core and mantle the upgoing material transmits part of its energy to the overlying rock followed by the material cooling. This results in the material lowering caused by the increases of its density. Additionally, the convection builds up as a result of light fractions floating up and heavy fractions falling off. Consolidation of fluid around the solid inner core followed by evolving of latent heat leads to an enhancement of the convection. It thus appears that the main source of the movements inside the outer core is the decrease of gravitational energy that is accompanied by the growth of the inner core. It is now unclear whether the convection is laminar or turbulent. At present the Earth's interior is not adequately explored to clear up this extremely important question.

The flow character is dependent on the Reynolds number $\mathrm{Re} = lV/\nu$, where l and V are characteristic space scale and velocity of the flow, and ν is the kinematical viscosity. So, this parameter is very important for the mechanism of the Earth's magnetic field generation. The numerical assessment of the flow velocity for both the laminar and turbulent convection gives approximately the same value $V = 0.1$–$0.2\,\mathrm{cm/s}$ (Golitsyn 1981; Stevenson 1983). Zeldovich and Ruzmaikin (1987) have used the higher estimate $V = 4\,\mathrm{cm/s}$. The characteristic size of the flow is of the order of the Earth's core radius, that is $l = 10^3\,\mathrm{km}$. The kinematical viscosity varies within interval of $10^{-7} < \nu < 10^5\,\mathrm{m^2/s}$, that presents great difficulties in estimating the Reynolds number. It is usually believed that the kinematical viscosity possesses the value $\nu = 10^{-6}\,\mathrm{m^2/s}$, which is close to the lower limit. The Reynolds number then exceeds 10^6; that means that the convection must be turbulent. Nevertheless, the large-scale flow inside the core flow gives no comprehensive evidence yet of having turbulent structure. In conclusion we note that the laminar flows have been studied in more detail and numerous dynamo-solutions for the steady-state velocity fields have been derived (e.g., see Brodsky 1983; Moffat 1968; Roberts 1971; and references therein).

1.1.2 Magnetohydrodynamic (MHD) Equations

To treat the electric and magnetic fields, Maxwell's electrodynamics equations are required. The first pair of these equations in their full form are given by

$$\nabla \times \mathbf{B} = \mu_0 \left(\mathbf{j} + \partial_t \mathbf{D} \right), \tag{1.1}$$

and

$$\nabla \times \mathbf{E} = -\partial_t \mathbf{B}, \tag{1.2}$$

where $\mathbf{B}$ is the induction of magnetic field, $\mathbf{E}$ is the electric field, $\mathbf{j}$ is the conduction current density, $\partial_t \mathbf{D}$ is the displacement current density, μ_0 denotes the magnetic constant/magnetic permeability of free space, and the symbol ∂_t stands for the partial time-derivative, that is $\partial_t = \partial/\partial t$. In what follows we only consider a non-magnetic medium so that the magnetic permeability of the medium is equal to unity.

The next pair of Maxwell's equations are given by

$$\nabla \cdot \mathbf{B} = 0, \tag{1.3}$$

and

$$\nabla \cdot \mathbf{D} = \rho_e, \tag{1.4}$$

where ρ_e is the electric charge density.

It is usually the case that the conduction current in the conducting fluid is much greater than the displacement current, i.e., $j \gg \partial_t D$, so that Eq. (1.1) is reduced to

$$\nabla \times \mathbf{B} = \mu_0 \mathbf{j}. \tag{1.5}$$

The electric and magnetic fields in the above equations are measured in a fixed or motionless coordinate system, K. Considering a conducting fluid moving at mass velocity $\mathbf{V}$, we first use the reference frame, K', moving with mass velocity $\mathbf{V}$ of the fluid. Transformation between two coordinate systems moving at relative velocity $\mathbf{V}$ causes the transformation of the electromagnetic field. According to Jackson (2001), in a nonrelativistic limit, while as $|\mathbf{V}| \ll c$, where c stands for the light speed in the vacuum, the electromagnetic field in the moving frame becomes

$$\mathbf{E}' = \mathbf{E} + \mathbf{V} \times \mathbf{B}, \tag{1.6}$$

$$\mathbf{B}' = \mathbf{B} \text{ and } \mathbf{j}' = \mathbf{j}, \tag{1.7}$$

where the primed variables are those measured in the moving frame, K', and the unprimed variables are measured in the motionless frame, K. Transformation between two reference frames leaves *Maxwell*'s equation invariant while the *Ohm*'s law has the form

$$\mathbf{j}' = \sigma \mathbf{E}', \tag{1.8}$$

which is valid only for the K' coordinate system. Here σ denotes the conductivity coefficient. It should be noted that the Ohm's law given by Eq. (1.8) is valid if the temporal and spatial dispersion in the medium can be neglected.

Substituting Eqs. (1.6)–(1.8) into Eq. (1.5) yields

$$\nabla \times \mathbf{B} = \mu_0 \sigma \left(\mathbf{E} + \mathbf{V} \times \mathbf{B} \right). \tag{1.9}$$

This implies that there is an electric field only if the conductor moves or the term $\partial_t \mathbf{B}$ in Eq. (1.2) is nonzero.

Now we consider the basic equations describing the dynamics of conducting fluid. The principle of conservation of fluid mass means that the fluid flux into or out of a volume through its surface must be equal to the rate of mass variations inside the volume. This mathematical statement in the integral form can be converted to the differential equation known as the continuity equation, which relates the fluid mass density ρ and the fluid velocity $\mathbf{V}$ through (e.g., see Kelley 1989)

$$\partial_t \rho + \nabla \cdot (\rho \mathbf{V}) = 0. \tag{1.10}$$

This equation can be reduced to the form

$$\frac{d\rho}{dt} + \rho\left(\nabla \cdot \mathbf{V}\right) = 0, \tag{1.11}$$

where the total time derivative

$$d/dt = \partial_t + \mathbf{V} \cdot \nabla. \tag{1.12}$$

The principle of conservation of momentum relates the fluid velocity to the forces acting on the fluid. This equation must include the familiar terms such as the pressure gradient, the gravitational and "viscous" forces. The presence of an electrical conductivity of the fluid requires the inclusion of volume force on the fluid given by $\mathbf{j} \times \mathbf{B}$; that is, by Ampére force. In the reference frame fixed to the Earth one should take into account the internal terms such as Coriolis force and centrifugal force resulted from the Earth spin. The equation of momentum in its general form can be written as

$$\rho d\mathbf{V}/dt = -\nabla P + \rho\mathbf{g} + \eta\nabla^2\mathbf{V} + \mathbf{j} \times \mathbf{B} + \mathbf{F}, \tag{1.13}$$

where $\rho\mathbf{g}$ describes the gravitational field, η is called the dynamic viscosity coefficient, and $\mathbf{F}$ stands for all the inertial forces acting on the fluid.

As the conducting fluid is immersed in an external magnetic field, the electric currents and fields can be developed from the hydrodynamic motion of the medium. The magnetic field forces the electric currents and thus may greatly affect the medium motion. The electric currents in turn change the magnetic field. In this notation, the interaction between the magnetic and hydrodynamic fields results in a complicated picture of Magnetohydrodynamic (MHD) flow. The set of MHD equations (1.2)–(1.4), (1.9), (1.10), and (1.13) can be supplemented by the equation of state and by the equation which is derived from the principle of conservation of energy. The interested reader is referred to the text by Kelley (1989) for details about these equations and for a more complete treatise on dissipative and viscous processes in conducting fluids.

The MHD approximation can be applied to a variety of physical objects such as a flow of conducting molten matter inside the Earth core, cosmic plasma in the solar wind, solar flares, the Earth's magnetosphere electrodynamics, and etc.

Applying a curl operator to Eq. (1.9) and substituting Eq. (1.2) for $\nabla \times \mathbf{E}$ into Eq. (1.9) yields

$$\partial_t \mathbf{B} = v_m \nabla^2 \mathbf{B} + \nabla \times (\mathbf{V} \times \mathbf{B}), \tag{1.14}$$

where $v_m = \left(\mu_0 \sigma\right)^{-1}$ is the magnetic diffusion coefficient measured in m^2/s.

Consider first the case of a still fluid so that Eq. (1.14) simplifies to

$$\partial_t \mathbf{B} = v_m \nabla^2 \mathbf{B}. \tag{1.15}$$

The implication of this equation, which is similar to the diffusion equation, is that the magnetic field can diffuse in the conducting media. In order to obtain the order-of-magnitude estimate of the characteristic time τ_d of the diffusion process one should give a rough assessment of the derivatives in Eq. (1.15)

$$\frac{\Delta B}{\tau_d} \sim v_m \frac{\Delta B}{l^2}. \tag{1.16}$$

Here ΔB is the perturbation of magnetic field and l is the characteristic space scale of perturbed region. Whence we get

$$\tau_d \sim l^2/v_m = \mu_0 \sigma l^2. \tag{1.17}$$

This estimate is representative for the diffusion processes since the time scale, τ_d, of magnetic field perturbation increases with the scale size l squared.

1.1.3   The Concept of "Frozen-In" Magnetic Field

It is usually the case that the magnetic field falls off with distance from the source due to Joule dissipation in a conducting medium. However the movement of conducting fluid may result in the amplification of original/inoculating magnetic field. The possibility of such phenomena follows from the concept of "frozen-in" magnetic field lines that have been discovered by *Alfvén* (e.g., see monograph by Alfvén 1950; Alfvén and Falthammar 1963, and references therein). The effect of "frozen-in" magnetic field arises when the temporal scale of magnetic field perturbations, τ_d, is large as compared to the variation time/period, T, of the velocity field. In the strict sense, this effect holds in an extreme case of infinite conductivity. This implies that the Joule dissipation can be neglected. Taking $\sigma \to \infty$ in Eq. (1.14) we get

$$\partial_t \mathbf{B} = \nabla \times (\mathbf{V} \times \mathbf{B}). \tag{1.18}$$

Expanding the curl operator and taking into account that $\nabla \cdot \mathbf{B} = \mathbf{0}$, Eq. (1.18) can also be written

$$\partial_t \mathbf{B} = (\mathbf{B} \cdot \nabla)\,\mathbf{V} - (\mathbf{V} \cdot \nabla)\,\mathbf{B} - \mathbf{B}\,(\nabla \cdot \mathbf{V}). \tag{1.19}$$

Now we use the principle of conservation of fluid mass given in Eq. (1.11). Eliminating $\nabla \cdot \mathbf{V}$ from Eqs. (1.19) and (1.11) gives

$$\partial_t \mathbf{B} - \frac{\mathbf{B}}{\rho}\partial_t \rho + (\mathbf{V} \cdot \nabla)\,\mathbf{B} - \frac{\mathbf{B}}{\rho}\,(\mathbf{V} \cdot \nabla)\,\rho = (\mathbf{B} \cdot \nabla)\,\mathbf{V}. \tag{1.20}$$

This equation can be rearranged to the form

$$\frac{d}{dt}\left(\frac{\mathbf{B}}{\rho}\right) = \left(\frac{\mathbf{B}}{\rho}\cdot\nabla\right)\mathbf{V},\qquad(1.21)$$

where the total time derivative is defined in Eq. (1.12).

Consider for the moment the "fluid line," which is labeled by "fluid particles" on the line (Landau and Lifshitz 1982). The "fluid line" moves together with the "fluid particles," which compose this line. Let $\delta\mathbf{l}$ be the length element of this line. We now study the temporal variations of this small element. If $\mathbf{V}$ is the mass velocity of the fluid at the end of the "fluid line," the mass velocity at another end of the "fluid line" must be $\mathbf{V} + (\delta\mathbf{l}\cdot\nabla)\,\mathbf{V}$. For the small time dt the increment of the length element achieves the value $dt\,(\delta\mathbf{l}\cdot\nabla)\,\mathbf{V}$, so the total time derivative of the length element is given by

$$\frac{d}{dt}\delta\mathbf{l} = (\delta\mathbf{l}\cdot\nabla)\,\mathbf{V}.\qquad(1.22)$$

As can be seen from Eqs. (1.21) and (1.22) the change of the vectors $\delta\mathbf{l}$ and $\mathbf{B}/\rho$ are determined by the same equation. In this sense, if these vectors are parallel at initial time, they are still parallel at any other time. Furthermore, the variations of their lengths are proportional to each other. This means that if two infinitely near "fluid particles" are situated at the same magnetic field line at initial time, they are still at the same field line at any other time. Moreover, the value B/ρ changes proportionally to the distance between these two "fluid particles."

Notice that this conclusion is valid not only for infinitely near points but also for the distant points situated at the same field line. This means that the magnetic field line moves together with the "fluid particles" related to this line. We usually say that if $\sigma \rightarrow \infty$, the magnetic field is "frozen in" and can be considered to move with the fluid. The quantity B/ρ varies in direct proportion to a tension of the "fluid line." In the case of incompressible flow, the mass density of moving "fluid particles" is unchanged. Whence it appears that the magnetic field B itself varies in direct proportion to the tension of the "fluid line."

When the frozen-in field takes place, the magnetic flux across an arbitrary surface S moving with the fluid velocity, that is a product of magnetic induction B and S, keeps a constant value. This may result in an enhancement of the magnetic field due to deformation or compression of the surface S. On the other hand, the flow may complicate the magnetic field lines thereby decreasing its scale size, which makes for the amplification of energy dissipation.

The concept of frozen-in field, which is very useful for the visualization of complex flow and field pattern, will be dealt with in a future study. As has already been stated, the concept of frozen-in field is valid for an extreme case of infinite conductivity. Nonetheless, the scale sizes of cosmic bodies are so large that the characteristic damping time τ_d of the magnetic field may be enormous, even though the medium conductivity is small or moderate. To illustrate this, consider the outer core of the Earth. Substituting the mean conductivity of the core $\sigma_g = 7 \times 10^5\,\mathrm{S/m}$ and the scale sizes $l = 10^3\,\mathrm{km}$ (Stacey 1969) into Eq. (1.17) gives the magnetic viscosity/diffusion coefficient $\nu_m \approx 1.1\,\mathrm{m^2/s}$ and the damping time of the Earth's magnetic field $\tau_d \approx 3 \times 10^4$ years. When the turbulent magnetic viscosity of about $4\,\mathrm{m^2/s}$ is taken into account, the damping time decreases up to 10^4 years (Parker 1979). If the characteristic time of the geomagnetic and mass velocity variations inside the core is smaller than 10^4 years, the concept of frozen-in field can be applied to the outer core.

1.1.4 A Simple Model of Hydromagnetic Dynamo

The geomagnetic field is mainly generated due to currents in the electrically conducting core of the Earth. The currents are in turn driven by the convection of molten matter in the core. This process is often called a dynamo in analogy to a motor-driven electric generator, which is capable of producing the electric current without wire and winding. Pioneering investigations of the dynamo mechanism were provided by Larmor (1919) who treated the nature of terrestrial and solar magnetism. The generally accepted theory states that the hydromagnetic dynamo is a principal origin of the Earth's magnetic field and that this mechanism is capable to sustain generation and amplification of the magnetic field due to hydrodynamical flow in the earth interior. It should be noted that the hydromagnetic dynamo is a self-excited generator, which can operate without maintaining the external supplies.

The effect of the conducting fluid flow on the magnetic field generation is controlled by the magnetic Reynolds number $\mathrm{Re}_m = lV/\nu_m$. This dimensionless parameter is on the order of ratio of the magnetic energy rate to the Joule dissipation of energy. As $\mathrm{Re}_m \gg 1$ then the magnetic field generation prevails over the energy dissipation due to the medium heating. Assuming the velocity of western drift of the matter in the Earth core $V = 0.3\,\mathrm{mm/s}$ and substituting the numerical parameters alluded to above we obtain that inside the core $\mathrm{Re}_m \sim 10^3$ (Parker 1979). Nevertheless, the condition $\mathrm{Re}_m \gg 1$ is insufficient for excitation of the hydromagnetic dynamo. In some sense, we need a topological complexity of the flow pattern as well. The turbulent flow is tangled enough so that the above requirement of topological complexity holds automatically.

A laminar flow must be nontrivial to produce a dynamo effect. For example, it was proved that the high symmetric flows of conducting media such as axially symmetric, centrally symmetric, and two-dimensional flows are unable to generate magnetic fields (Cowling 1934, 1953, 1976; Zeldovich 1956; Braginsky 1964; Zeldovich et al. 1983).

Surprisingly, however, the turbulent dynamo is simpler and more evident than the laminar dynamo. In the extreme case of large magnetic Reynolds number the analytic dynamo solutions to the equation for mean field and correlation function have been found (Moffat 1968; Parker 1979; Krause and Rädler 1980; Molchanov et al. 1985). In the case of turbulent flow the mean magnetic field is treated as a result

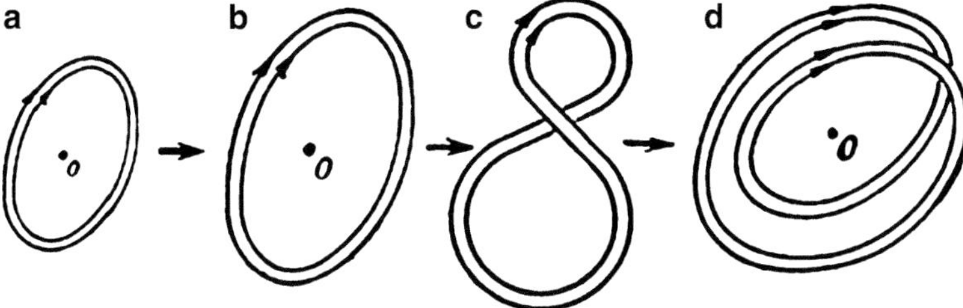

Fig. 1.2 A simple model of magnetic field amplification due to stretch (**a** → **b**), twisting (**b** → **c**) and doubling (**c** → **d**) of the magnetic loops

of the random pulsation averaging while in the case of laminar flow the azimuth and time averaging is considered. It is interesting to note also that in both of these cases the hydromagnetic dynamo is mainly due to the same characteristics of the flow, which are the mean helicity and nonuniform/differential rotation.

Some understanding of the hydromagnetic dynamo can be achieved by considering the following idealized descriptive model (Davies 1958; Vaynshtein et al. 1980). Imagine first a closed tube in perfectly conducting fluid as shown in Fig. 1.2a. There is a toroidal magnetic field inside the tube. In the framework of ideal MHD approach the magnetic field is frozen into the conducting fluid, so that the frozen-in field lines move with the fluid. Suppose that the fluid flow results in tension of the tube so that the tube length is doubled whereas its cross area is decreased roughly two times (Fig. 1.2b). Owing to the local flow rotation or for some other reason, the tube takes a shape of figure-of-eight (Fig. 1.2c) and then one loop of the figure-of-eight combines with another loop (Fig. 1.2d). The magnetic field is therefore doubled. Each iteration of this procedure results in duplication of the magnetic field. If this procedure is repeated n times, the magnetic field becomes 2^n as great as the initial magnetic field. This means that the magnetic field and magnetic flux increase exponentially. This mechanism of field amplification can arise in a turbulent flow. Since in the model the rate of field increase does not depend on the magnetic diffusion, this mechanism is referred to as fast dynamo.

1.1.5   Turbulent Diffusion and Mean Helicity

Random fluctuations of the conducting fluid flow give rise to fluctuations of currents and magnetic field. In what follows we study the flow pattern and conditions, which are required in order to produce a nonzero mean magnetic field in the conducting media.

Let us consider first the flow pattern where the eddies with certain sign, either right-handed or left-screw, are predominant. This means that a pronounced direction

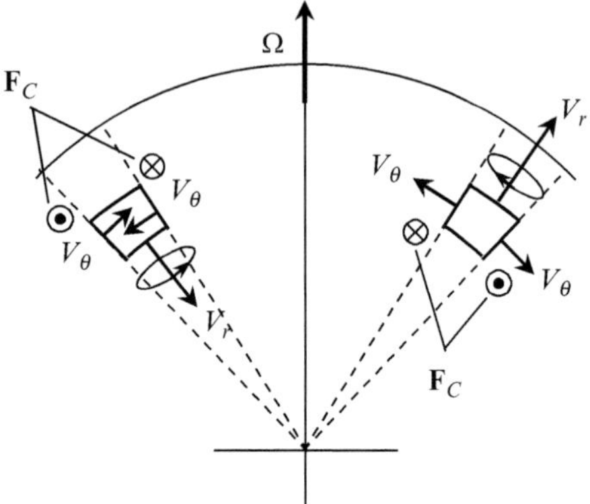

Fig. 1.3 A schematic plot of the helicity origination due to the convection in stratified media

of the helical curling/motion dominates in the flow. In such a way we say that the flow has the nonzero mean helicity, which is defined as a value which is proportional to $\langle \mathbf{V} \cdot (\nabla \times \mathbf{V}) \rangle$, where $\mathbf{V}$ is the mass velocity. Formally correct definition of the helicity α is (e.g., see monographs by Parker 1979; Vaynshtein et al. 1980, and references therein)

$$\alpha = -\frac{\tau}{3} \langle \mathbf{V} \cdot (\nabla \times \mathbf{V}) \rangle . \tag{1.23}$$

Here the correlation time is $\tau \sim l / V_0$, where l is scale size and V_0 is characteristic velocity.

Such helicity arises in outer space due to the spin of celestial bodies whereas in laboratory environment the helicity is atypical. The theoretical study has shown that the mean helicity is due to the violation of reflection symmetry of flow (Parker 1979). The reflection asymmetrical random movements are unable to generate a large-scaled magnetic field.

To understand the mechanism of helicity appearance in a little more detail we study the convection of matter in a rotating globe with radially stratified density. Suppose that a small convective fluid element with mass m ascends or is dropped along the radius. Owing to the density and pressure gradients, the element undergoes deformation and acquires the additional constituents of the velocity V_θ and V_ϕ perpendicular to the radius. These constituents result from dilatation or compression of the moving element. If the element moves up and the density falls off with increase of radius, the element volume dilates and whence the perpendicular velocity is directed so as shown on the right side of Fig. 1.3. The inverse case, when the element moves down, is shown on the left side of this figure. In the reference frame fixed to the globe, that is in the reference frame rotating with constant angular velocity $\mathbf{\Omega}$, the moving element undergoes the Coriolis inertia force

$$\mathbf{F}_C = 2m \, (\mathbf{V} \times \mathbf{\Omega}). \tag{1.24}$$

where $\mathbf{V}$ is given in that frame.

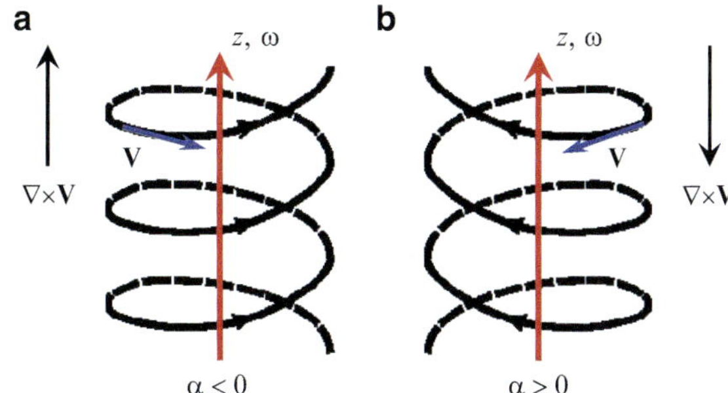

Fig. 1.4 Right-handed (**a**) and left-handed (**b**) spiral-helix motion

When the radial component of the velocity is taken into account, it produces the azimuthal component of the Coriolis force. However we now consider only the components of the Coriolis force due to the perpendicular velocity constituent V_θ shown in Fig. 1.3. As is seen from this figure, one of the components of the Coriolis force is pointed "out of paper" (the circle with point) whereas another component is oppositely directed (the circle with cross). These two forces create a moment, which rotates the convective fluid element around the radius as shown in Fig. 1.3. The same is true for the convective element located at an arbitrary point of the globe; that is, the Coriolis forces create a moment which rotates the element around the axis/position vector drawn from the center of the globe. Similarly, the Coriolis force can turn the oceanic flows, hurricanes, and tropical cyclones as it is seen from satellite pictures.

It is obvious that applying a curl operator to this rotary motion gives the vector pointed radially. We recall that the main movement of the element is radially directed whence it follows that $\nabla \times V_r \hat{\mathbf{r}} = 0$, where $\hat{\mathbf{r}}$ stands for unit vector. However, the mean helicity is nonzero, i.e., $\langle \mathbf{V} \cdot (\nabla \times \mathbf{V}) \rangle \neq 0$, due to the presence of the perpendicular components of the velocity.

Once the convective element is dropped, the radial velocity becomes negative. The perpendicular velocities V_θ, V_ϕ, and $\nabla \times \mathbf{V}$ change their directions as well due to compression of the element. In this case the Coriolis force moment rotates the element in the opposite direction. This means that the vector $\nabla \times \mathbf{V}$ changes sign whereas the helicity keeps the sign unchanged.

For illustrative purposes, a right-handed spiral motion is shown in Fig. 1.4a. Consider, for example, a progressive motion with constant velocity $V_z > 0$ positive parallel to z axis that combines with rotation around this axis with constant angular velocity $\boldsymbol{\omega} = \omega \hat{\mathbf{z}}$, where $\hat{\mathbf{z}}$ is a unit vector. Hence the azimuthal component of the velocity is $V_\phi = \omega r$. Applying a curl operator to the velocity of this helical motion yields $\nabla \times \mathbf{V} = 2\omega \hat{\mathbf{z}}$. This vector is vertically upward as shown in Fig. 1.4a, so that $\langle \mathbf{V} \cdot (\nabla \times \mathbf{V}) \rangle > 0$ while the helicity α in Eq. (1.23) is negative. If there is a counter-rotating helical motion as shown in Fig. 1.4b, the vector $\nabla \times \mathbf{V}$ is downward-directed so that $\alpha > 0$.

The turbulent hydrodynamic and magnetic fields are random in character that results in an extremely complicated pattern of MHD flow. In order to find the average overall realization of random function $\mathbf{B}$ and $\mathbf{V}$ in Eq. (1.14), it is necessary to calculate the correlation function $\langle \mathbf{V} \times \mathbf{B} \rangle$. In the course of this text we cannot come close to the exploration of this extremely complicated mathematical apparatus used for the study of MHD turbulence. The interested reader is referred to the text by Moffat (1968) for a more complete treatise on MHD turbulent processes (see also the treatise by Krause and Rädler 1980; Parker 1979; Vaynshtein et al. 1980).

Here we only reproduce the main equation describing the generation of mean large-scaled magnetic field $\langle \mathbf{B} \rangle$ in turbulent conducting media with nonzero mean helicity

$$\partial_t \langle \mathbf{B} \rangle = (\nu_m + \nu_t)\, \nabla^2 \langle \mathbf{B} \rangle + \nabla \times \alpha \langle \mathbf{B} \rangle + \nabla \times (\langle \mathbf{V} \rangle \times \langle \mathbf{B} \rangle). \tag{1.25}$$

Here α is the mean helicity given by Eq. (1.23), and $\nu_t \approx \langle V \rangle^2 \tau/3$ denotes the coefficient of turbulent diffusion where τ is the correlation time. The last term on the right-hand side of Eq. (1.25) describes the large-scaled motion of conducting media at the mean velocity $\langle \mathbf{V} \rangle$, including the so-called nonuniform/differential rotation.

In practice, the turbulent diffusion may be so much that $\nu_t \gg \nu_m$. Some understanding of the turbulent diffusion can be achieved by considering the following example. As has already been stated, the magnetic field is able to diffuse in a fixed conducting space in accordance with Eq. (1.15). A close analogy exists with the diffusion of impurity molecules in still air. Once the intermixing or turbulence occurs in the air, the impurity will propagate rapidly as compared to still air. The magnetic field propagates in moving conducting media analogously to the impurity in the air because of the "frozen-in" effect. Therefore, the magnetic field can propagate much more rapidly in turbulent media. Certainly, this analogy is incomplete since the diffusion of the impurity is described by a scalar equation whereas the magnetic field diffusion follows a vector equation. Other difference between diffusion of the magnetic field and of the impurity is that in contrast to the impurity, the field, if it is not weak, may greatly affect the flow.

1.1.6   Magnetic Field Generation

The effects of large-scaled magnetic field generation due to the helicity and nonuniform rotation of turbulent flow are the underlying physical principles on which the theory of the turbulent MHD is based. Since the gyrotropic turbulence due to the helicity is believed to play a major role in MHD dynamo excitation, we focus our attention on the following equation:

$$\partial_t \langle \mathbf{B} \rangle = \nu \nabla^2 \langle \mathbf{B} \rangle + \nabla \times \alpha \langle \mathbf{B} \rangle, \tag{1.26}$$

where $\nu = \nu_m + \nu_t$. Here we have dropped the term describing a large-scaled motion in Eq. (1.25). This equation has been studied repeatedly to demonstrate the possibility of magnetic field generation due to nonzero helicity. Following Vaynshtein et al. (1980) we now consider a simple one-dimensional case in order to derive a straightforward analytical solution of Eq. (1.26). Suppose first that the mean magnetic field $\langle \mathbf{B} \rangle$ is a function of only z and t so that the derivatives with respect to x and y are equal to zero. It follows from Eq. (1.3) that $\partial_z \langle B_z \rangle = 0$ and hence $\langle B_z \rangle = 0$. The components $\langle B_x \rangle$ and $\langle B_y \rangle$ are considered to vary as $\exp(\gamma t)$, where γ is the increment of growth. In this case Eq. (1.26) is reduced to

$$\gamma \langle B_x \rangle = -\alpha \frac{d \langle B_y \rangle}{dz} + \nu \frac{d^2 \langle B_x \rangle}{dz^2}, \tag{1.27}$$

$$\gamma \langle B_y \rangle = \alpha \frac{d \langle B_x \rangle}{dz} + \nu \frac{d^2 \langle B_y \rangle}{dz^2}. \tag{1.28}$$

In the case of infinite medium the set of Eqs. (1.27)–(1.28) has a particular solution (Vaynshtein et al. 1980)

$$\langle B_x \rangle = B_0 \exp(\gamma t) \sin(kz), \tag{1.29}$$

$$\langle B_y \rangle = B_0 \exp(\gamma t) \cos(kz), \tag{1.30}$$

where B_0 is a constant, and the increment of growth is related to the constant k by

$$\gamma = \alpha k - \nu k^2. \tag{1.31}$$

As is seen from Eq. (1.31) the magnetic field will increase exponentially in time under the requirement that the helicity is positive and $0 < k < \alpha/\nu$. The magnetic field thus can enhance if its characteristic scale $\lambda = k^{-1} > \nu/\alpha$. The increment of growth reaches a peak value $\gamma_{\max} = \alpha^2 / (4\nu)$, which corresponds to the spatial scale $\lambda = 2\nu/\alpha$. The rate of field enhancement is a maximum at this spatial scale. In contrast to the left-handed spiral field ($\alpha > 0$), the right-handed field ($\alpha < 0$) exponentially decreases with time at positive k since $\gamma < 0$.

The lack of a reflectional symmetry of the fluid flow means that the number of right-handed eddies in the flow is not equal to that of left-handed eddies. Notice that a single eddy has not the reflectional symmetry. Indeed, the reflection of eddy off a mirror plane, which is parallel to the spin axis, results in transformation of the right-screw eddy to the left-screw one. However, if the numbers of the right-screw eddies and left-screw ones are equal to each other, the flow has a mirror symmetry on average. Only if the eddies of certain sign are predominant in the turbulent flow, it can produce nonzero mean helicity in a way that the flow loses the mirror symmetry. In such a case the small-scale turbulence is able to produce a large-scale magnetic field. The credit for the discovery of this fact is given to Parker (1955) and Steenbeck et al. (1966).

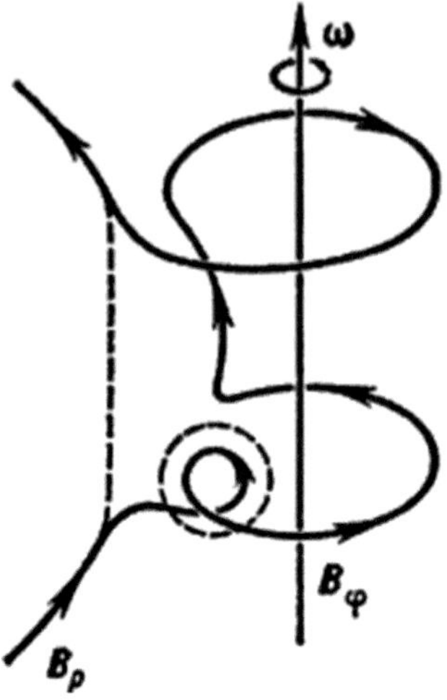

Fig. 1.5 A scenario of the generation of a toroidal magnetic field from a poloidal one for the case of nonuniform rotation of a conducting medium. Taken from the site http://www.astronet.ru/db/msg/1191490

1.1.7   Inhomogeneous Rotation

An astronomical observation has shown that stars, interstellar material, and galaxies rotate nonuniformly (differentially) that means that the angular velocity of the material rotation depends on the coordinates. The same phenomenon is believed to be a necessary accompaniment of the Earth rotation. Irregularities of the Earth spin are due to such factors as the presence of liquid core (Fig. 1.1) and the precession of the Earth rotation axis. This latter factor can play a role of driving force. The angular velocity of the precession depends on dynamical compression of a solid and thus the angular velocity in the mantle differs from that in the outer core. This difference may give rise to a motion of the molten material filling the space between the mantle and inner solid core. As a result the core will rotate slowly compared to the mantle.

The nonuniform rotation of a conducting medium may greatly affect the generation of magnetic field. To illustrate the mechanism of the phenomenon, consider the magnetic field frozen in the conducting media. This implies that the energy dissipation can be neglected. We now show that the nonuniform rotation generates a toroidal (azimuthal) magnetic field $\mathbf{B}_t = B_\varphi \hat{\varphi}$ from the poloidal (meridional) field $\mathbf{B}_p = B_r \hat{\mathbf{r}} + B_\theta \hat{\theta}$, where $\hat{\mathbf{r}}$, $\hat{\theta}$, and $\hat{\varphi}$ stand for unit vectors. Magnetic field lines of the poloidal field are in meridional planes which contain the spin axis, while the toroidal field lines are in the orthogonal planes. Since the different parts of the "frozen-in" field line rotate at different angular velocities $\omega = \omega(r, \theta)$, the magnetic field line stretches in the azimuthal direction. As shown in Fig. 1.5, the heterogenous rotation of the conducting medium results in generation of the toroidal magnetic field $\mathbf{B}_t$ in a way that the field lines of the toroidal field are perpendicular to the meridional planes. It should be noted that the generation of toroidal field from poloidal one due to the heterogeneous rotation is described, in a mathematical sense, via the term $\nabla \times (\langle \mathbf{V} \rangle \times \langle \mathbf{B} \rangle)$ on the right-hand side of Eq. (1.25).

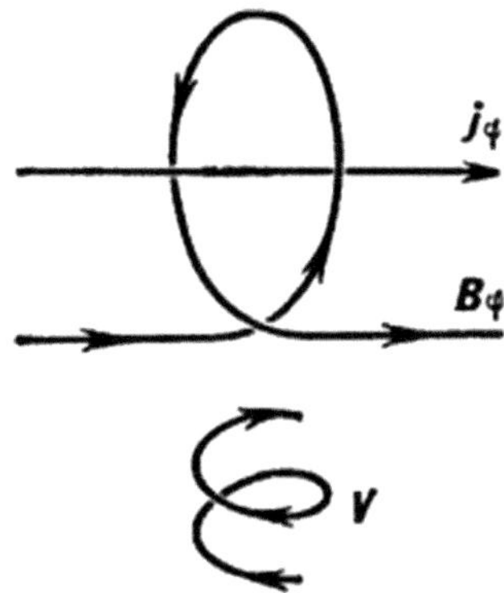

Fig. 1.6 A mechanism of hydromagnetic dynamo due to the mean helicity of a conducting medium. Taken from the site http://www.astronet.ru/db/msg/1191490

The turbulence can create and twist the loops of toroidal/azimuthal field $\langle B_\varphi \rangle$. One of such loops is highlighted in Fig. 1.5 with the dotted circle. The same loop is shown in Fig. 1.6 in a large scale. If the motion of conducting medium has the mean helicity, there is a predominant direction of twisting of the loops. We recall that the helicity is described by the term $\nabla \times \alpha \langle \mathbf{B} \rangle$ on the right-hand side of Eq. (1.25). The implication of this term is that there are not only the conduction current $\sigma \langle \mathbf{E} \rangle$ and the fluid-driven current $\sigma (\langle \mathbf{V} \rangle \times \langle \mathbf{B} \rangle)$ but also the current proportional to $\alpha \langle \mathbf{B} \rangle$ that results from the turbulence and helicity of the flow. This means that the azimuthal component $\langle B_\varphi \rangle$ of the mean magnetic field gives rise to the generation of azimuthal component of mean current since $\langle j_\varphi \rangle \propto \alpha \langle B_\varphi \rangle$. If the left-handed ($\alpha > 0$) helicity prevails over the right-handed one, the azimuthal current and magnetic field $\langle B_\varphi \rangle$ have the same sign as shown in Fig. 1.6 and vice versa in the inverse case of the right-handed helicity ($\alpha < 0$).

As is seen from Fig. 1.6, the azimuthal current in turn originates a new component poloidal/meridional field $\mathbf{B}_p$. Notice that the generation of the poloidal field from the toroidal one follows from Eq. (1.25). It is clear that substituting the toroidal field $\mathbf{B}_t$ into $\nabla \times \alpha \langle \mathbf{B} \rangle$ gives the term $\nabla \times \alpha \langle \mathbf{B}_t \rangle$ which plays a role of source of the poloidal field $\mathbf{B}_p$. To summarize, the poloidal magnetic field in a rotating conductive medium with nonzero mean helicity induces the toroidal field, which in turn generates the poloidal field and so on.

The combined action of the mechanisms treated here, that are the inhomogeneous rotation and the mean helicity, can result in self-excitation of the magnetic field. Moreover the dynamo-effect may arise at proper conditions even though only the latter mechanism is operative.

1.1.8 Magnetic Field Structure on the Earth Surface

To first order the Earth magnetic field in the atmosphere and ionosphere can be approximated by a dipole field. At present the Earth dipole magnetic moment is about $M_e \approx 8.3 \times 10^{22}$ A·m^2 (Stacey 1969; Yanovsky 1978). The dipole is slightly

Fig. 1.7 A sketch of dipole
magnetic field of the Earth

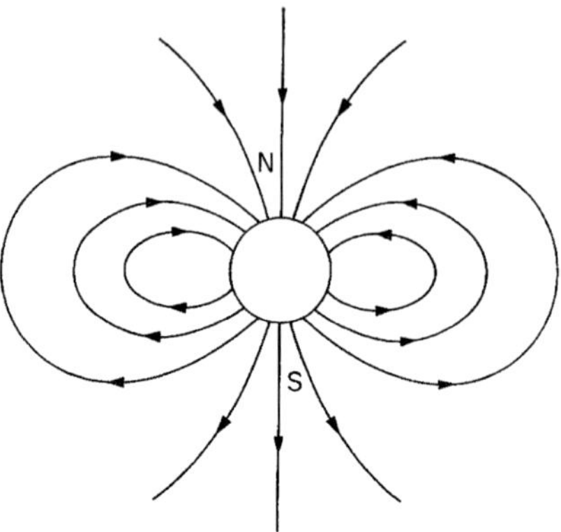

shifted with respect to the Earth center and the dipole vector makes an angle about
11.5° with respect to the Earth spin. In addition, the south magnetic pole is situated
in the northern hemisphere. The Earth's dipole magnetic field, $\mathbf{B}_0$, is given by

$$\mathbf{B}_0 = \frac{\mu_0}{4\pi r^3} \left(\frac{3\,(\mathbf{M}_e \cdot \mathbf{r})\,\mathbf{r}}{r^2} - \mathbf{M}_e \right), \tag{1.32}$$

where r is the distance from the dipole/Earth center. The reference frame used in
this study assumes that the z axis is positive parallel to the Earth's magnetic moment
$\mathbf{M}$ and the origin of the coordinate system is in the Earth center. Using the spherical
coordinates r, θ, and ϕ, the geomagnetic field components in Eq. (1.32) can be
written as

$$B_r = \frac{2\mu_0 M_e \cos\theta}{4\pi r^3}, \quad B_\theta = \frac{\mu_0 M_e \sin\theta}{4\pi r^3}. \tag{1.33}$$

The component $B_\varphi = 0$ because the vector $\mathbf{B}_0$ lies in a meridional plane.

As illustrated in Fig. 1.7, the field lines in the dipole approximation would extend
in loops of ever increasing dimension while the magnitude of the magnetic field
decreases with distance as r^{-3}. The mean value of the magnetic induction on the
ground surface is about 5×10^{-5} T, while at the aclinic line ($\theta = \pi/2$) the Earth
field is 3×10^{-5} T.

If the polar angle θ is expressed through the magnetic latitude λ (northern
hemisphere) via $\lambda = \theta - \pi/2$, then one should substitute $\cos\theta = -\sin\lambda$ and
$\sin\theta = \cos\lambda$ in Eq. (1.33). Applying the modified spherical coordinate system, the
equation for the magnetic field lines can be written as

$$r\,(\lambda) = R_e L \cos^2 \lambda, \tag{1.34}$$

where $R_e \approx 6{,}371$ km is the mean Earth radius, L is the so-called McIllwain
parameter defined as the radius of the equatorial crossing point of the field line.

Notice that the parameter L is measured in Earth radii R_e. For example, if the field line exits the Earth's surface ($r = R_e$) at a magnetic latitude $\lambda = 60°$, then $L = 4$. This implies that the field line crosses the equatorial plane at $r = 4R_e$.

The field near the ground surface, however, is not quite dipole since it contains abnormal regions. Such a region can be as high as several hundreds or thousands kilometers. There is known a number of the so-called world magnetic anomalies such as Brazilian, Siberian, and Canadian anomalies. Moreover, there are a lot of local anomalies on the Earth surface. The magnitude of the local anomalies is of the order of 2×10^{-7} T, and their sizes vary from unity up to several hundreds kilometers. For the Kursk magnetic anomaly (Russia) the deviation from the main geomagnetic field reaches a value of about 10^{-5} T. The influence of magnetized rocks, which contain ferrimagnetic minerals on the basis of iron oxide, can be the major cause of such local anomalies.

Empirical corrections to the field of the main dipole are described through a number of additional dipoles situated inside the core and at the boundary between the core and mantle. The observed magnitudes of the dipole harmonics decrease in accordance with a power law.

The geomagnetic field exhibits slow random variations with characteristic period from 10 to 10^4 years that is referred to as the secular variations. During the main period of the secular variations, that is 8×10^3 years, the main dipole moment can change by 1.5–2 times in magnitude. On average the dipole moment vector is approximately directed along the Earth spin axis. Therefore, the Earth rotation may greatly affect the evolution of the geomagnetic field. The result of the secular variations is that the geomagnetic pole exhibits precession around the geographic pole with the period of the order of 1.2×10^2 years.

Study of the residual magnetization of sedimentary which contain ferrimagnetic minerals has shown that the geomagnetic field keeps the magnitude close to the modern value at least during 2.5 billion years, that is comparable with the geologic age of the Earth, that is 4.6 billion years. On average during each 10^5–10^7 years the random fluctuations of the geomagnetic field reach the critical levels in a way that the geomagnetic field changes the initial direction to inverse one. When the northern and southern poles are traded their places the period of the poles inversion continues 10^3–10^4 years. The numerical modeling has shown that during this period the structure of the geomagnetic field is far from the dipole and the field pattern becomes complicated. It should be emphasized that these changes cannot reduce to the simple rotation of the vector of magnetic dipole.

It is interesting to note that such inversion of the geomagnetic field has not been observed during the last 7.8×10^5 years. The geomagnetic field was first measured at the beginning of nineteenth century and since then it has decreased by 10 %! Maybe the next inversion of magnetic poles is already close?

1.2   The Earth Magnetosphere

1.2.1   Solar Wind

The Earth magnetic field is immersed in the atmosphere of the Sun. As far as the temperature of the upper atmosphere/corona of the Sun is as high as 1.5×10^6 K, the corona consists of a fully ionized plasma. The gravitational pressure of the upper layers of the corona is unable to equilibrate the gas pressure of the corona material so that the Sun corona expands into outer space. Because of high temperature of the corona, the electrons, protons, alpha particles, and other ions can thus escape the Sun gravitational attraction and form a steadily streaming outflow of solar plasma called the solar wind. The solar wind expands radially from the Sun at speeds of about 300–800 km/s and fill the solar system up to the heliocentric distance of about 100 AU (astronomical units). The solar wind is also pervaded by a large-scale interplanetary magnetic field. Since the solar wind plasma conductivity is high enough, the magnetic field is frozen into the plasma. This yields that the solar magnetic field is transported outward into the solar system by the solar wind plasma.

Owing to the combination of plasma heating, compression, and subsequent expansion, the solar wind becomes a supersonic flow above a few solar radii (e.g., see monograph by Parker (1979), and references therein). The solar wind speed relative to the Earth is much greater than the speed of Alfvén, magnetosonic and other kind of MHD waves providing energy transfer inside the solar wind. The solar wind energy flux incident upon the cross section of Earth's magnetosphere is estimated as 2×10^{13} W. The solar wind flowing to the Earth brings the aurora, heats the polar upper atmosphere, and provides the plasma on the Earth's magnetic field lines thereby energizing a spacious revolving system of magnetospheric plasma. When the solar flares give rise to significant perturbations of the solar wind it produces magnetic storms and substorms which result in interferences in the communication system.

1.2.2   Interaction Between the Solar Wind and Earth's Magnetic Field

The basic concept of the Earth's magnetosphere was first deduced by Chapman and Ferraro (1931, 1932, 1933, 1940) in early 1930s. The magnetosphere is the near-earth space confined by the solar wind plasma blowing outward from the Sun. The Earth's magnetosphere is formed due to the interaction between the Earth's magnetic field and the electrically conducting plasma of the solar wind. This interaction brings the solar wind deform the Earth's dipolar magnetic field, compressing the field lines on the day side and stretching them out to form the elongated magnetotail on the night side.

Fig. 1.8 The deflection of solar wind by the geomagnetic field shown in an ecliptic plane. *1*—solar wind, *2*—bow shock, *3*—magnetopause, *4*—magnetopause current, *5*—the Earth's magnetic field, *6*—secondary magnetic field originated from the magnetopause current. The symbols e^- and p^+ denote electrons and protons/positive ions; z axis and the unit vectors **n** are both perpendicular to the bow shock, and A is the bow shock point nearest the Sun

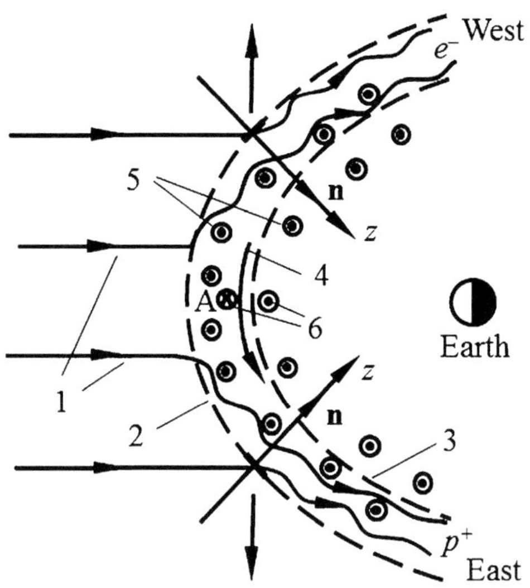

The solar wind particles of charge q moving at velocity **V** are subject to the magnetic force

$$\mathbf{F} = q\mathbf{V} \times \mathbf{B}.$$

The Earth's magnetic field will therefore deflect the positive-charged and negative-charged particles to the different sides of the magnetosphere. The deflected particles will spiral around the fields lines and drift perpendicularly to the meridian around the Earth.

Owing to the fact that the solar wind is a supersonic flow at high speed the Mach number of Alfvén and magnetosonic waves of the supersonic solar wind reach a value of about 7 at the distance equal to the Earth orbit radius. Therefore, a shock wave must form when the supersonic and super-Alfvén solar wind fall on the Earth magnetic field. At the front of the so-called collisionless bow shock the velocities of the solar wind particles change their direction and the solar wind flow around the Earth magnetic field forming the Earth's magnetosphere as shown in Fig. 1.8. The outer boundary of the magnetosphere is termed the magnetopause.

To study the electrodynamics of this region in a little more detail we note that the magnetoplasma itself is so tenuous that it can be treated to first order as a collisionless magnetized plasma with an MHD approach. The plasma conductivity is so high that the plasma is assumed to be a single fluid having infinite conductivity. This implies that in a reference frame moving at the plasma velocity **V** the electrical field $\mathbf{E}' = \mathbf{E} + \mathbf{V} \times \mathbf{B}$ vanishes both parallel and perpendicular to **B**. In a reference frame fixed to the Earth

$$\mathbf{E} = \mathbf{B} \times \mathbf{V}. \tag{1.35}$$

Notice that in this approximation the plasma is considered only as a conducting fluid, which does not exhibit anisotropic properties.

As usual in a fluid description, we use the continuity equation (1.11) and equation of the fluid motion (1.13). In the first approximation the gravity, viscosity, and inertial terms can be neglected when considering the magnetospheric plasma dynamics. In such a case the dynamical equation of plasma motion is given by

$$\rho d\mathbf{V}/dt = -\nabla P + \mathbf{j} \times \mathbf{B}, \tag{1.36}$$

where as before d/dt is the total time derivative. Substituting Eq. (1.5) for $\mathbf{j}$ into Eq. (1.36) yields

$$\rho \frac{d\mathbf{V}}{dt} = -\nabla P + \frac{1}{\mu_0} (\nabla \times \mathbf{B}) \times \mathbf{B}. \tag{1.37}$$

Taking into account the vector identity

$$\mathbf{B} \times (\nabla \times \mathbf{B}) = \nabla B^2/2 - (\mathbf{B} \cdot \nabla) \mathbf{B}, \tag{1.38}$$

we arrive at

$$\rho \frac{d\mathbf{V}}{dt} = -\nabla \left(P + \frac{B^2}{2\mu_0} \right) + \frac{1}{\mu_0} (\mathbf{B} \cdot \nabla) \mathbf{B}. \tag{1.39}$$

Across the bow shock the plasma is slowed, is compressed, and is heated. The velocity of the flow decreases whereas the thermal velocity of the particles enhances so much that the plasma is heated up to several million Kelvin degree. As a result a layer of turbulent plasma, called the magnetosheath, is formed between the shock and magnetopause.

To give some insight into the electrodynamics of this region we first leave out of account the interplanetary magnetic field and approximate the actual situation with a simple model shown in Fig. 1.8. The arrowed solid lines 1 indicate plasma streamlines, and the heavy long-dashed lines are the principal boundaries, that is, the bow shock 2 and magnetopause 3. The density, mass velocity, and other parameters of the flow may have a jump discontinuity across the bow shock surface whereas the flux of mass, momentum, and energy must be continuous at the shock.

Let z be the axis perpendicular to a small piece of the discontinuity surface. In order to derive the boundary conditions at the bow shock one should first integrate Eq. (1.10) with respect to z across the discontinuity surface in the vicinity of the interception point $z = 0$. Applying the operator

$$\lim_{\varepsilon \to 0} \int_{-\varepsilon}^{\varepsilon} dz \tag{1.40}$$

to Eq. (1.10) and making formally $\varepsilon \to 0$ yields

$$[\rho V_n] = 0, \qquad (1.41)$$

where square brackets denote the jump of the corresponding value and $V_n = V_z$ stands for the normal component of the fluid velocity. The implication here is that the flux of mass is continuous across the shock surface.

Similarly, applying the operator (1.40) to Eq. (1.39) and taking into account of Eq. (1.12) gives the boundary conditions

$$\left[\rho V_n \mathbf{V} + \left(P + \frac{B^2}{2\mu_0}\right)\mathbf{n} - \frac{B_n \mathbf{B}}{\mu_0}\right] = 0, \qquad (1.42)$$

where $\mathbf{n}$ is the unit vector perpendicular to the shock surface. Integrating of Eq. (1.3) across the discontinuity surface yields

$$[B_n] = 0. \qquad (1.43)$$

One more boundary equation describes the energy balance at the bow shock. We refer the reader to Akasofu and Chapman (1972) and Landau and Lifshitz (1982) for a more complete treatise on the boundary conditions on the bow shock.

To study the magnetic field at the bow shock in a little more detail, it is necessary at this point to consider the electric current flowing in the vicinity of the shock. Taking the cross product of Eq. (1.36) with $\mathbf{B}$, the current density perpendicular to $\mathbf{B}$ is given by

$$\mathbf{j}_\perp = \frac{1}{B^2}\left(\rho \mathbf{B} \times \frac{d\mathbf{V}}{dt} + \mathbf{B} \times \nabla P\right). \qquad (1.44)$$

As illustrated in Fig. 1.8, because of the deceleration of the solar wind in the bow shock, the eastward current arises in the region between the bow shock and magnetopause. The current flowing along the magnetopause is shown with an arrow in Fig. 1.8, which shows a view looking down onto the ecliptic plane. Notice that this current, called the Chapman-Ferraro/magnetopause current, exists in a thin plane sheet/magnetosheath extending also out of the paper. This surface current is due to both the solar wind ions and electrons since the planetary magnetic field deflects the solar wind ions to the east and electrons to the west as they move to the Earth. The current flowing around the magnetopause results in the generation of the secondary magnetic field. As is seen from Fig. 1.8, this field is parallel to the Earth's field on the earthward side of the current sheet. Conversely, on the sunward side of the current the secondary magnetic field is directed in opposition to the Earth's field. To the first order the secondary magnetic field cancels nearly the Earth's field in the solar wind whereas the value of magnetic field is approximately doubled on the earthward side of the current sheet.

Now we estimate the solar wind pressure. Equation of state of an ideal gas can be applied to the solar wind particles to relate the number density n_j and the pressure P_j of each species

$$P_j = n_j k_B T_j, \tag{1.45}$$

where the subscript "j" stands for the j-th ionized species, i.e., ions or electrons. In this text we use k_B to represent Boltzmann's constant. The mass density ρ_j relates to the number density n_j through $\rho_j = n_j m_j$ where m_j is the particle mass of each species. Now we will estimate the ratio between the pressure P due to the thermal motion and the dynamic pressure ρV^2 resulted from the solar wind motion. Taking the numerical values of the solar wind parameters $T = 10$–$40\,\text{eV}$, and $V = 300$–$800\,\text{km/s}$ we get $P/\rho V^2 \sim k_B T/m_P V^2 \approx 0.04$–$0.004$, where m_P is the proton/neutron mass. This means that the dynamic pressure dominates over the thermal one in the solar wind.

To estimate scale size of the magnetosphere on its upstream/sunlit side, we consider the bow shock at the ecliptic plane and the point A (Fig. 1.8), which is nearest to the Sun. Since the magnetic field is perpendicular to this plane, Eq. (1.42) is reduced to

$$\left[\rho V_n^2 + P + \frac{B^2}{2\mu_0} \right] = 0. \tag{1.46}$$

We set $B = 0$ in the solar wind because the interplanetary magnetic field is neglected. In the magnetosphere the dynamic pressure of plasma can be dropped since the flow is slowed across the shock. Additionally P is everywhere negligible compared with ρV^2. In this notation, the pressure balance at the boundary between the solar wind and the magnetosphere yields $\rho V^2 \approx B^2/(2\mu_0)$. Below we slightly specify this rough estimate.

The magnetosphere thus serves as an obstacle in such a way that the solar plasma flows around the magnetosphere. Assuming a perfect/mirror reflection of the solar wind from the boundary, then the dynamic pressure of the solar wind is estimated as $2\rho V^2$ that is twice as great as that under inelastic reflection. Since the Earth' magnetic field (1.32) is approximately doubled inside the magnetosphere, we get the assessment

$$2\rho V^2 \approx \frac{B^2}{2\mu_0} = \frac{\mu_0}{2} \left(\frac{M_e}{4\pi r_0^3} \right)^2, \tag{1.47}$$

where M_e is the dipole moment of the Earth magnetic field, and r_0 is geocentric distance of the magnetopause on the sunward side of the magnetosphere. Substituting the numerical parameters into Eq. (1.47) gives the value $r_0 \approx 10 R_e$. This estimate is compatible with observations, which show that the bow shock forms at about 13 Earth radii on the sunlit side of our planet.

In summary, an important aspect of the interaction between the solar wind and the Earth magnetic field follows from the fact that the solar wind plasma is frozen to the interplanetary magnetic field whereas the terrestrial/ionospheric plasma is frozen to the Earth's field. These plasmas form distinct regions separated by a thin boundary, that is the magnetopause, and they practically do not mix.

1.2.3 Structure of the Earth Magnetosphere

Although the solar wind plasma is frozen to the large-scale interplanetary magnetic field, the "frozen-in" picture can break down at the magnetopause boundary, where high current densities occur in the plasma. As a result the interplanetary magnetic field diffuses through the plasma in the magnetopause. In such a case the interplanetary and terrestrial field lines will connect through the dayside magnetopause, as shown in Fig. 1.9a with line 3. Following Dungey (1961) this process has been termed magnetic reconnection. The essential breakdown of frozen-in flow occurs not only at the dayside magnetopause but also in the magnetotail that results in the appearance of reconnection in that region. The concept of magnetic reconnection is a relatively new phenomenon, which frequently occurs in the magnetospheres of planets, stars, and other cosmic objects. The origin of this phenomenon has been the subject of a great deal of research during recent years.

A cutaway of an actual Earth magnetosphere illustrating the solar wind/ magnetosphere interaction is shown in Fig. 1.9b. The dipole magnetic field dominates only in the inner magnetosphere in the region with radius about $3R_e$. Across the magnetopause the magnetic field usually undergoes a sharp change in both strength and direction. An energy and momentum transfer from the solar wind into magnetosphere is basically due to reconnection and quasi-viscous interaction between the magnetic field frozen in the solar wind and the Earth's magnetic field (e.g., Mishin and Bazarzhapov 2002). The solar wind plasma is deflected at the bow shock, flows along the magnetopause, pulling the terrestrial magnetic field into a long magnetospheric tail/magnetotail on the night side thereby providing the magnetotail with antiparallel magnetic fields in North and South parts of the magnetotail. The magnetotail extends several hundred Earth radii in the antisunward direction. The diameter of the tail is about $40R_e$.

The solar wind flows around the Earth into the magnetic tail and is then injected back toward the Earth within the region called the plasma sheet, as shown in Fig. 1.9b. The plasma sheet is sandwiched between two bundles of magnetic field lines. One bundle, to the north of the equator, consists of field lines that enter the north magnetic pole, and the other bundle, to the south of the equator, contains field lines directed out of the southern polar cap. There are open field lines that cross the magnetopause and then connect interplanetary magnetic fields to the Earth's field.

The northern and southern polar cusps are narrow funnel-shaped regions of recently "opened" or merged magnetic field lines connected with those of the interplanetary magnetic field rather than the magnetosheath magnetic field. The hot

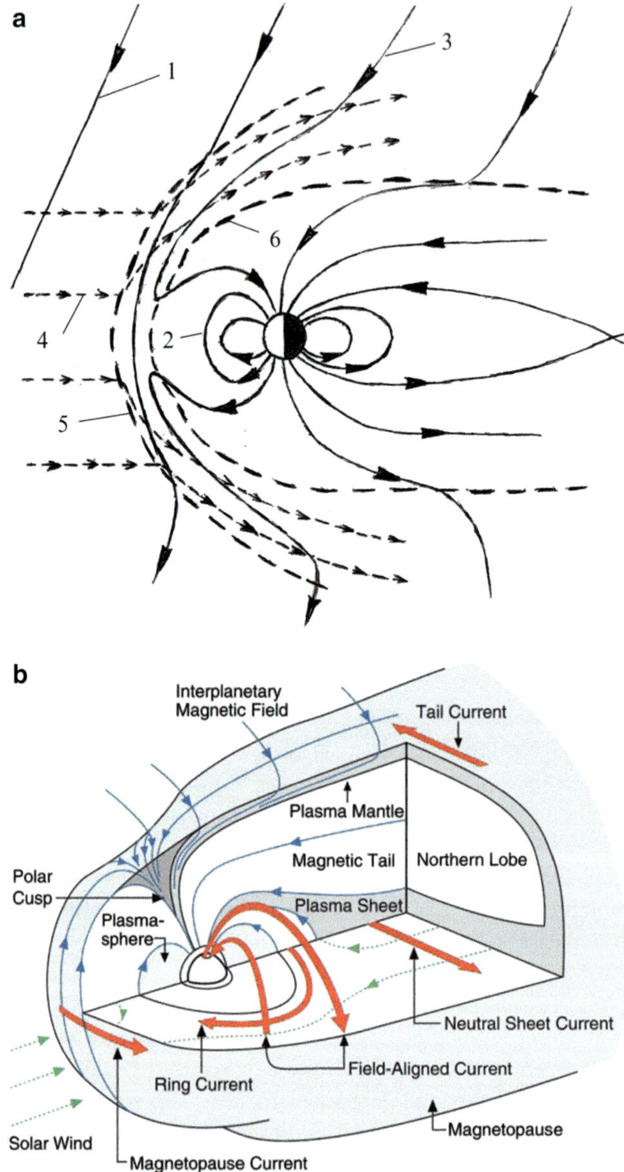

Fig. 1.9 (**a**) Sketch of the structure of the Earth's magnetosphere in the noon–midnight meridian plane. The *arrowed solid lines 1 and 2* indicate interplanetary and terrestrial field lines, connection between the interplanetary and terrestrial field is shown with line 3, the *arrowed dashed lines 4* are plasma streamlines, and the *heavy long-dashed lines 5 and 6* are the bow shock and magnetopause, correspondingly. (**b**) A cutaway view of the Earth magnetosphere. Taken from the site http://denji102.geo.kyushu-u.ac.jp/stp/study/study.html

electrons and ions of the solar wind plasma directly penetrate into the ionosphere through the cusp that heat the ionospheric plasma in the cusp and contributes to the dayside part of the auroral oval. The injection of magnetosheath plasma into the cusp has been found to trigger the outflow of ionospheric plasma into the magnetosphere.

The longitudinal/field-aligned currents flowing from the magnetosphere along magnetic field lines are closed through the high-conducting ionosphere. These persistent currents enter the morning ionosphere and go out from the evening ionosphere. The region of the current inflow and issue produces practically continuous strip along the auroral oval, which results from the plasma sheet and polar cusp mapping to the high-latitude ionosphere (100–200 km over the Earth surface). In this region the aurora borealism and magnetic storms are frequently observed due to the energetic particles precipitating along field lines from the magnetosphere into the ionosphere.

The plasmasphere is the region, which contains the ionospheric plasma with enhanced number density $n \sim 10^3 \, \text{cm}^{-3}$ and thermal energy $\sim 1.0 \, \text{eV}$. The plasmasphere is terminated by the so-called plasmapause (Nishida 1978). At this sharp boundary with geocentric distance about $4R_e$ the number density of the ionospheric plasma decreases abruptly to 0.1–$1.0 \, \text{cm}^{-3}$. Inside this region the ionospheric plasma approximately rotates with the Earth and the Earth's magnetic field because the magnetic field lines are frozen into the ionospheric plasma. At high magnetic latitudes the field lines are practically normal to the lower ionosphere. The terrestrial plasma flows from the ionosphere into the plasmasphere along these lines forming the so-called polar wind. The sources of the plasmas that occupy these regions are thus the solar wind and the Earth's ionosphere. The relative contributions of these two sources to the magnetospheric plasma depend on the level of geomagnetic activity.

The magnetosphere also contains two Van Allen radiation belts, ring and field-aligned currents and various large-scale regions depending on plasma parameters, which vary in time and space. The radiation belts are inner regions of the Earth's magnetosphere where the Earth's magnetic field maintains the charged particles, i.e., electrons, protons, ions, and so on (Nishida 1978). The kinetic energy of the trapped species varies from tens keV to hundreds MeV.

Under the influence of Lorentz force the trapped species spiral along the magnetic field lines from Northern pole towards Southern one and vise versa. The dynamics of the particle motion in magnetic field is described by

$$qV_{\perp}B = mV_{\perp}^2/R_g,$$

where q and m are the charge and mass of a particle, correspondingly, and $V_{\perp}$ is the component of particle velocity perpendicular to the magnetic field B. Whence it follows that the radius R_g of the circle or "gyration radius" is

$$R_g = \frac{mV_{\perp}}{qB}.$$

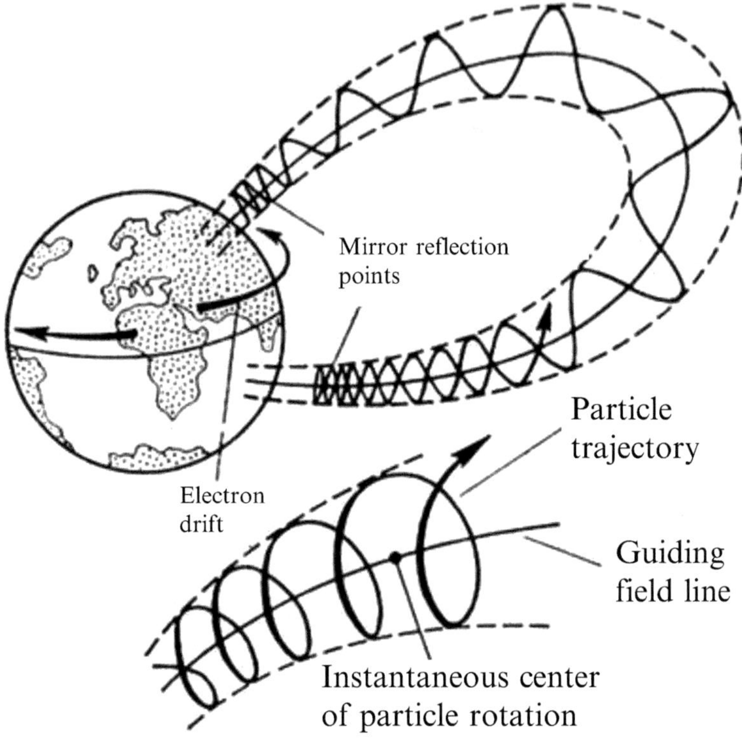

Fig. 1.10 Motion of charged particles trapped in the geomagnetic field. The particles spiral around the geomagnetic field line concurrently with eastward and westward drift. Taken from the site http://dic.academic.ru/dic.nsf/bse/

The particle velocity $V_\parallel$ parallel to the magnetic field is not influenced by the field because the Lorentz force is zero in that direction. The center of the particle rotation moves at this velocity along the guiding field line as shown in Fig. 1.10.

In addition to the rapid spiraling around field lines the trapped particles also undergo a slow eastward and westward drift around the Earth's magnetic axis. Positive ions drift to the West while the electrons drift to the East. The charged particles move in the Earth's magnetic field in such a way that its guiding center belongs to the same magnetic shell/L-shell the whole time. As the gyrating particles move closer to regions of stronger magnetic field, where field lines converge, it produces the particle reflection from this region, which has been termed the magnetic mirroring. At this point the particle velocity drops to zero and then reverses that results in the particle to move back towards the conjugate point in another hemisphere (Fig. 1.10). A proton with energy $\sim 100\,\mathrm{MeV}$ makes one vibration along the field line from Northern hemisphere to Southern one for the time $\sim 0.3\,\mathrm{s}$. The proton captured by the Earth's magnetic trap can make about 10^{10} vibrations during its lifetime, which can reach $\sim 3 \cdot 10^9\,\mathrm{s}$ (about 100 years).

The regions of trapped fast particles consist of two radiation belts. The inner radiation belt discovered by Van Allen in 1958 is a region sandwiched between magnetic shells at $L \sim 1.5$ and $L \sim 2$. This region contains very energetic protons resulted from collisions between cosmic ray ions and the atmospheric atoms. The outer radiation belt is bounded by two magnetic shells at $L \sim 3$ and $L \sim 6$. The flux density of the particles moving in this region has a peak value at $L \sim 4.5$. It should be noted that the radiation belts have not, in essence, distinct boundaries because each species has individual "radiation belt" depending on their energy.

1.3 Magnetic Storms

1.3.1 Solar-Quiet-Time Magnetic Variations

Magnetic variations with periods from several seconds to several years are generated by ionospheric and magnetospheric currents which in turn are influenced by the solar wind and radiation. The pattern and intensity of these variations depend on solar wind parameters, latitude, time, and season. The magnetic variations can be split into three main groups, i.e., a solar-quiet-time magnetic variations, perturbed variations, and short period oscillations (e.g., Gonzalez et al. 1994).

The solar-quiet-time variations are a strictly periodic phenomenon, which follows the Earth spin, the Earth orbiting around the Sun, and the moon location with respect to the horizon (lunar tides). These kinds of magnetic variations result predominantly from the ionospheric neutral winds and from the solar wind permanently flowing around the magnetosphere. The solar radiation is responsible for ionization of the upper atmosphere and for heating of the thermosphere, which in turn results in the diurnal generation of large-scale system of neutral winds at the ionospheric altitudes. When the neutral winds drag the ions, they produce the motion of the conducting media through the Earth's magnetic field followed by the generation of electric currents at altitude range 90–150 km. At middle latitude these currents give rise to solar-quiet-time magnetic variations (S_q-variations) with amplitude ~ 50 nT. At magnetic equator the amplitude of S_q-variations can enhance up to 2×10^2 nT due to the presence of equatorial electrojet flowing in the anisotropically conducting ionospheric plasma (Surkov et al. 1997; Fedorov et al. 1999).

The interaction of solar wind with the magnetosphere produces the eastward electric current at the magnetopause (Fig. 1.8). On the Earth surface in the vicinity of magnetic equator this current increases the noon magnetic field by the value ~ 25 nT, which can vary with amplitude ~ 4 nT for 24 h. Another result of the interaction between the solar wind and magnetosphere is the large-scale convection of plasma inside the magnetosphere that causes the current generation at high-latitude ionosphere and the magnetic S_q^p-variations with amplitude $\sim 10^2$ nT for summer season.

Interaction of the solar wind and frozen-in interplanetary magnetic field with the Earth's magnetic field in the auroral region can serve as another source mechanism for the solar-quiet-time magnetic variations. In the daytime the amplitude of magnetic variations reaches a peak value $\sim 1.5 \times 10^2$ nT at the magnetic latitude $\sim 80°$ in the vicinity of polar electrojet.

1.3.2   Storm Sudden Commencement

When the solar flares give rise to significant perturbations of the solar wind they produce the magnetic storms and substorms which result in interferences in communication systems. The solar wind energy flowing to the Earth brings about the aurora, heats the polar upper atmosphere and the plasma on the Earth's magnetic field lines thereby energizing a spacious revolving system of magnetospheric plasma.

During the most active period on the Sun, the solar flares are accompanied by the energy release as much as 10^{25}–10^{27} J for the short time interval about 2×10^3 s, which causes an enhancement of Roentgen and ultraviolet (UV) radiations, generation of shock waves. The strong solar energy release allows plasma clouds to escape into the space outside of the Earth's orbit. The sudden increase in Roentgen and UV radiations produces an excess ionization at the bottom ionosphere followed by an increase in the ionospheric currents and S_q-variations in sunlit hemisphere. The magnitude and duration of the S_q-variations on the ground surface is about 10 nT and 30 min, correspondingly.

The density and velocity of the solar wind plasma increase significantly behind the front of interplanetary shock wave so that the shock wave arrival results in compression of the magnetosphere from the day side and in strengthening the electric currents at the magnetopause. This has an effect of increasing the Earth's magnetic field, which has been termed the storm sudden commencement (SSC). The effect can be observed everywhere reaching a maximum value of several tens nT at the equator. Sometimes the SSC may initiate a magnetic storm.

1.3.3   Magnetic Storm and Substorms

A magnetic/geomagnetic storm is a temporary perturbation of the Earth's magnetic field caused by irregular disturbances in the solar wind and on the Sun that may greatly affect space weather (e.g., see McPherron 1979). The magnetic storm consists of three phases, i.e., an initial/growth phase, a main/active phase, and a recovery phase. The initial phase is the SSC, which may last from 10 min to 6 h and more. At this stage the strengthening field is mainly due to the increase in the electric currents at the magnetopause while the magnetic field is weakly perturbed.

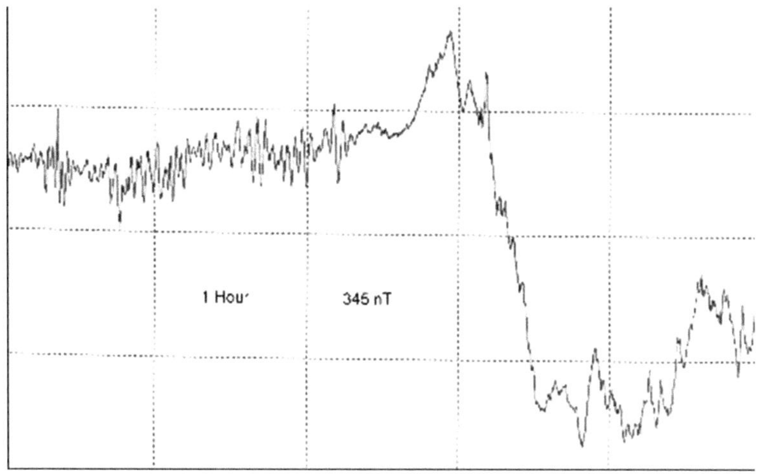

Fig. 1.11 Magnetic storm activity, Wednesday October 29, 2003—11:00 a.m. to 4:00 p.m. EST. Data acquired with GEM's Potassium (GSMP-40) ultra-sensitive magnetometer. (Taken from the site www.gemsys.ca/news/geomagnetic%20storm.htm)

The main phase begins when the solar plasma clouds stream out from the Sun and then reach the Earth's magnetosphere. This phase with duration from 3 to 20 h brings a sequence of violent processes referred to as substorms, which result from the solar wind energy intrusion into the magnetosphere. The magnetic storm develops because of the substorm superposition with time.

At middle and low latitudes the main phase of magnetic storm manifests itself as a decrease in H-component of the Earth's field by a value from 40 to 400 nT during several hours or day and more. Ground-based observations have shown that the magnetic substorm pattern characteristically exhibits a creek-shaped profile (magnetic bay) similar to the indentation of a coastline. The magnetic variations due to the substorm typically lasts for 1–2 h with a peak value about 30–300 nT while the irregular variations on the ground surface can reach a value from 5×10^2 to 3×10^3 nT. One of the most intensive magnetic storms happened on October 29, 2003 as shown in Fig. 1.11 (e.g., see Panasyuk et al. 2004). As is seen from this plot, the amplitude of a positive anomaly is as much as 500 nT and a subsequent negative anomaly is of the order of 900 nT while the change from maximum to minimum occurred over the period of approximately 1 h.

During the magnetic storm the energy flux incident upon the magnetosphere increases by 1–2 order of magnitude. It can reach a value $\sim 10^{12}$–10^{13} W that is close to the power requirements of the humanity but only 1–5 % of this energy enter the magnetosphere. The most part of this energy penetrates into the magnetosphere due to viscous friction at the magnetopause and via reconnection of the interplanetary and Earth's field lines. As a result, the magnetic energy piles up at the magnetotail.

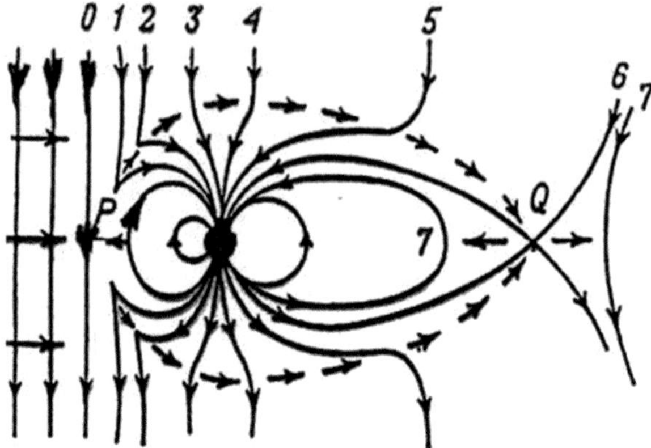

Fig. 1.12 Dungey model for the reconnection between magnetotail field lines and interplanetary magnetic field frozen in solar wind (Dungey 1961). The reconnection is assumed to occur at sunlit magnetopause during magnetic storms and substorms. Lines 0–7 illustrate convection of magnetic field lines in antisunward direction

The energy transfer caused by the magnetic reconnection is predominant at the boundary of the magnetosphere while the dissipation of the transmitted energy prevails at the magnetotail due to reconnection of the antiparallel magnetic fields in North and South portions of the tail.

For the solar-quiet time the magnetotail field lines are closed through the equatorial plane. The magnetic storm and substorms are accompanied by reconnection between the solar wind field and the Earth magnetic field at the sunlit magnetopause. For example, Fig. 1.12 shows a schematic plot of Dungey model (Dungey 1961) for the reconnection process under the requirement that the interplanetary magnetic field is directed from the North to the South. At first the undisturbed interplanetary field shown with line 0 interacts with the magnetosphere providing the North and the South portions of the opened field line shown with line 1. The high conducting plasma of the solar wind drags that part of the opened field line which is frozen in the solar wind. The result is that this part of the field line moves in the antisunward direction towards the magnetotail whereas the bottom of the line is fixed at the non-conducting atmosphere. As is seen from Fig. 1.12, the Northern and Southern portions of the "frozen in" field lines slide over the lower boundary of the ionospheric current layer, as shown with lines 1–5, until their interception at the point Q of the magnetotail. The upper portions of the field line joins the lower one at this point thereby producing the reverse convection of the fields lines and plasma towards the neutral point P as shown in Fig. 1.12 with field lines 6 and 7. So the reconnection mechanism generates plasma convective motion in the antisunward direction in outer part of the magnetotail whereas the reverse convection arises in inner part of the magnetotail.

Additionally, the similar effect of plasma convection at the magnetotail builds up as a result of the quasiviscous interaction between the solar wind and the magnetosphere although this mechanism is assumed to be not so intensive as compared to the reconnection mechanism.

The opened field lines 1–6, which converge over the polar caps, can occur simultaneously as the southward-directed interplanetary magnetic field exists for a long time. Since the opened lines cannot serve as the magnetic trap for the charged particles, the plasma does not pile up at these lines and hence the plasma density is lowered in this region. Moreover, the intensity of particle precipitation at the polar cap is much smaller than that at auroral region.

During the main phase of a magnetic storm the magnetic reconnection and changes in the magnetosphere size are responsible for the penetration of new particles into the magnetosphere. This tends to accelerate the magnetospheric plasma thereby exciting the westward ring current in the magnetosphere within the radii from $3R_e$ to $5R_e$. For the quiet Sun period the ring current is caused by the westward proton drift and eastward electron drift in the Earth's dipole magnetic field. Recent satellite measurements have shown that the energy of trapped protons can increase up to 10–120 keV during the main phase. The enhanced ring current generates a magnetic field in opposition to the geomagnetic field that causes the decreases of the geomagnetic field in accordance with ground-based observations.

The recovery phase starts with diffusion of the trapped particles in the magnetosphere followed by the decay of the ring current. The Coulomb scattering and reactions of protons and neutral hydrogens of the following type $H^+ + H \longleftrightarrow H + H^+$ are assumed to be the key mechanism for energy dissipation.

Now consider the changes in spacial pattern of magnetic field lines at magnetotail during a typical substorm. Figure 1.13 illustrates the so-called near-Earth neutral line (NENL) model of the substorm (e.g., Russell and McPherron 1973; Hones 1979; Baker et al. 1996). At the initial stage of substorm the "frozen in" field lines are dragged tailward. As a result the magnetotail is also stretched more and more along the magnetopause that gives rise to the energy accumulation at the tail. When the open magnetic flux is transported from the dayside magnetosphere towards the magnetotail lobes, it produces the pressure enhancement, neutral sheet sharpening, and decrease of the neutral sheet current. This process leads to the progressive break of the tail current and to the reconnection of the antiparallel magnetic fields of the lobes. In response to the reconnection the most part of the magnetotail is shaped into a huge plasmoid, which breaks away from the magnetosphere, and subsequently escapes in antisunward direction. The result is that the length of magnetotail decreases abruptly but it will be restored to its original size at subsequent relaxation.

The current disruption in the neutral sheet of magnetotail and generation of field-aligned currents flowing into the ionosphere is sketched in Fig. 1.14 (McPherron 1979). These currents flowing along the magnetic field lines close through the night sector of auroral oval thereby producing the so-called West auroral electrojet (AEJ-W). Thus, the field-aligned currents connect the magnetotail with auroral region of the ionosphere during a magnetospheric substorm. At this point the field-aligned current amplitude increases several times while the AEJ-W reaches a value $\sim (1–2) \times 10^6$ A. When the high-energy particles associated with the field-aligned currents precipitate into the ionosphere, they dramatically increase aurora borealis. The particle energy absorption in the atmosphere reaches a peak value

Fig. 1.13 Near-Earth Neutral
Line (NENL) substorm
model based on the field line
reconnection at the middle
portion of magnetotail.
Evolutions of the magnetotail
during substorm development
are shown with panels
numbered by 1–9. A huge
plasmoid builds up as a result
of the reconnection as shown
in panel 6, then it breaks
away from the magnetosphere
and subsequently escapes in
antisunward direction.
Adapted from McPherron
(1979)

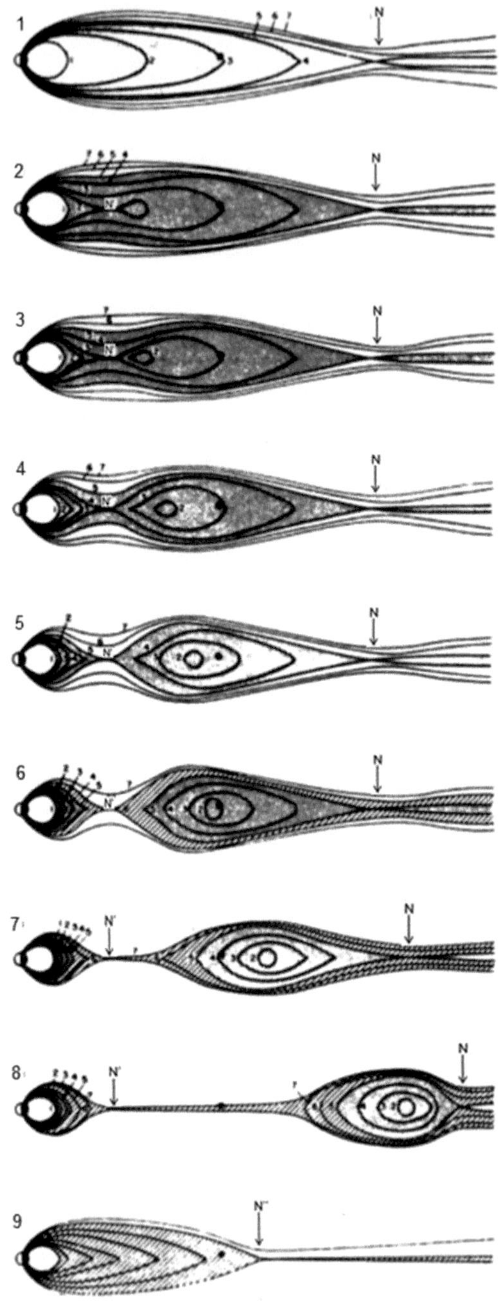

10^{-6}–10^{-5} W/cm^2 in the altitude range 100–200 km over the auroral oval where the
most bright aurora occurs. The permanent ground-based measurements of magnetic
variations caused by the auroral electrojet provide us with the so-called AE-index
which serves as one of the most important characteristics of space weather.

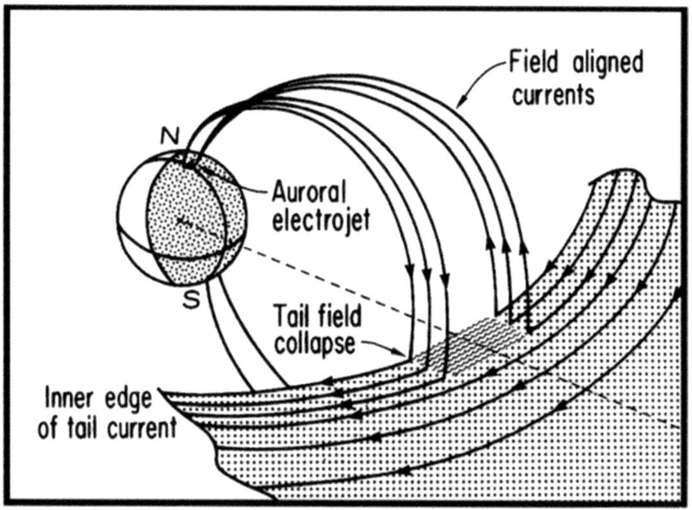

Fig. 1.14 Currents in the neutral sheet, field-aligned currents and ionospheric Hall currents that create eastward (AEJ-E) and westward (AEJ-W) auroral electrojets. A region of current break in the neutral sheet is also shown. Taken from McPherron et al. (1973)

The above-mentioned reconnection of the magnetic field lines in the neutral sheet of magnetotail constitutes the dominant bulk of the subject matter of the classic NENL model (Russell and McPherron 1973; Baker et al. 1996). In the last decade considerable attention has been focused on the alternative models for the active phase of the magnetic substorm. In these models the reconnection of the open magnetic field flux does not play a vital role. The key to the alternative or CD (Current Disruption) models is the nonlinear magnetosphere-ionosphere coupling provided by the magnetic energy which was accumulated in the closed magnetotail (e.g., see Erickson et al. 2000; Lui 2001). In the framework of the CD models the substorms are basically due to small-scale current disruptions in the near and closed magnetotail whereas the NENL model assumes that the substorms are caused by large-scale processes in the middle tail including the reconnection in the open lobes of the tail.

The concept of the self-organized criticality has been the subject of a great deal of research during recent years. In these models the substorm is considered as a stochastic process in the small scale cellular structure with random distribution of the cell parameters such as the magnetic field. Notice that the observation is actually indicative of the small-scale structure of the magnetotail field and neutral sheet current (e.g., Milovanov et al. 1996; Uritsky and Pudovkin 1998 and references herein; Klimas et al. 2000).

The numerical simulation of MHD processes in the magnetosphere provides us with a more comprehensive picture of the substorm development. The interested reader is referred to the text by Raeder and Maynard (2001) and Sonnerup et al. (2001) for a more complete treatise on numerical simulation of magnetospheric processes.

The magnetic substorms and storms trigger ULF MHD waves and the cosmic particle precipitation, which in turn influences the ionosphere conductivity and radiowave reflectivity allowing for disruption of radio communications. The enhancement of trapped particles may greatly affect the satellites even making the electronic equipment and solar batteries inoperative. The magnetic storm may have an impact on the power lines on the Earth, the atmosphere and biosphere and etc.

In relation to the next sections, it is pertinent to note that the interaction of solar wind with the Earth's magnetosphere is similar in part to the global seismotectonic phenomena such as earthquakes. The energy of solar wind plasma first tends to pile up at the magnetotail followed by sudden energy release that in turn gives rise to power fluxes of energetic particles and global redistribution of the magnetospheric currents. At this point the magnetic storm and substorms play a role of "magnetospherequake" while the major mechanism of these dramatic phenomena has been something of a mystery.

1.4   MHD Waves

1.4.1   Basic Equations for MHD Waves in a Homogeneous Conducting Medium

Variations in the solar wind dynamic pressure result in changes of plasma and field properties throughout the excitation of MHD waves propagating in the solar wind and the Earth's magnetosphere. A variety of MHD waves can be split into several classes including the field line resonances, cavity modes, waveguide modes, and so on.

Alfvén (1950) was the first who studied the MHD waves, which can propagate in a conducting medium immersed in the external magnetic field. The MHD approach can be applied to the waves propagating in both the plasma of solar wind and magnetospheric plasma. To study these waves in a little more detail we consider a homogeneous single fluid/plasma. The quiet state of the fluid is described by the constant mass density ρ_0, the pressure P_0, zero mass velocity, and the uniform magnetic field $\mathbf{B}_0$. So, we use the subscript zero to describe the unperturbed values of the medium parameters. Let $\delta\rho$ be the small perturbation of the mass density, so that $\delta\rho \ll \rho_0$. The continuity equation (1.10) can thus be linearized, and we obtain

$$\partial_t \delta\rho + \rho_0 \nabla \cdot \mathbf{V} = 0, \tag{1.48}$$

where $\mathbf{V}$ is the mass velocity of the conducting medium.

Let $\delta\mathbf{B}$ and δP be the small perturbations of the magnetic field $\mathbf{B}_0$ and of the pressure P_0, respectively. Similarly, in the first approximation Eq. (1.37) of the conducting fluid motion can be reduced to

$$\rho_0\partial_t\mathbf{V} = -\nabla\delta P + \frac{1}{\mu_0}\left(\nabla\times\delta\mathbf{B}\right)\times\mathbf{B}_0. \tag{1.49}$$

To simplify the problem, we assume that the fluid conductivity $\sigma\to\infty$, so that the magnetic field lines are frozen into the conducting fluid. This means that the electric field can be derivable from the velocity through Eq. (1.35), that is $\mathbf{E} = \mathbf{B}_0\times\mathbf{V}$. In such a case Eq. (1.18) is reduced to

$$\partial_t\delta\mathbf{B} = \nabla\times\left(\mathbf{V}\times\mathbf{B}_0\right). \tag{1.50}$$

Since the flow is isoentropic, the changes of pressure are related to the changes of density through

$$\delta P = c_s^2\delta\rho, \tag{1.51}$$

where $c_s^2 = (\partial P/\partial\rho)_S$ is the squared sound velocity taken at the constant entropy. We seek for the solution of the set of Eqs. (1.48)–(1.51) in the form of harmonic wave. All the quantities are assumed to vary as $\exp(i\mathbf{k}\cdot\mathbf{r} - i\omega t)$, where $\mathbf{k}$ is the wave vector and ω is the frequency. The following combined set of dynamic and electrodynamic equations for the conducting fluid remains after these simplifications:

$$\omega\delta\rho = \rho_0\mathbf{k}\cdot\mathbf{V}, \tag{1.52}$$

$$\omega\rho_0\mathbf{V} = c_s^2\delta\rho\mathbf{k} + \mu_0^{-1}\mathbf{B}_0\times(\mathbf{k}\times\delta\mathbf{B}), \tag{1.53}$$

$$-\omega\delta\mathbf{B} = \mathbf{k}\times(\mathbf{V}\times\mathbf{B}_0). \tag{1.54}$$

In addition, the equation $\nabla\cdot\mathbf{B} = 0$ is reduced to $\mathbf{k}\cdot\delta\mathbf{B} = 0$. The former equation holds automatically since $\delta\mathbf{B}$ is perpendicular to $\mathbf{k}$ as it follows from Eq. (1.54).

1.4.2 Shear Alfvén Waves

The set of Eqs. (1.52)–(1.54) can be split into two independent sets of variables (e.g., see Landau and Lifshitz 1982). The first one consists of the perpendicular components of magnetic perturbation $\delta\mathbf{B}_\perp$ and the velocity $\mathbf{V}_\perp$ as shown in Fig. 1.15 with the arrows parallel to z-axis. Both of these vectors are perpendicular to that plane in which the undisturbed field $\mathbf{B}_0$ and wave vector $\mathbf{k}$ are situated. As is seen from Eq. (1.52), this means that $\delta\rho = 0$, i.e., the medium density does not vary. Combining Eqs. (1.53) and (1.54), carrying out the triple cross product

$$\mathbf{A}_1\times(\mathbf{A}_2\times\mathbf{A}_3) = \mathbf{A}_2\left(\mathbf{A}_1\cdot\mathbf{A}_3\right) - \mathbf{A}_3\left(\mathbf{A}_1\cdot\mathbf{A}_2\right), \tag{1.55}$$

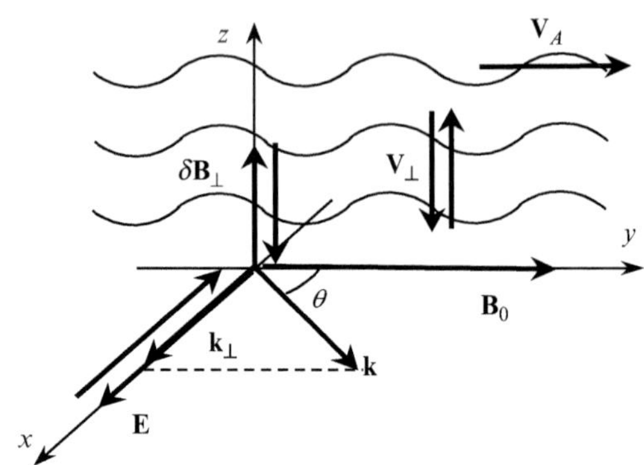

Fig. 1.15 Electric, **E**, magnetic, $\delta\mathbf{B}_\perp$, perturbations and mass velocity, $\mathbf{V}_\perp$, caused by the shear Alfvén wave propagation. Magnetic field lines disturbed by linearly polarized Alfvén waves are shown with *wavy lines*. $\mathbf{V}_A$ denotes the group velocity of the Alfvén wave

and taking into account that $\delta\rho = 0$, $\mathbf{B}_0 \cdot \delta\mathbf{B}_\perp = 0$ and $\mathbf{k} \cdot \mathbf{V}_\perp = 0$, yields

$$\omega\mu_0\rho_0\mathbf{V}_\perp + (\mathbf{k} \cdot \mathbf{B}_0)\,\delta\mathbf{B}_\perp = 0, \tag{1.56}$$

$$(\mathbf{k} \cdot \mathbf{B}_0)\,\mathbf{V}_\perp + \omega\delta\mathbf{B}_\perp = 0. \tag{1.57}$$

Here we have taken into account only projection of the equations on the vertical axes z. The set of Eqs. (1.56)–(1.57) has a nontrivial solution for $\mathbf{V}_\perp$ and $\delta\mathbf{B}_\perp$ only if the determinant of the set equals zero. Hence we get the following dispersion relation

$$\omega\,(\mu_0\rho_0)^{1/2} = \pm\mathbf{k} \cdot \mathbf{B}_0. \tag{1.58}$$

The phase velocity of the wave is thus given by

$$\omega/k = \pm V_A \cos\theta, \tag{1.59}$$

where θ is the angle included between the vectors $\mathbf{k}$ and $\mathbf{B}_0$, and V_A denotes the Alfvén speed

$$V_A = \frac{B_0}{(\mu_0\rho_0)^{1/2}}. \tag{1.60}$$

The group velocity $\mathbf{V}_A$ is pointed along $\mathbf{B}_0$

$$\mathbf{V}_A = \frac{d\omega}{d\mathbf{k}} = \pm\frac{\mathbf{B}_0}{(\mu_0\rho_0)^{1/2}}. \tag{1.61}$$

This wave mode has been termed the shear Alfvén wave because the mass velocity of the conducting fluid/plasma is perpendicular to both the group and phase velocities of the wave. Moreover, once the shear Alfvén wave propagates in the medium, the mass density remains unchanged; that is, the material can be considered as incompressible.

Magnetic field lines disturbed by the linearly polarized Alfvén wave are shown in Fig. 1.15 with wavy lines. It follows from Eq. (1.57) that for waves propagating along $\mathbf{B}_0$ ($\mathbf{k} \cdot \mathbf{B}_0 > 0$) the vectors $\mathbf{V}_\perp$ and $\delta\mathbf{B}_\perp$ are antiparallel, while for waves propagating antiparallel to $\mathbf{B}_0$ these vectors are parallel. This relation makes it possible to detect a direction of the Alfvén wave energy propagation, using only a single-point measurement (Glassmeier 1995).

It should be emphasized that the transverse displacements in the plasma can propagate along the magnetic field lines because the conducting plasma is frozen into the magnetic field. The transverse displacements in plasma result in curvature and stretching of the field lines shown in Fig. 1.15. Magnetic forces in a conducting medium act on the field lines in analogy to the quasielastic forces; that is, the magnetic forces act in such a way to tighten the field lines. In some sense, the Alfvén oscillations of the field lines are similar to oscillations of stretched strings.

To summarize, we note that the shear Alfvén wave has a field-aligned electric current parallel to the undisturbed field $\mathbf{B}_0$, an electric field $\mathbf{E}$ parallel to the perpendicular wave vector $\mathbf{k}_\perp$, and a magnetic component $\delta\mathbf{B}_\perp$ perpendicular to $\mathbf{B}_0$ and the vector $\mathbf{k}$. This shows that the Alfvén wave is a purely transverse wave with respect to the magnetic field $\mathbf{B}_0$. These properties of the Alfvén wave have an important role in magnetospheric plasma dynamics. The arbitrary perturbations of the plasma parameters can thus propagate along the magnetic field lines at the Alfvén velocity [Eq. (1.60)], which in turn depends on the magnetic field induction and the plasma density. The guiding by field lines, which is probably the major property of the Alfvén waves, holds in inhomogeneous plasmas and even under the finite curvature of the field lines.

1.4.3   Fast and Slow Magnetosonic Waves

Contrary to the Alfvén wave, the next wave mode can be excited in the plasma if both the magnetic perturbation $\delta\mathbf{B}$ and the mass velocity $\mathbf{V}$ are in that plane which contains the undisturbed field $\mathbf{B}_0$ and wave vector $\mathbf{k}$. A schematic representation of the field components is shown in Fig. 1.16. As is seen from this figure, all the vectors, i.e., $\delta\mathbf{B}$, $\mathbf{V}$, $\mathbf{B}_0$, and $\mathbf{k}$ are situated in the x, y plane whereas the electric field is normal to that plane. As before $\delta\mathbf{B}$ is perpendicular to $\mathbf{k}$. Taking into account the field polarization, Eq. (1.54) is reduced to

$$\omega\delta\mathbf{B} = -\left(\mathbf{V} \times \mathbf{B}_0\right)_z (\mathbf{k} \times \hat{\mathbf{z}}). \tag{1.62}$$

where the subscript z denotes the vector projection on z axis. Notice that the vector $\mathbf{k} \times \hat{\mathbf{z}}$ is anti-parallel to y axis as shown in Fig. 1.16. We may also eliminate $\delta\rho$ from Eqs. (1.52) and (1.53) to yield

$$\mu_0\rho_0\omega^2\mathbf{V} = \mu_0\rho_0 c_s^2 \left(\mathbf{k} \cdot \mathbf{V}\right)\mathbf{k} + \omega\mathbf{B}_0 \times (\mathbf{k} \times \delta\mathbf{B}). \tag{1.63}$$

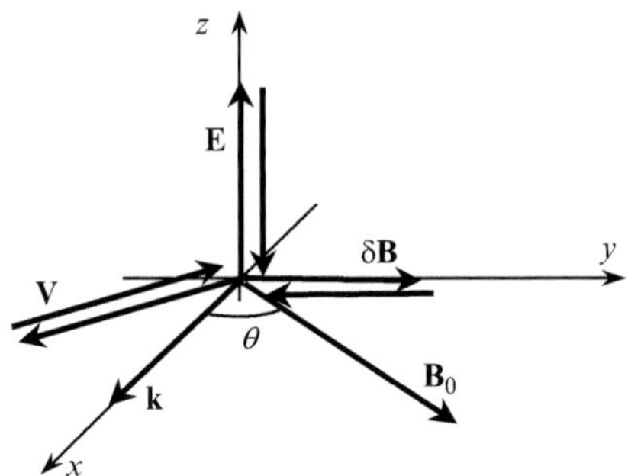

Fig. 1.16 The same field components as shown in Fig. 1.15 but for fast and slow magnetosonic (FMS and SMS) waves

Substituting Eq. (1.62) for $\delta\mathbf{B}$ into Eq. (1.63), taking into account the vector identity (1.55), and rearranging we arrive at

$$\left(\omega^2 - k^2 V_A^2\right)\mathbf{V} = c_s^2\left(\mathbf{k}\cdot\mathbf{V}\right)\mathbf{k} - k^2\left(\mathbf{V}\cdot\mathbf{V}_A\right)\mathbf{V}_A, \tag{1.64}$$

where V_A and $\mathbf{V}_A$ are given by Eqs. (1.60) and (1.61). Projections of this vector equation onto x and y axes can be written as

$$a_{xx}V_x + a_{xy}V_y = 0, \tag{1.65}$$

$$a_{yx}V_x + a_{yy}V_y = 0, \tag{1.66}$$

where the matrix coefficients are given by

$$a_{xx,yy} = \omega^2 - k^2 V_A^2 - c_s^2 k_{x,y}^2 + k^2 V_{Ax,y}, \tag{1.67}$$

$$a_{xy} = a_{yx} = k^2 V_{Ax} V_{Ay} - c_s^2 k_x k_y, \tag{1.68}$$

and V_{Ax} and V_{Ay} denotes the projections of the vector $\mathbf{V}_A$ onto the coordinate axes x and y.

The set of Eqs. (1.65)–(1.66) has a nontrivial solution for V_x and V_y only if the determinant of the set is equal to zero. Hence we come to the following dispersion relation

$$\frac{\omega^2}{k^2} = \frac{1}{2}\left\{V_A^2 + c_s^2 \pm \left[\left(V_A^2 + c_s^2\right)^2 - 4c_s^2\left(\hat{\mathbf{x}}\cdot\mathbf{V}_A\right)^2\right]^{1/2}\right\}, \tag{1.69}$$

where $\hat{\mathbf{x}}$ stands for the unit vector parallel to $\mathbf{k}$. This equation describes two different modes. The sign plus in Eq. (1.69) determines the mode, which is referred to as the fast magnetosonic (FMS) wave while the sign minus corresponds to the so-called slow magnetosonic (SMS) wave. Below we show that the fast wave propagates almost isotropically, while the slow wave is a strongly anisotropic mode. Both the magnetosonic modes readily carry a field-aligned magnetic perturbation; that is, they describe a magnetically compressive mode.

As long as the magnetic field $\mathbf{B}_0$ is weak or the fluid is practically incompressible ($c_s \to \infty$), the Alfvén velocity V_A can be neglected compared to the sound velocity c_s. In this extreme case the dispersion relation for the FMS wave is reduced to $\omega/k = c_s$. The FMS wave propagation brings the changes in distances between the field lines that in turn is accompanied by plasma compressions and rarefaction, that is, by plasma density variations. The implication here is that the FMS wave can transform into the conventional sound wave, which propagates isotropically. The FMS wave is often called a compressional mode in analogy to the acoustic longitudinal wave. In the same limit the dispersion relation of the SMS wave coincides with Eq. (1.59) for the Alfvén wave.

Conversely, if $V_A^2 \gg c_s^2$, that is $B_0^2 \gg \mu_0 \rho_0 c_s^2$, then the dispersion relation for the FMS wave is reduced to $\omega/k = V_A$, while the SMS mode is described by the relation $\omega/k = c_s \cos\theta$. This extreme case is of special interest in magnetospheric plasma dynamics because the magnetic pressure of the plasma is much greater than the thermal pressure in most regions of the Earth's magnetosphere. However, in regions of the ring current the thermal pressure might not be negligible. In practice, the phase velocity of the slow wave is rather small and therefore this wave is subject to Landau attenuation. The SMS wave is usually not observed in space plasmas so that this mode is of minor importance. There are therefore two most important MHD modes in the homogeneous magnetized plasma. The first mode is the shear Alfvén wave, which is guided by magnetic field lines according to Eqs. (1.59)–(1.61). The next one is the FMS/compressional mode, which propagates isotropically at the Alfvén velocity [Eq. (1.60)].

References

Akasofu S-I, Chapman S (1972) Solar-terrestrial physics, Pt. 1 and 2. Clarendon Press, Oxford

Alfvén H (1950) Cosmical electrodynamics. Clarendon Press, Oxford

Alfvén H, Falthammar C-G (1963) Cosmical electrodynamics, 2nd edn. Clarendon Press, Oxford

Baker DN, Pulkkinen TJ, Angelopoulos V, Baumjohann W, McPherron RL (1996) Neutral line model of substorms: past results and present view. J Geophys Res 101:12975–13010

Bott MHP (1982) The interior of the earth: its structure, constitution, and evolution, 2nd edn. Elsevier, London, 403 p

Braginsky SI (1964) Self-excitation of a magnetic field during the motion of a highly conducting fluid. J Exp Theor Phys (Zhurnal Eksperimental'noy i Teoreticheskoy Fiziki). Engl. Transl. 48:1084–1098

Brodsky YuA (1983) Analysis of geodynamo equations by means of perturbation theory. Geophys Astrophys Fluid Dyn 22:281–304

Brown GC, Mussett AE (1993) The inaccessible earth: an integrated view to its structure and composition, 2nd edn. Chapman and Hall, London, 278 p

Chapman S, Ferraro VCA (1931) A new theory of magnetic storms. Terr Magn Atmos Elect 36:77–97; 171–186

Chapman S, Ferraro VCA (1932) A new theory of magnetic storms. Terr Magn Atmos Elect 37:147–156

Chapman S, Ferraro VCA (1933) A new theory of magnetic storms. Terr Magn Atmos Elect 38:79–96

Chapman S, Ferraro VCA (1940) A new theory of magnetic storms. Terr Magn Atmos Elect 45:245–268

Cowling TG (1934) The magnetic field of sunspots. Mon Notices Roy Astron Soc 94:39–48

Cowling TG (1953) Solar electrodynamics. In: Kuiper GP (ed) The Sun. University of Chicago Press, Chicago

Cowling TG (1976) Magnetohydrodynamics, 2nd edn. Adam Hilger, London

Davies L (1958) In: Lehnert B (ed) Electromagnetic phenomena in cosmical physics. Cambridge University Press, Cambridge, p 27

Dungey JW (1961) Interplanetary magnetic field and the auroral zone. Phys Rev Lett. 6:47–49

Erickson GM, Maynard NC, Burke WJ, Wilson GR, Heinemann MA (2000) Electromagnetics of substorm onsets in the near-geosynchronous plasma sheet. J Geophys Res 105(A11):25265–25290

Fedorov E, Pilipenko V, Surkov V, Rao DRK, Yumoto K (1999) Ionospheric propagation of magnetohydrodynamic disturbances from the equatorial electrojet. J Geophys Res Space Phys 104:4329–4336

Glassmeier K-H (1995) ULF Pulsations. In: Volland H (ed) Handbook of atmospheric electrodynamics, vol 2. CRC Press, Boca Raton, pp 463–502

Golitsyn GS (1981) Convection structure under fast rotation. Rep USSR Acad Sci (Dokl Akad Nauk SSSR) 261(2):317–320 (in Russian)

Gonzalez, WD, Joselyn JA, Kamide Y, Kroehl HW, Rostoker G, Tsurutani BT, Vasyliunas VM (1994) What is a geomagnetic storm? J Geophys Res 99(A4):5771–5792

Gutenberg B (1914) Über Erdbebenwellen, VIIA. Beobachtungen an Registrierungen von Fernbeben in Göttingen und Folgerungen über die Konstitution des Erdköerpers. Nachr Ges Wiss Göttingen Math Phys Kl. 1:1–52

Hones EW Jr (1978) Plasma flow in magnetotail and its implications for substorm theories. In: Akasofu S-I (ed) Dynamics of the magnetosphere. Proceedings of the A.G.U. Chapman Conference 'Magnetospheric substorms and related plasma processes' held at Los Alamos Scientific Laboratory, Los Alamos, NM, USA, October 9–13, 1978. Astrophysics and space science library, vol 78. D. Reidel, Dordrecht, Netherlands; Boston, USA, pp 545–562

Jackson JD (2001) Classical electrodynamics, 3rd edn. Wiley, Hoboken

Jeffreys H (1970) The Earth, its origin, history and physical constitution, 5th edn. Cambridge University Press, Cambridge

Kelley MC (1989) The earth's ionosphere. Academic, New York

Klimas AJ, Valdivia JA, Vassiliadis D, Baker DN, Hesse M, Takalo J (2000) Self-organized criticality in the substorm phenomenon and its relation to localized reconnection in the magnetospheric plasma sheet. J Geophys Res 105:18765–18780

Krause F, Rädler K-H (1980) Mean-field magneto-hydrodynamics and dynamo theory. Pergamon Press, Oxford

Landau LD, Lifshitz EM (1982) Electrodynamics of continuous media, 2nd edn. Nauka, Moscow (in Russian)

Larmor J (1919) How could a rotating body such as the sun become a magnet? Rep Brit Assoc Adv Sci 87:159–160

Liu ATY (2001) Current controversies in magnetospheric physics. Rev Geophys 39:535–563

McPherron RL (1979) Magnetospheric substorms. Rev Geophys Space Phys 17:657–681

McPherron RL, Russell CT, Aubry MP (1973) Satellite studies of magnetospheric substorms on August 15, 1968: 9. Phenomenological model for substorms. J Geophys Res 78(16):3131–3149. doi:10.1029/JA078i016p03131

Milovanov AV, Zelenyi LM, Zimbardo G (1996) Fractal structures and power law spectra in the distant Earth's magnetotail. J Geophys Res 101:19903–19910

Mishin VM, Bazarzhapov AD (2002) Introduction into physics of the Earth magnetosphere: basic problems and some results of the team TIM ISZF. In: "Solar and Geophysical Studies", Proceedings of V session of young scientists, Baykal youth scientific school on fundamental physics, Irkutsk, pp 16–30 (in Russian)

Moffat H (1968) Magnetic field generation in electrically conducting fluids. Cambridge University Press, Cambridge

Molchanov SA, Ruzmaykin AN, Sokolov DD (1985) The kinematic dynamo in a random flux. Adv Phys Sci (Uspekhi Fizicheskikh Nauk) 145:593–628

Nikolaevskiy VN (1996) Geomechanics and fluidodynamics: with applications to reservoir engineering. Theory and applications of transport in porous media, vol 8. Kluwer Academic, Dordrecht

Nishida A (1978) Geomagnetic diagnosis of the magnetosphere. Springer, New York

Panasyuk MI, Kuznetsov SN, Lazutin LL, Avdyushinn SI, Alexeev II, Ammosov PP, Antonova AE, Baishev DG, Belenkaya ES, Beletsky AB, Belov AV, Benghin VV, Bobrovnikov SYu, Bondarenko VA, Boyarchuk KA, Veselovsky IS, Vyushkova TYu, Gavrilieva GA, Gaidash SP, Ginzburg EA, Denisov YuI, Dmitriev AV, Zherebtsov GA, Zelenyi LM, Ivanov-Kholodny GS, Kalegaev VV, Kanonidi KhD, Kleimenova NG, Kozyreva OV, Kolomiitsev OP, Krasheninnikov IA, Krivolutsky AA, Kropotkin AP, Kuminov AA, Leshchenko LN, Mar'in BV, Mitrikas VG, Mikhalev AV, Mullayarov VA, Muravieva EA, Myagkova IN, Petrov VM, Petrukovich AA, Podorolsky AN, Pudovkin MI, Samsonov SN, Sakharov YaA, Svidsky PM, Sokolov VD, Soloviev SI, Sosnovets EN, Starkov GV, Starostin LI, Tverskaya LV, Teltsov MV, Troshichev OA, Tsetlin VV, Yushkov BYu (2004) Magnetic storms in October 2003. Cosmic Res 42(5):489–534

Parker E (1955) Hydromagnetic dynamo models. Astrophys J122:293–314

Parker EN (1979) Cosmical magnetic fields. Their origin and their activity. Clarendon Press, Oxford

Raeder J, Maynard NC (2001) Foreword [to Special Section on Proton Precipitation Into the Atmosphere]. J Geophys Res 106(A1):345–348. doi:10.1029/2000JA000600

Roberts PH (1971) Dynamo theory. In: Reid WH (ed) Mathematical problems in the geophysical sciences. Lecturers in applied mathematics, vol 14. American Mathematical Society, Providence, pp 129–206

Russell CT, McPherron RL (1973) The magnetotail and substorms. Space Sci Rev 15:205–266

Sonnerup BUO, Siebert KD, White WD et al (2001) Simulation of the magnetosphere for zero interplanetary magnetic field: the ground state. J Geophys Res 106:24419–24434

Stacey FD (1969) Physics of the earth. Wiley, New York

Steenbeck M, Krause F, Rädler K-H (1966) Berechnung der mittlerer Lorentz-FeldStärke für ein elektrisch leitendes Medium in turbulenter, durch Coriolis-Kräfte beeinflusster Bewegung. Z Naturforsch 21:369–376

Stevenson DJ (1983) Planetary magnetic fields. Rep Prog Phys 46:555–620

Surkov, VV, Fedorov EN, Pilipenko VA, Rao DRI (1997) Ionospheric propagation of geomagnetic perturbations caused by equatorial electrojet. Geomagn Aeron 37:61–70

Uritsky VM, Pudovkin MI (1998) Low frequency l/f-like fluctuations of the AE-index as a possible manifestation of self-organized criticality in the magnetosphere. Ann Geophys 16(12): 1580–1588

Vaynshtein SI, Zeldovich YaB, Ruzmaykin AA (1980) Turbulent dynamo in astrophysics. Nauka, Moscow, 352 p (in Russian)

Yanovsky BM (1978) Earth Magnetism. Leningrad University Publisher, Leningrad, 592 pp (in Russian)

Zeldovich YaB (1956) The magnetic field in the two-dimensional motion of a conducting turbulent fluid. J Exp Theor Phys (Zhurnal Eksperimental'noy i Teoreticheskoy Fiziki) 31:154–156 [Engl. transl. Sov Phys JETP (1957) 4:460–462]

Zeldovich YaB, Ruzmaikin AA, Sokoloff DD (1983) Magnetic fields in astrophysics. Gordon & Breach Sci. Publ. Inc., NY

Zeldovich YaB, Ruzmaikin AA (1987) The hydromagnetic dynamo as the source of planetary, solar, and galactic magnetism. Sov Phys Usp 30:494–506. doi:10.3367/UFNr.0152.198706d. 0285

Zharkov VN (1983) Internal structure of earth and planets. Nauka Publishers, Moscow, USSR (in Russian)

Chapter 2
The Ionosphere and Atmosphere

Abstract This chapter contains some introductory materials on the formation, structure, and composition of the ionosphere and atmosphere. The altitude dependence of electrical conductivity is extremely important in the ionospheric study and is thus discussed in this chapter. In the remainder of this chapter we use the tensor of plasma conductivity to derive basic properties of shear Alfvén and compressional waves in a homogeneous magnetized plasma.

Keywords Compressional/magnetosonic wave • Ionosphere conductivity • Magnetized plasma • Shear Alfvén wave

2.1 Structure of the Ionosphere and Atmosphere

2.1.1 Formation and Composition of the Ionosphere

The Earth's ionosphere is a partly ionized and tenuous gas that surrounds the Earth. In this region the collisions between the ionized and neutral particles cannot be ignored since the number density of the neutral gas exceeds that of the ionospheric plasma. The lower margin of the ionosphere corresponds to the height about 50 km while the upper limit is the outer boundary of the magnetosphere. The ionosphere plays an important role in the formation of the Earth's magnetosphere.

Photoionization and photodissociation by solar UV radiation, ionization by energetic particle impact on the neutral gas along with recombination processes occur in the ionosphere continuously. In the dayside ionosphere the main source of the gas ionization is shortwave solar radiation with wavelength smaller than 103.8 nm. This process, which is responsible for the production of plasma, competes with recombination at which ions and electrons combine to form neutral molecules or atoms. The so-called dissociative recombination of the molecular ions is the most

V. Surkov and M. Hayakawa, *Ultra and Extremely Low Frequency Electromagnetic Fields*, Springer Geophysics, DOI 10.1007/978-4-431-54367-1_2,
© Springer Japan 2014

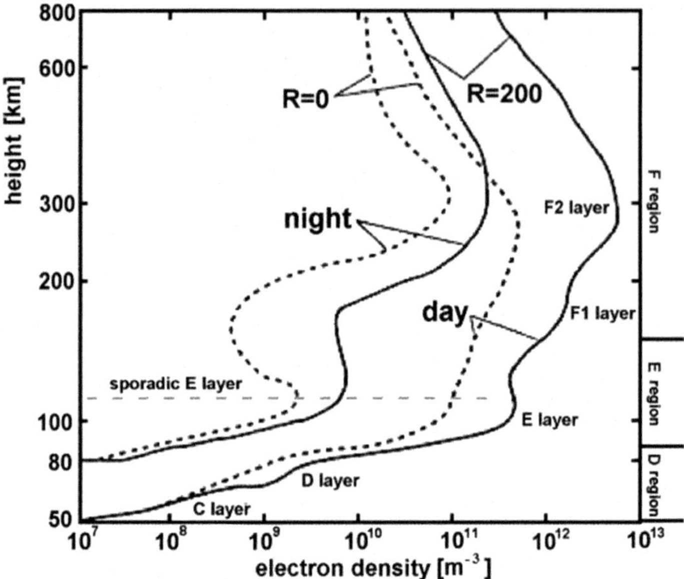

Fig. 2.1 Typical daytime (*solid line*) and nighttime (*dotted line*) distributions of electron number density with various layers designated. In this figure R represents the monthly median solar index. Taken from the site http://roma2.rm.ingv.it/en/research_areas/4/ionosphere

essential type of recombination, which dominates in the main part of the ionosphere. An example of such a reaction, where the molecule breaks apart, is given by

$$NO^+ + e^- \rightarrow N + O. \tag{2.1}$$

The reaction rate of this process is nearly 10^3 times that of usual radiative recombination of the atomic ions followed by the radiation of photons. For example, the radiative reaction

$$O^+ + e^- \rightarrow O + \gamma, \tag{2.2}$$

becomes effective just above 1,000 km where the number density of the atomic ions is 10^5 times that of the molecular ions. It should be noted that there are a variety of different reactions between the molecules and particles that have different reaction rates depending on the height, plasma composition, temperature, etc. Here we omit a complete treatise on atomic ion chemistry and do not go into the detail of numerous reactions.

In the first approximation the ionosphere is horizontally stratified. The ionospheric structure can be nearly organized by a representative altitude-dependence of the plasma number density. Typical mid-latitude profiles of electron number density versus altitude are shown in Fig. 2.1. This profile is controlled by the

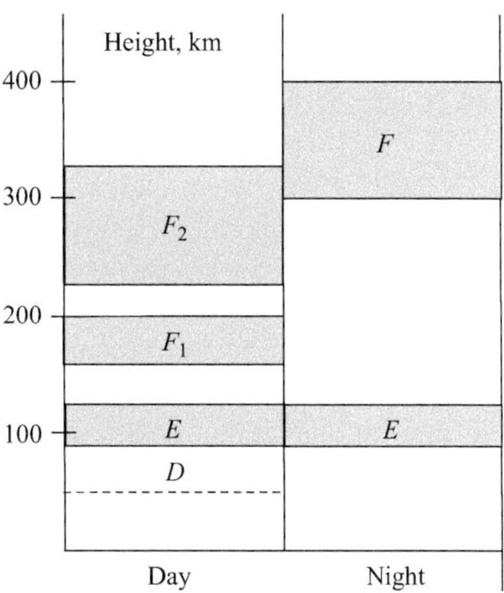

Fig. 2.2 A schematic representation of vertical structure of the ionosphere

equilibrium between the rate of production and recombination of the ions and electrons. The ion number density has approximately the same profile because of plasma quasineutrality. This means that the total number density of all the ions, n_i, must be nearly equal to the number density of electrons, n_e, that is the plasma density $n \simeq n_e \simeq n_i$. As is seen from Fig. 2.1, the electron density exhibits a number of peaks, which occur in several large-scale horizontal layers of ionization. It is widely accepted to divide the ionosphere into several layers shown in Fig. 2.2. The main peak of the electron number density occurs in the so-called F layer. The altitude for which the electron density reaches a maximum is termed the F peak. At night the F layer occupies the altitude range 300–400 km. In the daytime this layer lowers and is split (typically in summer) into two regions. The first density maximum is situated within the F_2 layer in the altitude range 220–320 km and the next maximum lies in the F_1 layer in the range of 160–200 km.

The ion and neutral composition varies with altitude. The atomic oxygen is the dominant neutral gas above about 250 km due to the photodissociation of O_2 by solar UV radiation, that in turn results in the dominance of ions O^+ in the plasma. Not surprisingly, at these altitudes the profiles of the electron and O^+ number densities are similar due to the quasineutrality of plasma. The peak value of the number density of both plasma constituents attain values as high as 10^6 cm^{-3} near noontime. The O^+ plasma at higher altitudes, on the other hand, is often sustained through the night at concentration between 10^4–10^5 cm^{-3}.

The ionizing shortwave radiation of the Sun is mostly absorbed at the altitude range about 100–200 km, which includes the F_1 layer and the E region that is usually located in the interval of 90–150 km. At this region the ionospheric parameters undergo the most regular diurnal and seasonal changes. The peak electron density about $(1–5) \times 10^5$ cm^{-3} occurs in this layer near noontime. At nighttime the plasma number density decreases dramatically in the region of altitude 125–160 km due to the absence of ionization source. Interestingly enough the plasma in the E

layer often survives the night at concentration about $(3\text{--}30) \times 10^3 \, \text{cm}^{-3}$. Unusually narrow layers with enhanced ionization have been occasionally observed in the E and D regions. This short-lived formation, the so-called sporadic E_s layer, is shown in Fig. 2.1 with a sharp peak at the nighttime profile of the electron number density. This layer often contains the ions of metals such as Mg^+, Fe^+, and Ca^+. It is assumed that the sporadic E_s layers can arise in those regions that have a zero wind velocity since these regions can accumulate the heavy ions of metals.

In the so-called D layer of the ionosphere, the height range below about 90 km, the plasma number density is below $10^3 \, \text{cm}^{-3}$. Owing to high values of the electron–neutral and ion–neutral collision frequency the dynamics of plasma in this region are mostly controlled by the neutrals.

The basic features of the ionosphere varies with latitude. It is customary to recognize equatorial, mid-latitude, auroral, and polar ionospheres. The structure of the ionosphere permanently undergoes the diurnal, seasonal variations and even the 11-years solar cycle.

2.1.2 Neutral Atmosphere

The atmosphere resembles a gas cocoon which surrounds the Earth. The horizontal stratification of the lower atmosphere is mostly due to the gravity. The atmospheric pressure P and the number density n_m of neutral gas decrease with height approximately exponentially

$$P = P_0 \exp\left(-z/H_a\right), \quad \text{and} \quad n_m = n_0 \exp\left(-z/H_a\right), \tag{2.3}$$

where the scale height of the atmosphere, H_a, is about several kilometers. In the isothermal atmosphere in hydrostatic equilibrium $H_a = RT/(Mg) \approx 8 \, \text{km}$, where R is the universal gas constant, T is temperature, and M is mass of air mole.

In contrast to the pressure and number density the temperature is a nonmonotonic function of height as shown in Fig. 2.3. The atmosphere is more sensibly organized by a representative temperature profile as compared with the ionosphere. In the so-called troposphere the temperature falls off with height in such a way that the temperature gradient is about 6 K/km. The temperature trend changes abruptly at altitude about 10 km within the transition layer, which is termed the tropopause. In the overlying layer, termed the stratosphere, the temperature increases with height. A maximum about 270 K is accomplished at the stratopause altitude ~ 55 km, where the temperature trend reverses again. In the so-called mesosphere, the height range 55–80 km, the temperature decreases to a minimum about 180 K. In the exosphere, the height range above about 10^3 km, the temperature increases drastically to values as high as 10^3 K and even more. This region is terminated by the mesopause from the thermosphere.

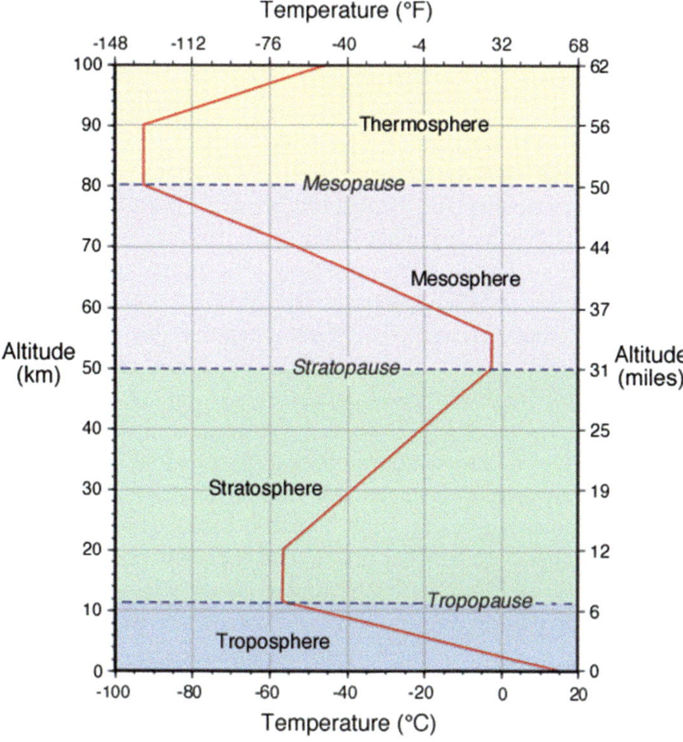

Fig. 2.3 A typical profile of neutral atmospheric temperature. Taken from the site http://www.physicalgeography.net/fundamentals/7b.html

Various physical mechanisms for the atmospheric gas motion result in permanent motions of the upper atmosphere (e.g., see Hines 1974). The basic long-period motions of the lower thermospheric dynamics are the average daily circulation, including both the zonal and meridional circulation, the diurnal and semidiurnal atmospheric tides, the internal gravity and acoustic waves and the turbulence. The oscillation period of the semidiurnal and diurnal tides are 12 and 24 h, respectively, while the motions with period approximately several hours are usually referred to as internal gravity waves (IGWs). Notice that due to a variety of turbulent mixing phenomena the atmosphere is relatively uniform in composition below about 100 km.

The diurnal tides are due to both the solar heating of the sunlit atmosphere, which results in rarefaction of the atmosphere, and the cooling of the nightside atmosphere is followed by its compression. The diurnal variations of the atmosphere heating and cooling excite the tide waves, which in turn produce the atmospheric gas motion in the horizontal direction. The tides generated at tropospheric and stratospheric heights can excite the lower E-region winds. The amplitude of the diurnal variations of the wind velocity increases from 10–30 m/s at the altitude 95 km up to

100–150 m/s above 200 km. It should be noted that the upper thermospheric winds are due to not only the solar heating but also the Joule heating and momentum transfer with the plasma.

The seasonal variations may greatly affect the wind pattern. For example, below 80 km the average daily wind blows from the west to the east in the winter hemisphere. The wind velocity reaches a peak value about 80 m/s at the altitude 60 km. In the summer hemisphere the wind blows to the west with a peak value of the order of 60 m/s at the altitude about 70 km.

It is customary to relate the IGW formation with the presence of tropospheric weather front, tornadoes and thunderstorms. The impulsive auroral zone momentum injection, volcanic eruptions, and other phenomena that can be accompanied by heating events are able to excite the IGW. The appearance of the IGW is due to the fact that the mass density in the atmosphere varies in the vertical direction as described by Eq. (2.3). It can be shown that the dispersion relation for waves in the stratified media is split into two different branches/modes. The so-called Brunt–Väisälä frequency, ω_b, serve as the boundary between these modes. The implication of ω_b is that if a small parcel of air is initially displaced from its equilibrium position it will oscillate around this position in the vertical direction at Brunt–Väisälä frequency. In the isothermal atmosphere $\omega_b = g (\gamma - 1)^{1/2} / c_s$, where γ is adiabatic index or ratio of specific heat at constant pressure to specific heat at constant volume. A typical value for the buoyancy period is $T_b = 2\pi / \omega_b \approx 5$ min. The waves with period less than T_b correspond to the acoustic branch, while the waves with period longer than T_b correspond to the IGW branch. The acoustic waves propagate in the atmosphere in all directions at constant sound velocity c_s, whereas the IGWs undergo dispersion, that is, the IGW velocity is a function of frequency.

The IGW and tides can play an important role in the ionospheric dynamics since their amplitude increases with altitude as $\exp (z/2H)$, at least in the theory (e.g., see Hines 1974), and thus can grow to significant amplitudes during the time they reach ionospheric altitudes.

2.2 Ionospheric Plasma

2.2.1 Tensor of Plasma Conductivity

In Sect. 1.4 we have studied the MHD waves propagating in homogeneous plasmas immersed in a uniform background magnetic field. This approach has a limited area of application in magnetospheric physics since the planetary magnetic field and plasma density are strongly inhomogeneous in space. Moreover, the Earth's magnetic field may affect the partly ionized ionospheric plasma in such a way that the field-aligned plasma conductivity differs from that perpendicular to the magnetic field lines. This implies that the electric current density, $\mathbf{j}$, is related to the electric field, $\mathbf{E}$, through the tensor of the plasma conductivity

$$\mathbf{j} = \hat{\boldsymbol{\sigma}} \cdot \mathbf{E}' \tag{2.4}$$

where $\mathbf{E}' = \mathbf{E} + \mathbf{V} \times \mathbf{B}$ is the electric field in a reference frame moving at the neutral flow velocity $\mathbf{V}$, and $\mathbf{E}$ is the electric field in a reference frame fixed at the earth.

Thus, the Ohm's law in a collisional magnetized plasma differs from its usual form (1.8) in homogeneous conducting media since the plasma conductivity is not a scalar. The tensor components depend on both the gyrofrequencies of the ionized particles and the collision frequencies between the plasma constituents and neutrals. This tensor can be deduced from the equations that describe the electrodynamics of a partially ionized plasma immersed in the external magnetic field. If we use a local coordinate system in which the Earth magnetic field is in the direction of the z axis, the tensor of plasma conductivity can be written as

$$\hat{\boldsymbol{\sigma}} = \begin{pmatrix} \sigma_P & -\sigma_H & 0 \\ \sigma_H & \sigma_P & 0 \\ 0 & 0 & \sigma_{\parallel} \end{pmatrix}, \tag{2.5}$$

where $\sigma_{\parallel}$ is the field-aligned/parallel plasma conductivity, σ_P and σ_H are called the Pedersen and Hall conductivities, respectively. So, the scalar conductivity in the Ohm's law (1.8) should be replaced by the tensor of plasma conductivity (2.5). As a result the Ohm's law (2.4) can be written as follows:

$$\mathbf{j} = \sigma_{\parallel} \mathbf{E}_{\parallel} + \sigma_P \mathbf{E}'_{\perp} + \sigma_H \left(\hat{\mathbf{z}} \times \mathbf{E}'_{\perp} \right), \tag{2.6}$$

where $\mathbf{E}_{\parallel}$ and $\mathbf{E}'_{\perp}$ are the electric field components parallel and perpendicular to the undisturbed magnetic field $\mathbf{B}$, and $\hat{\mathbf{z}} = \mathbf{B}/B$ stands for the unit vector parallel to $\mathbf{B}$.

The tensor components are given by (e.g., see texts by Ginzburg (1970) and Ginzburg and Rukhadze (1972), Kelley (1989), and references therein)

$$\sigma_{\parallel} = e^2 n \left[\frac{1}{m_e \left(v_e - i\omega \right)} + \frac{1}{m_i \left(v_i - i\omega \right)} \right], \tag{2.7}$$

$$\sigma_P = e^2 n \left[\frac{v_e - i\omega}{m_e \left\{ (v_e - i\omega)^2 + \omega_H^2 \right\}} + \frac{v_i - i\omega}{m_i \left\{ (v_i - i\omega)^2 + \Omega_H^2 \right\}} \right], \tag{2.8}$$

$$\sigma_H = e^2 n \left[\frac{\omega_H}{m_e \left\{ (v_e - i\omega)^2 + \omega_H^2 \right\}} - \frac{\Omega_H}{m_i \left\{ (v_i - i\omega)^2 + \Omega_H^2 \right\}} \right], \tag{2.9}$$

Here $n = n_e = n_i$ is the number densities of electrons and ions, ω is wave frequency, $\omega_H = eB_0/m_e$ and $\Omega_H = eB_0/m_i$ are gyrofrequencies/cyclotron frequencies of the electrons and ions, v_{in}, v_{ie}, v_{en}, and v_{ei} are the ion–neutral, ion–electron, electron–neutral, and electron–ion collision frequencies, respectively, $v_i = v_{ie} + v_{in}$ and $v_e = v_{ei} + v_{en}$ are the total ion and electron collision frequencies.

For simplicity we have considered a single ion species of mean mass $\overline{m_i}$, using the symbol m_e for the electron mass and the symbol e for the elemental charge, which is taken to be positive.

To estimate the collision frequencies, which play a crucial role in a partially ionized plasma, a standard atmosphere model is necessary (e.g., see Johnson 1961). The ion–electron collisions at ionospheric heights are of minor importance since the mass of electrons is too small. This implies that the total ion collision frequency $\nu_i = \nu_{in} + \nu_{ie} \approx \nu_{in}$. Notice that in the magnetosphere the ion–electron collisions cannot be dropped since there are no neutrals. The ion–neutral collision frequency derived from the standard parameters of the atmosphere and ionosphere is given by the following approximate formula (Chapman 1956),

$$\nu_{in} = 2.6 \times 10^{-9} \left(n_n + n_i \right) M_n^{-1/2} \ \mathrm{s}^{-1}, \tag{2.10}$$

where n_n and n_i are the neutral and ion number densities in cm^{-3}, respectively, and M_n is the mean molecular mass of the neutrals/ions measured in atomic mass units. Equation (2.10) is valid for both daytime and nighttime ionosphere since $n_i \ll n_n$.

The electron–neutral and the electron–ion collision frequencies are given by (Kelley 1989)

$$\nu_{en} = 5.4 \times 10^{-10} n_n T_e^{1/2} \ \mathrm{s}^{-1}, \tag{2.11}$$

$$\nu_{ei} = \left[34 + 4.18 \ln \left(T_e^3 / n_e \right) \right] n_e T_e^{-3/2} \ \mathrm{s}^{-1}, \tag{2.12}$$

where both the electron and neutral number densities, n_e and n_n, are measured in cm^{-3} and the electron temperature, T_e, is measured in K. The sum of the two collision frequencies yields the total electron collision frequency, i.e., $\nu_e = \nu_{ei} + \nu_{en}$.

All the collision frequencies are highly dependent on the altitude, whereas the gyrofrequencies do not so undergo the changes in the altitude range of interest. For an equatorial ionosphere with $B_0 = 2.5 \times 10^{-5}\,\mathrm{T}$ the gyrofrequencies are $\omega_H = 0.44 \times 10^7\,\mathrm{Hz}$ and $\Omega_H = 0.8 \times 10^2\,\mathrm{Hz}$. The ratio of gyro to collision frequencies for electrons and ions, that is $\omega_H / \nu_e = e B_0 / (\nu_e m_e)$ and $\Omega_H / \nu_{in} = e B_0 / (\overline{m_i} \nu_{in})$, is plotted in Fig. 2.4. The ratio ω_H / ν_e passes through unity near 75 km. Inside the E layer, the altitude range 90–120 km, the value of ω_H / ν_e becomes much greater than unity, whereas the ratio Ω_H / ν_{in} keeps this value small. This means that in a reference frame moving with the neutral wind velocity the electron does exhibit many cycles around the magnetic field line before a collision takes place. In the extreme collisionless limit the drift velocity of the electrons is equal to $\mathbf{E} \times \mathbf{B}_0 / B_0^2$, which is perpendicular to both the electric and magnetic fields. In contrast, the ion does undergo many collisions and predominantly moves parallel to the electric field as if there were no influence of the magnetic field. Considering this kind of plasmas, we say that the electrons are magnetized whereas the ions are not magnetized. In the F layer, that is above 130 km, both the electrons and ions are magnetized and thus can move at the identical velocity $\mathbf{E} \times \mathbf{B}_0 / B_0^2$.

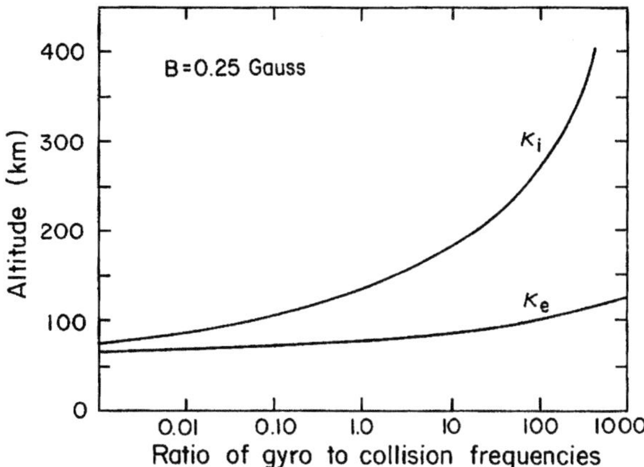

Fig. 2.4 Ratio of gyro to collision frequencies versus altitude. The indices i and e stand for ions and electrons, respectively. Taken from Kelley (1989)

At the altitudes below 150 km the ion collision frequency, v_i, is no more than 30–40 s^{-1}. Notice that v_e is much greater than v_i for any altitude of interest. This implies that in the ULF frequency range ($f = \omega/2\pi < 3$ Hz) the frequency ω is negligible compared with both v_i and v_e, so that the components of the plasma conductivity tensor given by Eqs. (2.7)–(2.9) can be considered as practically constant values, that is

$$\sigma_\parallel = e^2 n \left(\frac{1}{m_e v_e} + \frac{1}{m_i v_i} \right), \tag{2.13}$$

$$\sigma_P = e^2 n \left[\frac{v_e}{m_e \left(v_e^2 + \omega_H^2 \right)} + \frac{v_i}{m_i \left(v_i^2 + \Omega_H^2 \right)} \right], \tag{2.14}$$

$$\sigma_H = e^2 n \left[\frac{\omega_H}{m_e \left(v_e^2 + \omega_H^2 \right)} - \frac{\Omega_H}{m_i \left(v_i^2 + \Omega_H^2 \right)} \right]. \tag{2.15}$$

The typical profiles of the parallel, Pedersen, and Hall conductivities for mid-latitude ionosphere and for $\omega = 0$ are plotted in Fig. 2.5. The parallel plasma conductivity is much more enhanced than σ_P and σ_H in the altitude range above 90 km both in the nighttime and daytime conditions. This conductivity is so high that the ratio $\sigma_\parallel/\sigma_P$ is greater than 10^4 above 130 km. This effect follows from the high electron mobility along the magnetic field lines. In this case the first term prevails in Eq. (2.7), so that the parallel plasma conductivity is equal to $e^2 n / (m_e v_e)$ as a good approximation.

To a great extent the Pedersen and Hall conductivities define the properties of the conducting gyrotropic E layer of the ionosphere. In the daytime the Pedersen conductivity reaches a peak value of about $(1.5–3) \times 10^{-4}$ S/m (mho/m) at the altitudes 130–135 km, while in the nighttime the peak value decreases up to $(0.4–3) \times 10^{-5}$ S/m. As is seen from Fig. 2.5 the peak of the Pedersen conductivity is smaller than that of the Hall conductivity.

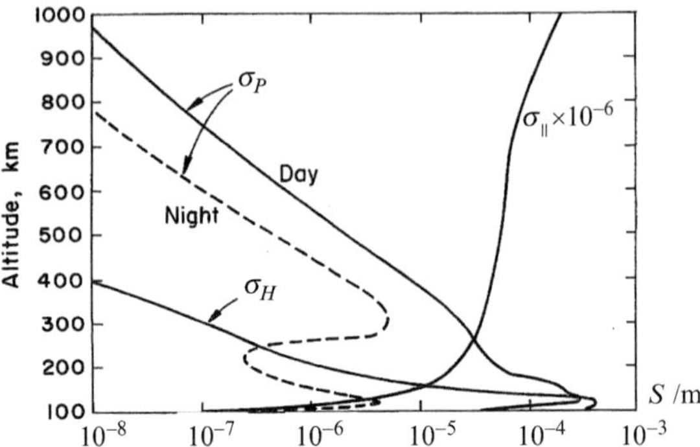

Fig. 2.5 Typical profiles of the parallel $\sigma_\parallel$, Pedersen σ_P, and Hall σ_H conductivities for the mid-latitude ionosphere. Adapted from Kelley (1989)

The Hall conductivity, σ_H, amounts to the value of about $(3\text{--}8) \times 10^{-4}$ S/m at the altitude range 100–110 km. In the nighttime the maximum value of σ_H becomes as much as $(0.8\text{--}1.5) \times 10^{-5}$ S/m. This conductivity is significant only in a narrow altitude range of the E layer since σ_H falls off more rapidly with altitude than does σ_P. The Hall conductivity is important due to its role in a mode coupling mechanism, which relates the shear Alfvén and compressional waves in the E layer.

2.2.2   Shear Alfvén and Compressional Waves in a Homogeneous Magnetized Plasma

It is customary to introduce the tensor of dielectric permittivity of the plasma via

$$\hat{\varepsilon} = \hat{\mathbf{1}} - i\hat{\sigma} / (\varepsilon_0 \omega). \tag{2.16}$$

Here $\hat{\mathbf{1}}$ denotes a unit matrix, ε_0 is the electric constant (dielectric permittivity of free space), and tensor $\hat{\sigma}$ is given by Eq. (2.5). The conduction and displacement currents entering the Maxwell's equation (1.1) can be expressed through the tensor of dielectric permittivity to yield

$$\nabla \times \mathbf{B} = -\frac{i\omega}{c^2} \hat{\varepsilon} \mathbf{E}. \tag{2.17}$$

At high altitudes above 200–300 km the collisionless approach is more appropriate to study the electrodynamics of plasma. In the extreme limit when the collision frequencies ν_e and ν_i are negligible compared with ω the parallel and Pedersen conductivities in Eqs. (2.7)–(2.9) are reduced to pure imaginary quantities $\sigma_\parallel = ie^2 n/(m_e \omega)$ and $\sigma_P = -i\omega \rho_0/B_0^2$, where $\rho_0 = n\overline{m_i}$ is the plasma mass density. The Hall conductivity becomes insignificant at higher altitudes and thus can be dropped. This means that in the reference frame with z axis parallel to the magnetic field the plasma dielectric permittivity tensor becomes diagonal

$$\hat{\boldsymbol{\varepsilon}} = \begin{pmatrix} \varepsilon_\perp & 0 & 0 \\ 0 & \varepsilon_\perp & 0 \\ 0 & 0 & \varepsilon_\parallel \end{pmatrix}, \tag{2.18}$$

with the components

$$\varepsilon_\perp = \frac{c^2}{V_A^2} \text{ and } \varepsilon_\parallel = -\frac{\omega_{pe}^2}{\omega^2}, \tag{2.19}$$

where c is the speed of light in vacuum and ω_{pe}^2 is the squared plasma frequency

$$\omega_{pe} = \left(\frac{ne^2}{m_e \varepsilon_0} \right)^{1/2}. \tag{2.20}$$

Now we study the plasma wave properties which come from the anisotropical character of the plasma conductivity. Consider a homogeneous plasma immersed in a uniform magnetic field $\mathbf{B}_0$. Let $\delta\mathbf{B}$ be the small perturbation of the magnetic field $\mathbf{B}_0$, so that $\delta B \ll B_0$. If all perturbed quantities are considered to vary as $\exp(i\mathbf{k}\cdot\mathbf{r} - i\omega t)$, where $\mathbf{k}$ denotes the wave vector, then Maxwell's equations (2.17) and (1.2) are reduced to the following equations

$$\mathbf{k} \times \delta\mathbf{B} = -\frac{\omega}{c^2} \hat{\boldsymbol{\varepsilon}}\mathbf{E}, \tag{2.21}$$

$$\mathbf{k} \times \mathbf{E} = \omega\delta\mathbf{B}, \tag{2.22}$$

whence it follows that

$$\mathbf{k} \times (\mathbf{k} \times \mathbf{E}) = -\frac{\omega^2}{c^2} \hat{\boldsymbol{\varepsilon}}\mathbf{E}. \tag{2.23}$$

As long as $\omega \ll \omega_{pe}$ the absolute value of the parallel plasma dielectric permittivity in Eq. (2.23) is much greater than unity and thus can be assumed to be infinite in this frequency range. The field-aligned component $\varepsilon_\parallel E_\parallel$ in Eq. (2.23) is finite, however, that means that the field-aligned electric field $E_\parallel$ must be zero.

Applying Eq. (1.55) for the triple cross product to Eq. (2.23) and rearranging yields

$$\mathbf{k}_\perp (\mathbf{k}_\perp \cdot \mathbf{E}_\perp) - \mathbf{E}_\perp \left(k^2 - \frac{\omega^2}{V_A^2} \right) = 0. \tag{2.24}$$

where $\mathbf{k}_\perp$ and $\mathbf{E}_\perp$ are the components perpendicular to the unperturbed magnetic field $\mathbf{B}_0$.

If the electric field $\mathbf{E}_\perp$ is parallel to $\mathbf{k}_\perp$, such a polarization is referred to as the shear Alfvén wave as shown in Fig. 1.15. The dispersion relation of the Alfvén mode can be derived from Eq. (2.24) under the requirement that the latter equation has a nonzero solution whence it follows that

$$k^2 - k_\perp^2 = \frac{\omega^2}{V_A^2}.$$ (2.25)

On account of the relation $k^2 - k_\perp^2 = k_\parallel^2 = k^2 \cos^2\theta$ this dispersion relation coincides with Eq. (1.59) for Alfvén mode of the MHD waves. As it follows from Eq. (2.22), the magnetic field perturbation, $\delta\mathbf{B} = \left(\mathbf{k}_\parallel \times \mathbf{E}_\perp\right)/\omega$, is perpendicular to the parallel wave vector that means $\delta\mathbf{B}$ is perpendicular to both the electric field and $\mathbf{B}_0$.

Conversely, if the electric field $\mathbf{E}_\perp$ is perpendicular to $\mathbf{k}_\perp$, Eq. (2.24) reduces to the dispersion relation

$$k^2 = \omega^2/V_A^2,$$ (2.26)

which describes the compressional/FMS wave. This equation is compatible with Eq. (1.69) describing the dispersion relation for the FMS waves in the extreme case $V_A^2 \gg c_s^2$. The polarization of the electric and magnetic components of the compressional wave is the same as shown in Fig. 1.16.

Finally we note that the dispersion relations for the shear Alfvén and the compressional waves in the collisionless magnetized plasma are practically the same as the dispersion relations under MHD approach.

References

Chapman S (1956) The electric conductivity of the ionosphere: a review. Nuovo Cimento 5(Suppl.):1585–1412

Ginzburg VL (1970) The propagation of electromagnetic waves in plasmas. Pergamon Press, Oxford

Ginzburg VL, Rukhadze AA (1972) Waves in magnetoactive plasma. Hand physics, vol. 49. Springer, New York

Hines CO (1974) The upper atmosphere in motion: a selection of papers with annotation. Geophysical monograph, vol 18. American Geophysical Union, Washington, DC

Johnson FS (ed) (1961) Satellite environment handbook. Stanford University Press, Stanford

Kelley MC (1989) The earth's ionosphere. Academic, New York

Chapter 3
Atmospheric Electricity

Abstract In this chapter the main attention is paid on the atmospheric electricity and on the global lightning activity as a machine supplying the negative charges to the Earth. Here we deal with the electric field and charge distribution in thunderstorm clouds and with the conventional mechanism for air breakdown. Lightning discharge parameters and global thunderstorm activity are discussed. In the remainder of this chapter the main emphasis is on the low frequency effects associated with recently documented evidences of previously unknown forms of upward propagating gigantic electric discharges, also known as transient luminous events (TLEs), which occur above a large thunderstorm system.

Keywords Global thunderstorm activity • Lightning return stroke • Runaway electron breakdown • Streamer • Transient luminous events (TLEs)

3.1 Global Electric Circuit

3.1.1 Electric Field and Conductivity of the Atmosphere

It is usually the case that the clouds, precipitation, fogs, and dust clouds contain a large amount of spatial electric charges. The electric field permanently exists in the atmosphere even though there is a fine weather condition. Near the ground surface the so-called fair weather electric field, i.e., the steady electric field, is vertically downward with mean value of about 100–130 V/m. The solid Earth is negatively charged with net charge about -3×10^5 C, while the positive charges are mainly concentrated in the lower atmosphere. Under fair weather conditions the atmospheric electric field falls off with height as shown in Fig. 3.1, so that its value is no more than several volts per meter at the height about 10 km. Owing to the presence of charged aerosols the electric field may increase in amplitude in the mixing layer over

V. Surkov and M. Hayakawa, *Ultra and Extremely Low Frequency Electromagnetic Fields*, Springer Geophysics, DOI 10.1007/978-4-431-54367-1_3,
© Springer Japan 2014

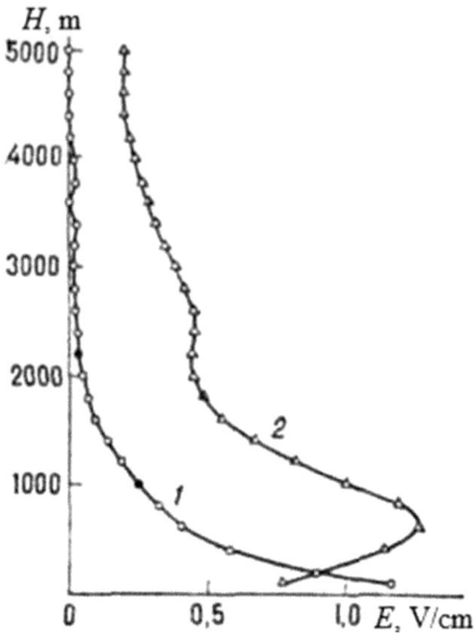

Fig. 3.1 Typical profile of the electric field in the atmosphere. *1*—fair weather conditions over ocean, arctic region and etc. *2*—over mainland. Adapted from the site http://www.femto.com. ua/articles/part_1/0217.html

the land, as shown in Fig. 3.1 with line 2. Above the mixing layer, whose depth is about 0.3–3 km, the field value decreases approximately exponentially. The voltage drop between the Earth and the ionosphere is about 200–250 kV.

The atmospheric conductivity at the ground level is about $\sigma_a = (2-3) \times 10^{-14}$ S/m which is smaller than that of the ionosphere by several orders of magnitude. In the mixing layer the atmospheric conductivity σ_a increases insignificantly and then it rises nearly exponentially with altitude, with a characteristic scale of 3–7 km, the value of which depends on altitude. For example, at the daytime conditions the approximate law for the conductivity as a function of altitude z can be written as (Chalmers 1967)

$$\sigma_a = \sigma_0 \exp\left(z/z_0\right), \quad 0 < z < 3.6\,\text{km},$$

$$\sigma_a = \sigma_1 \exp\left(z/z_1\right), \quad 3.6 < z < 17.7\,\text{km},$$

$$\sigma_a = \sigma_2 \exp\left(z/z_2\right), \quad 17.7 < z < 40\,\text{km}, \tag{3.1}$$

where $z_0 = 0.82$ km, $z_1 = 4.1$ km, $z_2 = 7.0$ km, $\sigma_0 = 1.14 \times 10^{-14}$ S/m, $\sigma_1 = 0.38 \times 10^{-12}$ S/m, and $\sigma_2 = 2.29 \times 10^{-12}$ S/m. This exponential tendency holds true for larger altitudes although the atmospheric conductivity depends on local time.

Above the lower edge of the E region at altitudes 75–90 km the ratio of the electron gyrofrequency to the electron–neutral collision frequency is no longer negligible, and the conductivity converts from a scalar quantity to a tensor one. Note that the atmospheric conductivity undergoes diurnal variations depending on latitude, local meteorological conditions, and so on.

The so-called fair weather current, that is, a weak background current flowing from the mesosphere to the ground plays an important role in the generation of global electric circuit. The mean value of the background atmospheric current density is about $(3.5-4) \times 10^{-12}$ A/m^2 (e.g., see Feynman et al. 1964). In stratus rainclouds the vertical atmospheric current increases up to $(0.5-1) \times 10^{-11}$ A/m^2, and in storm precipitation it enhances up to $10^{-10}-10^{-9}$ A/m^2.

This current is mostly due to the atmospheric conductivity although the diffusion and convective transfer of the electric charges may also be operative in the atmosphere. The convection and diffusion currents may be comparable to the conduction current within the mixing layer, whereas the sum of these currents is approximately the same as the conduction current at high altitude since the net current usually exhibits weak variations with altitude. In contrast to that the conductivity and electric field can be highly dependent on altitude. As an example, it is worthwhile to mention the case of mesospheric altitudes in which layered peaks of downward electric field and relatively low conductivity have been occasionally detected (Bragin et al. 1974; Hale et al. 1981; Maynard et al. 1981).

The permanent thunderstorm activity around the world is thought to be a major electric source for the global atmospheric electric circuit, which is formed by the lower ionosphere and terrestrial surface conducting layers (e.g., see Rakov and Uman 1998, 2003). The global electric circuit is closed via lightning discharge currents flowing basically upwards and the background atmospheric current flowing downward to the Earth's surface from the atmosphere and lower ionosphere. A typical negative cloud-to-ground ($-$CG) lightning flash carries about $q = -20$ C of the negative electricity. Taking into account that worldwide number of the lightning discharges per second is about $v = 10^2$ s^{-1} we obtain that the total current flowing to the Earth is $I = |q| \, v \approx 2 \times 10^3$ A. Based on Optical Transient Detector (OTD) satellite data one can specify this value since the global annual mean flash frequency has recently been estimated as 44 ± 5 s^{-1} (Nickolaenko and Hayakawa 2002, 2014; Christian et al. 2003; Hayakawa et al. 2005; Sato et al. 2008; Sátori et al. 2009).

The negative sign of the typical lightning charge means that the flash current points outward, that is, from the Earth to the ionosphere. The opposite-directed conduction current carries the positive charges from the upper atmosphere and troposphere to the Earth. Assuming for the moment that this background conduction current is approximately uniformly distributed around the Earth, the total current flowing from the ionosphere to the Earth surface can be estimated as $I_f = 4\pi j_f R_e^2$, where j_f is the mean density of the background current and R_e is Earth radius. Taking the numerical values of the parameters $j_f \approx (3-4) \times 10^{-12}$ A/m^2 and $R_e \approx 6.4 \times 10^3$ km we get the estimate $I_f \approx (1.5-2) \times 10^3$ A, which is consistent in magnitude with the inverse background current due to the global thunderstorm activity.

3.1.2   Electric Field and Charges in Thunderstorm Clouds

At the moment there are about 2,000 thunderstorms simultaneously operating on the Earth. As the global mean flash frequency is about 10^2 s^{-1}, the individual

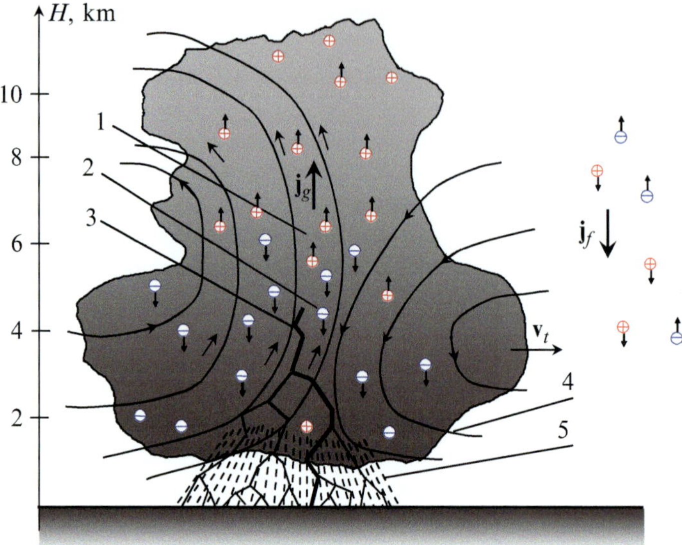

Fig. 3.2 A sketch of thundercloud structure. *1* and *2*—centers of positive and negative charge distributions. *3*—a lightning return stroke. *4*—air flow lines. *5*—a downpour. *Horizontal arrow* $\mathbf{v}_t$ shows the direction of the thundercloud motion

thunderstorm is characterized by the local mean flash frequency about $0.05\,\mathrm{s}^{-1}$. Whence we can estimate the mean interval between lightning flashes as $20\,\mathrm{s}$. The charge of thunderstorm clouds is thus renewed for this short interval of time.

We now raise an interesting question, what is the basic mechanism for electric charge formation in the thunderstorm cloud? The electric charges in the clouds are concentrated on the small particles such as rain-drops, snowflake, pieces of ice and aerosols. In stratus and stratocumulus clouds the charge of rain-drop reaches the value of $q_0 = (10$–$100)\,e$, where e is elementary charge, while in the nimbostratus the charges of separate drops amounts to $q_0 = (10^5$–$10^6)\,e$ and in the thunderstorm clouds the separate charges amounts to a very great value of the order of $(10^6$–$10^7)\,e$ (Israël 1970, 1973, Imyanitov et al. 1971, Muchnik and Fishman 1982).

Despite that the electrical structure of a typical thundercloud is rather a stratiform, the most part of positive charges tend to pile up at the upper portion of the thundercloud whereas most of negative charges predominantly accumulate at its bottom (Coroniti 1965; Nelson 1967; Bhartendu 1969; Wahlin 1973; Winn et al. 1974; Uman 1987; McGorman and Rust 1998; Rakov and Uman 2003). The simplest model of the spatial charge separation in the thundercloud is shown in Fig. 3.2. In the middle latitudes the thundercloud top amounts to 8–12 km, while in tropics the thundercloud top may be as high as 20 km. The thunderstorms are mainly formed in the zone of power convective fluxes. The separation of oppositely

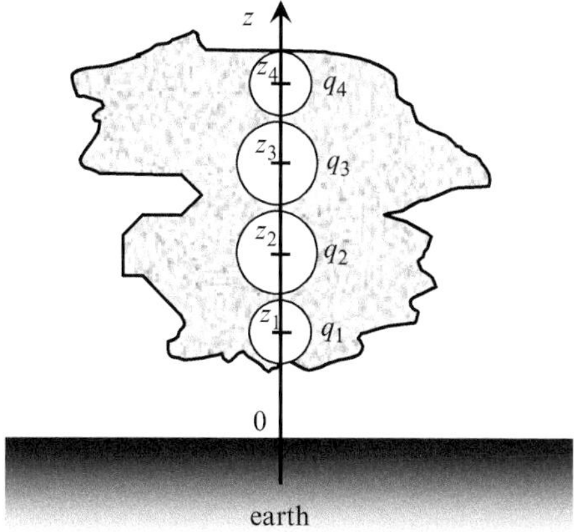

Fig. 3.3 A simplified model of electric charge distribution in a thundercloud. The spatial charge densities are constant inside the spherical regions whose centers are located on the z-axis at different altitudes and z_1, z_2, z_3, and z_4 (Surkov and Hayakawa 2012)

charged particles is thought to be due to slow hydrodynamics processes inside the thundercloud. It appears that upward air fluxes drag small and light positively charged ice fragments, whereas heavy negatively charged hailstones predominantly fall downward due to the gravity (Lyons et al. 2003; Lyons 2006; Krehbiel et al. 2008; Pasko 2010). Since the current is upward inside the thundercloud and approximately zero outside, charges pile up at the thundercloud boundaries as shown in Fig. 3.2. Here $\mathbf{j}_g$ is updrafts- and gravity-driven current density inside the thundercloud. This current causes the charge separation in the thundercloud and thus it plays a role of a battery/source for the generation of upward or downward-directed lightning discharges. $\mathbf{j}_f$ denotes the so-called fair weather current. The lightning can be operative as long as the current $\mathbf{j}_g$ can separate the charges and provide the top of the thundercloud with sufficient amount of positive charges.

The charge density, ρ, which is usually observed under the fine weather condition is about $0.01\,\mathrm{nC/m^3}$. In stratocumulus clouds the charge density increases up to $0.1\,\mathrm{nC/m^3}$. In cumulonimbus clouds under the downpour the mean charge density is about $\rho = 0.3$–$10\,\mathrm{nC/m^3}$ while in the thunderclouds $\rho = 3$–$30\,\mathrm{nC/m^3}$ (Imyanitov et al. 1971).

Notice that the electrical structure/charge distribution of actual thunderclouds is much more complicated as compared to the above model. Moreover a certain charge imbalance may persist in a thunderstorm, which leads to strong variations of the electric field with altitude. The electrical structure of a standard thundercloud can be described via a stratiform/multilayered thundercloud model in which the charged regions are situated at different altitudes (Krehbiel et al. 2008; Riousset et al. 2010a). The spatial distribution of these charges was assumed to obey a Gaussian law and was not spherically symmetric. To simplify the problem and to interpret this model, we assume that all the charges are uniformly distributed in spherical regions shown in Fig. 3.3. A normally electrified storm, which corresponds to a typical $-CG$ lightning, was characterized by the following numerical values $q_i = 12.5, -60, 40,$ and $-20\,\mathrm{C}$, where $i = 1, 2, 3, 4$ (Krehbiel et al. 2008). The Earth is considered to be a perfect conductor. This implies that the vertical component of the net electric field taken along the z-axis is given by $E_z = \sum_{i=1}^{4} E_{zi}$, where

$$E_{zi} = \frac{q_i \, (z - z_i)}{4\pi \varepsilon_0 r_i^3} + E_{zi}' \qquad (3.2)$$

inside the charged balls and

$$E_{zi} = \frac{q_i \, (z - z_i)}{4\pi \varepsilon_0 \, |z - z_i|^3} + E_{zi}' \qquad (3.3)$$

outside the balls. Here

$$E_{zi}' = \frac{q_i}{4\pi \varepsilon_0 \, (z + z_i)^2} \qquad (3.4)$$

stands for the electric field of mirror electric images of thundercloud charges, and z_i denote the coordinates of centers and radii of the charged spherical regions, respectively. The variations of the net electric field, E_z, as a function of altitude z can be easily calculated after these simplifications.

To illustrate the results of this simulation we made use of the following numerical parameters $z_i = 3.7, 6.9, 12.1, 15.7\,$km, $r_i = 1.0, 2.2, 2.3, 1.3\,$km and above-mentioned values of q_i taken from Krehbiel et al. (2008). The results of calculations shown in Fig. 3.4 with solid line are in qualitative agreement with the vertical profiles of thunderstorm electric field as measured by balloon equipment (Marshall et al. 1995). The first two peaks at the bottom of Fig. 3.4 are basically due to the field of a pair of charges $q_1 = 12.5\,$C and $q_2 = -60\,$C whereas the two peaks at the top of Fig. 3.4 are caused by the upper charges $q_3 = 40\,$C and $q_4 = -20\,$C. It should be noted that both the magnitude and location of the peaks are very sensitive to the distances between the charges.

3.1.3 Conventional Mechanism for Air Breakdown and Streamers

The conventional mechanism of air breakdown is due to the thermal ionization of the air by low energy electrons in the presence of strong electric fields, which occasionally occurs inside and around thunderclouds. Typical energy of electrons producing the ionization is about 10–20 eV while the mean electron energy is about 2 eV. According to laboratory tests the breakdown threshold E_c is approximately proportional to the gas pressure p at least under condition $pd > 10^6\,$Pa m, where d is the inter-electrode gap size (e.g., see Raizer 1991; Lieberman and Lichtenberg 1994). It follows from the state equation of perfect gas ($p = k_B n_m T$) that at constant

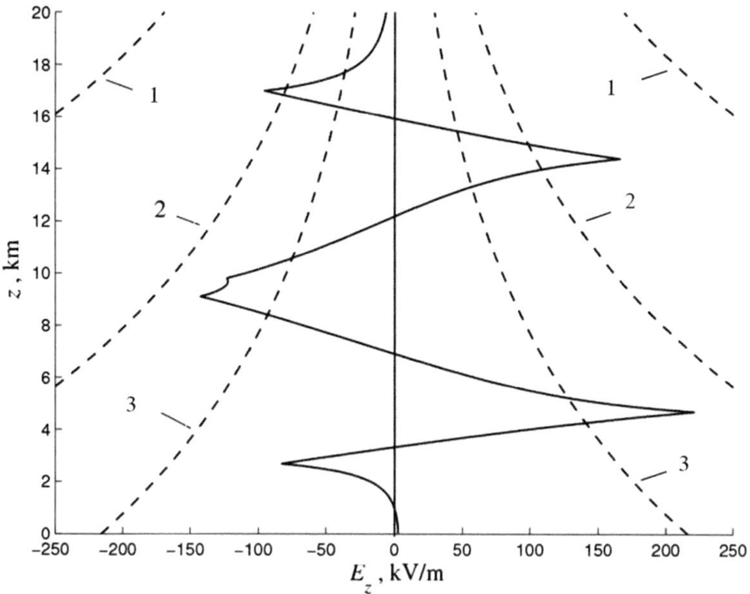

Fig. 3.4 Model calculations of thunderstorm quasielectrostatic (QE) field preceding a conventional −CG stroke. The vertical electrical field profile along z-axis as a function of altitude z is shown with solid line. The fields required for the propagation of negative and positive streamers in the air are shown with *dash lines* 1 and 2, respectively. A runaway breakdown field is shown with *dash line* 3. The numerical values of parameters are assumed to be typical for the generation of −CG strokes (Surkov and Hayakawa 2012)

gas temperature the value of E_c is proportional to number density n_m of the neutral gas. Taking into account of Eq. (2.3) for n_m we thus obtain that the conventional breakdown threshold falls off approximately exponentially with altitude

$$E_c = E_0 \exp\left(-z/H_a\right). \tag{3.5}$$

Here $E_0 \approx 32\,\text{kV/m}$ is the constant of the order of breakdown threshold at the ground level.

As the electric field exceeds the breakdown threshold (3.5), the streamer mechanism of air breakdown may develop (e.g., see Bazelyan and Raizer 1998). A typical streamer is the self-propagating narrow filament of cold low-conducting plasma which can propagate at the velocity 10^2–$10^4\,\text{km/s}$ as measured at the ground pressure. The electric field in the vicinity of the streamer head can be about 4–7 times larger than E_c due to the high charge density at streamer head. This results in the electron impact- and photo-ionization in the streamer head followed by an enhancement of ionization coefficient up to the value occurring at the streamer channel (e.g., Raizer et al. 1998; Pasko 2006; Celestin and Pasko 2010). In laboratory experiments such as point-to-plane corona discharges, the individual electron avalanches initiate the streamer in the vicinity of the sharp portion of an electrode thereby producing the streamer branching phenomena. The transition from an electron avalanche to a streamer generation also requires the critical number of avalanching electrons, a minimum radius of the avalanche region and many other factors (e.g., Raizer et al. 1998; Bazelyan and Raizer 1998).

The streamers can be divided into two types depending on the sign of the space charge in their heads. A negative streamer propagates due to ejection of electrons from its head into ambient air and vice versa, that is a positive streamer propagates due to injections of ambient/seed electron avalanches from surroundings. Along with the traditional channels of the seed electron production such as the cosmic rays and photo-ionization of air due to solar radiation, the effective source of the electron production is the strong electric field at the streamer head which gives rise to the high rate of impact- and photoionization around the head (Bazelyan and Raizer 1998). In electronegative gases such as air, the oxygen and nitrogen ions could be an additional source of seed electrons due to fast electron detachment in an electric field (Pancheshnyi 2005).

In some sense, the streamers can be considered as a kind of ionization waves which require the strong electric field for their initiation. However, once the steamer was generated it can propagate through the region where the electric field is smaller than E_c. The minimum value of electric field required for the propagation of positive streamers in the air at ground pressure is $E_s^+ = 4.4$ kV/cm, while the same value for negative streamers is $E_s^- = -12.5$ kV/cm (Raizer 1991; Allen and Ghaffar 1995; Babaeva and Naidis 1997; Pasko 2006). The dependence of these values on altitude is described by an equation analogous to Eq. (3.5). The altitude-dependences of electrical field required for breakdown in the air are shown in Fig. 3.4 with lines 1–3, which correspond to different mechanisms of the air breakdown. We shall discuss the runaway breakdown mechanisms (line 3) in more detail later on. Despite the calculated value of electric field does not exceed the breakdown threshold for streamer propagation, the lightning nucleation is the most probable in the regions where the peaks of thunderstorm electric field are close to the breakdown threshold. Random spatial inhomogeneities of the charge distribution may increase the local electric field. As is seen from Fig. 3.4, the lightning discharge between the cloud and ground may be initiated in the vicinity of the first peak at 4–6 km altitude range.

3.1.4 Lightning Discharge

The first studies have shown that the streak lightning originated from the thundercloud is a sort of spark discharge with length of several kilometers and with peak current of about 20 kA (e.g., see Stekolnikov 1940; Schonland 1956; Ishikawa 1961). The typical lightnings can be divided into two general classes: cloud-to-ground (CG) and intracloud (IC) lightning discharges. A newly discovered class of high-altitude gigantic discharges occurring above a large thunderstorm

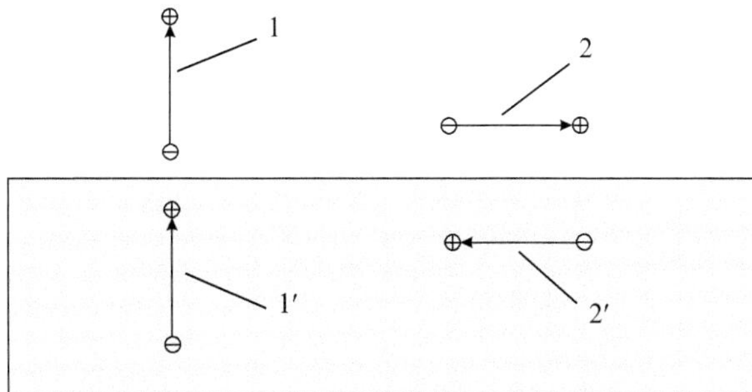

Fig. 3.5 Vertical and horizontal dipoles above the perfect conducting ground. The vertical dipole is approximately duplicated due to the electric images in the ground while the horizontal dipole is converted to quadrupole

at stratospheric and mesospheric altitudes (Franz et al. 1990) is discussed in the Sect. 1.3.2. The IC discharges are strongly prevalent over the CG and transient luminous events (TLEs). Note that the peak current of the IC discharges is one order of magnitude smaller than that for CG discharges. However the IC discharges may have a length up to 50–150 km.

Effectiveness of the vertical and horizontal discharges, considered as electromagnetic wave transmitter, is extremely different (Kudintseva et al. 2009; Hayakawa et al. 2012). To illustrate this, we approximate the actual discharge with the effective electric dipole/antenna shown in Fig. 3.5 with lines 1 (CG) and 2 (IC). If the ground is considered as a perfect conductor, the electric field of induction electric charges arising on the ground surface is equivalent to the electric field of dipole image, which is in the ground symmetrically with respect to the ground surface as shown in Fig. 3.5 with lines $1'$ and $2'$. In the vertical case it practically produces the duplication of the net dipole moment, whereas in the horizontal case the net dipole moment vanishes, so that the horizontal antenna is equivalent to a quadrupole. This implies that CG lightning can be a much stronger radiator as compared to IC especially in ULF/ELF (extremely low frequency, < 3 kHz) range. One more argument in favor of this statement is the presence of the so-called continuing current (CC) following the CG flash because the CC may greatly contribute to the ULF/ELF portion of the lightning spectrum. The IC lightning cannot, in general, be observed at distances as great as CG lightning (Heavner et al. 2003). The energetic intracloud (EIC) discharges are the most powerful source of lightning radiation in the HF (high frequency) and VHF (very high frequency) radio bands (e.g., see Smith et al. 1999, 2004; Jacobson 2003). Certainly the portion of the EIC with vertical channels may also contribute to the background ULF/ELF electromagnetic noise produced by the global lightning activity. However in what follows the main emphasis is on the CG lightning as the most creditable candidate for the excitation of global electromagnetic low-frequency resonances.

To a great extent our knowledge of the lightning parameters comes from the optical and electromagnetic measurements in the VLF (very low frequency) and other regions (e.g., see Uman 1987; Raizer 1991; Nickolaenko and Hayakawa 2002, 2014; Rakov and Uman 2003). As noted above, the majority of CG lightning are negative. This implies that the current of the lightning discharge is upward-directed which corresponds to the positive current moment. The $-$CG lightning discharge starts with the downward-propagating stepped leader, which creates the thin conducting plasma channel that connects the thundercloud to the ground. The leader motion can be divided into two phases. At the first phase the pilot-streamer, which arises in the thundercloud, begins to move jerkily. On the average, during the jerk the streamer travels a distance about 100 m at the mean velocity of the order of 5×10^2 km/s. Then the heavily ionized leader catches up with the streamer front for a short time of about $1 \, \mu$s. The leader velocity is believed to be about 7×10^4 km/s. After that a new pilot-streamer arises from the end of the ionized channel and a newly leader catches up with the streamer front and etc. step-by-step, so that the ionized channel makes longer. The mean pause between the leader steps is about $50 \, \mu$s. It is usually the case that the channel branches out when it moves downward to the ground. The stepped leader carries the negative charge of the order of 5 C with the mean vertical velocity of 1.5×10^2 km/s and the leader current is as high as 300 A.

The subsequent upward-propagating return stroke produces a main breakdown of the CG interval. The return stroke is an upgoing wave propagating rapidly along the warm conducting channel. The stroke begins to move with the velocity $(0.5–1) \times 10^5$ km/s and further it decelerates gradually. During an interval of $5–10 \, \mu$s the return stroke current amounts to the peak value 10–20 kA and thereafter the current decreases up to half the peak value for the interval about $20–50 \, \mu$s. The net result of a leader/return-stroke pair is that negative charge of about 10 C is lowered from the thundercloud to the ground.

The $-$CG lightning flash usually contains return strokes and the mean interval between the strokes is 40 ms, so that the net duration of the flash is about 0.2 s. Note that leaders preceding the second and subsequent strokes propagate continuously without any steps and pauses. Such leaders termed as dart leaders move at the velocity $10^3–10^4$ km/s that is much higher than that of the stepped leader. A highly branched system of the streamer channels arises around the leader (Uman 1987; Rakov and Uman 2003).

A portion of $-$CG return strokes is accompanied by the CC that immediately follows the return strokes. This current of several tens to hundreds of amperes flows in the same channel to the ground for tens to hundreds of milliseconds (e.g., see Rakov and Uman 2003). Thus the CC manifests itself as a slowly varying current flowing between the thundercloud and ground along the path created by the preceding leader return stroke pairs. This current makes a significant contribution to the low-frequency portion of natural electromagnetic noise especially to ULF/ELF region which contains global electromagnetic resonances (see Chap. 2).

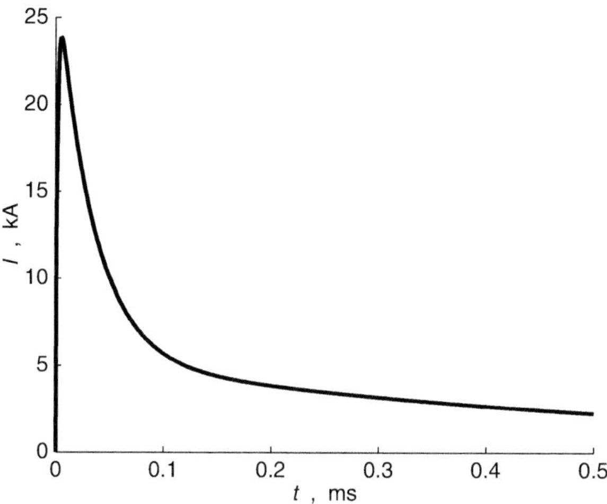

Fig. 3.6 Model calculation of the return stroke current versus time

It is thought that about 5–10 % of global CG lightning activity is composed of positive cloud-to-ground (+CG) lightning, which transfers the positive charge to the ground (Rakov 2003). A positive flash usually consists of a single stroke followed by a CC that typically lasts for several tens or hundreds ms. The amplitude of positive CC current varies from several kA to tens kA, an order of magnitude larger than that for the −CG (Rakov 2003; Rakov and Uman 2003). The reader is referred to the extensive special literature for details about CG and IC lightning discharges (e.g., see Krider 1986; Uman 1987; Raizer 1991; Lyons 1997; McGorman and Rust 1998; Rakov and Uman 2003 and references therein).

There are a number of relevant models, which can serve as "engineering" models of lightning stroke (Rakov and Uman 1998). According to the known models for a −CG lightning, the current at the base of the stroke is described by a combination of a power function and several exponents with different relaxation times (e.g., see Nickolaenko and Hayakawa 2002). More usually we choose four items in this model (e.g., see Jones 1970; Taylor 1972; Uman and Krider 1982; Uman 1987)

$$I\,(t) = \sum_{m=1}^{4} I_m \exp\left(-\omega_m t\right) \tag{3.6}$$

where ω_m are inverse time constants. Since the current given by Eq. (3.6) is equal to zero at the initial moment $t = 0$, the amplitudes I_m of individual current terms must satisfy the condition $\sum_{m=1}^{4} I_m = 0$. Following Jones (1970) we use the following values of the parameters, which are typical for the models of return strokes $I_{1-4} = -28.45, 23.0, 5.0, 0.45$ (in kA) and $\omega_{1-4} = 6.0 \times 10^5, 3.0 \times 10^4, 2.0 \times 10^3, 147.0$ (in s^{-1}). Figure 3.6 shows the temporal variation of the return stroke current for these parameters.

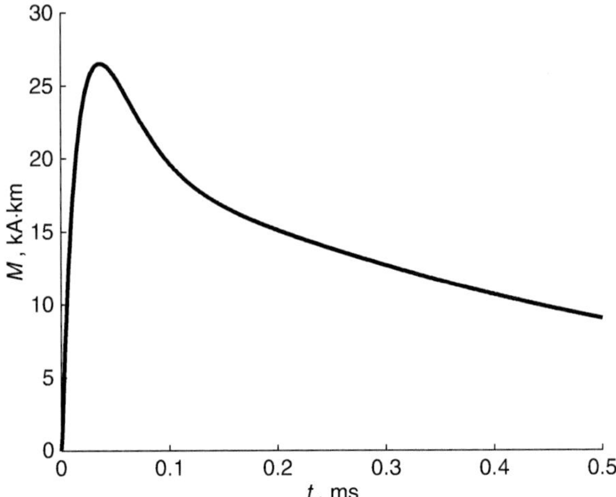

Fig. 3.7 Model calculation of current moment of a return stroke versus time

The two last terms on the right-hand side of Eq. (3.6), that is $I_3 \exp(-\omega_3 t) + I_4 \exp(-\omega_4 t)$, describe the CC, which is responsible for the final decay of the electrostatic field of the stroke. As we have noted above, a $-$CG stroke transfers the negative electric charge to the ground and the lightning current is thus pointed upward. It follows from the model that the total lightning charge equals $-6.3\,$C, and about 50% of this charge is carried by the weak CC, which is described by the component I_4 and, in part, by the component I_3. Below we show that these terms make a main contribution to the ULF range of the lightning spectrum.

Far from the lightning discharge the ULF electromagnetic field can be characterized by the current moment of the stroke as the product of the discharge current and the length of the current channel, i.e., $m(t) = I(t)\,l(t)$. One more important lightning characteristic is the charge moment, which is equal to the product of the total charge transferred from the thundercloud to the ground and the final length of the current channel.

In the theory the lightning channel length, l, is assumed to be governed by an exponential law $dl/dt = V_0 \exp(-\Omega t)$, where V_0 is the current wave velocity at the ground level (see Ogawa 1995). In another words, the vertical current channel grows upward with exponentially attenuated velocity. In this case

$$m(t) = M F_1(t)$$
$$F_1(t) = [1 - \exp(-\Omega t)] \sum_{m=1}^{4} \frac{I_m}{|I_1|} \exp(-\omega_m t) \qquad (3.7)$$

where $M = |I_1|\,l$ and $l = V_0/\Omega$ is the final channel length.

For illustrative purposes, a model calculation of the lightning current moment is shown in Fig. 3.7 as a function of time. In making the plot of $m(t)$ we have used the following numerical values of parameters: the maximum of lightning channel velocity is $V_0 = 8 \times 10^4\,$km/s and the inverse time parameter is $\Omega = 2 \times 10^4\,$s^{-1}.

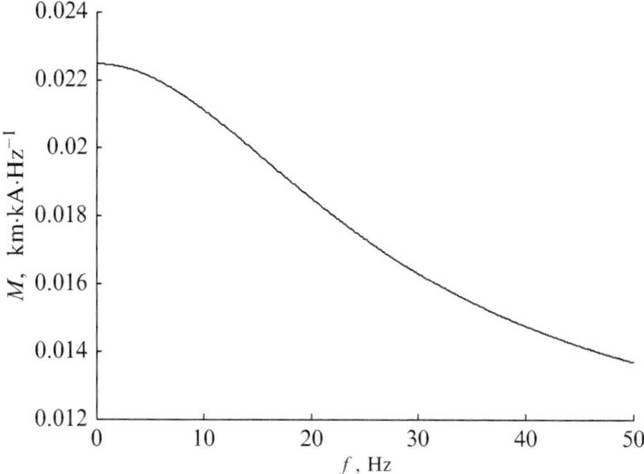

Fig. 3.8 Model calculation of the absolute value of current moment spectrum of an individual return stroke

Whence it follows that the final channel length is $l = V_0/\Omega = 4\,\mathrm{km}$ (see, e.g., Berger et al. 1975; Nickolaenko and Hayakawa 1998, 1999, 2002; Visacro et al. 2004). As is seen from Fig. 3.7, the current moment magnitude is about $27\,\mathrm{kA\,km}$ for these parameters.

The spectrum of the current moment $M\,(t)$ can be written as

$$m\,(\omega) = MF_1\,(\omega) = V_0 \sum_{m=1}^{4} \frac{I_m}{(\omega_m - i\,\omega)\,(\omega_m + \Omega - i\,\omega)} \qquad (3.8)$$

As illustrated in Fig. 3.8 the absolute value of the return stroke spectrum by Eq. (3.8) is practically constant within the ULF band. It is not surprising since in this case $\omega \ll \omega_m$ and thus the frequency-dependent terms in the denominator of Eq. (3.8) are negligible.

3.1.5   Multiple Return Stroke

As we have noted above, a typical $-$CG lightning discharge consists of $n = 2 - 6$ return strokes. More frequently there are 3–4 return strokes with characteristic duration of about $100\,\mu\mathrm{s}$ (Uman 1987; Ogawa 1995; Lyons 1997; McGorman and Rust 1998; Borovsky 1998). The mean interval between them is of the order of $t_0 = 40\,\mathrm{ms}$. The current profile of the individual strokes can be different.

Now we consider a simple model of multiple return stroke proposed by Jones (1970). The final/peak length of each return stroke is assumed to increase with its number n by the fixed value Δl, that is, as $l_n = l_1 + (n-1)\Delta l$. The increase in the lightning channel length results in gradual enhancement of the electric current moment.

Actually the current profiles of the individual strokes can be different whereas in this model all the current impulses are assumed to have the same shape. The inverse time parameter, Ω_n, of the return stroke is related to the channel length, l_n, as follows

$$\Omega_n^{-1} = \frac{l_n}{V_0} = \Omega_1^{-1} + \frac{\Delta l}{V_0}(n-1), \tag{3.9}$$

where $\Omega_1 = V_0/l_1$. The net magnetic moment of the flash which contains the multiple discharge can be described by an equation analogous to Eq. (3.7), that is $m(t) = MF(t)$, where $M = l_1|I_1|$ is the "magnitude" of the current moment while the dimensionless function $F(t)$ determines the shape of the multiple discharge

$$F(t) = \sum_{n=1}^{n_0} \frac{l_n}{l_1}\left[1 - \exp\left(-\Omega_n t_n'\right)\right]$$

$$\times \eta\left(t_n'\right) \sum_{m=1}^{4} \frac{I_m}{|I_1|}\exp\left(-\omega_m t_n'\right), \tag{3.10}$$

where $t_n' = t - (n-1)t_0$ and $\eta(x)$ denotes the step-function, i.e., $\eta = 1$ if $x \geq 0$ and $\eta = 0$ if $x < 0$.

Taking the numerical values $l_1 = 4\,\mathrm{km}$, $\Delta l = 1\,\mathrm{km}$ and above-mentioned parameters of the return stroke, one can estimate the typical magnitude of the magnetic moment of the multiple return stroke as $M \sim 10{-}10^2\,\mathrm{kA\,km}$. Model calculation of the flash current moment for $n_0 = 3$ is shown in Fig. 3.9.

The spectrum of the current moment $MF(t)$ is given by

$$MF(\omega) = V_0 \sum_{n=1}^{n_0}\sum_{m=1}^{4} \frac{I_m \exp\left[i\omega t_0(n-1)\right]}{(\omega_m - i\omega)(\Omega_n + \omega_m - i\omega)}. \tag{3.11}$$

Figure 3.10 shows model calculation of the absolute value of the spectrum $MF(\omega)$ originated from $n_0 = 3$ return strokes, which follow one by one with equal interval $t_0 = 40\,\mathrm{ms}$. In such a case the spectrum has approximately a quasi-oscillatory profile with maximum repetition period about $25\,\mathrm{Hz}$. This peculiarity of the spectrum is due to the presence of imaginary exponents in Eq. (3.11) and thus it follows from the interference between the fields of individual return strokes (e.g., see Surkov et al. 2010). Actually the number of strokes and intervals between their occurrence are rather random values that may result in randomizing of these oscillations.

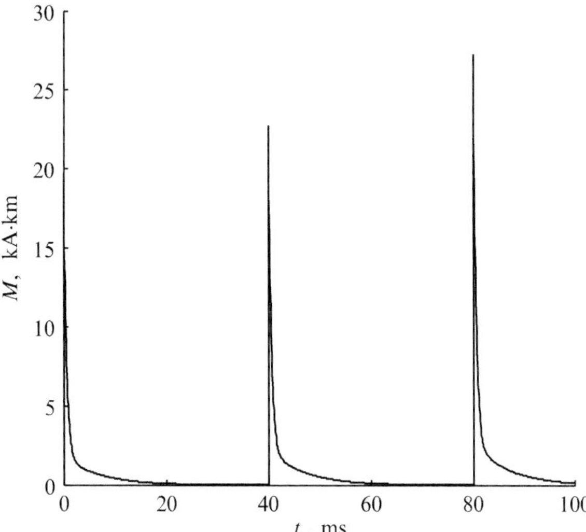

Fig. 3.9 Model calculation of current moment of a multiple return stroke versus time

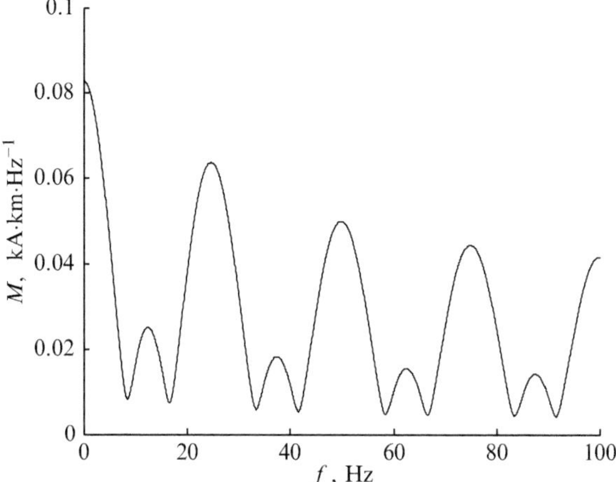

Fig. 3.10 Model calculation of the absolute value of current moment spectrum of multiple return strokes

3.1.6 Global Thunderstorm Activity

Considering that there are perhaps 2,000 thunderstorms in progress around the world at any time, the measurements show that slow diurnal variations of the Earth electric field are in a good agreement with the variations of the global thunderstorm activity. Worldwide, both the Earth field and thunderstorm activity vary within $\pm 15\,\%$ and reach the peak value approximately at the same period from 14 UT till 20 UT. In particular, the diurnal variations of the Earth electric field correlate with diurnal variation of the global thunderstorm activity (e.g., see Bering et al. 1998; Füllekrug et al. 1999; Rycroft et al. 2000).

The weather condition and climate peculiarities may greatly affect the distribution of lightning discharges over the Earth surface. The thunderstorm cells more

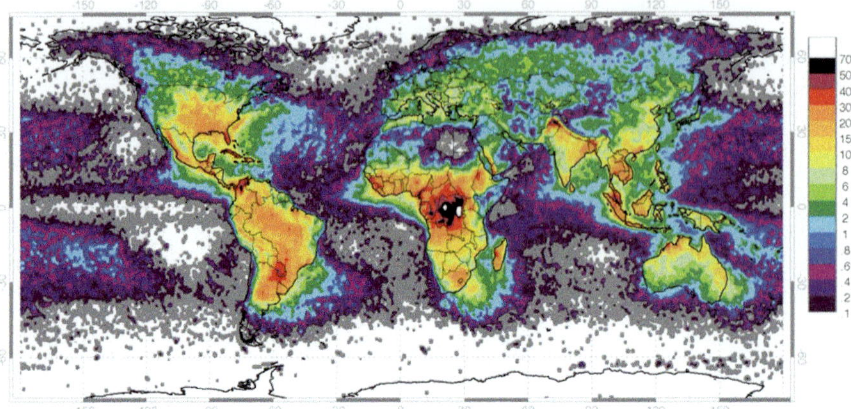

Fig. 3.11 NOAA satellite data of worldwide average annual lightning flashes per square kilometer. Taken from the site http://www.boqueteweather.com/lightning_year.htm

frequently occur in the regions covered with cyclones, typhoons, frontal zone of temperature inversion, climatological fronts and etc. (Watt 1967; Bhartendu 1969; Uman and Krider 1982; Uman 1987; Nickolaenko and Hayakawa 2002; Rakov and Uman 2003). The most favorable areas for the thunderstorm formation are usually associated with such regions as mountain ridges, which may affect the monsoon circulation, the river basins and bottom-lands, where the humid climate is predominant. Additionally, the group of islands in an ocean may influence the wind system in such a way that the wind system in this region is capable of sustaining the generation of thunderstorm clouds (Watt 1967). Based on observational data gathered for a long period one can characterize the lightning activity over the Earth surface through the mean number of flashes per unit of square in a year or month. An example of annual pattern of lightning activity over the globe as observed from space by the OTD is displayed in Fig. 3.11. The number of lightning per square kilometer in a year is shown with different colors. There is a small wonder that the main centers of lightning activity, termed global thunderstorm centers, are concentrated in tropics and around the equator. As is seen from Fig. 3.11, such centers cover three broad continental tropical regions, (1) the sub-Saharan Africa; (2) Central America and the Amazon basin in South America; and (3) the Malaysian Archipelago/Maritime continent extending from Southeast Asia across the Philippines, Indonesia, and Borneo into Northern Australia.

The number, v, of lightning discharges per one km squared in a year inside these global thunderstorm centers is greater than 25 $km^{-2} \cdot year^{-1}$ (Bliokh et al. 1980). The value $v = 2.5–7.5$ $km^{-2} \cdot year^{-1}$ is typically for the regions with an enhanced lightning activity, while the regions with a moderate thunderstorm activity can be characterized by the value $v = 1.5–2.5$ $km^{-2} \cdot year^{-1}$.

The global thunderstorm centers play a crucial role in the formation of global electric circuit of the Earth. It is commonly accepted that the total electric current arising from the global lightning activity must nearly cancel the inverse background atmospheric current, which is distributed around the whole globe. The global electric circuit is capable of sustaining both a constant potential difference between the Earth and the ionosphere and the fair weather electric field near the Earth surface.

3.2 Sprites, Blue Jets, and Other High Altitude Electric Discharges

3.2.1 Classification of TLEs

Gigantic electric discharges, also known as TLEs occur above a large thunderstorm at stratospheric and mesospheric altitudes. Since their recent discovery (Franz et al. 1990), much emphasis has been put into studies of these pleasing phenomena in the ground-based observations (e.g., Neubert et al. (2008) and references herein) as well as in the aircraft (Sentman and Wescott 1993; Sentman et al. 1995; Wescott et al. 1995), satellite (Chern et al. 2003; Mende et al. 2005; Cummer et al. 2006a; Chen et al. 2008), and space shuttle measurements (Boeck et al. 1992; Yair et al. 2004).

In a broad sense, the term TLEs includes not only the gigantic electric discharges but also a few extremely fast and highly dynamical electrical and optical phenomena which arise between the top of the thundercloud and the ionosphere. Depending on their properties the TLEs may be categorized by several types which are sprites/red sprites, blue jets (BJs), halos, elves and recently discovered blue starters and gigantic jets (GJs) (e.g., see Chen et al. 2008). The ISUAL satellite measurements have shown that the global occurrence rate of elves, sprites, halos, and GJs can be estimated as 3.23, 0.50, 0.39, and 0.01 events per minute, respectively. We cannot come close to exploring these topics in any detail since the main scope of our study is the low frequency effects associated with the TLEs. The reader is referred to the reviews by Ebert and Sentman (2008), Pasko (2010), Surkov and Hayakawa (2012) and Pasko et al. (2013) for the details on basic features of TLEs. However before discussing these effects, we need to understand a little about the underlying mechanisms of the TLEs.

The sprite is a luminous red glow occurring at 50–90 km altitude range with gradually changing to blue color below 50 km. As is seen from Fig. 3.12, the typical sprite consists of the upper diffuse region in red color and lower tendril-like filamentary structure in blue color with lateral dimension from 20–30 km to 50–100 km (e.g., Pasko 2006; Neubert et al. 2008; Stenbaek-Nielsen and McHarg 2008; Montanyà et al. 2010). Figure 3.13 shows that the visible inner structure of the tendrils and branches is very complicated. The bright streamer heads, shown with red color, vary in size from ~ 10 to ~ 100 m. This picture is highly dynamical since the streamer heads move in different directions at velocities about 10^3–10^4 km/s (Stenbaek-Nielsen et al. 2007). Typically, the sprite flash is lasted from a few to several tens of ms. The sprite halos have been occasionally observed approximately 1 ms prior to the sprite occurrence. The typical halo is visible as a ring area with 50 km diameter and 10 km thickness.

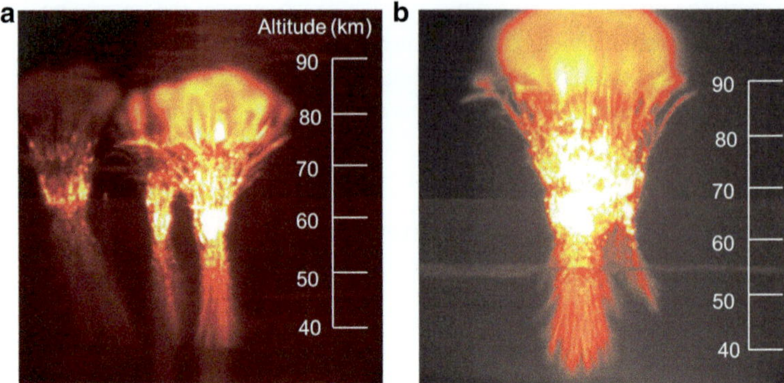

Fig. 3.12 The images illustrating the spatial sprite structure and transition between diffuse and streamer regions in the sprites as observed (**a**) 04:36:09.230 UT and (**b**) 05:24:22.804 UT on August 18, 1999. Taken from Stenbaek-Nielsen et al. (2000)

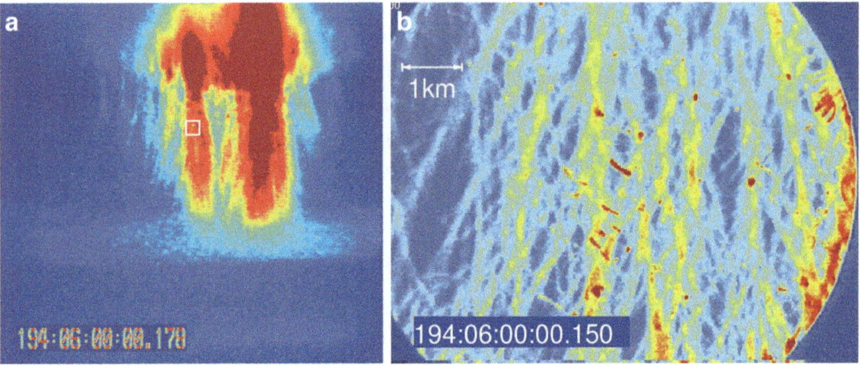

Fig. 3.13 Sprite telescopic images at low (**a**) and high (**b**) resolutions. The small area highlighted on the panel (**a**) is shown on the panel (**b**) at large scale. Taken from Gerken et al. (2000)

Blue jets (BJs) are beams of luminosity propagating upwards in narrow cones of about 15° from the tops of thunderclouds (e.g., Wescott et al. 1995; Boeck et al. 1995; Mishin and Milikh 2008). A color video imagery of the BJs has shown that this kind of TLEs exhibit primarily blue color. On average they are several km in diameter and typically brighter than sprites. BJs propagate with velocity of the order of 100 km/s, that is slower than sprites. They climb in the stratosphere up to 40–50 km altitude which implies a jet lifetime of 0.2–0.3 s. Blue starters are a kind of BJs which differ from them by a lower terminal altitude. They can develop upward from cloud tops at 17–18 km to terminal altitudes of about 25 km (Wescott et al. 1996; Heavner et al. 2000; Pasko 2006). Gigantic jets (GJs) are more intensive discharges with a much greater length than that of the BJs which results in the formation of electrical connection between thundercloud tops and the conducting E-layer of the ionosphere (Wescott et al. 2001; Su et al. 2003; van der Velde et al.

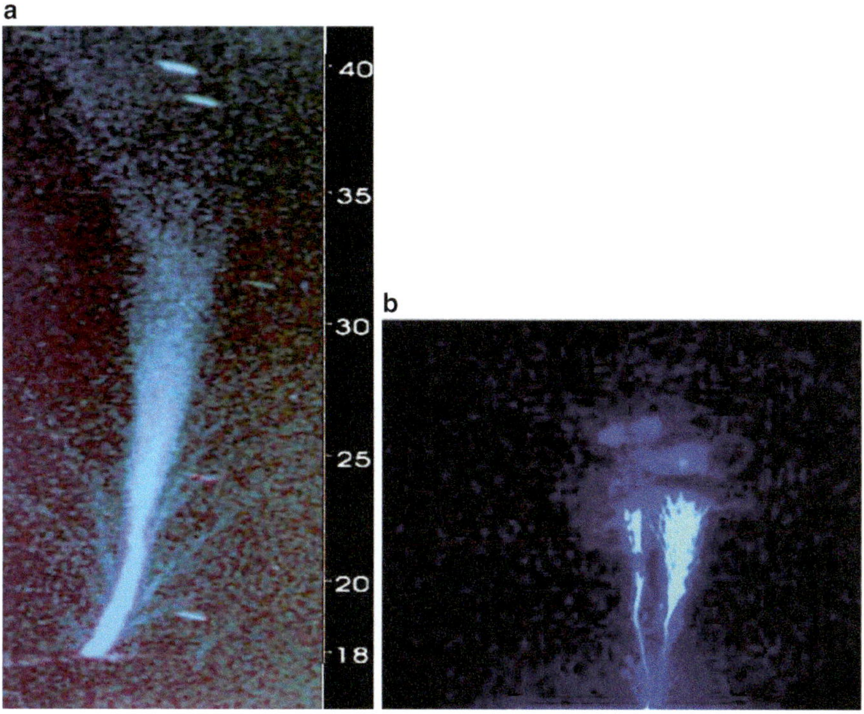

Fig. 3.14 Images of (**a**) blue jet (BJs) (Wescott et al. 2001), and (**b**) gigantic blue jet (GJ) (Pasko et al. 2002) discharged from the top of a thundercloud and upwards propagated to the lower ionosphere. The original images were recorded using a monochrome low-light video systems though the researchers observed the blue color flashes. To reproduce this effect, these images were enhanced with false color

2007; Kuo et al. 2009; Cummer et al. 2009). As illustrated in Fig. 3.14, the ground-based images of BJs and GJs exhibit a filamentary structure.

The predominance of red and blue colors in the optical emission of TLEs is believed to be due to the excitation of molecules of N_2 and O_2 by electron impact. At altitudes above 50 km the emissions of the first positive band of N_2 ($N_2 1P$) enhance a red optical region of red sprite emission whereas below 50 km the strong quenching of $B_3 \Pi_g$ state gives rise to the suppression of this emission. In the stratosphere the emission of the second positive band of N_2 ($N_2 2P$) becomes dominant, which results in predominance of blue color in the optical emission of BJs (e.g., see Vallance-Jones 1974; Pasko 2006, 2010). The recent high resolution measurements have shown that the most portion of the optical emission comes from the high-ionized streamer heads which manifest themselves as mobile bright compact balls (Liu and Pasko 2006; McHarg et al. 2007).

The elves are an abbreviation for Emission of Light and VLF perturbations due to EMP Sources (Fukunishi et al. 1996). The "elf" manifest itself as a divergent ring of optical emissions at the bottom of the ionosphere at $\sim$90 km altitude.

Typically the elves are visible for time interval less than 0.1 ms while their size can reach a value of 300–700 km in radius and 10–20 km in thickness. This short-term effect is associated with the ionospheric response to a strong electromagnetic pulse radiated by the CG discharge current of either polarity (e.g., see Boeck et al. 1992; Nickolaenko and Hayakawa 1995; Inan et al. 1996a, 1997; Cho and Rycroft 1998; Rowland 1998; Cheng et al. 2007).

Certainly, here we cannot come close to treating of other striking optical phenomena in the upper atmosphere such as trolls, gnomes, fairies, and so on.

3.2.2   Underlying Mechanisms for Blue Jets (BJs) and Gigantic Jets (GJs)

It is generally believed that BJs may occur under certain relatively rare conditions, when large amount of positive charge piles up at the top of thundercloud. Since the actual charge distribution in the cloud is very complicated we simplify the problem assuming as before that the charges are uniformly distributed in four spherical regions as shown in Fig. 3.3. According to this simplified model the dependence of the vertical field on altitude is described by the set of Eqs. (3.2)–(3.4). To simulate a quasielectrostatic (QE) field preceding a normal BJ discharge, Krehbiel et al. (2008) have suggested the set of parameters $q_i = 5, -40, 57.5$ and -20 C, where $i = 1, 2, 3, 4$. In Fig. 3.15, we plot the numerical calculation of the vertical field versus altitude based on the same parameters z_i and r_i as those used in making Fig. 3.4. It is obvious from this figure that the positive peak of E_z exceeds E_s^+ (line 2) around the altitude $z = 14$ km. On account of the positive field direction in this altitude range, we would expect the initiation of the upward-directed positive discharge which can propagate towards the ionosphere.

In this picture the BJ can be considered as upward-propagating positive leader with a streamer corona on the top (Petrov and Petrova 1999) as schematically displayed in Fig. 3.16. The streamer-to-leader transition is assumed to be accompanied by the Joule heating and subsequent electron detachment processes (Bondiou and Gallimberti 1994). So, more precisely, the BJs should be considered as hot leader-like discharges (Raizer et al. 2010) rather than cold streamer-like discharges, as the early modelers did. The leader bears a positive charge from the thundercloud top into the stratosphere up to altitude about several tens kilometers. A great number of the short-lived streamers are emitted from the leader thereby producing a streamer corona and a branching structure of BJ. However the leader velocity $\mathbf{V}_j$ is much less than that of individual streamers.

We have already discussed the scaling of critical breakdown field E_c which is proportional to the gas pressure and to the neutrals number density n_m at constant temperature. In a similar fashion the typical streamer parameters can be estimated on the basis of a similarity law and on dimension attributes (e.g., see Raizer et al. 1998; Pasko 2006; Surkov and Hayakawa 2012), according to which the typical

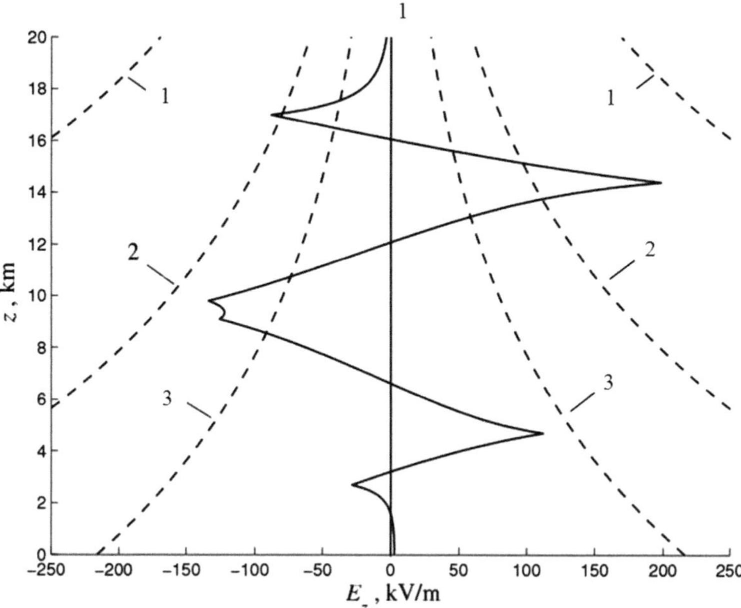

Fig. 3.15 The same as in Fig. 3.4 but for thunderstorm QE field preceding BJ discharge (Surkov and Hayakawa 2012)

discharge size L such as discharge tube length, streamer radius and etc. scales as $L \propto n_m^{-1} \propto \exp(z/H_a)$. Typical time interval T such as relaxation time, mean free time between collisions, two-body attachment time and etc. scales as $T \propto n_m^{-1}$. The typical velocity $V \propto L/T$ whence it follows that the streamer velocity, electron drift velocity V_d, and so on are independent of n_m whereas the electron mobility $\mu \sim V_d/E_c \propto n_m^{-1}$. Plasma and charge density inside the streamer body follows the scaling law: $n_e = n_i \propto n_m^2$ while the plasma conductivity scales as $\sigma \sim e n_e V_d \propto n_m$.

The increase in streamer size with altitude predicted by the scaling theory is compatible with the BJ observations. Although the similarity law for leader does not exist (Raizer 1991), with some care one may speculate that the conical shape of BJs and GJs (see Fig. 3.14) follows this similarity law since the scale of individual streamers and of the whole streamer zone has to increase with height. However, our calculations have demonstrated that the electric field produced by thundercloud charges is still smaller than that required for propagation of positive streamer (Fig. 3.15). Reasonable guesses as to the electrical inhomogeneity need to start breakdown ionization of the air. Another way of explaining this contradiction has been proposed by Raizer et al. (2006, 2007). In their model the BJ can be resulted from a bidirectional uncharged leader which in turn originates in the thundercloud area where the electric field reaches a maximum value. Owing to the exponential profile of the air density, the leader and the streamer corona are assumed to grow

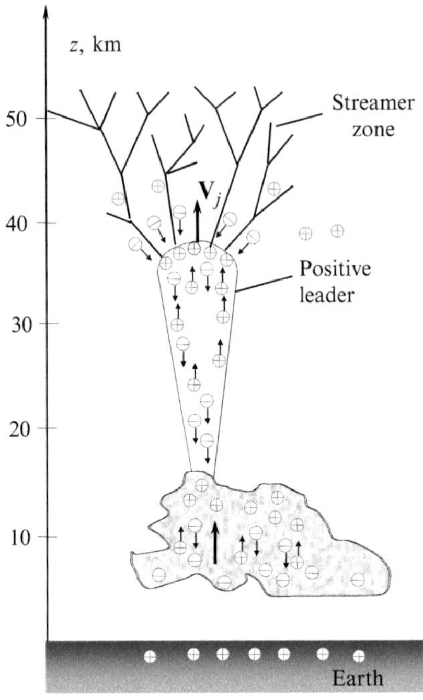

Fig. 3.16 A scenario of BJ development considered as a positive leader-like discharge propagating upwards at the velocity $\mathbf{V}_j$ (Surkov and Hayakawa 2012)

predominantly upward in contrast to laboratory conditions. Notice that the theory of streamers/leaders propagating at stratospheric and mesospheric altitudes is still far from being accurate.

It seems likely that the GJs can be associated with a large amount of negative electric charges accumulated at the middle region of thundercloud either by chance or as a result of another effect. One of the conceivable sets of the parameters for existence of this situation is as follows: $q_i = 25, -120, 82.5$ and -3 C (Krehbiel et al. 2008). A model calculation of the vertical QE field is presented in Fig. 3.17 as a function of height. The numerical values of other parameters used in making this plot are as follows: $z_i = 4.3, 8.0, 13.2, 15.4$ km, $r_i = 1.1, 2.6, 1.9, 0.3$ km. As is seen from this figure, the thunderstorm electric field is close to the breakdown threshold E_s^- (line 1) in the area above the charge $q_2 = -120$ C within the altitude range 10–12 km. This means that the GJ can be originated in this area as an upward-propagating IC discharge which transfers a negative charge of the order of 100 C through the region with upward/positive electric field towards the thundercloud top. Although the field is positive in a narrow region of 13–15 km above the charge $q_3 = 82.5$ C, the GJ can overcome this region to propagate out of the thundercloud towards the ionosphere (Krehbiel et al. 2008; Pasko 2010). However, we cannot explain in any detail why the GJs look more powerful than the BJs and why they can extend to higher altitudes.

It appears that the most of GJs develop in the form of upward-propagating negative leaders. In support of this conclusion it was noted that the visible

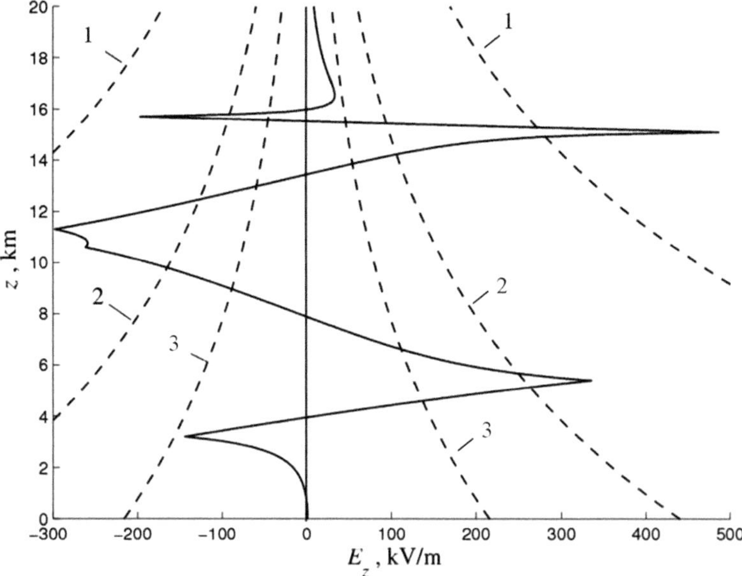

Fig. 3.17 The same as in Fig. 3.4 but for the thunderstorm QE field preceding a GJ discharge (Surkov and Hayakawa 2012)

patterns of GJs are similar to inverted images of conventional $-$CG (Pasko 2010; Neubert et al. 2011). Despite this similarity the other parameters of GJs differ significantly from those of standard $-$CG. For example, a GJ recently observed by Cummer et al. (2009) was estimated to transfer the negative charge of $-144\,$C from the thundercloud to the lower ionosphere. This value is much greater than a typical charge $\sim$5–10 C lowered to the ground by a normal $-$CG stroke. Additionally, the onset time of the GJ current was about 30 ms which is much greater than that ($\sim$5 μs) due to the stroke. This kind of GJ can be referred to as the class of negative cloud-to-ionosphere discharge ($-$CI)

The first documented event of positive cloud-to-ionosphere discharge ($+$CI) has been recently observed by van der Velde et al. (2010) during winter thunderstorms in the Mediterranean. This event is characterized by the current peak value of 3.3 kA and by short duration of 120–160 ms. This $+$CI discharge was estimated to lower negative charge $-136\,$C down from the ionosphere to the positively charged origins in the cloud only 6.5 km tall that result in a huge charge moment change of 11,600 C km. This event has also demonstrated that high altitudes are not a necessary condition for initiation of GJs.

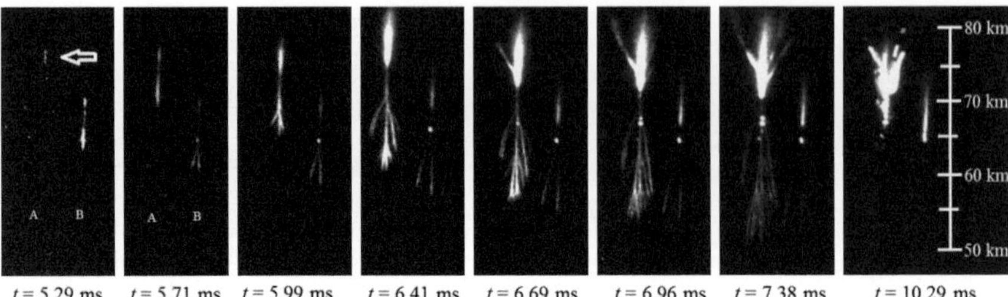

Fig. 3.18 High speed images of two sprites nucleation. The initiation point of the left sprite (*A*) is marked with an *arrow*. The images were recorded at Yucca Ridge Field Station on August 13, 2005 at 03:43:09.4 UT. The time was measured from the moment of lightning return stroke onset. The first image is contrast enhanced. Adapted from Cummer et al. (2006a)

3.2.3 Underlying Mechanisms for Sprites

A great deal of observations has shown that worldwide sprites and halos are triggered by large CG flashes almost exclusively with positive polarity (e.g., Boccippio et al. 1995; Williams et al. 2007). The charge moment change of the causative +CG was found to be greater than the critical value of the order of 500 C km in order to initiate the sprite discharge (Stanley et al. 2000; Cummer 2003; Cummer and Lyons 2005; Rycroft 2006; Hiraki and Fukunishi 2007). The high-speed video recording of sprites initiation has shown (Cummer et al. 2006b) that at first the downward streamer originates either spontaneously from a bright nucleolus between 70 and 75 km altitude (Fig. 3.18) or from brightening inhomogeneities at the bottom of a halo (Fig. 3.19). As is seen from the images shown in Fig. 3.18, the brighter column continues to expand upward and downward from the nucleation point followed by the generation of bright upward propagating streamers that branch and terminate in diffuse emissions. In the case shown in Fig. 3.19, at first the distinct bright nucleolus develop at the lower edge of the originally homogeneous halo. A downward streamer then initiates from that point thereby producing the bright column which in turn begins to expand upward and downward. The upward streamers propagate at velocity $(0.5–2) \times 10^4$ km/s and terminate in diffuse emissions as in the previous example (Stanley et al. 1999; Cummer et al. 2006b; Stenbaek-Nielsen and McHarg 2008).

It is generally accepted that there are two basic visible shapes of sprites: "carrot" or "jellyfish" configuration and columniform (e.g., Cho and Rycroft 1998; Matsudo et al. 2007; Myokei et al. 2009). The carrot type sprites are characterized by diffuse tops and lower tendrils extending down to altitudes of 30–40 km, while the columniform sprite has a very fine spatial structure as compared with the "carrot" sprite (e.g., Wescott et al. 1998; Hayakawa et al. 2004). It appears that these kinds of sprites differ in time delay with respect to a causative +CG. Winter thunderstorm observations in the Hokuriku area of Japan have shown that the "column" sprites

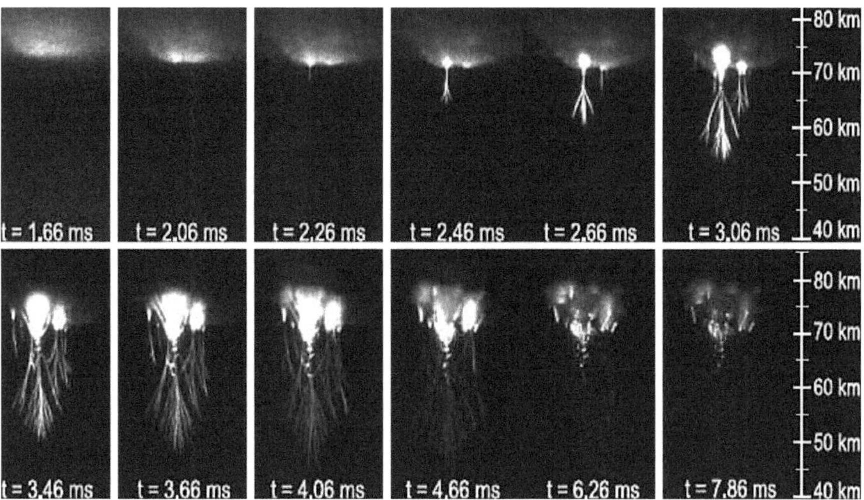

Fig. 3.19 High speed images of the sprite nucleation. The images were recorded at Yucca Ridge Field Station on August 13, 2005 at 03:12:32.0 UT. Adapted from Cummer et al. (2006a)

might be delayed by several ms after the causative $+$CG discharge while "carrot" sprites occur tens of ms after the $+$CG lightning (e.g., Matsudo et al. 2009; Suzuki et al. 2011).

The positive CG flashes may result in the extraordinary large charge transfer ($\sim$100 C) for the short time that gives rise to the strong QE field caused by uncompensated negative charges located in the thundercloud. To estimate this QE field at high altitudes, consider first a short period just after the causative $+$CG lightning. In the first approximation this implies that the conduction current, $\sigma_a E$, is much smaller than the displacement current, $\varepsilon_0 \partial_t E$, so that the air conductivity can be neglected. To simplify the problem, the thundercloud charge is assumed to be uniformly distributed inside the ball with radius r. The center of the charged ball is located on z-axis at the altitude h above the perfectly conducting ground. In this model the vertical component of electric fields E_z on z axis is described by equations similar to Eqs. (3.2)–(3.4). The thundercloud charge q and QE field can gradually increase just after the moment of main stroke due to strong CC in the sprite-associated $+$CG lightning. This CC is normally much greater than that due to negative stroke and its value amounts to 5–10 kA for a relatively long period of 10–100 ms (Rakov 2000). To illustrate this tendency, plots of E_z versus altitude z are shown in Fig. 3.20 with lines 1–3, which correspond to $q = 50$, 100, and 150 C, respectively. In making these plots the numerical values $h = 10$ km and $r = 2$ km are used.

We recall that all the threshold fields exponentially decrease with altitude whereas the QE field caused by the thundercloud charge and its electric image in the conducting ground falls off according to the dipole law; that is, inversely as the cube of the distance from the source. This means that the breakdown fields decrease

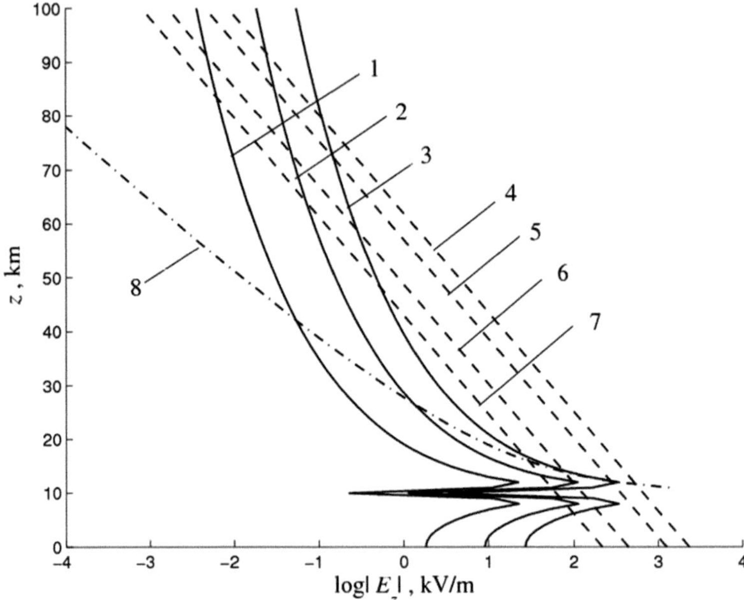

Fig. 3.20 Model calculations of thundercloud QE field just after a strong +CG which is able to trigger sprite discharge. The absolute value of the vertical electric field for different thundercloud charge q is plotted in this figure with lines 1–3 and 8 as a function of altitude: *1—q* $= 50\,$C; *2—q* $= 100\,$C; 3 and *8—q* $= 150\,$C, respectively. In making the plots 1–3 the air conductivity was ignored. The breakdown threshold electric fields which correspond to different air breakdown criteria are shown with dotted lines: *4*—conventional breakdown threshold, *5*—negative streamer propagation, *6*—positive streamer propagation, *7*—relativistic runaway breakdown. *Dash-and-dot line 8* illustrates the air conductivity effect on thunderstorm QE field (Surkov and Hayakawa 2012)

with altitude more rapidly than does the thunderstorm electric field. So, there may be a height above which the thundercloud electric field exceeds the breakdown threshold (Wilson 1925). As is seen from Fig. 3.20, this situation may exist at the mesospheric altitude range 50–80 km.

Actually, the generation of QE electric fields above a thundercloud may be greatly reduced due to the exponential increase of the atmospheric conductivity with altitude. The background atmospheric conductivity is a subject of a variety of factors: cosmic-ray ionization rate, ion–neutral collision rate, electron attachment and detachment and etc., which in turn vary with altitude due to changes of the air density. However in the first approximation the air conductivity can be approximated by Eq. (3.1); that is, as an exponential function of altitude z. The thundercloud charge variations, which follow primary +CG stroke, are basically due to the CC. However, if the time scale of the charge variations is much greater than the relaxation time $\tau = \varepsilon_0/\sigma_a$ due to air conductivity, then the problem is reduced to a stationary one. In this extreme case a distribution of electric potential

Φ is described by Poisson equation. Considering the thundercloud as a point current source located on z-axis at the altitude h, we come to the following equation:

$$\frac{\sigma_a}{r}\partial_r\left(r\partial_r\Phi\right) + \partial_z\left(\sigma_a\partial_z\Phi\right) = \frac{I\delta\left(r_-\right)}{4\pi r_-^2},\tag{3.12}$$

where I denotes the total current flowing from the source, δ stands for Dirac delta-function, and $r_\mp = \left\{r^2 + (z \mp h)^2\right\}^{1/2}$. This equation should be supplemented by the proper boundary conditions for the conducting ground and at the infinity, that is, $\Phi = 0$ at $z = 0$ and $\Phi \to 0$ when $z \to \infty$. Substituting the relationship $\sigma_a = \sigma_0\exp\left(\alpha z\right)$ into Eq. (3.12) and solving the problem gives (e.g., see Soloviev and Surkov 2000)

$$\Phi = \frac{I}{4\pi\sigma_0}\exp\left[-\frac{\alpha}{2}\left(z + h\right)\right]\left\{r_-^{-1}\exp\left(-\frac{\alpha r_-}{2}\right) - r_+^{-1}\exp\left(-\frac{\alpha r_+}{2}\right)\right\}.\tag{3.13}$$

The total charge q of the source/thundercloud can be related to the source current I through the Gauss theorem and Ohm law whence it follows that $I = (q\sigma_0/\varepsilon_0)\exp\left(\alpha h\right)$. When $\alpha = 0$, Eq. (3.13) describes a potential of point charge q and its mirror image in the perfectly conducting ground. In general case the exponential factors in Eq. (3.13) lead to the strong field attenuation with altitude due to air conductivity.

This model with the parameter $\alpha = 0.15\,\text{km}^{-1}$ and $q = 150\,\text{C}$ was used for numerical calculation of the vertical component of the thundercloud electric field $E_z = -\partial_z\Phi$ along z-axis in the presence of atmospheric conductivity as shown in Fig. 3.20 with dash-and-dot line 8. The role of the atmospheric conductivity comes into particular prominence when comparing this graph with that calculated at the same thundercloud charge and zero atmospheric conductivity (line 3). It is obvious from Fig. 3.20 that the conduction current due to the atmospheric conductivity may decrease the thunderstorm field to such an extent that it makes impossible the air breakdown in the mesosphere. In a more accurate model which takes into account both the time-dependent CC and the air conductivity, the altitude profile of the electric field be situated between lines 3 and 8 (Mareev and Trakhtengerts 2007). In this notation, the sprites must build up very quickly just after the causative lightning discharge for the short period limited by the relaxation time τ that varies within 1–100 ms in the altitude range 60–80 km.

The sprite initiation, visible evolution, streamer structure, and their relationship with IC process are so complex that any quantitative theory of the sprites has not been established yet except a number of numerical simulations (e.g., Pasko et al. 2000, 2001; van der Velde et al. 2006, 2007; Asano et al. 2009a,b; Ebert et al. 2010).

Luque and Ebert (2009, 2010, 2012) have recently developed a numerical model of sprite initiation that takes into account the photoionization effect and altitude-dependent transport and ionization parameters of electrons and neutrals. This model does not require any kind of seed electrons since the primary streamer is assumed to be due to drift of the background electrons subjected to thundercloud electric field.

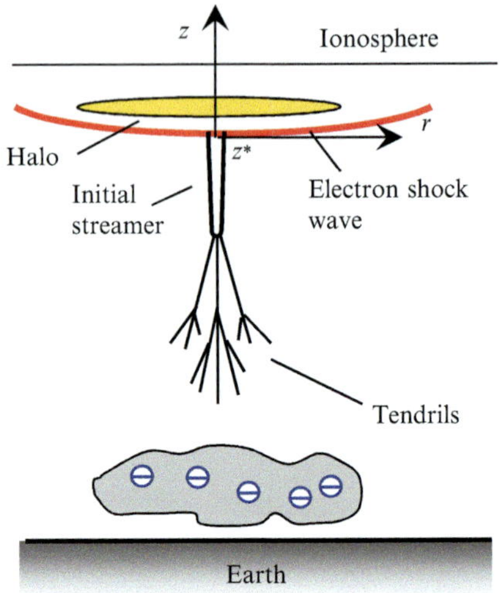

Fig. 3.21 A schematic plot of sprite originated from a sprite halo in the presence of QE field generated above a thundercloud after +CG lightning (Surkov and Hayakawa 2012)

Their numerical simulations show that (1) several ms after the powerful causative +CG lightning a downward-propagating electron density shock wave can develop in the lower ionosphere, (2) then this wave transforms into the downward self-propagating narrow filament which can serve as a positive sprite streamer, and (3) the impact and photo-ionization rates are the highest at the streamer head. In this model the electron density wave can be considered as a possible candidate for the visible sprite halo.

To give a qualitative interpretation of these results, we assume that a dipole approximation could be applied to mesospheric electric field of the thundercloud charges and of their mirror image in the perfectly conducting ground

$$E = \frac{d}{4\pi\varepsilon_0 \left(r^2 + z^2\right)^{3/2}} \left(1 + \frac{3z^2}{r^2 + z^2}\right)^{1/2}, \tag{3.14}$$

where r and z are cylindrical coordinates which are shown in Fig. 3.21. The electric dipole moment $d = 2qL$, where q is the thunderstorm charge and L is the distance from the thundercloud to the ground. We also assume that the thundercloud electric field at the front of the electron wave is close to the breakdown threshold. Substituting Eq. (3.5) for E into Eq. (3.14) gives the implicit dependence $r(z)$ or $z(r)$, which defines the surface of the wave front. This surface crosses z axis at the point z_* which can be found from the following equation

$$\frac{d}{2\pi\varepsilon_0 z_*^3} = E_0 \exp\left(-\frac{z_*}{H_a}\right). \tag{3.15}$$

In the vicinity of the axis of symmetry ($r \ll z$) the relationship between r and z can be simplified in such a way that we come to the explicit dependence $r\,(z)$ which is valid as $z \geq z_*$:

$$r = 4z \left\{ \frac{\pi \varepsilon_0 E_0}{15d} \left[z_*^3 \exp\left(-\frac{z_*}{H_a}\right) - z^3 \exp\left(-\frac{z}{H_a}\right) \right] \right\}^{1/2}. \qquad (3.16)$$

The wave front given by Eq. (3.16) is schematically shown in Fig. 3.21 with red line. It follows from Eq. (3.15) that an increase in dipole moment d results in a decrease of z_*. This means that the enhancement of thundercloud field is accompanied by downward propagation of the point z_* and the wave front given by Eq. (3.16). Although this qualitative analysis is consistent with the results of numerical simulations reported by Luque and Ebert (2009), the above approach cannot predict the sharp prominence arising in the center of the wave surface because we have ignored the electric field of charges accumulated at the wave front. This point requires the precise analytical analysis because this effect can be due to plasma or other kind of instabilities (e.g., Derks et al. 2008).

3.2.4 Runaway Electron Breakdown

As has already been discussed, the conventional streamer-leader mechanism for air breakdown can explain, in principle, the basic properties of the TLEs (e.g., see Riousset et al. 2010a,b; Raizer et al. 2010). An alternative approach assumes the relativistic runaway electron avalanches as the proper candidate for producing the air breakdown at stratospheric and mesospheric altitudes (e.g., see Gurevich et al. 1992, 1994; Roussel-Dupré and Gurevich 1996; Lehtinen et al. 1997, 1999; Babich et al. 1998, 2008; Gurevich and Zybin 2001; Lehtinen 2000; Füllekrug et al. 2010, 2011). One of the merits of this mechanism is that the electric field threshold required for air breakdown may be one order of magnitude lower than that due to the conventional breakdown.

In the course of this text, the runaway breakdown is only treated in a sketchy fashion. First of all we note that if the electron energy greater than $50\,\mathrm{eV}$ then there prevails the electron forward scattering at small angles. In this notation we consider the simple one-dimensional model in which all the high-energy electrons can move only along z axis parallel to the constant electric field $\mathbf{E}$. The electron collisions are taken into account by means of the so-called dynamical friction force F_{fr} which is pointed oppositely to the vector of electron momentum $\mathbf{p}$. In such a case the equation of electron motion is reduced to the following (Gurevich et al. 1992, 1994)

$$\frac{dp}{dt} = eE - F_{fr}. \qquad (3.17)$$

In a more accurate model one should take into account the angle included between the electric force eE and the electron momentum.

The dynamical friction force is equal to the electron energy loss due to the electron collisions per unit length, that is,

$$F_{fr}(\varepsilon) = \frac{d\varepsilon}{dz}. \tag{3.18}$$

A main contribution to the energy losses from high-energy electrons is caused by the ionization of air. A peculiarity of this ionization process is that the energy of the fast-moving electron is much greater than the energies of atomic electrons. This means that the high-energy electron interacts with atomic electrons and nuclei as with free particles. In such a case the friction force can be estimated as $F_{fr}(\varepsilon) \sim \varepsilon/\lambda$, where the free length of the electrons $\lambda \sim (Zn_m\sigma_c)^{-1}$ depends on the number density of molecules n_m, the mean number of electrons in molecule Z, and the scattering cross section σ_c. In the non-relativistic energy range the interaction between charged particles is governed by the Coulomb law through Rutherford scattering cross-section $\sigma_c \sim e^4/\varepsilon^2$ (e.g., see Gurevich and Zybin 2001). Combining the above relationships, we arrive at the following estimate $F_{fr}(\varepsilon) \sim e^2 Z n_m/\varepsilon$. Notice that this dependence is in good agreement with the equation derived by Bethe (1930) in a more accurate model:

$$F_{fr}(\varepsilon) = \frac{2\pi e^4 Z n_m}{\varepsilon} \ln \frac{\varepsilon}{J_z}. \tag{3.19}$$

Here $J_z \sim \varepsilon_i$, where ε_i is the energy of ionization.

A schematic plot of the dynamical friction force of electrons as a function of their kinetic energy is displayed in Fig. 3.22. Here we do not show a few resonance peaks in the low-energy region although on average the friction force approximately increases in this region as shown with dashed line. As is seen from this figure, the friction force reaches a maximum value at the energy ε_*. This maximum corresponds to the so-called thermal runaway breakdown threshold, which occurs at the electric field $E_{\text{th}} \approx 260\,\text{kV/cm}$. So large electric field does not occur at stratospheric and mesospheric altitudes.

Equation (3.19) can be applied to the 10^2–10^6 eV energy range where the friction force falls off with increasing the electron energy. At higher energies one should take into account relativistic effects which were ignored in deriving the above estimates. The contribution of the relativistic effects results in gradual changes in the above tendency in such a way that the friction force reaches a minimum F_{min} at the energy $\varepsilon_{\text{min}} \approx 1.4\,\text{MeV}$, and then a logarithmically slow increase begins to prevail at higher energies (Gurevich and Zybin 2001).

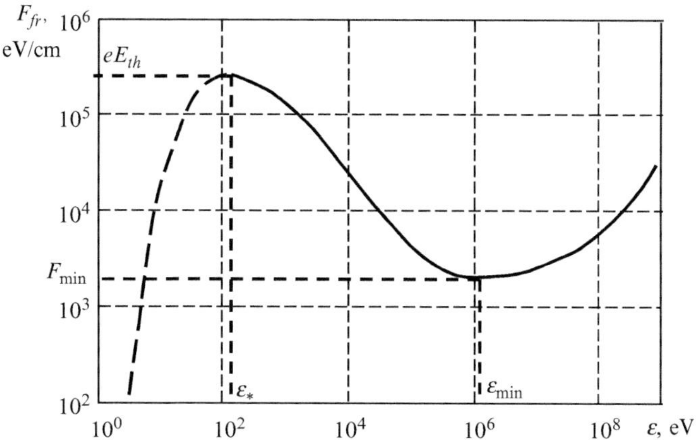

Fig. 3.22 A schematic plot of dynamical friction force of electrons in the air versus electron kinetic energy. The figure is partly adapted from Pasko (2006)

The generation of the runaway electrons is possible in the energy range from ε_* to $\varepsilon_{\min}$ where the fall off of the friction force dominates. It follows from Eq. (3.17) that the runaway electrons can appear under the requirement

$$eE > F_{fr}(\varepsilon).$$ (3.20)

which means that the electric field will accelerate the electrons with such energies continuously so that they become "runaway" electrons. The implication here is that the increase of the electron energy in the electric field prevails over the energy losses due to ionization of air. Conversely, if $eE < F_{fr}(\varepsilon)$, then the electron energy falls off quickly due to the ionization of air and other inelastic processes result in the energy losses. The requirement given by Eq. (3.20) can be satisfied for the ambient electric field $E > E_r = F_{\min}/e$. It follows from the detailed analysis that the minimal value of the threshold electric field is given by (Gurevich and Zybin 2001)

$$E_r = \frac{4\pi e^3 Z n_m a}{m_e c^2},$$ (3.21)

where the dimensionless parameter $a \approx 11$.

To satisfy Eq. (3.20), the runaway electron energy must be greater than the threshold value, ε_r, which depends on the ambient electric field. Combining Eqs. (3.19)–(3.21) we obtain that

$$\varepsilon > \varepsilon_r \approx \frac{m_e c^2 E_r}{2E}.$$ (3.22)

Runaway electron propagation through the air is accompanied by the generation of large amount of secondary low-energy electrons due to the neutral molecule ionization by runaway electron impact. Although a majority of secondary electrons have a small energy, a portion of such electrons may gain energy ε which is greater than the threshold value; that is, $\varepsilon > \varepsilon_r$. The ambient electric field will accelerate these energetic electrons, so that they may also become runaway electrons, which in turn results in additional ionization of air and the generation of a new portion of secondary and runaway electrons. The exponentially increasing avalanche of runaway electrons is a crucial factor in the development of air breakdown since a great deal of the secondary slow electrons is produced along with the runaway electrons (e.g., see Colman et al. 2010).

It follows from Eq. (3.21) that the runaway breakdown field E_r is proportional to the neutral number density. Taking the notice of Eq. (2.3) for n_m gives the following approximation (Gurevich and Zybin 2001):

$$E_r = 2.16 \exp\left(-z/H_a\right), \quad \text{kV/cm}. \tag{3.23}$$

It should be emphasized that the value of the runaway breakdown threshold E_r is one order of magnitude smaller than the conventional breakdown E_c. This important fact follows from a comparison of Eqs. (2.3) and (3.23). The runaway breakdown field E_r is shown in Figs. 3.4, 3.15 and 3.17 with line 3. On the other hand, the thermal runaway breakdown threshold $E_{th} \approx 260\,\text{kV/cm}$ is approximately 100 times greater than the runaway threshold E_r. Under such a strong electric field all the thermal electrons become runaway ones since the electric force acting on electrons becomes greater than the maximal dynamical friction force shown in Fig. 3.22. A more sophisticated treatment has shown that as the electric field E is close to the breakdown threshold then the characteristic length l_r of runaway electron avalanches is inversely proportional to n_m (Gurevich and Zybin 2001). Since the neutral number density decreases with altitude, the value of l_r, on the contrary, increases from several tens meters at the ground surface level to a few km at the mesospheric altitudes. It is obvious from Fig. 3.20 that the thundercloud electric field arising after a CG discharge (lines 1–3) can exceed the runaway breakdown threshold (line 7) in the tens km altitude range which is greater than the length l_r of exponential growth of runaway electron avalanche.

Radiations of relativistic electrons give rise to the generation of Roentgen and gamma quanta which in turn are able both to ionize the molecules and to generate electron–positron pairs when interacting with nuclei of molecules. In the course of this text, we cannot come close to exploring these topics owing to the complexity of this problem. In a more accurate theory, the Boltzmann transport equation is used to describe the runaway electron distribution function $f\left(\mathbf{r}, \mathbf{p}, t\right)$ in phase space (e.g., see recent reviews by Roussel-Dupré et al. (2008) and by Milikh and Roussel-Dupré (2010))

$$\partial_t f + \mathbf{V} \cdot \nabla f + e\mathbf{E} \cdot \nabla_{\mathbf{p}} f = S\left(f, f_n\right), \tag{3.24}$$

where $\nabla_{\mathbf{p}}$ denotes gradient with respect to components of the momentum $\mathbf{p}$. Here the collision integral S is dependent on the distribution functions of electrons f and neutral molecules f_n. The interested reader is referred to a discussion by Gurevich and Zybin (2001) and by Trakhtengerts et al. (2002, 2003) for details about solutions of this equation.

The models of runaway breakdown in the atmosphere are based on the assumption that cosmic rays generate a shower of secondary particles, called an extensive air shower (EAS), thereby producing seed/secondary relativistic electrons which are capable of initiating the runaway breakdown in the presence of a strong QE field of thundercloud (Gurevich et al. 1999; Lehtinen et al. 1999; Gurevich and Zybin 2001, 2004; Inan and Lehtinen 2005; Roussel-Dupré et al. 2008; Milikh and Roussel-Dupré 2010). The incident cosmic ray particle energy to initiate runaway breakdown was estimated to be greater than or of the order of 10^{15} eV (Gurevich et al. 1999). The EAS typically consists of 89 % photons, 10 % electrons with the energy up to 30 MeV and 1 % other particles, largely muons (e.g., Carlson et al. 2008).

Modeling of interference between electromagnetic wave radiated by the horizontal branch of the parent lightning discharge and the waves reflected from the night ionosphere and the ground has shown that the transient electric field in the mesosphere can exceed the runaway electron threshold that supports the idea of free electron bunching in the mesosphere by the pulsed electric field (Kudintseva et al. 2010).

The numerical simulation has shown that one more conceivable reason for the existence of runaway is the electron acceleration during the propagation of lightning streamers and stepped leaders (e.g., Gurevich et al. 2007; Carlson et al. 2010; Chanrion and Neubert 2010; Celestin and Pasko 2011). The secondary fast electrons with the MeV energies may come out from the radioactive decays of a rest muon after an IC lightning discharge (Paiva et al. 2009).

Despite the threshold E_c for conventional breakdown is approximately an order of magnitude greater than that for runway breakdown, the focusing of the electric field in the vicinity of any inhomogeneity could lower the value of E_c by a factor of 10 or 30 (Fernsler and Rowland 1996) and vice versa, the actual field needs to be two to three times the runaway threshold E_r to get sufficient ionization for starting of runaways (Rowland 1998). It may be suggested that the conventional and runaway breakdowns develop at different altitudes, which are separated by only a few kilometers. Once either process is triggered prior to the next one, it can suppress the other process from triggering because of fast electric field relaxation due to the air polarization caused by the increase of plasma density. This implies that there may be a hybrid sprite model in which both the breakdown mechanisms may occur simultaneously (Roussel-Dupré and Gurevich 1996; Yukhimuk et al. 1999; Li et al. 2009, 2010; Chanrion and Neubert 2010).

An observational hint toward the runaway electron mechanism does occur in the atmosphere is the observations of the so-called terrestrial gamma ray flashes (TGFs); that is, short bursts of gamma rays originating from Earth's atmosphere. These events are believed to be due to Bremsstrahlung emissions from energetic ($\sim$1 MeV) electrons interacting with neutral molecules (e.g., see; Dwyer et al. 2010;

Carlson et al. 2010). Since their experimental finding (Fishman et al. 1994), the TGFs have been studied intensively for the last decades and much is now known of their properties. Typically, the TGFs occur in the form of narrow beams with energies up to 20 MeV and duration from 0.2 to 3.5 ms. It is generally believed that the TGFs are associated with an individual lightning strike though the observed rate of TGF events is much smaller than that of lightning flashes (Fishman et al. 1994; Inan et al. 1996b; Inan and Lehtinen 2005; Cummer and Lyons 2005; Smith et al. 2005, 2010; Briggs et al. 2010). This fact can be due to difficulties in detecting the TGFs.

3.2.5 *VLF Probing of the Lower Ionosphere Above Thunderstorm: Early/Fast and Early/Slow Events*

A major part of our knowledge of sprite properties is based on optical and spectral measurements, video observations of sprite morphology, and much was done for improvement of the spatial and temporal resolution of the sprites structure. Other instrumentations and technique are necessary to study the IC processes associated with the sprite evolution. The distribution of sprite delay between a sprite and its causative +CGs is indicative of correlation between IC processes and sprite generation mechanisms. Simultaneous optical and ELF/VLF observations are believed to be an effective technique for discussing the relationship between the sprite and its causative lightning (Füllekrug and Constable 2000; Sato and Fukunishi 2003; Hobara et al. 2006; Cummer et al. 2006a,b; Neubert et al. 2008; Surkov et al. 2010).

The ground-based narrowband VLF transmitters and receivers are commonly used to detect the perturbations of ionospheric and mesospheric conductivity caused by lightning discharges. This effect can be observed by distant measuring of the changes in the amplitude and phase of VLF electromagnetic wave propagating in Earth-Ionosphere waveguide and passing over a thunderstorm region (e.g., Dowden et al. 1996; Neubert et al. 2008). Design of the experiment scheme is shown in Fig. 3.23. The so-called lightning-induced electron precipitation effects (LEPs) or Trimpi effect are considered as a possible cause for this phenomenon (Helliwell et al. 1973). A portion of electromagnetic energy radiated by lightning penetrates through the ionosphere thereby exciting a whistler mode wave in the magnetosphere. As the Doppler-shifted frequency of the whistler mode wave is close to the gyrofrequency of trapped radiation-belt electrons then a resonance wave-particle interaction occurs which results in changing the electron pitch angle sufficiently to reduce it below the loss cone (Trakhtengerts and Rycroft 2008). As a result, the precipitation of 0.1–0.3 MeV electrons occurs at the base of the field line causing the local increase in the ionization and conductivity in D region of the ionosphere. The typical lateral size of the ionization region is about 1,000 km. The lag time between LEPs and the lightning is $\sim$1 s, and the onset time is about several seconds, while the recovery time varies within 10–100 s.

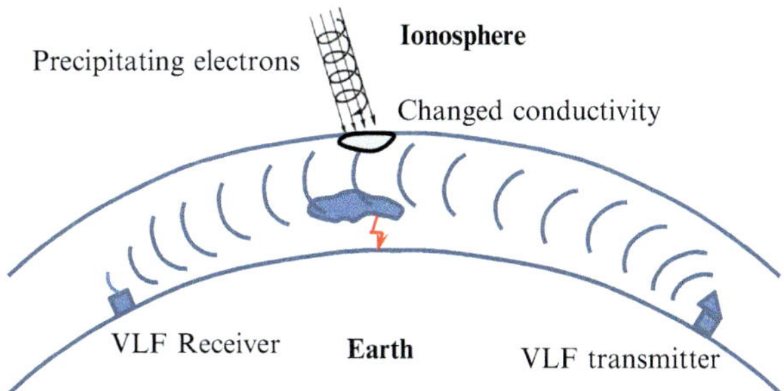

Fig. 3.23 Experimental scheme for remote measurement of the amplitude and phase changes of VLF electromagnetic wave propagating in the Earth-Ionosphere waveguide over a thunderstorm region. The local conductivity enhancement in the D region can be due to lightning-induced electron precipitation effects. The similar phenomena can be produced by the sprite ionization column and sprite halos in the D region

A similar effect associated with sprite-producing lightning is referred to as early/fast Trimpis (Inan et al. 1995). An example of the "early/fast" events observed in Crete during the 2003 EuroSprite campaign is displayed in Fig. 3.24. The observations have shown that sprites are nearly always accompanied by "early" VLF perturbations (Neubert et al. 2005, 2008). The lag time between the perturbations in signals from distant VLF transmitter and causative +CG lightning is less than 20 ms, that is shorter than the lag time observed during LEPs. Typically the onset time of the "early" perturbations is less than 50 ms while the recovery time varies within 10–300 s (Inan et al. 1995, 1996a, 2010; Hobara et al. 2001; Otsuyama et al. 2004; Neubert et al. 2008). Most of "early" VLF events are supposed to be due to the region of enhanced conductivity produced by the sprite ionization column and sprite halos in the upper D region. The typical lateral size of this region is about 100 km.

To treat early VLF events associated with sprite discharges in the D region, we need a combined set of Maxwell and continuity equations for charged particles. These equations govern the dynamics at least four kinds of particles; that is, electrons, positive ions, negative ions, and positive cluster ions (e.g., Glukhov et al. 1992; Haldoupis et al. 2009). As is seen from Fig. 3.24, the early VLF perturbation is characterized by an abrupt signal onset and a long recovery. In what follows we focus on a more slowly recovery process. Thus far, no consideration has been given to the short-term stage of electron production and ionization in the lower ionosphere caused by impact of a causative lightning and its sprite. Considering the plasma recombination, the continuity equation for electrons can be thus written as

$$\frac{dn_e}{dt} = -\alpha_d n_e n_i - \alpha_d^c n_e n_x, \tag{3.25}$$

where n_e, n_i, and n_x stand for electrons, positive ions, and positive cluster ions number densities, respectively. Here α_d denotes the coefficient of dissociative

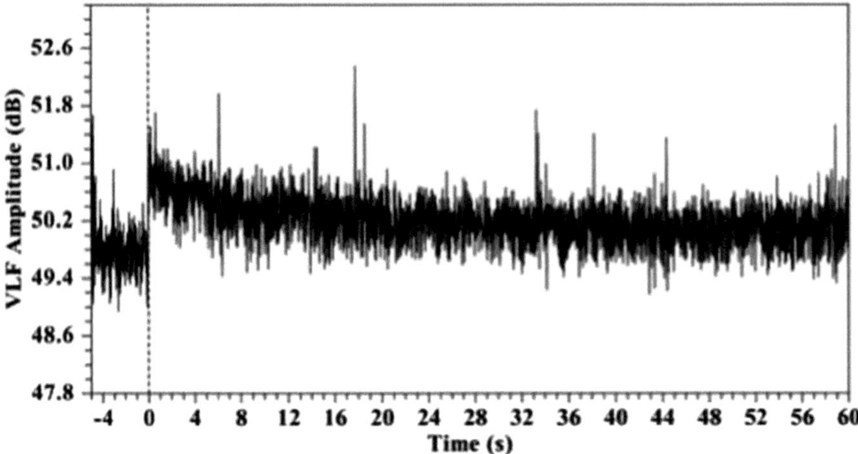

Fig. 3.24 A signature of "early/fast" VLF event associated with a sprite which was observed in Crete on August, 29 during the 2003 EuroSprite campaign (Neubert et al. 2005). Typically these events are characterized by an abrupt onset and a signal recovery ranging from 10–300 s. Adapted from Haldoupis et al. (2009)

recombination with positive ions, the majority of which compose O^{2+} and NO^+ while α_d^c is the effective coefficient of recombination of electrons with positive cluster ions which are produced from positive ions via a hydration chain reaction. Notice that the interaction between electrons and negative ions due to electron attachment and detachment does not enter this simplified equation.

Taking into account that the dissociative recombination of electrons and single positive ions dominates above 80–85 km (Glukhov et al. 1992; Haldoupis et al. 2009); that is, in the altitude range of interest, one may ignore the last term on the right-hand side of Eq. (3.25). Additionally, taking the notice of plasma quasi-neutrality one should substitute $n_i \approx n_e$ into Eq. (3.25). Let n_{e0} be the electron number density arising after the short-term stage of electron production. Taking this value as the initial one and performing integration of this equation we come to the following usual law for binary plasmas

$$n_e = \frac{n_{e0}}{1 + \alpha_d n_{e0} t}. \tag{3.26}$$

Substituting $n_e = n_{e0}/2$ into Eq. (3.26) we obtain the rough estimate of the plasma relaxation time $t_r \sim (\alpha_d n_{e0})^{-1}$. In the altitude range of interest the numerical values of the parameters are as follows: $\alpha_d = (1-3) \times 10^{-7}\,\mathrm{cm^3 s^{-1}}$ (Lehtinen and Inan 2007) and $n_{e0} = 6 \times 10^4\,\mathrm{cm^{-3}}$ (Haldoupis et al. 2009). Substituting these values into the above relationship gives the estimate $t_r \sim (2-5) \times 10^2\,\mathrm{s}$ which is compatible with the recovery time of early VLF events.

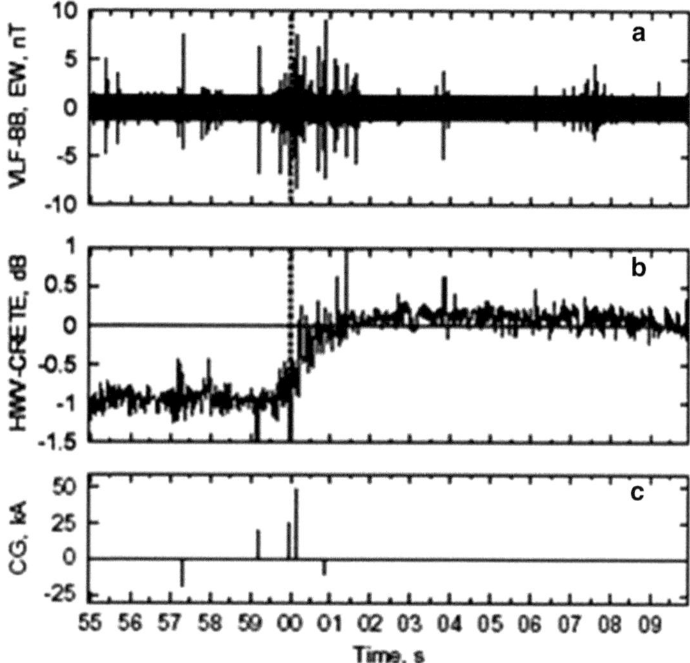

Fig. 3.25 An "early/slow" VLF event (*middle panel b*) which was observed in Crete at the distance about 2,000 km from a convective storm in central France. The sprite appearance was detected at the time marked by the *vertical dashed line*. A signature of the causative +CG discharge and other CGs is displayed in the (*bottom panel c*). The sferics possibly associated with intracloud lightning discharges were recorded by a broadband VLF receiver in Nançay, France, at about 200 km northeast of the thunderstorm (*upper panel a*). Adapted from Neubert et al. (2008)

Hence the early VLF events can be associated with the lightning and sprite-produced extra ionization in the D-region. In this picture the recovery time seems to be controlled by the plasma relaxation time.

A new type of the so-called early/slow VLF perturbations associated with sprites have been recently observed (Neubert et al. 2008). An experimental evidence of such events provided by the Crete receiver is shown in Fig. 3.25 (middle panel b). The sprite was observed over a convective storm in central France at the distance about 2,000 km from the receiver. This "early/slow" event is characterized by a relatively long onset time of ∼2 s after the moment of sprite appearance. As is seen from the bottom panel, this event can be associated with a few sequential CG lightning strokes. The broadband time series (upper panel a) provided by Nançay VLF receiver exhibits a number of sferics possibly associated with bursts of IC lightning discharges. Haldoupis et al. (2006) have assumed that the sequential electromagnetic pulses radiated upwards from horizontal IC discharges accelerate sprite-produced electrons which in turn can result in ionization of the lower ionosphere. One may also speculate that the electron impact upon the ionosphere causes the generation of secondary electron avalanche thereby ionizing the ionosphere. In this picture the onset of "early/slow" VLF perturbations and the period of sferics clusterization are correlated.

3.2.6   ELF Field Measurements of Sprite-Producing Events

Analysis of the ELF field measurements made it apparent that the sprite-associated events can be accompanied by appearance of two distinct peaks in the ELF recordings (Cummer et al. 1998, 2006a). Simultaneous optical and ELF observations have shown that the first peak corresponds to the causative lightning whereas the second one coincides in time with the moment of sprite luminosity. As is evident from the observations, the currents flowing inside the sprite body may generate 1–2 pulses comparable in amplitude with that produced by a causative CG flash and it appears that the peak amplitude is proportional to the sprite brightness. As one example, Fig. 3.26 shows the ELF field variations caused by +CG causative lightning and sprite which were detected by an interferometric optic system called SAFIR in the Hokuriku area (37.48° N, 136.76° E), Japan on February 03, 2007 during the 2006/2007 winter campaign. The first two peaks in this figure are assumed to be caused by two +CG return strokes while the third peak that follows the first ones can be resulted from the sprite current because this peak practically coincides with the sprite initiation moment (red vertical line in Fig. 3.26) which was found from the optical measurements.

As is seen from Fig. 3.26, the sprite delay between a sprite and its causative +CGs is about 50 ms although the lag time can reach a few hundred ms (Cummer 2003; Cummer et al. 2006a). The same order value seems to be typical as the duration of long-lasting intense CC in the positive causative lightning (Reising et al. 1996; Cummer and Füllekrug 2001; Lyons 2006; Hu et al. 2007). It may be suggested that the CC and possibly the higher frequency components (like M-component) in the CC play an important role in the initiation of the long delayed sprites (Yashunin et al. 2007; Asano et al. 2009a,b). The horizontal lightning currents between clouds and IC lightning discharges followed by nonuniform ionization of the upper atmosphere and an increase in mesospheric electric field can serve as the triggering events for the delayed sprite generation (Bell et al. 1998; Cho and Rycroft 1998, 2001; Ohkubo et al. 2005).

When the ELF recording and luminosity data are compared with that derived from model approximation of the sprite currents, this makes it possible to extract the sprite charge moment change (Boccippio et al. 1995; Hobara et al. 2001, 2006; Cummer 2003; Hayakawa et al. 2004; Matsudo et al. 2009) and also the sprite current moment waveform from the observations (Cummer and Inan 2000). Based on this approach, Cummer et al. (2006a) have estimated the sprite current moments as much as several hundred kA km for two case-studies.

It is notable that the dynamic spectrogram and ULF/ELF power spectrum of sprite-associated events can exhibit an approximately quasi-oscillatory pattern (Surkov et al. 2010). For example, the upper panel in Fig. 3.27 displays the dynamic

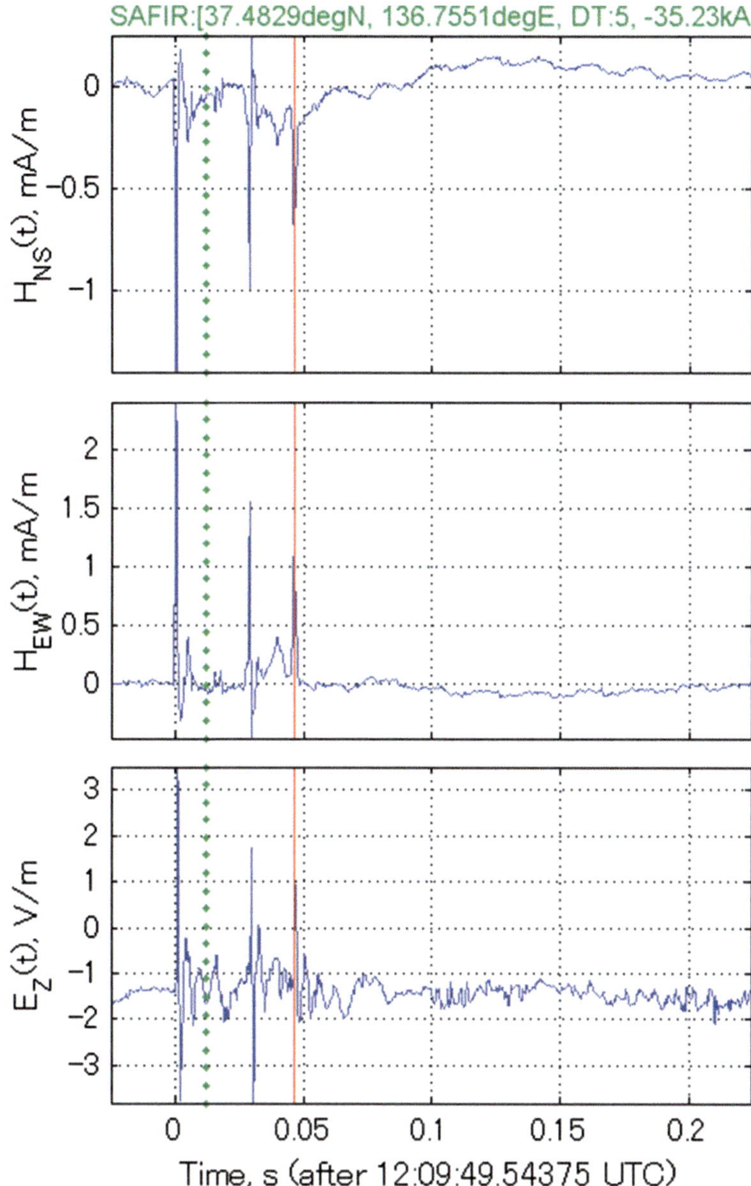

Fig. 3.26 ELF electromagnetic perturbations as observed in the Hokuriku area, Japan (37.48° N, 136.76° E) on February 03, 2007 during a sprite-associated event. The first and middle panels show north-south and east-west, H_{NS} and H_{EW}, magnetic field components while vertical electric field, E_z, is displayed in bottom panel. The vertical *red line* at approximately 0.05 s indicates a moment of sprite flash. Adapted from Surkov et al. (2010)

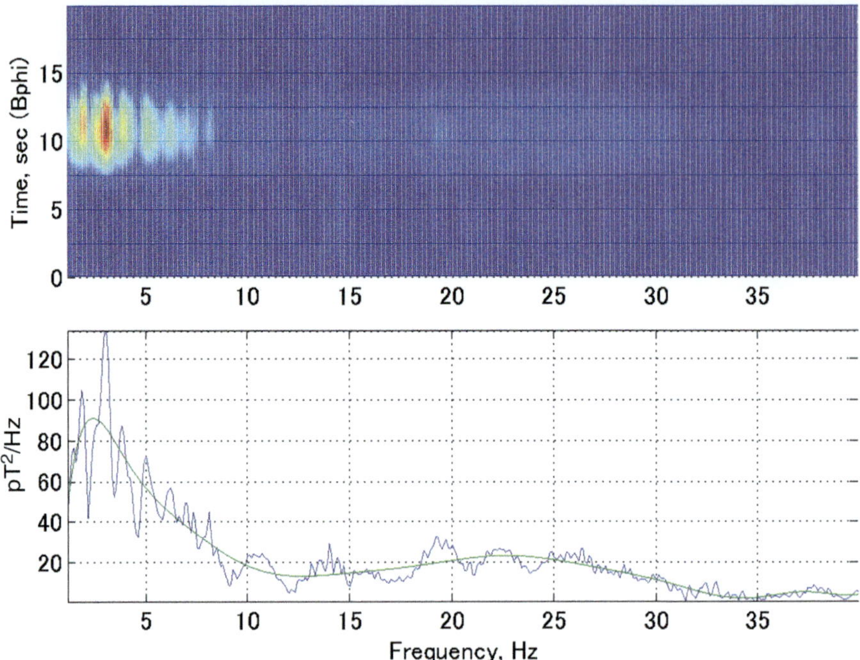

Fig. 3.27 Dynamic spectrogram (*upper panel*) and power spectral density of absolute value of the magnetic field variations shown in Fig. 3.26. Adapted from Surkov et al. (2010)

spectrogram which corresponds to the magnetic variations shown in Fig. 3.26. Every so often the same tendency for oscillating spectra has been observed during the 2006/2007 winter campaign. A distinct resonance structure below 7 Hz is assumed to be due to excitation of the so-called ionospheric Alfvén resonator (IAR) which will be described in more detail in the text section. What is more important, the power spectrum in Fig. 3.27 (lower panel) shows evidence of some vibrations with maximum repetition period of about 15–20 Hz. These oscillations are much more pronounced with the smooth envelope of the power spectrum shown in Fig. 3.27 with green line. To explain this peculiarity, one should note that the spectrum of net magnetic field variations resulted from the causative lightning and the delayed sprite is given by

$$\mathbf{B}\left(\omega\right) = \mathbf{B}_c\left(\mathbf{r}_c,\omega\right) + \mathbf{B}_s\left(\mathbf{r}_s,\omega\right)\exp\left(i\omega T\right), \tag{3.27}$$

where T is the lag of time between the causative lightning and sprite occurrences, and $\mathbf{r}_c$ and $\mathbf{r}_s$ are their position vectors, respectively. Since the power spectrum is proportional to $\left|\mathbf{B}\left(\omega\right)\right|^2$, it is evident that the amplitude modulation of the power spectrum shown in Fig. 3.27 with green line can be due to the oscillatory factor $\exp\left(i\omega T\right)$ in Eq. (3.27). This implies that the sprite lag time is inversely proportional to the "period" $\omega = 2\pi/T$ of spectrum oscillations. In the next section we shall study a model of the Earth-Ionosphere waveguide which makes it possible to calculate the ULF/ELF spectrum $\mathbf{B}(\omega)$ caused by the lightning discharge and sprite. Comparing the calculated power spectrum with the observation data permits us to find the sprite charge moment, the time lag of the sprite current and other parameters (Surkov et al. 2010). The ULF/ELF measurements thus provide with important information that will assist us in understanding the role played by long-lasting CC and IC processes in the delayed sprite generation.

In conclusion we note that despite much progress toward a comprehension of underlying mechanisms for recently discovered TLEs, the analytical theory of these phenomena is still far from accurate. The numerical modeling continues to be a basic instrument for theoretical study of TLEs behavior except for simple estimates which follow from the similarity law. This approach leaves unexplained why the charge moment change of the causative $+$CG must be greater than $\sim$500 C km to produce sprites (Stanley et al. 2000; Cummer 2003; Rycroft 2006). Moreover the relationship between the sprite structure and meteorological conditions, the effect of IC lightning activity, and etc. are far from being well understood. There are a lot of such problems to be solved.

References

Allen NL, Ghaffar A (1995) The conditions required for propagation of cathode-directed positive streamer in air. J Phys D Appl Phys 28:331–337

Asano T, Suzuki T, Hiraki Y, Mareev E, Cho MG, Hayakawa M (2009a) Computer simulations on sprite initiation for realistic lightning models with higher-frequency surges. J Geophys Res 114:A02310. doi:10.1029/2008JA013651

Asano T, Suzuki T, Hayakawa M, Cho MG (2009b) Three-dimensional EM computer simulation on sprite initiation above a horizontal lightning discharge. J Atmos Solar Terr Phys 71:983–990

Babaeva NY, Naidis GV (1997) Dynamics of positive and negative streamers in air in weak uniform electric fields. IEEE Trans Plasma Sci 25:375–379

Babich LP, Kutsyk IM, Donskoy EN, Kudryavtsev AY (1998) New data on space and time scales of relativistic runaway electron avalanche for thunderstorm environment: Monte Carlo calculations. Phys Lett A 245:460–470

Babich LP, Kudryavtsev AY, Kudryavtseva ML, Kutsyk IM (2008) Atmospheric gamma-ray and neutron flashes. J Exp Theor Phys 106(1):65–76. doi:10.1007/s11447-008-1005-4

Bazelyan EM, Raizer YP (1998) Spark discharge. CRC Press, Boca Raton

Bell TF, Reising SC, Inan US (1998) Intense continuing currents following positive cloud-cloud lightning associated with red sprites. Geophys Res Lett 25:1285–1288

Berger K, Anderson RB, Kröninger H (1975) Parameters of lightning flashes. Electra 41:23–27

Bering EA III, Few AA, Benbrook JR (1998) The global electric circuit. Phys. Today 51:24–30

Bethe HA (1930) Zur Theorie des Durchgans schneller Korpuskularstrahlen durch Materie. Ann Phys Leipzig B5:325–400

Bhartendu H (1969) Thunder-a survey. Naturaliste Can 96(4):671–682

Bliokh PV, Nickolaenko AP, Filippov YF (1980). In: Llanwyn-Jones D (ed) Schumann resonances in the Earth-ionosphere cavity. Peter Peregrines, London

Boccippio DJ, Williams ER, Heckman SJ, Lyons WA, Baker IT, Boldi R (1995) Sprites, ELF transients, and positive ground strokes. Science 269(5227):1088–1091

Boeck WL, Vaughan OH Jr, Blakeslee R, Vonnegut B, Brook M (1992) Lightning induced brightening in the airglow layer. Geophys Res Lett 19:99–102

Boeck WL, Vaughan OH, Blakeslee RJ, Vonnegut B, Brook M, McKune J (1995) Observations of lightning in the stratosphere. J Geophys Res 100:1465–1475

Bondiou A, Gallimberti I (1994) Theoretical modeling of the development of the positive spark in long gaps. J Phys D Appl Phys 27:1252–1266

Borovsky JE (1998) Lightning energetics: estimates of energy dissipation in channels, channel radii, and channel-heating risetimes. J Geophys Res 103:11537–11553

Bragin YA, Tyutin AA, Kocheev AA, Tyutin AA (1974) Direct measurement of the atmospheric vertical electric field strength up to 80 km. Cosmic Invest [Kosmicheskie Issledovaniya, English Translation] 12:279–280

Briggs MS, Fishman GJ, Connaughton V, Bhat PN, Paciesas WS, Preece RD, Wilson-Hodge C, Chaplin VL, Kippen RM, von Kienlin A, Meegan CA, Bissaldi E, Dwyer JR, Smith DM, Holzworth RH, Grove JE, Chekhtman A (2010) First results on terrestrial gamma ray flashes from the fermi gamma-ray burst monitor. J Geophys Res 115:A00E49. doi:10.1029/2009JA014853

Carlson BE, Lehtinen NG, Inan US (2008) Runaway relativistic electron avalanche seeding in the Earth's atmosphere. J Geophys Res 113:A10307. doi:10.1029/2008JA013210

Carlson BE, Lehtinen NG, Inan US (2010) Terrestrial gamma ray flash production by active lightning leader channels. J Geophys Res 115:A10324. doi:10.1029/2010JA015647

Celestin S, Pasko VP (2010) Effects of spatial non-uniformity of streamer discharges on spectroscopic diagnostics of peak electric fields in transient luminous events. Geophys Res Lett 37:L07804. doi:10.1029/2010GL042675

Celestin S, Pasko VP (2011) Energy and fluxes of thermal runaway electrons produced by exponential growth of streamers during the stepping of lightning leaders and in transient luminous events. J Geophys Res 116:A03315. doi:10.1029/2010JA016260

Chalmers JA (1967) Atmospheric electricity, 2nd edn. Pergamon Press, New York

Chanrion O, Neubert T (2010) Production of runaway electrons by negative streamer discharges. J Geophys Res 115:A00E32

Chen AB, Kuo C-L, Lee Y-J, Su H-T, Hsu R-R, Chern J-L, Frey HU, Mende SB, Takahashi Y, Fukunishi H, Chang Y-S, Liu T-Y, Lee L-C (2008) Global distributions and occurrence rates of transient luminous events. J Geophys Res 113:A08306. doi:10.1029/2008JA013101

Cheng Z, Cummer SA, Su HT, Hsu RR (2007) Broadband very low frequency measurement of D region ionospheric perturbations caused by lightning electromagnetic pulses. J Geophys Res 112:A06318. doi:10.1029/2006JA011840

Chern JL, Hsu RR, Su HT, Mende SB, Fukunishi H, Takahashi Y, Lee LC (2003) Global survey of upper atmospheric transient luminous events on the ROCSAT-2 satellite. J Atmos Solar Terr Phys 65(5):647–659

Cho M, Rycroft MJ (1998) Computer simulation of the electric field structure and optical emission from cloud-top to the ionosphere. J Atmos Solar Terr Phys 60:871–888

Cho M, Rycroft MJ (2001) Non-uniform ionization of the upper atmosphere due to the electromagnetic pulse from a horizontal lightning discharge. J Atmos Solar Terr Phys 63:559–580

Christian HJ, Blakeslee RJ, Boccippio DJ, Boeck WL, Buechler DE, Driscoll KT, Goodman SJ, Hall JM, Koshak WJ, Mach DM, Stewart MF (2003) Global frequency and distribution of lightning as observed from space by the optical transient detector. J Geophys Res 108(D1):4005. doi:10.1029/2002JD002347

Colman JJ, Roussel-Dupré RA, Triplett L (2010). Temporally self-similar electron distribution functions in atmospheric breakdown: the thermal runaway regime. J Geophys Res 115:A00E16. doi:10.1029/2009JA014509

Coroniti SC (1965) Theories of generation in thunderstorm. Problems of atmospheric and space electricity. Elsevier Publishing Co., New York

Cummer SA (2003) Current moment in sprite-producing lightning. J Atmos Solar Terr Phys 65:499–508

Cummer SA, Füllekrug M (2001) Unusually intense CC in lightning produces delayed mesospheric breakdown. Geophys Res Lett 28:495–498

Cummer SA, Inan US, Bell TF, Barrington-Leigh CP (1998) ELF radiation produced by electrical currents in sprites. Geophys Res Lett 25:1281–1284

Cummer SA, Inan US (2000) Modelling ELF radio atmospheric propagation and extracting lightning currents from ELF observations. Radio Sci 35:385–394

Cummer SA, Lyons WA (2005) Implications of lightning charge moment changes for sprite initiation. J Geophys Res 110:A04304. doi: 10.1029/2004JA010812

Cummer SA, Frey HU, Mende SB, Hsu R-R, Su H-T, Chen AB, Fukunishi H, Takahashi Y (2006a) Simultaneous radio and satellite optical measurements of high-altitude sprite current and lightning continuing current. J Geophys Res 111:A10315. doi:10.1029/2006JA011809

Cummer SA, Jaugey N, Li J, Lyons WA, Nelsen TE, Gerken EA (2006b) Submillisecond imaging of sprite development and structure. Geophys Res Lett 33:L04104. doi:10.1029/2005GL024969

Cummer SA, Li J, Han F, Lu G, Jaugey N, Lyons WA, Nelson TE (2009) Quantification of the troposphere-to-ionosphere charge transfer in a gigantic jet. Nature Geosci 2:1–4. doi:10.1038/NGEO607

Derks G, Ebert U, Meulenbroek B (2008) Laplacian instability of planar streamer ionization fronts: an example of pulled front analysis. J Nonlinear Sci 18:551–590

Dowden RL, Brundell JB, Lyons WA, Nelson T (1996) Detection and location of red sprites by VLF scattering of subionospheric transmissions. Geophys Res Lett 23:1737–1740. doi:10.1029/96GL01697

Dwyer J, Smith D, Uman M, Saleh Z, Grefenstette B, Hazelton B, Rassoul H (2010) Estimation of the influence of high-energy electron bursts produced by thunderclouds and the resulting radiation doses received in aircraft. J Geophys Res 115:D09206. doi:10.1029/2009JD012039

Ebert U, Sentman D (2008) Editorial review: streamers, sprites, leaders, lightning: from micro- to macroscales. J Phys D Appl Phys 41:230301

Ebert U, Nijdam S, Li C, Luque A, Briels T, van Veldhuizen E (2010) Review of recent results on streamer discharges and discussion of their relevance for sprites and lightning. J Geophys Res 115:A00E43. doi:10.1029/2009JA014867

Fernsler R, Rowland H (1996) Models of lightning-produced sprites and elves. J Geophys Res 101:29653–29662

Feynman RP, Leighton RB, Sands M (1964) The Feynman lectures on physics, vol 2. Addison-Wesley/Readings: Palo Alto/London

Fishman GJ, Bhat PN, Mallozzi R, Horack JM, Koshut T, Kouveliotou C, Pendleton GN, Meegan CA, Wilson RB, Paciesas WS, Goodman SJ, Christian HJ (1994) Discovery of intense gamma-ray flashes of atmospheric origin. Science 264:1313–1316

Franz RC, Nemzak RJ, Winkler JR (1990) Television image of a large upward electrical discharge above a thunderstorm system. Science 249:48–51

Fukunishi H, Takahashi Y, Fujito M, Wanatabe Y, Sakanoi S (1996) Fast imagining of elves and sprites using a framing/streak camera and a multi-anode array photometer. EOS Trans AGU 77(46, Fall Meeting Suppl.):F60

Füllekrug M, Constable S (2000) Global triangulation of intense lightning discharges. Geophys Res Lett 27(3):333–336

Füllekrug M, Fraser-Smith AC, Bering EA, Few AA (1999) On the hourly contribution of global cloud-to-ground lightning activity to the atmospheric electric field in the Antarctic during December 1992. J Atmos Solar Terr Phys 61:745–750

Füllekrug M, Roussel-Dupré RA, Symbalisty EMD, Chanrion O, Odzimek A, van der Velde O, Neubert T (2010) Relativistic runaway breakdown in low-frequency radio. J Geophys Res 115:A00E09. doi:10.1029/2009JA014468

Füllekrug M, Roussel-Dupré R, Symbalisty EMD, Colman JJ, Chanrion O, Soula S, van der Velde O, Odzimek A, Bennett AJ, Pasko VP, Neubert T (2011) Relativistic electron beams above thunderclouds. Atmos Chem Phys 11:7747–7754

Gerken EA, Inan US, Barrington-Leigh CP (2000) Telescopic imaging of sprites. Geophys Res Lett 27(17):2637–2640

Glukhov V, Pasko V, Inan U (1992) Relaxation of transient lower ionospheric disturbances caused by lightning-whistler-induced electron precipitation bursts. J Geophys Res 97:16971–16979

Gurevich AV, Milikh GM, Roussel-Dupré R (1992) Runaway electron mechanism of air breakdown and preconditioning during a thunderstorm. Phys Lett A 165:463–468

Gurevich AV, Milikh GM, Roussel-Dupré R (1994) Nonuniform runaway air-breakdown. Phys Lett A 187:197–203

Gurevich AV, Zybin KP (2001) Runaway breakdown and electric discharges in thunderstorm. Phys. Uspekhi 44:1119–1140

Gurevich AV, Zybin KP (2004) High energy cosmic ray particles and the most powerful discharges in thunderstorm atmosphere. Phys Lett A 329:341–347. doi:10.1016/j.physleta.2004.06.094

Gurevich AV, Zybin KP, Roussel-Dupré RA (1999) Lightning initiation by simultaneous effect of runway breakdown and cosmic ray shower. Phys Lett A 254:79–97

Gurevich A, Zybin K, Medvedev Y (2007) Runaway breakdown in strong electric field as a source of terrestrial gamma flashes and gamma bursts in lightning leader steps. Phys Lett A361:119–125

Haldoupis C, Steiner RJ, Mika Á, Shalimov S, Marshall RA, Inan US, Bösinger T, Neubert T (2006) "Early/slow" events: a new category of VLF perturbations observed in relation with sprites. J Geophys Res 111:A11321. doi:10.1029/2006JA011960

Haldoupis C, Mika Á, Shalimov S (2009) Modeling the relaxation of early VLF perturbations associated with transient luminous events. J Geophys Res 114:A00E04. doi:10.1029/2009JA014313

Hale LC, Croskey CL, Mitchell JD (1981) Measurements of middle-atmosphere electric fields and associated electrical conductivities. Geophys Res Lett 8:927–930

Hayakawa M, Nakamura T, Hobara Y, Williams E (2004) Observation of sprites over the sea of Japan and conditions for lightning-induced sprites in winter. J Geophys Res 105:4689–4697

Hayakawa M, Sekiguchi M, Nickolaenko AP (2005) Diurnal variations of electric activity of global thunderstorms deduced from OTD data. J Atmos Electr 25(2):55–68

Hayakawa M, Hobara Y, Suzuki T (2012) Lightning effects in the mesosphere and ionosphere. In: Cooray V (ed) Lightning electromagnetics. Institution of Engineering and Technology, London, pp 611–646

Heavner MJ, Sentman DD, Moudry DR, Wescott EM (2000) Sprites, blue jets, and elves: optical evidence of energy transport across the stratopause. In: Siskind DE, Eckermann SD, Summers ME (eds) Atmospheric science across the stratosphere. Geophysical monograph series, vol 123. AGU, Washington, DC, pp 69–82

Heavner MJ, Suszcynsky DM, Smith DA (2003) LF/VLF intracloud waveform classification. Paper presented at International Conference on Atmospheric Electricity, ICAE, Versailes, France

Helliwell R, Katsufrakis JP, Trimpi M (1973) Whistler-induced amplitude perturbation in VLF propagation. J Geophys Res 78:4679–4688

Hiraki Y, Fukunishi H (2007) Theoretical criterion of charge moment change by lightning for initiation of sprites. J Geophys Res 111:A11305

Hobara Y, Iwasaki N, Hayashida T, Hayakawa M, Ohta K, Fukunishi H (2001) Interrelation between ELF transients and ionospheric disturbances in association with sprites and elves. Geophys Res Lett 28:935–938

Hobara Y, Hayakawa M, Williams E, Boldi R, Downes E (2006) Location and electrical properties of sprite-producing lightning from a single ELF site. In: Füllekrug M, Mareev EA, Rycroft MJ (eds) Sprites, elves and intense lightning discharges. NATO science series II, vol 225. Springer, Dordrecht, pp 211–235

Hu W, Cummer SA, Lyons WA (2007) Testing sprite initiation theory using lightning measurements and modeled EM fields. J Geophys Res 112:D13115

Imyanitov IM, Chubarina EV, Schwartz YM (1971) Electricity of clouds. Gydrometeoizdat, Leningrad, p 239 (in Russian)

Inan US, Lehtinen NG (2005) Production of terrestrial gammaray flashes by an electromagnetic pulse from a lightning return stroke. Geophys Res Lett 32:L19818. doi:10.1029/2005GL023702

Inan US, Bell TF, Pasko VP, Sentman DD, Wescott EM, Lyons WA (1995) VLF signatures of ionospheric disturbances associated with sprites. Geophys Res Lett 22:3461–3464. doi:10.1029/95GL03507

Inan US, Sampson WA, Taranenko YN (1996a) Space-time structure of optical flashes and ionization changes produced by lightning-EMP. Geophys Res Lett 23:133–136. doi:10.1029/95GL03816

Inan US, Reising SC, Fishman GJ, Horack JM (1996b) On the association of terrestrial gamma-ray bursts with lightning and implications for sprites. Geophys Res Lett 23:1017–1020

Inan US, Barrington-Leigh C, Hansen S, Glukhov VS, Bell TF, Rairden R (1997) Rapid lateral expansion of optical luminosity in lightning-induced ionospheric flashes referred to as "elves". Geophys Res Lett 24(5):583–586

Inan US, Cummer SA, Marshall RA (2010) A survey of ELF and VLF research on lightning-ionosphere interactions and causative discharges, J Geophys Res 115, A00E36. doi:10.1029/2009JA014775

Ishikawa H (1961) Nature of lightning discharge as origin of atmospherics. Proc. Res. Inst. Atmos Nagoya Univ 8A:1–274

Israël H (1970) Atmospheric electricity, vol 1, 2nd edn. Israel Program for Scientific Translations, Jerusalem

Israël H (1973) Atmospheric electricity, vol 2, 2nd edn. Israel Program for Scientific Translations, Jerusalem

Jacobson AR (2003) Relationship of intracloud lightning radiofrequency power to lightning storm height, as observed by the FORTE satellite. J Geophys Res 108(D7):4204. doi:10.1029/2002JD002956

Jones DLl (1970) Electromagnetic radiation from multiple return stroke of lightning. J Atmos Terr Phys 32:1077–1093

Krehbiel PR, Riousset JA, Pasko VP, Thomas RJ, Rison W, Stanley MA, Edens HE (2008) Upward electrical discharges from thunderstorms. Nature Geosci 1(4):233–237. doi:10.1038/ngeo162

Krider EP (1986) Physics of lightning. In: Earth's electrical environment. National Academy Press, Washington, DC, pp 30–40

Kudintseva IG, Nickolaenko AP, Hayakawa M (2009) Spatial fine structure of model electric pulses in mesosphere above Γ -shaped stroke of lightning. Surv Geophys 71:603–608

Kudintseva IG, Nickolaenko AP, Hayakawa M (2010) Transient electric field in the mesosphere above a Γ-shape lightning stroke. Surv Geophys 31:427–448. doi:10.1007/s10712-0010-9095-x

Kuo C, Chou JK, Tsai LY, Chen AB, Su HT, Hsu RR, Cummer SA, Frey HU, Mende SB, Takahashi Y, Lee LC (2009) Discharge process, electric field, and electron energy in ISUAL-recorded gigantic jets. J Geophys Res 114:A04314. doi:10.1029/2008JA013791

Lehtinen N (2000) Relativistic runaway electrons above thunderstorms. Ph.D. dissertation, Stanford University

Lehtinen NG, Inan US (2007) Possible persistent ionization caused by giant blue jets. Geophys Res Lett 34:L08804. doi:10.1029/2006GL029051

Lehtinen N, Bell T, Pasko V, Inan U (1997) A two-dimensional model of runaway electron beams driven by quasi-electrostatic thundercloud fields. Geophys Res Lett 24:2639–2642

Lehtinen N, Bell T, Inan U (1999) Monte Carlo simulation of runaway MeV electron breakdown with application to red sprites and terrestrial gamma ray flashes. J Geophys Res 104:24699–24712

Li C, Ebert U, Hundsdorfer W (2009) 3D hybrid computations for streamer discharges and production of run-away electrons. J Phys D Appl Phys 42:202003

Li C, Ebert U, Hundsdorfer W (2010) Spatially hybrid computations for streamer discharges with generic features of pulled fronts: I. Planar fronts. J Comput Phys 229:200

Lieberman MA, Lichtenberg AJ (1994) Principles of plasma discharges and material processing. Wiley, New York

Liu NY, Pasko VP (2006) Effects of photoionization on similarity properties of streamers at various pressures in air. J Phys D Appl Phys 39:327–334. doi:10.1088/0022-3727/39/2/013

Luque A, Ebert U (2009) Emergence of sprite streamers from screening-ionization waves in the lower ionosphere. Nature Geosci 2:757–760. doi:10.1038/ngeo662

Luque A, Ebert U (2010) Sprites in varying air density: charge conservation, glowing negative trails and changing velocity. Geophys Res Lett 37:L06806. doi:10.1029/2009GL041982

Luque A, Ebert U (2012) Density models for streamer discharges: beyond cylindrical symmetry and homogeneous media. J Comput Phys 231:904–918. doi:10.1016/j.jcp.2011.04.019

Lyons WA (1997) The handy weather answer book. Accord Publishing, Detroit

Lyons WA (2006) The meteorology of transient luminous events. An introduction and overview. In: Füllekrug M, Mareev EA, Rycroft MJ (eds) Sprites, elves and intense lightning discharges. NATO Science Series II, vol 225. Springer, Dordrecht, pp 19–44

Lyons WA, Nelson T, Williams ER, Cummer SA, Stanley MA (2003) Characteristics of sprite-producing positive cloud-to-ground lightning during the 19 July STEPS mesoscale convective systems. Mon Weather Rev 131:2417–2427

Mareev EA, Trakhtengerts VY (2007) Puzzles of atmospheric electricity. Priroda 3:24–33 (in Russian)

Marshall T, McCarthy MP, Rust WD (1995) Electric field magnitudes and lightning initiation in thunderstorms. J Geophys Res 100(D4):7097–7103. doi:10.1029/95JD00020

Matsudo Y, Suzuki T, Hayakawa M, Yamashita K, Ando Y, Michimoto K, Korepanov V (2007) Characteristics of Japanese winter sprites and their parent lightning as estimated by VHF lightning and ELF transients. J Atmos Solar Terr Phys 69:1431–1446

Matsudo Y, Suzuki T, Michimoto K, Myokei K, Hayakawa M (2009) Comparison of time delays of sprites induced by winter lightning flashes in the Japan Sea with those in the Pacific Ocean. J Atmos Solar Terr Phys 71:101–111

Maynard NC, Croskey CL, Mitchell JD, Hale LC (1981) Measurement of volt/meter vertical electric fields in the middle atmosphere. Geophys Res Lett 8:923–926

McGorman DR, Rust WD (1998) The electrical nature of storms. Oxford University Press, Oxford, p 422

McHarg MG, Stenbaek-Nielsen HC, Kanmae T (2007) Observation of streamer formation in sprites. Geophys Res Lett 34:L06804. doi:10.1029/2006GL027854

Mende, SB, Frey HU, Hsu RR, Su HT, Chen AB, Lee LC, Sentman DD, Takahashi Y, Fukunishi H (2005) D region ionization by lightning-induced electromagnetic pulses. J Geophys Res 110:11312. doi:10.1029/2005JA011064

Milikh G, Roussel-Dupré R (2010) Runaway breakdown and electrical discharges in thunderstorms. J Geophys Res 115:A00E60. doi:10.1029/2009JA014818

Mishin EV, Milikh GM (2008) Blue jets: upward lightning. Space Sci Rev 137(1–4):473–488

Montanyà J, van der Velde O, Romero D, March V, Solá G, Pineda N, Arrayas M, Trueba JL, Reglero V, Soula S (2010) High-speed intensified video recordings of sprites and elves over the western Mediterranean Sea during winter thunderstorms. J Geophys Res 115:A00E18. doi:10.1029/2009JA014508

Muchnik VM, Fishman BE (1982) Electrization of coarse-dispersion aerosols in the atmosphere. Gidrometeoizdat, Leningrad (in Russian)

Myokei K, Matsudo Y, Asano T, Suzuki T, Hobara Y, Michimoto K, Hayakawa M (2009) A study of the morphology of winter sprites in the Hokuriku area of Japan in relation to cloud height. J Atmos Solar Terr Phys 71:579–602

Nelson PH (1967) Ionospheric perturbations and Schumann resonances data. Ph.D. thesis, MIT, Cambridge

Neubert T, Allin TH, Blanc E, Farges T, Haldoupis C, Mika A, Soula S, Knutsson L, van der Velde O, Marshall RA, Inan U, Sátori G, Bór J, Hughes A, Collier A, Laursen S, Rasmussen IL (2005) Co-ordinated observations of transient luminous events during the Eurosprite2003 campaign. J Atmos Solar Terr Phys 67:807–820. doi:10.1016/j.jastp.2005.02.004

Neubert T, Rycroft M, Farges T, Blanc E, Chanrion O, Arnone E, Odzimek A, Arnold N, Enell C-F, Turunen E, Bösinger T, Mika Á, Haldoupis C, Steiner RJ, van der Velde O, Soula S, Berg P, Boberg F, Thejll T, Cristiansen B, Ignaccolo M, Füllekrug M, Verronen PT, Montanyá J,

Crosby N (2008) Recent results from studies of electric discharges in the mesosphere. Surv Geophys 29:71–137. doi:10.1007/s10712-008-9043-1

Neubert T, Chanrion O, Arnone E, Zanotti F, Cummer S, Li J, Füllekrug M, Soula S, van der Velde O (2011) The properties of a gigantic jet reflected in a simultaneous sprite: observations interpreted by a model. J Geophys Res 116:A12329. doi:10.1029/2011JA016928

Nickolaenko AP, Hayakawa M (1995) Heating of the lower ionosphere electrons by electromagnetic radiation of lightning discharges. Geophys Res Lett 22:3015–3018

Nickolaenko AP, Hayakawa M (1998) Electric fields produced by lightning discharges. J Geophys Res 103:17175–17189

Nickolaenko AP, Hayakawa M (1999) A model for causative discharge of ELF-transients. J Atmos Electr 19:11–24

Nickolaenko AP, Hayakawa M (2002) Resonances in the earth-ionosphere cavity. Kluwer Academic Publishers, Dordrecht, p 380

Nickolaenko AP, Hayakawa M (2014) Schumann resonances for tyros. Springer, Tokyo, p 348

Ogawa T (1995) Lightning currents. In: Volland H (ed) Handbook of atmospheric electrodynamics, 1. CRC Press, Florida, pp 93–136

Ohkubo A, Fukunishi H, Takahashi Y, Adachi T (2005) VLF/ELF sferic evidence for in-cloud discharge activity producing sprites. Geophys Res Lett 32:L04812. doi:10.1029/2004GL021943

Otsuyama T, Manaba J, Hayakawa M, Nishimura M (2004) Characteristics of subionospheric VLF perturbation associated with winter lightning around Japan. Geophys Res Lett 31:L04117. doi:10.1029/2003GL019064

Paiva GS, Pavão AC, Bastos CC (2009) "Seed" electrons from muon decay for runaway mechanism in the terrestrial gamma ray flash production. J Geophys Res 114:D03205. doi:10.1029/2008JD010468

Pancheshnyi S (2005) Role of electronegative gas admixtures in streamer start, propagation and branching phenomena. Plasma Sources Sci Technol 14:645–653

Pasko VP (2006) Theoretical modeling of sprites and jets. In: Füllekrug M, Mareev EA, Rycroft MJ (eds) Sprites, elves and intense lightning discharges. NATO Science Series II, vol 225. Springer, Dordrecht, pp 255–311

Pasko VP (2010) Recent advances in theory of transient luminous events. J Geophys Res 115:A00E35. doi:10.1029/2009JA014860

Pasko VP, Inan US, Bell TF (2000) Fractal structure of sprites. Geophys Res Lett 27:497–500

Pasko VP, Inan US, Bell TF (2001) Mesosphere-ionosphere coupling due to sprites. Geophys Res Lett 28:3821–3824

Pasko VP, Stanley MA, Matheus JD, Inan US, Wood TG (2002) Electrical discharge from a thunderstorm top to the lower ionosphere. Nature 416:152–154

Pasko VP, Qin J, Celestin S (2013) Toward better understanding of sprite streamers: initiation, morphology, and polarity asymmetry. Surv Geophys 34(6):797–830. doi:10.1007/s10712-013-9246-y

Petrov NI, Petrova GN (1999) Physical mechanism for the development of lightning discharges between a thundercloud and the ionosphere. Tech Phys 44:472–475

Raizer YP (1991) Gas discharge physics. Springer, Berlin

Raizer YP, Milikh GM, Shneider MN, Novakovski SV (1998) Long streamer in the upper atmosphere above thundercloud. J Phys D Appl Phys 31:3255–3264

Raizer YP, Milikh GM, Shneider MN (2006) On the mechanism of blue jet formation and propagation. Geophys Res Lett 33(23):L23801. doi:10.1029/2006GL027697

Raizer YP, Milikh GM, Shneider MN (2007) Leader-streamers nature of blue jets. J Atmos Solar Terr Phys 69(8):925–938. doi:10.1016/j.jastp.2007.02.007

Raizer YP, Milikh GM, Shneider MN (2010) Streamer-and leader-like processes in the upper atmosphere: models of red sprites and blue jets. J Geophys Res 115:A00E42. doi:10.1029/2009JA014645

Rakov VA (2000) Positive and bipolar lightning discharges: a review. In: Proceedings of the 25th International Conference on Lightning Protection, Rhodos, pp 103–108

Rakov VA (2003) A review of positive and bipolar lightning discharges. Bull Am Meteorol Soc
    84:767–775
Rakov VA, Uman MA (1998) Review and evaluation of lightning return stroke models including
    some aspect of their application. IEEE Trans Electromagn Compatibility 40:403–426
Rakov VA, Uman MA (2003) Lightning: physics and effects. Cambridge University Press,
    Cambridge, p 687
Reising SC, Inan US, Bell TF, Lyons WA (1996) Evidence for CCs in sprite-producing cloud-to-
    ground lightning. Geophys Res Lett 23:3639–3642
Riousset JA, Pasko WA, Krehbiel PR, Rison W, Stanley MA (2010a) Modeling of thundercloud
    screening charges: implications for blue and gigantic jets. J Geophys Res 115:A00E10.
    doi:10.1029/2009JA014286
Riousset JA, Pasko VP, Bourdon A (2010b) Air-density-dependent model for analysis of air heating
    associated with streamers, leaders, and transient luminous events. J Geophys Res 115:A12321.
    10.1029/2010JA015918
Roussel-Dupré R, Gurevich A (1996) On runaway breakdown and upward propagating discharges.
    J Geophys Res 101:2297–2311
Roussel-Dupré R, Colman JJ, Symbalisty E, Sentman D, Pasko VP (2008) Physical processes
    related to discharges in planetary atmospheres. Space Sci Rev 137:51–82. doi:10.1007/s11214-
    008-9385-5
Rowland HL (1998) Theories and simulations of elves, sprites and blue jets. J Atmos Solar Terr
    Phys 60:831–844
Rycroft MJ (2006) Introduction to the physics of sprites, elves and intense lightning discharges.
    In: Füllekrug M, Mareev EA, Rycroft MJ (eds) Sprites, elves and intense lightning discharges.
    NATO Science Series II, vol 225. Springer, Dordrecht, pp 1–18
Rycroft MJ, Israelsson S, Price C (2000) The global atmospheric electric circuit, solar activity and
    climate change. J Atmos Solar Terr Phys 62:1563–1576
Sato M, Fukunishi H (2003) Global sprite occurrence locations and rates derived from
    triangulation of transient Schumann resonances events. Geophys Res Lett 30(16):1859.
    doi:10.1029/2003GL017291
Sato M, Takahashi Y, Yoshida A, Adachi T (2008) Global distribution of intense lightning
    discharges and their seasonal variations. J Phys D Appl Phys 41:234011. doi:10.1088/0022-
    3727/41/23/234011
Sátori G, Williams E, Lemperger I (2009) Variability of global lightning activity on the ENSO time
    scale. Atmos Res 91:500–507
Schonland BEJ (1956) The lightning discharge. In: Handbuch der physik, Bd22. Springer, Berlin
Sentman DD, Wescott EM (1993) Observations of upper atmosphere optical flashes recorded from
    an aircraft. Geophys Res Lett 20:2857–2860
Sentman DD, Wescott EM, Osborne DL, Hampton DL, Heavner MJ (1995) Preliminary results
    from the Sprites94 aircraft campaign: 1. Red sprites. Geophys Res Lett 22:1205–1208
Smith DA, Shao XM, Holden DN, Rhodes CT, Brook M, Krehbiel PR, Stanley M, Rison
    W, Thomas RJ (1999) A distinct class of isolated intracloud lightning discharges and their
    associated radio emissions. J Geophys Res 104:4189–4212
Smith DA, Heavner MJ, Jacobson AR, Shao XM, Massey RS, Sheldon RJ, Wines KC (2004) A
    method for determining intracloud lightning and ionospheric heights from VLF/LF electric field
    records. Radio Sci. 39:RS1010. doi:10.1029/2002RS002790
Smith DM, Lopez LI, Lin LI, Barrington-Leigh LI (2005) Terrestrial gamma-ray flashes observed
    up to 20 MeV. Science 307:1085–1088
Smith DM, Hazelton BJ, Grefenstette BW, Dwyer BW, Holzworth RH, Lay EH (2010)
    Terrestrial gamma ray flashes correlated to storm phase and tropopause height. J Geophys Res
    115:A00E49. doi:10.1029/2009JA014853
Soloviev SP, Surkov VV (2000) Electrostatic field and lightning generated in the gaseous dust
    cloud of explosive products. Geomagn Aeron 40:61–69
Stanley M, Krehbiel P, Brook M, Moore C, Rison W, Abrahams B (1999) High speed video of
    initial sprite development. Geophys Res Lett 26:3201–3204

Stanley M, Brook M, Krehbiel P, Cummer S (2000) Detection of daytime sprites via a unique sprite ELF signature. Geophys Res Lett 27:871–874

Stekolnikov IS (1940) Lightning. USSR Academy of Science, Leningrad (in Russian)

Stenbaek-Nielsen HC, Moudry DR, Wescott EM, Sentman DD, Sao Sabbas FT (2000) Sprites and possible mesospheric effects. Geophys Res Lett 27:3829–3832

Stenbaek-Nielsen HC, McHarg MG, Kanmae T, Sentman DD (2007) Observed emission rates in sprite streamer heads. Geophys Res Lett 34:L11105. doi:10.1029/2007GL029881

Stenbaek-Nielsen HC, McHarg MG (2008) High time-resolution sprite imagining: observation and implications. J Phys D Appl Phys 41:234009

Su HT, Hsu RR, Chen RR, Wang YC, Hsiao WS, Lai WC, Lee LC, Sato M, Fukunishi H (2003) Gigantic jets between a thundercloud and the ionosphere. Nature 423:974–976

Surkov VV, Hayakawa M (2012) Underlying mechanisms of transient luminous events: a review. Ann Geophys 30:1185–2012. doi:10.5194/angeo-30-1185-2012

Surkov VV, Matsudo Y, Hayakawa M, Goncharov SV (2010) Estimation of lightning and sprite parameters based on observation of sprite-producing lightning power spectra. J Atmos Solar Terr Phys 72:448–456

Suzuki T, Matsudo Y, Asano T, Hayakawa M, Michimoto K (2011) Meteorological and electrical aspects of several winter thunderstorms with sprites in the Hokuriku area of Japan. J Geophys Res 116:D06205. doi:10.1029/2009JD013358

Taylor WL (1972) Atmospherics and severe storms. In: Deer VE (ed) Remote sensing of the troposphere. Chapter 17, US Government Printing Office, Washington, DC

Trakhtengerts VY, Rycroft MJ (2008) Whistler and Alfvén Mode Cyclotron Masers in Plasma. Cambridge University Press, Cambridge, p 354

Trakhtengerts VY, Iudin DI, Kulchitsky AV, Hayakawa M (2002) Kinetics of runaway electrons in a stochastic electric field. Phys Plasmas 9:2762–2766

Trakhtengerts VY, Iudin DI, Kulchitsky AV, Hayakawa M (2003) Electron acceleration by a stochastic electric field on the atmospheric layer. Phys Plasmas 10:3290–3296

Vallance-Jones A (1974) Aurora. D. Reidel, Dordrecht

van der Velde OA, Mika A, Soula S, Haldoupis C, Neubert T, Inan US (2006) Observations of the relationship between sprite morphology and in-cloud lightning processes. J Geophys Res 111:D15203. doi:10.1029/2005JD006879

van der Velde OA, Lyons WA, Nelson TE, Cummer SA, Li J, Bunnell J (2007) Analysis of the first gigantic jet recorded over continental North America. J Geophys Res 112:D20104. doi:10.1029/2007JD008575

van der Velde OA, Bór J, Li J, Cummer SA, Arnone E, Zanotti F, Füllekrug M, Haldoupis C, NaitAmor S, Farges T (2010) Multi-instrumental observations of a positive gigantic jet produced by a winter thunderstorm in Europe. J Geophys Res 115:D24301. doi:10.1029/2010JD014442

Visacro S, Soares A Jr, Schroeder MAO, Cherchiglia LCL, de Sousa VJ (2004) Statistical analysis of lightning current parameters: measurements at Morro do Cachimbo Station. J Geophys Res 109:D01105. doi: 10.1029/2003JD003662

Uman MA (1987) The lightning discharge. Academic Press, New York

Uman MA, Krider EP (1982) A review of natural lightnings: experimental data and modeling. IEEE Trans Electromagn Compatibility 24:79

Wahlin L (1973) A possible origin of atmospheric electricity. Found Phys 3:450–472

Watt AD (1967) VLF radio engineering. Pergamon Press, New York

Wescott EM, Sentman D, Osborne D, Hampton D, Heavner M (1995) Preliminary results from the Sprites94 aircraft campaign: 2. Blue jets. Geophys Res Lett 22:209–1212

Wescott EM, Sentman DD, Heavner MJ, Hampton DL, Osborne DL, Vaugham OH Jr (1996) Blue starters: brief upward discharges from an intense Arkansas thunderstorm. Geophys Res Lett 23:2153–2156

Wescott EM, Sentman DD, Heavner MJ, Hampton DL, Lyons WA, Nelson T (1998) Observations of columniform sprites. J Atmos Solar Terr Phys 60:733–740

Wescott EM, Sentman DD, Stenbaek-Nielsen HC, Huet P, Heavner MJ, Moudry DR (2001) New evidence for the brightness and ionization of blue starters and blue jets. J Geophys Res 106:21549–21554

Williams E, Downes E, Boldi R, Lyons W, Heckman S (2007) Polarity asymmetry of sprite-producing lightning: a paradox? Radio Sci 42:RS2S17. doi:10.1029/2006RS003488

Wilson CTR (1925) The electric field of a thundercloud and some of its effects. Proc Phys Soc London 37:32D–37D

Winn WP, Schwede GW, Moore CB (1974) Measurements of electrical fields in thunderclouds. J Geophys Res 79(12):1761–1767

Yair Y, Israelevich P, Devir AD, Moalem M, Price C, Joseph JH, Levin Z, Ziv B, Sternlieb A, Teller A (2004) New observations of sprites from space shuttle. J Geophys Res 109:D15201. doi:10.1029/2003JD004497

Yashunin SA, Mareev EA, Rakov VA (2007) Are lightning M components capable of initiating sprites and sprite halo? J Geophys Res 112:D10109. doi:10.1029/2006JD007631

Yukhimuk V, Roussel-Dupré R, Symbalisty EMD (1999) On the temporal evolution of red sprites: runaway theory versus data. Geophys Res Lett 26:679–682

Part II
Global Electromagnetic Resonances and ULF Noises

Chapter 4
Earth-Ionosphere Cavity Resonator

Abstract This chapter covers both the global ULF/ELF electromagnetic resonances and background noises. We choose first to study Schumann resonances excited in the Earth-Ionosphere cavity resonator by the global lightning activity treated as stochastic process. In the analysis that follows, we first examine models of the Earth-Ionosphere cavity and then derive eigenfrequencies of the Schumann resonances. In this chapter we estimate the contribution of positive and negative cloud-to-ground (CG) lightning discharges to the low-frequency portion of power spectra and to the Schumann resonances.

Keywords Earth-Ionosphere cavity resonator • Negative cloud-to-ground (CG) lightning • Positive CG lightning • Quality/energy-factor • Random lightning process • Schumann resonances

4.1 Structure and Models of the Earth-Ionosphere Cavity Resonator

4.1.1 Model of the Earth-Ionosphere Cavity and Basic Equations

A large fraction of the ground surface is covered by sea water with conductivity $\sigma_g \approx 5\,\text{S/m}$, while the continental conductivity of the Earth crust varies in the range $\sigma_g = 10^{-4}\text{–}10^{-2}\,\text{S/m}$. Both of these values are much greater than the atmospheric conductivity at the sea level. The conductivity of the air increases exponentially with altitude according to Eq. (3.1). In the ELF range of 3–30 Hz the displacement current in the atmosphere is much greater than the conduction one up to the altitudes 45–50 km. This implies that in the ELF range the lower atmosphere can be considered as a perfect insulator sandwiched between two conducting layers.

V. Surkov and M. Hayakawa, *Ultra and Extremely Low Frequency Electromagnetic Fields*, Springer Geophysics, DOI 10.1007/978-4-431-54367-1_4, © Springer Japan 2014

The atmospheric layer between the highly conducting terrestrial surface boundary and the conducting but dissipative ionosphere forms a dissipative spherically concentric cavity, the Earth-Ionosphere cavity, that can serve as a resonator for electromagnetic waves with wavelengths comparable to the Earth radius. As shown below, the resonant spectra of these waves fall in the ELF range.

The main mechanism for excitation of the resonance spectrum is the electromagnetic energy stemming from the global thunderstorm activity (e.g., see Sentman 1995; Nickolaenko and Hayakawa 2002; Hayakawa et al. 2011). The lightning discharges radiate electromagnetic impulses with broadband spectrum. The lowest portion of such a spectrum is responsible for excitation of the quasi-steady electromagnetic waves in the Earth-Ionosphere resonance cavity. The total resonant spectrum is an incoherent superposition of contributions from the individual lightning discharges occurring over the globe. This phenomenon, called Schumann resonances, constitutes a prevailing part of the natural background electromagnetic spectrum over the frequency range 5–50 Hz.

It was Schumann who first studied and predicted the resonance properties of the Earth-Ionosphere cavity. The early study of the Schumann resonances was stimulated in part by the U.S. Navy's interest in investigating the ELF band for possible use in submarine communications (Wait 1974; Sentman 1995). A large amount of experimental study on this problem was performed between 1965 and 1982 when Schumann resonances research was at its peak (e.g., see review by Bliokh et al. (1980) and references therein). The overview of these studies and recent results, including the theory of relevance to Schumann resonances, have been summarized in the book by Nickolaenko and Hayakawa (2002). The reader is referred to that book and to the texts by Budden (1962), Galejs (1972), Wait (1972), and Bliokh et al. (1980) for details about more complete theory of the Schumann resonances.

A number of different theoretical models have been used to describe the resonance spectrum and sources of the cavity excitation. Following Schumann (1952a,b, 1957) we consider a simplified model of the resonance cavity. In this model the atmosphere is considered as an insulator while the Earth and the ionosphere are supposed to be perfect conductors, so that the Earth-Ionosphere forms a spherical capacitor, which contains the resonance cavity bounded from below and from above by the perfectly conductive walls.

CG lightning discharges are believed to be the principal source for excitation of the Schumann resonances (Sentman 1995). Since the vertical CG return stroke is the most effective emitter of the electromagnetic waves, consider first a single vertical current impulse as a source for the cavity excitation. All the values are supposed to vary as $\exp(-i\omega t)$, where ω is the frequency. The Maxwell equations (1.1) and (1.2) for the atmosphere can thus be rewritten as

$$\nabla \times \mathbf{B} = \mu_0 \mathbf{j}_s - \frac{i\omega}{c^2}\mathbf{E}, \qquad (4.1)$$

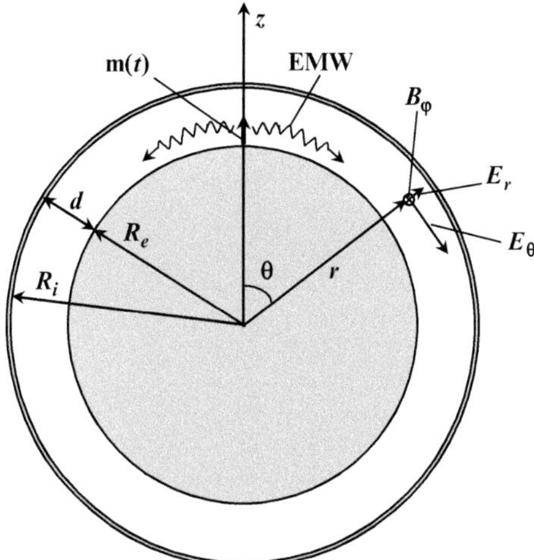

Fig. 4.1 A simple model of the Earth-Ionosphere resonance cavity. Here R_e and R_i are the radii of the Earth and the ionosphere, d is the height of the non-conductive atmosphere, EMW denotes electromagnetic wave radiated by the vertical current moment $\mathbf{m}(t)$, and B_ϕ, E_r, and E_θ are the components of the TM mode

and

$$\nabla \times \mathbf{E} = i\omega\mathbf{B}. \tag{4.2}$$

The current density $\mathbf{j}_s$ due to the vertical lightning discharge can serve as a source function for the electromagnetic waves inside the resonance cavity.

The conduction current in the atmosphere is much smaller than the displacement one under the requirement of $\sigma_a E \ll \varepsilon_0 \omega E$. In what follows we show that the Schumann resonances lie in the frequency range $f > 7.5\,\text{Hz}$, that is $\omega = 2\pi f > 47\,\text{Hz}$. Whence it follows that the above requirement reduces to the following $\sigma_a < \varepsilon_0 \omega = 4 \times 10^{-10}\,\text{S/m}$. Taking into account that the atmospheric conductivity σ_a increases exponentially with altitude, one can find that this requirement is valid in the lower atmosphere up to the altitude 40–50 km at the daytime and up to 60–70 km at night. These altitudes may serve as an estimate of the upper boundary of the "spherical capacitor." In order to make our consideration as transparent as possible, we assume a constant distance between the resonator walls thereby disregarding the difference between daytime and nighttime conditions.

We shall use spherical variables r, θ, and φ and a coordinate system in which the vertical current of the return stroke is in the direction of the polar axis z. By symmetry of the problem the vertical current produces the so-called transverse magnetic mode (TM mode), which contains only three components B_φ, E_r, and E_θ, as shown in Fig. 4.1. The transverse electric mode (TE mode), E_φ, B_r, and B_θ, can be excited by virtue of the mode coupling at the boundary between the atmosphere and the ionosphere. The mode coupling to the shear Alfvén and FMS wave mode in the bottom E region ionosphere is due to the Hall conductivity of the ionospheric plasma. In our model the ionospheric conductivity is assumed to be a scalar, so the TE mode cannot be excited since we have ignored the gyrotropic properties of the ionosphere. On account of symmetry of the problem, all the quantities are independent of the coordinate φ. In such a case Eqs. (4.1) and (4.2) for TM mode reduce to the following form

$$\frac{\partial_\theta \left(B_\varphi \sin \theta\right)}{r \sin \theta} = -\frac{i\omega}{c^2} E_r + \mu_0 j_s, \tag{4.3}$$

$$\frac{1}{r}\partial_r \left(rB_\varphi\right) = \frac{i\omega}{c^2} E_\theta, \tag{4.4}$$

$$\partial_r \left(rE_\theta\right) - \partial_\theta E_r = i\omega r B_\varphi. \tag{4.5}$$

where the symbols ∂_r and ∂_θ stand for partial derivatives with respect to r and θ. Following Wait (1962) we now introduce the scalar potential, U, which corresponds to the radial component of the Hertz vector. The field components can be expressed via the potential U as follows:

$$B_\varphi = \frac{i\omega}{c^2}\partial_\theta U, \tag{4.6}$$

$$E_r = \left(\partial_r^2 + k^2\right)(Ur), \quad E_\theta = r^{-1}\partial_{r\theta}^2 (Ur), \tag{4.7}$$

where $k = \omega/c$ is the wave number. Substituting these quantities into the set of Eqs. (4.3)–(4.5) one can check that Eqs. (4.4) and (4.5) become the identities, while Eq. (4.3) yields

$$\left(\partial_r^2 + k^2\right)(Ur) + \frac{\partial_\theta \left(\sin \theta \partial_\theta U\right)}{r \sin \theta} = \frac{j_s}{i\omega\varepsilon_0}. \tag{4.8}$$

To proceed analytically, it is necessary at this point to construct a suitably idealized model of the source that is a reasonable approximation to the lightning discharge parameters.

4.1.2 Model of Lightning Discharge and Boundary Conditions

For now, we approximate the actual CG return stroke by a lumped source with the current moment $m(t) = Il$, where $I(t)$ is the lightning current and $l(t)$ denotes the length of the current channel. Let $m(\omega)$ be a Fourier transform of the current moment. The $-$CG current is upward directed. In standard meteorological practice this current corresponds to negative current moment. In our study the $-$CG current moment is positive since we use the upward directed vertical axis z, as shown in Fig. 4.1.

Let R_e and $R_i = R_e + d$ be the radii of the Earth and the ionosphere, where d is the height of the non-conductive atmosphere. The upward-directed lumped element **m** situated in the atmosphere has the coordinates $r = R$ and $\theta = 0$ as shown in Fig. 4.1. In such a case the Fourier transform of the current density j_s produced by the lumped source can be written as follows:

$$j_s (r, \theta, \omega) = \frac{m (\omega)}{2\pi r^2 \sin \theta} \delta (r - R) \delta (\theta), \qquad (4.9)$$

where $\delta (x)$ is the so-called Dirac's function/delta-function, which equals zero for all the x, except for $x = 0$, and by definition

$$\int_{-\infty}^{\infty} \delta (x) \, dx = 1. \qquad (4.10)$$

The arrangement of the formula (4.9) for j_s is in accord with the requirement that the integral of j_s over the space gives the total current moment $m (\omega)$. Indeed, the elementary current moment is $j_s dV$, where $dV = r^2 \sin \theta d\theta d\varphi dr$ is the volume element/differential in the spherical coordinates. Multiplying Eq. (4.9) by dV and integrating the elementary current moment over the whole space yields the value $m (\omega)$, which is required to be proved.

Substituting Eq. (4.9) for j_s into Eq. (4.8) gives a differential equation which contains the delta-function. In order to derive a boundary condition at point $r = R$ one should integrate Eq. (4.8) over r between $R - \alpha$ and $R + \alpha$, and then formally take $\alpha \to 0$. Taking into account the continuity of U at $r = R$, we thus obtain the following condition at $r = R$

$$[\partial_r (rU)] = \frac{m (\omega) \delta (\theta)}{2\pi i \varepsilon_0 \omega R^2 \sin \theta}, \qquad (4.11)$$

where the square bracket denotes the jump of the function, that is, $[f (x)] = f (x + 0) - f (x - 0)$. In Eq. (4.11) the square bracket stands for the jump of the function $\partial_r (rU)$ at $r = R$.

Boundary conditions (4.11) may be simplified because the point current element occurs at the small altitude above the ground, that is, in the case $R - R_e \ll d$. If the Earth is supposed to be a perfect conductor, the tangential component of the electric field E_θ becomes zero at the terrestrial surface. Whence it follows from Eq. (4.7) that $\partial_r (rU) = 0$ at $r = R_e$. Combining this condition with Eq. (4.11) in the extreme limit case $R \longrightarrow R_e$ we come to the following boundary condition at $r = R_e$

$$\partial_r (rU) = \frac{m(\omega)\,\delta(\theta)}{2\pi i\,\varepsilon_0 \omega R_e^2 \sin\theta}. \qquad (4.12)$$

Similarly, if the ionosphere is supposed to be a perfect conductor, the tangential component of the electric field E_θ becomes zero at the boundary with the ionosphere. Whence it follows from Eq. (4.7) the boundary condition at $r = R_i$:

$$\partial_r (rU) = 0. \qquad (4.13)$$

4.1.3   Solution of the Problem

Finally, we have arrived at the second order differential Eq. (4.8), where the right-hand side is equal to zero. This equation for the unknown potential function U should be supplemented by two boundary conditions of Eqs. (4.12) and (4.13). In standard mathematical technique, it is recommended to seek for the solution of Eq. (4.8) in the form

$$U = \sum_{n=0}^{\infty} \left\{ A_n h_n^{(1)}(kr) + B_n h_n^{(2)}(kr) \right\} P_n(\cos\theta), \qquad (4.14)$$

where A_n and B_n are the undetermined coefficients, $h_n^{(1)}(x)$ and $h_n^{(2)}(x)$ are the spherical Bessel functions of the third kind, and $P_n(x)$ stands for Legendre polynomials. The definitions of these functions are found in Appendix A. To construct the solution of the problem, the boundary condition (4.12) can be expanded in a series of the Legendre polynomials. The representation of the delta-function $\delta(\theta)$ through Legendre polynomials is given by Eqs. (4.59) and (4.60). Combining these equations with Eq. (4.12), we come to the following boundary condition at $r = R_e$

$$\partial_r (rU) = \frac{m(\omega)}{2\pi i\,\varepsilon_0 \omega R_e^2} \sum_{n=0}^{\infty} \left(n + \frac{1}{2}\right) P_n(\cos\theta). \qquad (4.15)$$

Substituting Eq. (4.14) for U into Eqs. (4.13) and (4.15) we get the boundary condition expanded in a series of the Legendre polynomials. This representation of the boundary conditions gives a set of equations for undetermined coefficients A_n and B_n

$$A_n u_n^{(1)}(kR_i) + B_n u_n^{(2)}(kR_i) = 0, \qquad (4.16)$$

$$A_n u_n^{(1)}(kR_e) + B_n u_n^{(2)}(kR_e) = \frac{m(\omega)}{2\pi i\,\varepsilon_0 \omega R_e^2} \left(n + \frac{1}{2}\right). \qquad (4.17)$$

Here we made use of the following abbreviations:

$$u_n^{(1,2)} = \frac{d}{dr} \left\{ r h_n^{(1,2)} (kr) \right\}. \tag{4.18}$$

The set of Eqs. (4.16) and (4.17) can be solved for A_n and B_n to yield the potential U

$$U = \frac{m(\omega)}{2\pi i \varepsilon_0 \omega R_e^2} \sum_{n=0}^{\infty} C_n \left(n + \frac{1}{2} \right) P_n (\cos \theta), \tag{4.19}$$

where

$$C_n = \frac{u_n^{(2)} (kR_i) h_n^{(1)} (kr) - u_n^{(1)} (kR_i) h_n^{(2)} (kr)}{u_n^{(1)} (kR_e) u_n^{(2)} (kR_i) - u_n^{(1)} (kR_i) u_n^{(2)} (kR_e)}. \tag{4.20}$$

Substituting Eq. (4.19) for U into Eqs. (4.6) and (4.7) gives the solution of problem.

The structure of the solution given by these equations is very complicated despite the fact that the simplest model of Earth-Ionosphere resonance cavity has been used. In what follows we simplify this solution to extract the eigenfrequencies of the resonator.

4.2 Schumann Resonances

4.2.1 Eigenfrequencies of the Schumann Resonances

It should be emphasized that the set of eigenfrequencies of the Earth-Ionosphere cavity is its inner property, which depends on the sizes and inner structure of the resonant cavity. The eigenfrequencies are affected by neither a type of source nor the way of the cavity excitation. Eigenvalues of the frequency are determined by singular points of the solution given by Eqs. (4.19) and (4.20). To find the eigenfrequencies, it is necessary at this point to examine the poles of the functions C_n, that is, the points in complex ω-plane where the denominator of Eq. (4.20) vanishes. First, we note that the set of corresponding wave numbers $k = \omega/c$ is derivable from the following equation set

$$u_n^{(1)} (kR_e) u_n^{(2)} (kR_i) - u_n^{(1)} (kR_i) u_n^{(2)} (kR_e) = 0, \tag{4.21}$$

where $n = 1, 2, 3 \ldots$

When performing the integration of U in complex ω-plane, the area surrounding the poles $\omega = \omega_n$, which are commonly complex, makes a contribution to the integral via residues of the functions C_n in the poles. From physical viewpoint, the real part of the poles ω_n defines eigenfrequencies of the resonator while the imaginary part of ω_n defines the damping factors of the eigenmodes. As we shall see, these eigenfrequencies typically lie in the range of $f = 7.5$–50 Hz. In such a case Eq. (4.21) contains the small parameter kd, where $d = R_i - R_e$ is the altitude/thickness of the nonconducting atmospheric layer. Indeed, taking $\omega = 2\pi f = 50$ Hz and $d = 40$–90 km, one obtains $kd = (0.67$–$1.5) \times 10^{-2}$. Taking into account that $kd \ll 1$, we use the approximation

$$u_n^{(1,2)}(kR_i) \approx u_n^{(1,2)}(kR_e) + \frac{du_n^{(1,2)}(kR_e)}{dr}kd. \tag{4.22}$$

Substituting Eq. (4.22) into Eq. (4.21) yields

$$u_n^{(1)}\frac{du_n^{(2)}}{dr} - u_n^{(2)}\frac{du_n^{(1)}}{dr} = 0, \tag{4.23}$$

where all the functions are taken at $r = kR_e$. Substituting Eq. (4.18) for $u_n^{(1)}$ and $u_n^{(2)}$ into Eq. (4.23), we come to the equation, which contains both the functions $h_n^{(1)}$ and $h_n^{(2)}$ and their derivatives. Eliminating from this equation the second order derivatives with the help of Eq. (4.53) and rearranging, we obtain

$$\{k^2 R_e^2 - n(n+1)\}\left(h_n^{(1)}\frac{dh_n^{(2)}}{dr} - h_n^{(2)}\frac{dh_n^{(1)}}{dr}\right) = 0. \tag{4.24}$$

The factor in the round brackets is Wronskian of the functions $h_n^{(1)}$ and $h_n^{(2)}$, which is equal to $-2i(kR_e)^{-2}$ (Abramowitz and Stegun 1964). Since this factor is nonzero, the first factor in Eq. (4.24) must be equal to zero. Hence, we arrive at the following result

$$\omega^2 - \frac{c^2 n(n+1)}{R_e^2} = 0, \tag{4.25}$$

whence it follows that (Schumann 1952a,b, 1957; Nickolaenko and Hayakawa 2002)

$$f_n = \frac{\omega_n}{2\pi} = \frac{c}{2\pi R_e}[n(n+1)]^{1/2}, \tag{4.26}$$

where $n = 1, 2, 3\ldots$ enumerates the mode number. Substituting $R_e = 6{,}370$ km into Eq. (4.26) gives the following frequencies of the first Schumann's resonances $f_1 = 10.6$, $f_2 = 18.3$, $f_3 = 25.9$, $f_4 = 33.5$, $f_5 = 41.1,\ldots$, $f_n = f_1[n(n+1)/2]^{1/2},\ldots$ (in Hz).

Schumann was the first who predicted the spectrum of the eigenfrequencies defined by Eq. (4.26), and this resonance spectrum has been termed Schumann spectrum. In the first approximation this spectrum is independent of the atmosphere altitude d.

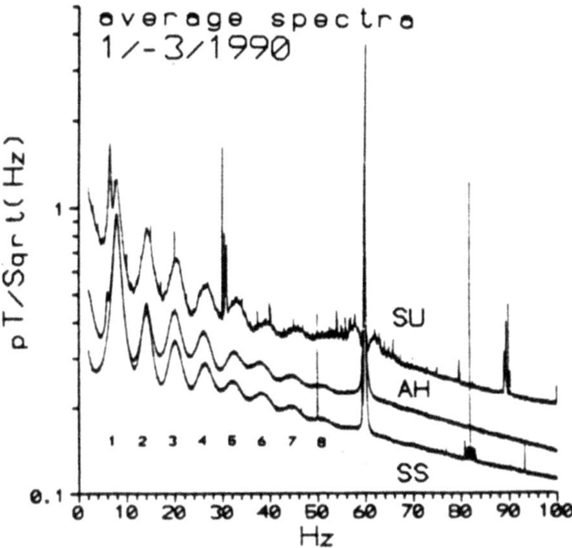

Fig. 4.2 Experimental power spectra, which contain the Schumann resonances. Arrival Heights, Antarctica (AH), Sondrestromfjord, Greenland (SS), Stanford, California, USA (SU). Taken from www.crucible.org/equip_lifeforce.htm

To interpret this phenomenon one may assume that the Schumann resonances arise from the large-scaled electromagnetic waves propagating inside the atmospheric waveguide around the Earth. Typical wavelength of these waves is of the order of the Earth's radius and thus is much greater than the atmospheric altitude d. The eigenfrequencies, f_n, of the Schumann resonances can be roughly estimated in terms of the fact that whole numbers of electromagnetic wavelength must keep within a circle $2\pi R_e$, where R_e is the Earth's radius. Hence we get the estimate $f_n \sim cn/(2\pi R_e) \approx 7.5n$ (in Hz), which is close to that given by Eq. (4.26), especially as for the large numbers n.

The tangential component, E_θ, of the electromagnetic field caused by a CG lightning discharge is small inside the resonator since it becomes zero at the resonator sides. The radial electric component E_r is perpendicular to the direction of wave propagation. This component of the quasi-transverse electromagnetic normal mode dominates in the resonance cavity. As far as $d \ll R_e$, the resonator is partly similar to a flat waveguide between two infinite parallel conductive plates. In such a case the perpendicular component E_r is practically insensitive to the waveguide altitude.

Worldwide thunderstorm activity is believed to be the main source/mechanism for excitation of the Schumann's spectra. The Schumann resonances were originally measured by Balser and Wagner (1960) and have been extensively studied by a number of authors (e.g., see Galejs 1965, 1972; Bliokh et al. 1980; Nickolaenko and Hayakawa 2002). In Fig. 4.2 the Schumann's spectrum measured in $pT/Hz^{1/2}$ is plotted. As is seen from Fig. 4.2, the observed resonance frequencies are somewhat smaller than that predicted by Eq. (4.26). Numerous data obtained by a number of researchers may be summarized to give the following mean annual values: $f_1 = 7.8$, $f_2 = 14.0$–14.1, $f_3 = 20.0$–20.3, $f_4 = 26.0$–26.4, $f_5 = 31.8$–32.5 (in Hz, Nickolaenko and Hayakawa 2002). More usually, the highest order harmonics ($n \geq 6$) are hardly distinguished from the background since they may be below the signal-to-noise threshold.

In the first place the small discrepancy between the theory and observations results from the fact that we have used the idealized resonator model that ignores actual ionospheric and atmospheric conductivity which in turn leads to the absorption of wave energy. The atmospheric conductivity becomes significant at high altitude since it exponentially increases with altitude. More accurate models of the conductivity profiles, including exponential and two-layer models have been developed to account for actual distribution of the electromagnetic field at the bottom ionosphere (Wait 1960, 1962; Galejs 1961; Jones 1964). In this case some complication arises due to the Earth magnetic field impact on conductivity of the ionospheric plasma. Inside the gyrotropic E-layer the plasma conductivity becomes anisotropic so that the proper boundary condition at the lower ionosphere should be applied instead of the simple boundary condition of Eq. (4.13). Additionally, the plasma conductivity tensor depends on both dip angle of the geomagnetic field and the ionosphere status, which is highly dependent on the current time. For instance, the conductivity of sunlit ionosphere is much greater than that of nighttime ionosphere (see Fig. 2.5). This implies that the boundary conditions at the ionosphere depend on both polar θ and azimuthal φ angles, and local time. The reader is referred to the books by Wait (1972), Bliokh et al. (1980), and Nickolaenko and Hayakawa (2002) for details about more accurate calculation of the resonator eigenfrequencies.

It is worth mentioning that another branch of Schumann resonances can be excited in the Earth–Ionosphere cavity. These resonances are due to TE mode propagation, which result in the formation of standing electromagnetic waves perpendicular to the resonator sides, that is perpendicular to ground surface and lower boundary of the ionosphere. Assuming that the integer number of electromagnetic wavelengths must keep within the atmospheric altitude d, we get the rough estimate of the transverse normal mode eigenfrequencies $f_n = cn/(2d)$, where $n = 1, 2, 3 \ldots$ (Outsu 1960; Hayakawa et al. 1994; Shvets and Hayakawa 1998). This estimate shows that the corresponding values of eigenfrequencies must be greater than $2\,$kHz. Due to strong damping of the electromagnetic waves in this frequency range, such resonances are practically indistinguishable from the background variations and thus they are of minor interest.

4.2.2 Quality/Energy-Factor

Further complexities of the theory of Schumann resonances arise from the attenuation of electromagnetic waves due to Joule dissipation of the wave energy at the conducting sides of the Earth–Ionosphere cavity. Furthermore, some electromagnetic energy can penetrate through the conductive E-layer thereby exciting the shear and magnetosonic Alfvén waves which are radiated into the magnetosphere.

Let Q be the quality/energy-factor, which is defined as

$$Q = \frac{f_{\max}}{\Delta f},$$
(4.27)

where $f_{\max}$ is the central frequency corresponding to the local maximum/resonance of power spectrum and Δf is its full width at semi-height/half of the maximum. The Q-factor can be defined as

$$Q = \frac{\pi \tau_e}{T} = \pi N_e,$$
(4.28)

where τ_e is the relaxation time or the time which is necessary to decrease the amplitude of damping oscillation by the factor e (base of natural logarithm), T is the period of damping oscillations, and N_e denotes the number of wave periods keeping within a relaxation time. Furthermore, the Q-factor is inversely proportional to energy dissipation for a period of oscillations. Typical values of the Q-factors for various Schumann resonances observed at mid-latitudes are as follows: $Q_1 = 4.6$, $Q_2 = 6.0$, $Q_3 = 6.6$, $Q_4 = 6.8$, and $Q_5 = 7.0$ (e.g., see Bliokh et al. (1980)).

For the fundamental mode ($n = 1$) the relaxation time $\tau_e = Q_1 / (\pi f_1) \approx 0.2$ s. On average approximately 10^2 strokes per second occur worldwide due to the global lightning activity, and thus within 0.2 s of damping interval of the fundamental mode about 20 lightnings may occur. The vast majority of individual lightning strokes are randomly distributed in time, so that their electromagnetic fields are added incoherently thereby producing quasi-permanent excitation of the normal modes in the Earth-Ionosphere cavity.

In a more complete theory complications due to the field attenuation in conducting layers of the ionosphere and of the Earth may be included. Numerical modeling permits of fitting adequately the theoretical and experimental values of both the eigenfrequencies and Q-factors.

4.2.3 Solution of the Problem in a More Accurate Model

Before leaving this section, it is useful to revise the solution of the problem derived above in order to take into account the effect of wave damping. Substituting Eq. (4.19) for U into Eqs. (4.6) and (4.7) yields

$$B_\varphi = \frac{\mu_0 m\,(\omega)}{2\pi R_e^2} \sum_{n=1}^{\infty} C_n \left(n + \frac{1}{2} \right) \partial_\theta P_n\,(\cos\theta),$$
(4.29)

$$E_r = \frac{\mu_0 m(\omega)}{2\pi i \omega r} \sum_{n=0}^{\infty} C_n \omega_n^2 \left(n + \frac{1}{2}\right) P_n(\cos\theta), \qquad (4.30)$$

where the coefficients C_n are given by Eq. (4.20). Here we have taken into account that the summand with number $n = 0$ in Eq. (4.29) is equal to zero since $\partial_\theta P_0(\cos\theta) = 0$.

The implication here is that the electromagnetic field inside the Earth-Ionosphere cavity is represented as a linear superposition of normal modes excited by the vertical lightning discharge with the current moment $m(\omega)$. Each normal mode has a specific angular structure, which is determined by the Legendre polynomial associated with the mode number. In practice, the angular dependence determines the decrease of the normal mode amplitude with distance between the source and observation point.

The coefficient C_n in Eqs. (4.29) and (4.30) can be expanded in a power series of a small parameter kd. Using the first order approximation similar to that given by Eq. (4.22), one can reduce Eq. (4.20) to the following:

$$C_n \approx \frac{R_e}{d\left[k^2 R_e^2 - n(n+1)\right]} = \frac{c^2}{R_e d\left(\omega^2 - \omega_n^2\right)}. \qquad (4.31)$$

On the ground surface $r = R_e$ we thus get

$$B_\varphi = \frac{m(\omega)}{4\pi\varepsilon_0 R_e^3 d} \sum_{n=1}^{\infty} \frac{(2n+1)}{\left(\omega^2 - \omega_n^2\right)} \partial_\theta P_n(\cos\theta), \qquad (4.32)$$

$$E_r = \frac{im(\omega)}{4\pi\varepsilon_0 \omega R_e^2 d} \sum_{n=0}^{\infty} \frac{\omega_n^2(2n+1)}{\left(\omega_n^2 - \omega^2\right)} P_n(\cos\theta). \qquad (4.33)$$

We recall that this form of solution is based on the idealized model of the Earth-Ionosphere resonance cavity which includes the perfect conducting walls.

The finite conductivity of the ionospheric plasma is believed to be the major cause for wave energy losses. At the altitudes of the gyrotropic E layer the ionospheric plasma is described by the conductivity tensor of Eq. (2.5) or by the plasma dielectric permittivity tensor of Eq. (2.16). In the D layer, the altitude range 50–70 km, the total electron and ion collision frequencies, i.e., $\nu_e = \nu_{ei} + \nu_{en}$ and $\nu_i = \nu_{in} + \nu_{ie}$, are much greater than the corresponding gyrofrequencies, ω_H and Ω_H, of the electrons and ions. This implies that all diagonal components of the conductivity/dielectric permittivity tensor [see Eqs. (2.7)–(2.9)] become equal to each other in such a way that this tensor can be replaced by the scalar value

$$\varepsilon_p = 1 - \frac{\omega_p^2}{\omega(\omega + i\nu_e)}, \qquad (4.34)$$

where ω_p is the plasma frequency given by Eq. (2.20). Here we have neglected the small terms of the order of $m_e/\overline{m_i}$.

In the lower ionosphere and D layer the total electron collision frequency, ν_e, is mainly determined by the electron–neutral collisions. Note that ν_e and ω_p are much larger than ω for all wave frequencies of interest here, so that we obtain the pure imaginary value $\varepsilon_p \approx i\omega_p^2/(\omega\nu_e)$. At the height about 60 km the typical daytime parameters are as follows (Nickolaenko and Hayakawa 2002) $\omega_p \sim 5 \times 10^5\,\mathrm{s}^{-1}$ and $\nu_e \sim 5 \times 10^7\,\mathrm{s}^{-1}$, while $\omega \sim 50\,\mathrm{s}^{-1}$. Substituting these values into Eq. (4.34) yields $\varepsilon_p \sim 10^2 i$.

Now we chose the simplest way to estimate the effect of wave energy absorption, aiming at physical intuition rather than detailed analysis. In order to take into account the damping factor of the electromagnetic waves, one should formally remove the poles $\omega = \omega_n$ from real axes in the complex plane ω. From physical viewpoint, the real part of the poles ω_n defines eigenfrequencies of the resonator while the imaginary part of ω_n defines the damping factors of the eigenmodes. In a more accurate model of dissipative resonator the eigenfrequencies can be found from the following equation (e.g., see Nickolaenko and Hayakawa (2002), and references therein)

$$\omega^2 + \frac{ic\omega Z(\omega)}{d} - \frac{c^2 n(n+1)}{R_e^2} = 0, \tag{4.35}$$

instead of Eq. (4.25). Here $Z(\omega)$ stands for the surface impedance of the ionosphere. In the simple model of the isotropic ionosphere and perfectly conducting Earth this impedance is given by $Z = \varepsilon_p^{-1/2}$.

When comparing the rigorous solution of the problem (e.g., see monographs by Wait (1972), Galejs (1972)) with the approximate solution given by Eq. (4.32), the difference in the denominators of these solutions comes into particular prominence. To make Eq. (4.32) identical to the rigorous solution, one should replace the term ω^2 in the denominator of the sum in Eq. (4.32) by the imaginary expression $\omega^2 + ic\omega Z(\omega)/d$. As a result we obtain that

$$B_\varphi = M g_\varphi(\theta, \omega). \tag{4.36}$$

Here we made use of the following abbreviation:

$$g_\varphi(\theta, \omega) = \frac{F(\omega)}{4\pi\varepsilon_0 R_e^3 d} \sum_{n=1}^{\infty} \frac{(2n+1)\,\partial_\theta P_n(\cos\theta)}{(\omega^2 + ic\omega Z(\omega)/d - \omega_n^2)}. \tag{4.37}$$

where ω_n is given by Eq. (4.26). As before, the spectrum of magnetic variations produced by the lightning discharge is proportional to the spectrum of the lightning current moment $m(\omega) = M F(\omega)$.

A similar rearranging of approximate solution for the electric field is found in Appendix A. In this case Eq. (4.33) should be replaced by the following equation:

$$E_r = -\frac{iMF(\omega)}{4\pi\varepsilon_0 R_e^2 d} \sum_{n=0}^{\infty} \frac{(\omega + icZ/d)(2n+1)}{(\omega^2 + ic\omega Z(\omega)/d - \omega_n^2)} P_n(\cos\theta). \qquad (4.38)$$

By symmetry of the problem the electromagnetic field of the vertical CG lightning discharge is azimuthally symmetric in the Earth-Ionosphere cavity. Before discussing these solutions, we note that Eq. (4.36) may be simplified in an extreme case $kR_e \ll 1$ or $\omega \ll \omega_1$. Letting formally $\omega \to 0$, we come to

$$B_\varphi = -\frac{\mu_0 MF(\omega)}{4\pi R_e d} \partial_\theta \sum_{n=1}^{\infty} \frac{(2n+1)}{n(n+1)} P_n(\cos\theta). \qquad (4.39)$$

The sum of infinite series in Eq. (4.39) is given by Eq. (4.61). Taking this sum and performing differentiation over θ yields

$$B_\varphi = \frac{\mu_0 MF(\omega)}{4\pi R_e d} \cot\left(\frac{\theta}{2}\right). \qquad (4.40)$$

This equation coincides with the known solution for quasi-steady magnetic field of the radial point current element located inside the thin ($d \ll R_e$) spherical condenser (e.g., see Belyaev et al. 1989). In the same approximation one can simplify Eq. (4.38) for E_r.

The measured resonance frequencies, i.e., the frequencies at which the spectra tend to maximize, undergo a weak diurnal variation with magnitude about $\pm 0.5\,\mathrm{Hz}$ about their average values (Balser and Wagner 1962). Furthermore, the resonance frequencies of the magnetic variations are observed to depend on the orientation of the antenna; that is, these values depend on whether the measurement is made along the north-south or east-west direction (Sentman 1987). The shift of the eigenfrequencies of the Earth-Ionosphere cavity may occur due to changes in dissipative properties of the ionosphere. The solar Roentgen radiation and energetic particle precipitation from the magnetosphere may greatly affect the ionosphere conductivity due to ionization of the neutral particles (Mitra 1974; Rodger 1999).

It should be noted that the shift in the eigenfrequencies is also expected for some kinds of field asymmetry in the actual Earth-Ionosphere cavity. The electromagnetic field of the vertical CG lightning discharge given by Eqs. (4.36) and (4.38) is azimuthally symmetrical in the Earth-Ionosphere cavity. The n-th normal mode of this field is described by $(2n + 1)$-fold degenerate eigenfunction. This implies the degeneracy of the eigenfunction with respect to azimuthal variable φ. The observed effect can be due to the removal of this degeneracy of the normal modes of the resonator possibly due to some kind of accidental factors (e.g., see a book by Nickolaenko and Hayakawa (2002)).

4.3 Sources of Resonator Excitation

4.3.1 Lightning Discharges Treated as a Stochastic Process

To study the Earth-Ionosphere cavity resonator in a little more detail, it is necessary at this point to construct a suitably idealized model of the spatial and temporal source distribution that is a reasonable approximation to the variation of the global thunderstorm parameters. In the simple model the global thunderstorm activity is modeled as a point constant-amplitude emitter located in the vicinity of equator near the nighttime terminator at 17–18 LT (Galejs 1972). Following the sun this emitter runs around the Earth for the day. Bliokh et al. (1980) have approximated the actual situation of the lightning activity with the configuration, which includes three or more fixed world thunderstorm centers. In a more complete theory, however, thunderstorm distribution on the Earth surface in accordance with climatological data may be included (Ogawa and Murakami 1973), that results in some complications of the theory.

Actually, the lightning discharge in the thunderstorm region should be considered most likely as random events described by a sequence of randomly occurring pulses of random amplitudes (Raemer 1961a,b; Galejs 1965; Polk 1969; Nickolaenko 1981; Surkov et al. 2005, 2006; Surkov and Hayakawa 2010). In what follows we consider the global lightning activity, whose electromagnetic signals are being recorded by a ground-based station which is far away from thunderstorms. Let N be the number of thunderstorms that are in progress around the world at the moment. Typical thunderstorm area is of the order of $10^3\,\mathrm{km}^2$. This means that a typical size of thunderstorm area is smaller than the distance from the recording station so that we can ignore the lightning spatial distribution within the thunderstorm area. In this notation all the lightning related to the given thunderstorm will have the same coordinates as those of the given thunderstorm.

We introduce a coordinate system fixed to the center of the Earth in such a way that the recording station is situated on the polar axis z, as shown in Fig. 4.3. Let r_μ, θ_μ, and φ_μ be the spherical coordinates of a thunderstorm, where the subscript μ stands for the number of thunderstorms, i.e., $\mu = 1, 2, \ldots, N$. Notice that in fact r_μ equals to the Earth's radius R_e.

Let $\mathbf{b}\left(\mathbf{r}_\mu, t - t_{n_\mu}\right)$ be the random magnetic field at the ground-based station generated by an individual CG lightning discharge occurring at the point $\mathbf{r}_\mu$ at a random moment t_{n_μ}, where $n_\mu = 1, 2, \ldots$ is a number of the lightning discharge happened at this point. Here $\mathbf{r}_\mu$ denotes a set of the thunderstorm's spherical coordinates r_μ, θ_μ, and φ_μ. Notice that $\mathbf{b}\left(\mathbf{r}_\mu, t - t_{n_\mu}\right)$ vanishes at $t < t_{n_\mu}$.

In standard meteorological practice a local coordinate system fixed at the recording station has the x axis eastward, the y axis to the north, and z axis vertically upward. For convenience, this local coordinate system is shown in Fig. 4.3 from the right of the main image.

In Sects. 4.1 and 4.2 the electromagnetic field of vertical CG lightning has been expressed through the spherical components, b_r, b_θ, and b_φ, in the source-local

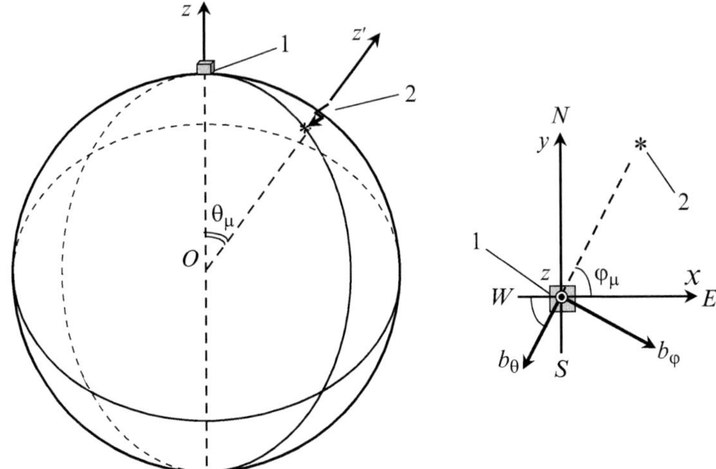

Fig. 4.3 Coordinate systems used in this study. *1*—ground-based recording station. *2*—thunderstorm. A base coordinate system is fixed to the center O of the Earth in such a way that the recording station is situated on the polar axis z. The same picture from a viewpoint looking down on the Earth from above is shown from the right of the main image. Presented here are the station, the local Cartesian coordinate system, and horizontal components, b_θ and b_ϕ

system of coordinates with polar axes z', which coincides with the lightning current, as shown in Fig. 4.3. It is usually the case that the far-field/wave zone includes only the perpendicular component, b_φ, whereas the near-field consists of all the components due to nonuniformity and asymmetry of the actual ionosphere (Surkov et al. 2005, 2006). More usually we measure the field components in the Cartesian system of coordinates fixed to the ground recording station. In this case the components of the magnetic field can be expressed through b_r, b_θ, and b_φ as follows:

$$\begin{pmatrix} b_x \\ b_y \\ b_z \end{pmatrix} = \begin{pmatrix} 0 & -\cos\varphi_\mu & \sin\varphi_\mu \\ 0 & -\sin\varphi_\mu & -\cos\varphi_\mu \\ 1 & 0 & 0 \end{pmatrix} \begin{pmatrix} b_r \\ b_\theta \\ b_\varphi \end{pmatrix}. \tag{4.41}$$

In the text we shall use the abbreviation $\hat{\mathbf{A}}_\mu$ for the transformation matrix in Eq. (4.41).

Far from the site of lightning discharge occurrence, the electromagnetic field generated by the lightning discharge can be expressed via only lightning parameter, that is, via the current moment of the stroke $m(t)$. As before we approximate the actual current moments with the function $m(t) = MF(t)$, where the magnitude, M of the current moment is assumed to be a random value whereas the function $F(t)$ describing the current moment profile is a deterministic function of time (Surkov et al. 2005, 2006). For example, the current moment of individual and multiple return strokes can be approximated by Eqs. (3.7) and (3.10), respectively. As we have noted above, the $+$CG lightning discharges which transfer the positive charge to ground differ from the $-$CG lightning discharges in amplitude and duration of current, especially their long-lasting continuing current. In the case of high amplitude, the statistical distributions of the negative and positive charge moments are quite different (Williams et al. 2007). In this notation, we introduce two types of deterministic functions $F(t)$, that is, $F_n(t)$ and $F_p(t)$ and two independent random values M_n and M_p for the negative and positive lightning, respectively.

Considering either of these current moments, the magnetic field of the single lightning discharge can be written as

$$\mathbf{b}\left(\mathbf{r}_\mu, t - t_{n_\mu}\right) = M_{n_\mu}\mathbf{G}_\mu\left(\mathbf{r}_\mu, t - t_{n_\mu}\right), \tag{4.42}$$

where the propagation factors $\mathbf{G}_\mu$ is supposed to be equal to zero as $t < t_{n_\mu}$. These functions are derivable from Maxwell equations for the Earth-Ionosphere resonance cavity that should be supplemented by proper boundary conditions at the ground and the ionosphere. In the simple model of the Earth-Ionosphere cavity considered in Sects. 4.1 and 4.2, the components of the function $\mathbf{G}_\mu$ can be extracted from Eq. (4.36).

The net magnetic perturbation at the ground-recording station is also a random quantity, $\mathbf{B}$, which equals the sum of the magnetic perturbations caused by individual lightning discharges. In the Cartesian reference frame fixed to the ground-recording station, the net magnetic field of all the lightning due to global thunderstorm activity is then

$$\mathbf{B}(t) = \sum_{\mu=1}^{N}\mathbf{b}_\mu\left(\mathbf{r}_\mu, t\right), \quad \mathbf{b}_\mu\left(\mathbf{r}_\mu, t\right) = \sum_{n_\mu}\mathbf{b}\left(\mathbf{r}_\mu, t - t_{n_\mu}\right), \tag{4.43}$$

where $\mathbf{b}_\mu\left(\mathbf{r}_\mu, t\right)$ is the magnetic field due to a thunderstorm with number μ.

It is common practice to describe the stochastic processes via mean value/mathematical expectation and correlation functions of the stochastic process. In the analysis that follows, we first introduce the basic characteristic of a random process and then extend them to estimate the so-called power spectrum of electromagnetic variations. The detailed calculations are found in Appendix B.

4.3.2   Correlation Matrix of Random Field Variations

The comprehensive analysis of random vector fields is mainly based on correlation methods. We next compute the correlation matrix of the magnetic variations defined as

$$\Psi_{jk}\left(t, t'\right) = \left\langle B_j(t) B_k(t')\right\rangle - \left\langle B_j(t)\right\rangle\left\langle B_k(t')\right\rangle, \tag{4.44}$$

where $j, k = x, y, z$ and the angular brackets denote the averaging over all available/possible realizations of the random process. This matrix is defined in such a way that the component Ψ_{jk} may vanish in the case of statistically independent functions B_j and B_k. Below we focus only on magnetic variations while the same technique can be applied to electric components as well.

The field fluctuation is assumed to be a steady stochastic process. This means that the correlation matrix depends on t and t' in such a way that it is a function of only the time difference $\tau = t' - t$. In such a case the spectral density of this random process must be delta correlated (Rytov et al. 1978; Weissman 1988), that is,

$$\Psi_{jk}\left(\omega, \omega'\right) = \delta\left(\omega - \omega'\right)\psi_{jk}\left(\omega\right). \tag{4.45}$$

The spectral matrix $\psi_{jk}\left(\omega\right)$ is of our prime interest in this section.

In what follows the lightning discharge occurrence is treated as a sequence of independent random events, which obeys the Poisson random distribution. This implies that the elementary probability, dP, of the lightning origin from the moment t to $t + dt$ is proportional to dt and does not depend on t, that is $dP = vdt$, where v stands for the mean number of the lightning discharges per unit time. We refer the reader to Appendix B for details about the Poisson probability law. In reality, just after the end of the lightning discharge the probability for the next discharges slightly decreases since it is necessary about $5\,\text{s}$ to reconstruct the total charge of thunderstorm cloud (Uman and Krider 1982; Uman 1987). We shall ignore this fact assuming that the mean time v^{-1} between two adjacent discharges is much greater than the last value.

Lightning activity in each thunderstorm is considered as an independent random process. Consider first the random process associated with the thunderstorm center with number μ. The moments of lightning discharge appearance, t_{n_μ}, and the current moment magnitudes, M_{n_μ}, are supposed to be statistically independent of each other and their probability distributions are independent of the impulse number n_μ. Recall that the propagation factor $\mathbf{G}_\mu\left(\mathbf{r}_\mu, t\right)$ describing the shape of magnetic field generated by individual lightning discharge is considered as deterministic functions.

In the simple model where the Earth-Ionosphere forms a spherically symmetric resonance cavity, the functions $\mathbf{G}_\mu$ for the vertical CG lightning can be considered as a universal function of the angular distance θ_μ between the lightning and the ground-based station, that is, $\mathbf{G}_\mu = \mathbf{G}\left(\theta_\mu, t\right)$. The influence of the solar wind and radiation on the ionospheric plasma and the presence of the Earth's magnetic field bring the actual ionosphere is nonuniform and asymmetric. This means that the shape of the functions $\mathbf{G}_\mu$ can depend on the lightning/thunderstorm coordinates.

The mean value of $\langle\mathbf{b}_\mu\left(t\right)\rangle$ under above requirements is obtained in Appendix B. Assuming that the lightning discharge processes at each thunderstorm are independent of each other and taking into account Eqs. (4.71) and (4.43), we obtain the mean value of net magnetic field generated by all the thunderstorms

$$\langle\mathbf{B}\rangle = \sum_{\mu=1}^{N} v_\mu\langle M_\mu\rangle \int_{-\infty}^{\infty} \mathbf{G}\left(\mathbf{r}_\mu, t'\right) dt'. \tag{4.46}$$

Here the mean number of the lightning discharges per unit time, ν_μ, and the mean magnitude of current moment, $\langle M_\mu \rangle$, are related to an individual thunderstorm with number μ. Notice that Eq. (4.46) for the mean value of the magnetic field variations has been derived for the stationary random process. Not surprisingly, this value is independent of time.

In the measurements the magnetic variations observed on the ground have an alternating-sign shape due to the presence of a variety of natural and man-made noises. The mean value of such variations is normally close to zero and therefore is of little importance. The correlation matrix at this point is a more suitable value for adequate description of the magnetic variations. However, for the sake of generality, in the analysis that follows we shall keep the mean values $\langle B_j\,(t) \rangle$ and $\langle B_k\,(t') \rangle$ in Eq. (4.44) despite these values are close to zero.

Far away from the lightning the perpendicular field b_φ dominates over other magnetic components. As the lightning far-field includes only the perpendicular component, the spectral density given by Eqs. (4.90) and (4.91) is simplified to

$$\psi_{xx}\,(\omega) = 2\pi \sum_{\mu=1}^{N} \nu_\mu \left\langle M_\mu^2 \right\rangle \sin^2 \varphi_\mu \left| g_\varphi\,(\mathbf{r}_\mu, \omega) \right|^2 \tag{4.47}$$

and

$$\psi_{yy}\,(\omega) = 2\pi \sum_{\mu=1}^{N} \nu_\mu \left\langle M_\mu^2 \right\rangle \cos^2 \varphi_\mu \left| g_\varphi\,(\mathbf{r}_\mu, \omega) \right|^2, \tag{4.48}$$

where $g_\varphi\,(\mathbf{r}_\mu, \omega)$ is the Fourier transform of the propagation factor $G_\varphi\,(\mathbf{r}_\mu, t)$, that is

$$g_\varphi\,(\mathbf{r}_\mu, \omega) = \int_{-\infty}^{\infty} G_\varphi\,(\mathbf{r}_\mu, t) \exp\,(-i\omega t)\,d\omega. \tag{4.49}$$

More usually we measure the so-called power spectrum of magnetic/electric variations, which is proportional to $|\mathbf{B}\,(\omega)|^2$ or $|\mathbf{E}\,(\omega)|^2$. This spectrum determines the magnetic/electric energy distribution over frequency. In a theory the power spectrum can be correlated with the matrix of spectral densities of the stochastic process, which are defined through Fourier integral of the correlation matrix (4.44). More exactly, the power spectrum can be expressed through the sum $\psi = \psi_{xx} + \psi_{yy}$ or $\psi = \psi_{xx} + \psi_{yy} + \psi_{zz}$ depending on whether the total magnetic field is measured or only its horizontal components. As is seen from these equations, the sum $\psi = \psi_{xx} + \psi_{yy}$ is independent of azimuthal angles φ_μ. This means that this value, as well as the power spectrum, depends only on the parameters θ_μ which in turn are the functions of distances from the ground-based station to the sites of thunderstorms.

It follows from different models of the Earth-Ionosphere resonator that the set of functions g_φ contains singularities, which correspond to the Schumann resonances (e.g., see Sentman 1995). Given the functions g_φ and the lightning distributions over the globe, the set of Eqs. (4.47)–(4.49) describes the spectral correlation matrix and power spectra of the electromagnetic variations due to global thunderstorm activity. One may take into account only the most active world thunderstorm centers situated at tropics in order to give a rough estimate of this power spectrum.

One further comment should be made that the IC (intra-cloud) lightning discharges cannot, in general, be observed at distances as great as CG lightning (see Chap. 3). Although the IC lightning discharges of both positive and negative polarities are strongly prevalent over CG ones, it appears that the IC lightning contribution in the Schumann resonance frequency range is smaller as compared to CG because the CG lightning is a much stronger radiator at these frequencies due to the presence of continuing current.

There should be emphasized three important properties followed from Eqs. (4.47) and (4.48). (1) The total resonant power spectrum is an incoherent superposition of contributions from individual lightning discharges occurring over the globe. (2) The spectral matrices produced by individual thunderstorms are summed independently to provide the total spectral matrix of random process. In a similar fashion we may add variances produced by independent random variables. (3) The spectral matrix produced by an individual thunderstorm is directly proportional to $\nu_\mu \left\langle M_\mu^2 \right\rangle$.

In conclusion we note that the functions $g_\varphi \left(\theta_\mu, \omega \right)$ given by Eq. (4.37) are proportional to the function $F(\omega)$ describing the profile of individual or multiple return strokes. This implies that the spectral densities of Eqs. (4.47) and (4.48) are directly proportional to $|F(\omega)|^2$; that is, to the power spectrum generated by a single lightning discharge. In a more accurate model, one should distinguish between $F_n(\omega)$ and $F_p(\omega)$ as well as between M_n and M_p when describing the negative and positive lightning discharges, respectively.

4.3.3 A Role of Positive and Negative Cloud-to-Ground (CG) Lightning

As has already been stated, most CG lightning discharges are negative, that is, the lightning current transfer negative charges from the thundercloud to the ground. The $-$CG current is thus upward directed. Only about 5–10 % of global CG lightning activity results from $+$CG lightning, which transfer the positive charge to ground. Despite their infrequent occurrence compared to the negative flashes, the global $+$CG lightning can be of primary importance in generating the ULF/ELF power spectrum and Schumann resonances since their charge moment and continuing current are, on average, larger than those of the $-$CG lightning.

The positive ground flashes usually stand well above the global lightning population for their large current/charge moment. With the charge moment change more than 250–350 C km, that is beyond its mean value, the ratio between +CG occurrence rate and −CG occurrence rate increases in a gradual manner. Considering, as an example, the charge moment distribution for lightning flashes located in Africa (Williams et al. 2007), one can find that the ratio of the positive lightning number to negative lightning number changes from 1.7 to more than 10, as the lightning charge moment increases from 500 to 2,000 C km. It is of interest at this point to compare the contribution of +CG and −CG global lightning activities to the excitation of the Schumann resonances.

It is obvious from Fig. 3.8 that in the low-frequency band and, in particular, in the range of the Schumann resonances, the spectral density $m(\omega)$ of the current moment generated by −CG return stroke is a slowly varying function of frequency. In this frequency region Eq. (3.8) can be simplified to the following equation:

$$m(\omega) = MF(\omega) = l\left(\frac{I_3}{\omega_3} + \frac{I_4}{\omega_4 - i\omega}\right). \tag{4.50}$$

Here the parameters I_3 and I_4 determine the magnitude of the low-frequency portion of the lightning spectrum, ω_3 and ω_4 are inverse time parameters which define the total duration of the signal, l is approximately equal to the maximal lightning channel length, and $\omega = 2\pi f$. In the frequency range of the first Schumann resonances, that is $f = 7$–22 Hz, the function (4.50) varies within 20 %. In the subsequent discussion we will neglect this change; that is, the function $F(\omega)$ is considered as a constant in the frequency range of interest.

From Eq. (3.6) for the stroke current, it is clear that the function (4.50) describes the long-lasting CC (continuing current) that immediately follows the −CG peak current. In this picture the CC makes a considerable contribution to ULF/ELF region and thus can play a significant role in the generation of Schumann resonances.

According to Ballarotti et al. (2005), only 28 % of the strokes in −CG flash are followed by some kind of CC whereas almost every +CG lightning is accompanied by the CC that can reach a value of about $I_p = 5$–10 kA for periods up to $\tau_p = 5$–10 ms (Brook et al. 1982). Typically the CCs last for ten to hundreds of milliseconds (Rakov and Uman 2003). So large a CC may result in the extraordinary large charge transfer of the order of tens coulombs. The low-frequency spectrum of the +CG lightning can be written similarly to Eq. (4.50) where one should replace the parameter $I_3/\omega_3 + I_4/\omega_4$ by the mean parameter I_p/ω_p $\left(\omega_p \sim \tau_p^{-1}\right)$.

The charge moment distributions for hundreds of thousands of positive and negative lightning flashes have been measured over the globe including world thunderstorm centers located in Africa, North and South America (e.g., see Williams et al. 2007). In practice, with decreasing charge moment the lightning waveforms overlap in such a way that individual flashes cannot be resolved (Füllekrug et al. 2002). So, the empirical distribution of the distant flashes is bounded from below by the magnitude about 50 C km (Williams et al. 2007). In this study we do not come close to exploring the statistical distributions of the lightning parameters in any detail, since we focus on a rough estimate based on the mean parameters of the −CG and $+$CG lightning. To compare the contribution to the global lightning spectrum due to $+$CG lightning with that due to $-$CG, we first consider a certain thunderstorm center. In this case all the lightning is assumed to have the same propagation factor g and the same arrival angle φ so that the amplitude of the power spectrum given by Eqs. (4.47) and (4.48) can be estimated as $|\psi| \sim v \langle M^2 \rangle |F|^2$. Whence it follows that (Surkov and Hayakawa 2010)

$$\frac{|\psi_n|}{|\psi_p|} \sim \frac{v_n \langle M_n^2 \rangle |F_n|^2}{v_p \langle M_p^2 \rangle |F_p|^2} \sim \frac{v_n I_n^2 \omega_p^2 l_n^2}{v_p I_p^2 \omega_n^2 l_p^2}, \tag{4.51}$$

where $I_n/\omega_n = I_3/\omega_3 + I_4/\omega_4$, and the subscripts n and p stand for the $-$CG and $+$CG lightning, correspondingly. The same equation can serve as a rough estimate of the ratio $|\psi_n|/|\psi_p|$ as the global lightning activity is considered.

On account of about an order-of-magnitude dominance in occurrence rate of negative discharges over positive ones we can estimate a ratio between global positive and global negative mean frequencies of the strokes as $v_p/v_n \approx 0.05$ (Shalimov and Bösinger 2008). As we have noted above, the numerical values of other parameters are as follows: $I_p/\omega_p \approx 25$–$100\,\mathrm{C}$, $I_n/\omega_n \approx 5.6\,\mathrm{C}$, and $l_n \approx l_p$. Substituting the above numerical parameters of the global CG lightning into Eq. (4.51) yields $|\psi_n|/|\psi_p| \sim (0.06$–$1)$.

This numerical estimate has shown that the $+$CG lightning can make a considerable contribution to the low-frequency part of the background spectra provided by the global lightning activity. Furthermore this contribution can be even greater than that due to $-$CG lightning in spite of the infrequent occurrence of the $+$CG compared to the $-$CG lightning. This also implies that $+$CG lightning can play a crucial role in the generation of Schumann resonances (Surkov and Hayakawa 2010).

Before discussing this, we must point out that the above estimate of Eq. (4.51) basically depends on the parameters of the CCs which greatly affect the low-frequency part of the lightning spectra. Taking into account that the CCs carry the negative/positive charge of the order of $q_n = -I_n/\omega_n$ or $q_p = I_p/\omega_p$, we can reduce Eq. (4.51) to

$$\frac{|\psi_n|}{|\psi_p|} \sim \frac{v_n (q_n l_n)^2}{v_p (q_p l_p)^2}, \tag{4.52}$$

where the parameters $q_n l_n$ and $q_p l_p$ play a role of mean value of the charge moment change due to the CC. As we have noted above, the positive CC typically transfers the total charge as high as $q_p \approx 25$–$100\,\mathrm{C}$ (Rakov 2003), an order of magnitude greater than that due to the negative CC, that results in $q_p l_p \gg q_n l_n$. Although the flash rate $v_n \gg v_p$, the parameters of global lightning are such that $|\psi_n|$ may be even smaller than that of $|\psi_p|$.

As would be expected, considering the important role played by the CC in the generation of Schumann resonances, the parameter v_n in Eqs. (4.51) and (4.52) can depend only on those negative strokes that have a CC. As we have noted above, only 28 % of the negative strokes in a flash are followed by some kind of CC whereas practically each of positive flashes is followed by the CC. Considering this fact, the ratio $|\psi_n| / |\psi_p|$ can be even smaller than above estimate.

This suggests that our present insight into the major source of the Schumann resonances is far from complete. In particular, we may assume that the diurnal and seasonal variations in intensity of the Schumann resonances can be dependent on the variations of the global +CG lightning activity.

4.3.4   Observations of Schumann Resonances

The Schumann resonances dominate the natural background of the electromagnetic noise spectrum over the frequency range from 6 to 50 Hz and can, in principle, be detected from any place of the Earth away from thunderstorms and man-made electromagnetic sources. This fact is consistent with the assumption that the global lightning activity of the planet is the major source of excitation of the Schumann resonances. The Schumann resonances can therefore serve as a planetwide thunderstorm monitor, which makes it possible to interpret the Schumann resonances as a unique signature of global lightning activity.

A typical power density spectrum which contains the Schumann resonances is shown in Fig. 4.2. As is seen from this figure, the amplitudes of the resonances are very weak and can be easily masked by nearby lightning or other unrelated sources of the ULF electromagnetic noise. Since the amplitude of the resonances is of the order of several pT/Hz$^{1/2}$, the low noise and highly sensitive induction coils are typically used.

Diurnal variations in modulation of the Schumann resonances is approximately consistent with the observed variations of the global lightning activity, which tends to maximize in the late afternoon in each of world thunderstorm centers. The lightning activity reaches a peak value in Caribbean Basin and South-East Asia, in Central Africa, and the Americas, at approximately 1,000, 1,600, and 2,200 UT, respectively, that results in a corresponding amplification of the resonance power spectra at this time. The electric and magnetic field spectra that are displayed in Fig. 4.4a,b illustrate the diurnal variations of the Schumann resonances (Sentman 1995). The average power spectra during the daytime maximum and during the nighttime minimum are shown with dotted lines. The end-to-end system noise spectra for (a) a ball-over-plane antenna and (b) a magnetic induction coil are shown at the bottom.

The power spectra of the Schumann resonances exhibit day-to-day variations in the diurnal profiles and there may be seasonal effect as well. It is generally believed that these variations reflect a corresponding variability in the totality of the

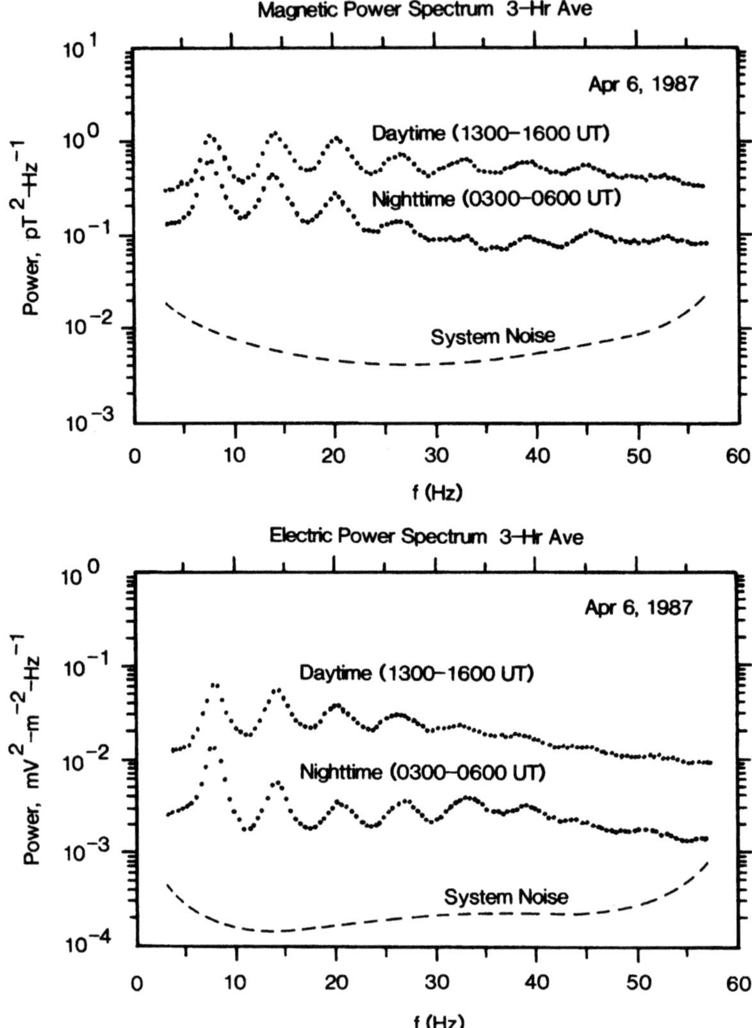

Fig. 4.4 Typical average 3-h power spectra of the vertical electric and horizontal magnetic components over the passband from 3 to 57 Hz. The Schumann resonances were recorded in the measurements at Table Mountain, California during intervals of diurnal maximum intensity (1,300–1,600 UT) and minimum intensity (0300–0600 UT). Taken from Sentman (1995)

global lightning activity (e.g., Clayton and Polk 1974). In spite of global character of the Schumann resonances excited by the totality of global lightning, the striking difference may occur between the diurnal spectra simultaneously recorded at widely separated locations. Sentman and Fraser (1991) have demonstrated such a difference between diurnal profiles recorded in California and Western Australia (distance about 14,800 km) during 8-day interval in April 1990. It has been hypothesized that the difference is due to a modulation related to the local height of the ionosphere at respective observing sites. It is known that the day–night asymmetry in the height of the D region is mainly due to diurnal variations in the solar UV and Roentgen radiation. On the basis of this assumption, Sentman and Fraser (1991) have shown that the function describing the global lightning may be determined once the diurnal modulation factor is removed. Further improvement has been developed by Nickolaenko et al. (2010) in order to separate the UT (universal time) and LT (local time) variations in Schumann resonance data.

As would be expected, considering the major role played by the global lightning in the excitation of Earth-Ionosphere resonance cavity, the Schumann resonances exhibit a very high degree of spatial and temporal coherence. For one example, Holtham and McAskill (1988) and Sweeney (1989) have examined the cross-spectra of the lowest Schumann resonances recorded across baselines of 1,100 and 480 km. The results show the coherence in the lowest modes at a level about 90–98 %, which means that the lightning over oceans and in the regions outside the world thunderstorm centers plays a relatively minor role in the excitation of Schumann resonances as compared to the contributions of these three continental thunderstorm centers.

A single impulse from a very large discharge, however, can sporadically excite the Schumann resonances to an amplitude greater than that due to incoherent sum of random fields caused by the global lightning activity. Such large discharges can serve as the sources of transients called Q-bursts, which can be simultaneously detected at very far distances (Ogawa et al. 1967; Sentman 1989; Nickolaenko et al. 2010). It is usually the case that a Q-burst is a damping quasi-periodic oscillation at one of the eigenfrequencies, more usually at the lowest frequency near 8 Hz. Figure 4.5 shows typical recordings of the Q-bursts recorded in California in 1985 (Sentman 1989). The power spectrum of the Q-bursts is shown to be concentrated at the Schumann resonance frequencies. As is seen from Fig. 4.5 the amplitude of the oscillations decreases exponentially for ~ 0.5 s, that is, at a rate corresponding to the Q-factor of the Earth-Ionosphere cavity.

As we have noted above, the probability of +CG lightning dominates that of −CG lightning when the charge moment magnitude exceeds several hundreds C km. These huge +CG flashes can play a significant role in sporadically exciting the Schumann resonances since this kind of flashes stands well above the global lightning population for short periods of time. Perhaps, the same conclusion can be applied to the variations of atmospheric electric field. Recently Füllekrug (2004) has reported that the intense positive lightning discharges result in a weak decrease in intensity of the global atmospheric electric field.

Recently a distinctive pattern of Schumann resonances has been measured on the low earth orbiting C/NOFS satellite within the altitude region of 400–850 km sampled by the satellite (Simoes et al. 2011). The C/NOFS satellite was equipped with instrumentation for measuring three-component electric field. Typical amplitudes of the first peak was $\sim 0.25\,\mu\mathrm{Vm}^{-1}\,\mathrm{Hz}^{-1/2}$ whereas the sensitivity of the electric field onboard measurements was $\sim 10\,\mathrm{nVm}^{-1}\,\mathrm{Hz}^{-1/2}$ in ELF range. These amplitudes are about three orders of magnitude lower than that observed

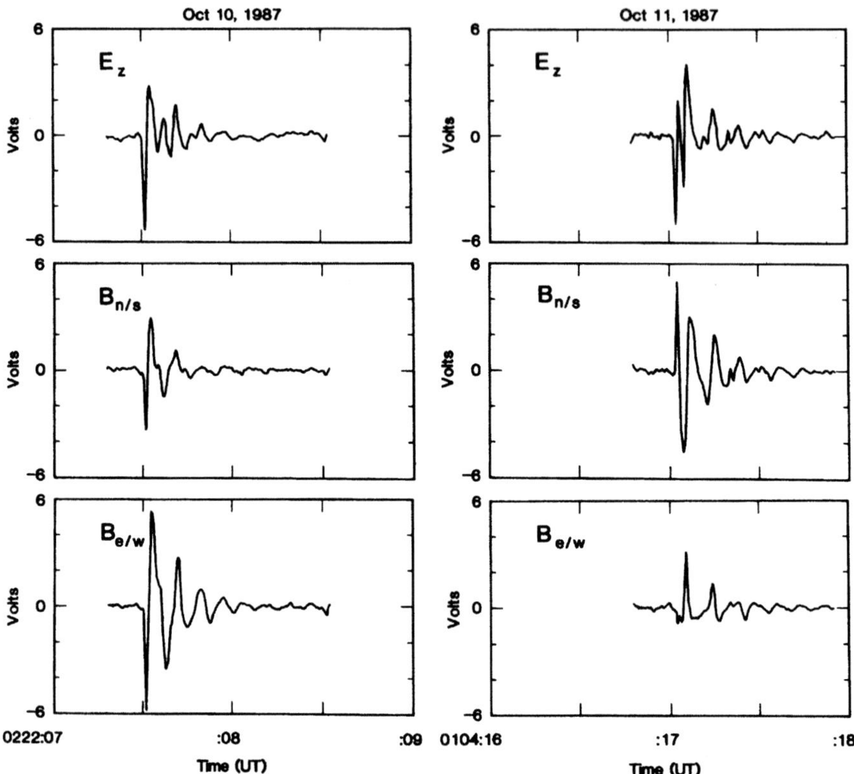

Fig. 4.5 Typical Q-bursts that are believed to be signature of the Earth-Ionosphere cavity excitation due to extremely large lightning flashes. The temporal dependencies of the vertical electric component E_z and orthogonal magnetic components $B_{n/s}$ and $B_{e/w}$ were recorded on October 10 (*left*) and October 11 (*right*), 1987. Taken from Sentman (1989)

by the ground-based station. The resonant frequencies measured on the satellite in the range 8–34 Hz differ from that measured on the ground by no more than several percent. Interestingly enough, the amplitudes of the nighttime spectra were on average one order of magnitude greater than the amplitudes of the daytime spectra. To explain this effect, one may suppose that the ionosphere under nighttime condition is more transparent to the resonant ELF radiation penetrating through a conductive layer of the ionosphere and coming into the outer space (Surkov et al. 2013). This result testifies to the fact that the satellites can be used for the study of lightning parameters in the range of Schumann resonances.

Immediately below the frequency interval covered by the Schumann resonances are the spectra of the ionospheric and magnetospheric Alfvén resonators. This kind of resonances is referred to as the class of field-line resonances for shear Alfvén waves. In the next sections we study in earnest the physical mechanism of this kind of oscillations.

Appendix A: Spherical Bessel Functions

The spherical Bessel functions are the solutions of the second order differential equation (Abramowitz and Stegun 1964)

$$z^2 w_n'' + 2z w_n' + \left[z^2 - n\,(n+1) \right] w_n = 0, \tag{4.53}$$

where z and $w\,(z)$ are generally complex and the prime denotes the derivative with respect to z. The spherical Bessel functions of the third kind are the partial and linearly independent solutions of Eq. (4.53). They can be expressed as follows:

$$h_n^{(1)}\,(z) = -i\,z^n \left(-\frac{d}{z\,dz} \right)^n \left(\frac{\exp\,(i\,z)}{z} \right),$$

$$h_n^{(2)}\,(z) = i\,z^n \left(-\frac{d}{z\,dz} \right)^n \left(\frac{\exp\,(-i\,z)}{z} \right), \tag{4.54}$$

where $n = 0, 1, 2, \ldots$

Legendre Polynomials

The Legendre polynomials, $P_n\,(\cos\theta)$, are the solutions of the differential equation

$$\frac{d_\theta\,(\sin\theta\,d_\theta\,P_n)}{\sin\theta} + n\,(n+1)\,P_n = 0, \tag{4.55}$$

where $d_\theta = d/d\theta$ denotes the derivative with respect to θ. These polynomials are defined as follows:

$$P_n\,(\cos\theta) = \frac{1}{2^n n!} \frac{d^n}{(d\cos\theta)^n} \left(\cos^2\theta - 1 \right)^n. \tag{4.56}$$

The Legendre polynomials are mutually orthogonal functions on the segment $[-\pi, \pi]$, that means that

$$\int_0^\pi P_n\,(\cos\theta)\,P_m\,(\cos\theta)\,\sin\theta\,d\theta = 0, \quad \text{if } n \neq m, \tag{4.57}$$

and

$$\int_0^\pi [P_n\,(\cos\theta)]^2 \sin\theta\,d\theta = \frac{2}{2n+1}. \tag{4.58}$$

In the text we need to develop the boundary condition (4.12) as a series in Legendre polynomials. For that one should use the following representation of the delta-function

$$\frac{\delta(\theta)}{\sin\theta} = \sum_{n=0}^{\infty} C_n P_n(\cos\theta). \qquad (4.59)$$

In order to obtain the undetermined coefficients C_n, one should multiply Eq. (4.59) by $\sin\theta P_m(\cos\theta)$ and then integrate both sides of this equation between zero and π. On account of Eqs. (4.57) and (4.58) one can find

$$C_n = \left(n + \frac{1}{2}\right) P_n(1) = n + \frac{1}{2}. \qquad (4.60)$$

In the course of this text, we also need the following sum

$$\sum_{n=1}^{\infty} \frac{(2n+1)}{n(n+1)} P_n(\cos\theta) = -\ln\left(\sin^2\frac{\theta}{2}\right) - 1. \qquad (4.61)$$

Rearrangement of Solution

Before rearranging Eq. (4.33), it should be noted that this equation contains the delta-function of θ because the source function, that is, the current density j_s includes the same factor, i.e., $\delta(\theta)$. In order to eliminate the delta-function, one should rearrange the sum in Eq. (4.33). Taking the notice of

$$\frac{\omega_n^2}{\omega_n^2 - \omega^2} = \frac{\omega^2}{\omega_n^2 - \omega^2} + 1 \qquad (4.62)$$

and accounting of Eqs. (4.59) and (4.60), we obtain

$$\sum_{n=0}^{\infty} \frac{\omega_n^2(2n+1)}{(\omega_n^2 - \omega^2)} P_n(\cos\theta) = \sum_{n=0}^{\infty} \frac{\omega^2(2n+1)}{(\omega_n^2 - \omega^2)} P_n(\cos\theta) + \frac{2\delta(\theta)}{\sin\theta}. \qquad (4.63)$$

If the delta-function is omitted and ω^2 in the sums is replaced by the $\omega^2 + ic\omega Z(\omega)/d$, Eq. (4.33) is reduced to (4.38).

Appendix B: Mean Value and Correlation Function of Random Process

In this appendix the process of lightning discharges is treated as a sequence of independent random events, which obeys the Poisson random distribution. This implies that the elementary probability, dP, of the lightning origin from the moment t till $t + dt$ is proportional to dt and does not depend on t, that is $dP = v dt$, where v stands for the mean number of the lightning discharges per unit time. Hence the probability, $P(n)$, of appearance of n lightning discharges during a time interval $(0, T)$ is given by a Poisson distribution

$$P(n) = \frac{\langle n \rangle^n \exp(-\langle n \rangle)}{n!}, \tag{4.64}$$

where $\langle n \rangle = vT$ stands for the mean number of the lightning discharges per time T.

We first consider a single thunderstorm as a source of lightning activity. For simplicity, we shall omit the subscript μ for the number of this thunderstorm. Let $\mathbf{b}(\mathbf{r}, t)$ be the net magnetic field at the point $\mathbf{r}$ originated from the lightning discharges happened at random moments. Here we ignore the spatial distribution of the lightning discharges in a thunderstorm area since the magnetic field is measured far away form the recording station. For reasons of convenience, we shall therefore omit the argument $\mathbf{r}$ of the function. The mean value of the random value $\mathbf{b}(t)$ can thus be written as

$$\langle \mathbf{b}(t) \rangle = \sum_{n=0}^{\infty} \left\langle {}^{(n)}\mathbf{b}(t) \right\rangle P(n), \tag{4.65}$$

where $P(n)$ is the Poisson distribution (4.64). The angular brackets on the right-hand side of (4.65) denote a conditional mean of the function $\mathbf{b}(t)$, which is the mean value of $\mathbf{b}(t)$ under the condition that there were n lightning flashes during the interval $(0, T)$ (Rytov et al. 1978):

$$\left\langle {}^{(n)}\mathbf{b}(t) \right\rangle = \frac{1}{T} \sum_{m=1}^{n} \int_0^T \mathbf{b}(t - t_m) \, dt. \tag{4.66}$$

Suppose now that the random moment of the lightning discharge/impulse occurrence, t_m, and the magnitude of current moment M_m, are statistically independent and their probability distributions are independent of the impulse number n. On account of Eq. (4.42) we get

$$\left\langle {}^{(n)}\mathbf{b}(t) \right\rangle = \sum_{m=1}^{n} \langle M_m \rangle \langle \mathbf{G}(t - t_m) \rangle, \tag{4.67}$$

where

$$\langle \mathbf{G}\,(t - t_m)\rangle = \frac{1}{T}\int_0^T \mathbf{G}\,(t - t_m)\,dt_m. \tag{4.68}$$

Assuming for the moment that the time interval T in Eq. (4.68) is much greater than the duration of the lightning discharge and making allowance for $\langle M_m\rangle = \langle M\rangle$, yields

$$\left\langle {}^{(n)}\mathbf{b}\,(t)\right\rangle = \frac{n\,\langle M\rangle}{T}\int_{-\infty}^{\infty} \mathbf{G}\,(t')\,dt'. \tag{4.69}$$

It should be noted that the mean value (4.69) is independent of time. This is due to the fact that the stationary random process has been assumed. Substituting Eq. (4.69) for $\left\langle {}^{(n)}\mathbf{b}\,(t)\right\rangle$ into Eq. (4.65), and taking into account that

$$\sum_{n=0}^{\infty} nP\,(n) = \langle n\rangle = vT, \tag{4.70}$$

gives the mean value of $\langle \mathbf{b}\,(t)\rangle$

$$\langle \mathbf{b}\,(t)\rangle = v\,\langle M\rangle \int_{-\infty}^{\infty} \mathbf{G}\,(t')\,dt'. \tag{4.71}$$

To treat a correlation matrix of the random process in Eq. (4.44), we consider a steady stochastic process such as Poisson random process. This means that the correlation matrix (4.44) depends only on the time difference $\tau = t' - t$, that is,

$$\Psi_{jk}\,(\tau) = \left\langle B_j\,(t)\,B_k\,(t + \tau)\right\rangle - \left\langle B_j\,(t)\right\rangle \left\langle B_k\,(t)\right\rangle, \tag{4.72}$$

where the mean values $\left\langle B_j\,(t)\right\rangle$ are given by Eq. (4.46) and $j, k = x, y, z$. Substituting Eq. (4.43) for $\mathbf{B}\,(t)$ into Eq. (4.72) leads to

$$\left\langle B_j\,(t)\,B_k\,(t + \tau)\right\rangle = \sum_{\mu=1}^{N}\sum_{\rho=1}^{N}\left\langle b_{j,\mu}\,(\mathbf{r}_\mu, t)\,b_{k,\rho}\,(\mathbf{r}_\rho, t + \tau)\right\rangle, \tag{4.73}$$

where $\mathbf{r}_\mu$ and $\mathbf{r}_\rho$ denote the coordinates of the corresponding thunderstorms.

If $\mu \neq \rho$, then

$$\left\langle b_{j,\mu}\,(\mathbf{r}_\mu, t)\,b_{k,\rho}\,(\mathbf{r}_\rho, t + \tau)\right\rangle = \left\langle b_{j,\mu}\,(\mathbf{r}_\mu, t)\right\rangle \left\langle b_{k,\rho}\,(\mathbf{r}_\rho, t)\right\rangle, \tag{4.74}$$

since the lightning discharges processes associated with different thunderstorms are supposed to be statistically independent of each other.

Consider now the case of $\mu = \rho$. This implies that both magnetic variations, $b_{j,\mu}\left(\mathbf{r}_\mu, t\right)$ and $b_{k,\mu}\left(\mathbf{r}_\mu, t + \tau\right)$ are taken at the same thunderstorm. For the reason of convenience, we can therefore omit the inferior index μ and argument $\mathbf{r}_\mu$. Since we assume that the lightning discharges form a random Poisson process, the corresponding terms on the right-hand side of Eq. (4.73) can be written analogous to Eq. (4.65), that is

$$\langle b_j\left(t\right) b_k\left(t + \tau\right)\rangle = \sum_{n=0}^{\infty} \left\langle {}^{(n)}b_j\left(t\right) b_k\left(t + \tau\right)\right\rangle P\left(n\right), \tag{4.75}$$

where the angular brackets denote a conditional mean of the function

$$b_j\left(t\right) b_k\left(t + \tau\right) = \sum_{m_1=1}^{n} \sum_{m_2=1}^{n} M_{m_1} M_{m_2} G_j\left(t - t_{m_1}\right) G_k\left(t + \tau - t_{m_2}\right), \tag{4.76}$$

that is, the mean value of this function under the condition that there were n flashes during the interval $(0, T)$. Here we made use of Eq. (4.42).

As the components of magnetic field are represented in the spherical coordinate system in which the source is located on the polar axis z' shown in Fig. 4.3 from the right of the main image, the components of the magnetic field in the local Cartesian coordinate system fixed at a ground-based station can be found through the transformation matrix $\hat{\mathbf{A}}_\mu$ given by Eq. (4.41). For example, if we choose first to study the component $\Psi_{xx}\left(\tau\right)$ the function G_x is given by

$$G_x = G_\varphi \sin\varphi - G_\theta \cos\varphi. \tag{4.77}$$

For the sake of generality, we shall consider all the components b_r, b_θ, and b_φ though only the component b_φ given by Eq. (4.36) is nonzero in the simple model of the waveguide treated above. As we shall see subsequently the TE mode, which contains the components b_r and b_θ, can be excited as well due to the generation of Hall currents in the E-layer of the ionosphere.

Taking into account the definition of a conditional mean (4.66) and assuming that the moment T is much greater than the duration of the lightning discharge, we obtain

$$\langle G_{k_1}\left(t - t_{m_1}\right) G_{k_2}\left(t + \tau - t_{m_2}\right)\rangle = \begin{cases} \frac{1}{T^2} h_{k_1} h_{k_2}, & \text{if } m_1 \neq m_2 \\ \frac{1}{T} \xi_{k_1 k_2}\left(\tau\right), & \text{if } m_1 = m_2 \end{cases}, \tag{4.78}$$

where the subscripts k_1 and k_2 denote θ or φ. Here the cross-correlation function $\xi_{k_1 k_2}\left(\tau\right)$ takes the following form

$$\xi_{k_1 k_2}(\tau) = \int_{-\infty}^{\infty} G_{k_1}(t') \, G_{k_2}(t' + \tau) \, dt' \tag{4.79}$$

and the constants $h_{k_1 k_2}$ are given by

$$h_{k_1 k_2} = \int_{-\infty}^{\infty} G_{k_1, k_2}(t') \, dt'. \tag{4.80}$$

In practice, the mean value of the magnetic variations generated by the lightning discharge is of little importance since it is close to zero. This means that integrals in Eq. (4.80) tend to zero and thus the constants h_θ and h_φ are negligible.

Using the statistical independence of random values M_m yields

$$\langle M_{m_1} M_{m_2} \rangle = \begin{cases} \langle M_{m_1} \rangle \langle M_{m_2} \rangle = \langle M_m \rangle^2, & \text{if } m_1 \neq m_2 \\ \langle M_{m_1}^2 \rangle = \langle M_m^2 \rangle, & \text{if } m_1 = m_2 \end{cases} \tag{4.81}$$

Taking into account Eqs. (4.78)–(4.81) and making allowance for $\langle M_m \rangle = \langle M \rangle$ and $\langle M_m^2 \rangle = \langle M^2 \rangle$, we arrive at

$$\left\langle {}^{(n)}b_x(t) \, b_x(t + \tau) \right\rangle = \frac{n \langle M^2 \rangle}{T} \left[\xi_{\theta\theta}(\tau) \cos^2 \varphi + \xi_{\varphi\varphi}(\tau) \sin^2 \varphi - \xi_c(\tau) \sin 2\varphi \right]$$
$$+ \frac{\left(n^2 - n\right) \langle M \rangle^2}{T^2} \left[h_\theta \cos \varphi - h_\varphi \sin \varphi \right]^2. \tag{4.82}$$

where

$$\xi_c(\tau) = \frac{1}{2} \left\{ \xi_{\theta\varphi}(\tau) + \xi_{\varphi\theta}(\tau) \right\} \tag{4.83}$$

Taking the notice of a useful relation followed from the Poisson distribution:

$$\langle n^2 - n \rangle = \langle n \rangle^2 = v^2 T^2, \tag{4.84}$$

and substituting Eq. (4.82) into Eq. (4.75) yields

$$\langle b_{x,\mu}(t) \, b_{x,\mu}(t + \tau) \rangle = v_\mu \left\langle M_\mu^2 \right\rangle \left[\xi_{\theta\theta}(\mathbf{r}_\mu, \tau) \cos^2 \varphi_\mu + \xi_{\varphi\varphi}(\mathbf{r}_\mu, \tau) \sin^2 \varphi_\mu \right.$$
$$\left. - \xi_c(\mathbf{r}_\mu, \tau) \sin 2\varphi_\mu \right] + \langle b_{x,\mu}(\mathbf{r}_\mu, t) \rangle^2. \tag{4.85}$$

Here we have restored the index μ and arguments of the functions.

Substituting Eqs. (4.74) and (4.85) into (4.73) and rearranging, we can find the mean value $\langle B_x(t) B_x(t+\tau)\rangle$. Substituting the former value into Eq. (4.72), we come to

$$\Psi_{xx}(\tau) = \sum_{\mu=1}^{N} v_\mu \left\langle M_\mu^2 \right\rangle \{ \xi_{\theta\theta}(\mathbf{r}_\mu, \tau) \cos^2 \varphi_\mu$$

$$+ \xi_{\varphi\varphi}(\mathbf{r}_\mu, \tau) \sin^2 \varphi_\mu - \xi_c(\mathbf{r}_\mu, \tau) \sin 2\varphi_\mu \} , \tag{4.86}$$

Similarly, one can use an analogous procedure to obtain the matrix component $\Psi_{yy}(\tau)$

$$\Psi_{yy}(\tau) = \sum_{\mu=1}^{N} v_\mu \left\langle M_\mu^2 \right\rangle \{ \xi_{\theta\theta}(\mathbf{r}_\mu, \tau) \sin^2 \varphi_\mu$$

$$+ \xi_{\varphi\varphi}(\mathbf{r}_\mu, \tau) \cos^2 \varphi_\mu + \xi_c(\mathbf{r}_\mu, \tau) \sin 2\varphi_\mu \} . \tag{4.87}$$

In a similar fashion one can find other components of the correlation matrix.

Fourier integral of the correlation functions (4.86) and (4.87) defines the spectral densities, $\psi_{xx}(\omega)$ and $\psi_{yy}(\omega)$, of the stochastic process

$$\Psi_{xx,yy}(\tau) = \int_{-\infty}^{\infty} \psi_{xx,yy}(\omega) \exp(i\omega\tau) \, d\omega, \tag{4.88}$$

Taking Fourier integrals of the functions G_θ and G_ϕ in the same form, using the convolution transform/faltung theorem

$$\int_{-\infty}^{\infty} G_{k_1}(t) G_{k_2}(t+\tau) \, dt$$

$$= 2\pi \int_{-\infty}^{\infty} g_{k_1}(\omega) g_{k_2}^*(\omega) \exp(i\omega\tau) \, d\omega, \tag{4.89}$$

and combining Eqs. (4.79), (4.83), (4.88), and (4.89), we finally arrive at

$$\psi_{xx}(\omega) = 2\pi \sum_{\mu=1}^{N} v_\mu \left\langle M_\mu^2 \right\rangle \left[\cos^2 \varphi_\mu \left| g_\theta(\mathbf{r}_\mu, \omega) \right|^2 \right.$$

$$\left. + \sin^2 \varphi_\mu \left| g_\varphi(\mathbf{r}_\mu, \omega) \right|^2 - g_c(\mathbf{r}_\mu, \omega) \sin 2\varphi_\mu \right], \tag{4.90}$$

and

$$\psi_{yy}(\omega) = 2\pi \sum_{\mu=1}^{N} v_\mu \left\langle M_\mu^2 \right\rangle \left[\sin^2 \varphi_\mu \left| g_\theta \left(\mathbf{r}_\mu, \omega \right) \right|^2 \right.$$

$$\left. + \cos^2 \varphi_\mu \left| g_\varphi \left(\mathbf{r}_\mu, \omega \right) \right|^2 + g_c \left(\mathbf{r}_\mu, \omega \right) \sin 2\varphi_\mu \right], \qquad (4.91)$$

where $g_{\theta,\mu}$ and $g_{\phi,\mu}$ are Fourier transforms of the functions $G_{\theta,\mu}$ and $G_{\phi,\mu}$, respectively, and

$$g_c = \frac{1}{2} \left\{ g_\varphi \left(\mathbf{r}_\mu, \omega \right) g_\theta^* \left(\mathbf{r}_\mu, \omega \right) + g_\theta \left(\mathbf{r}_\mu, \omega \right) g_\varphi^* \left(\mathbf{r}_\mu, \omega \right) \right\}. \qquad (4.92)$$

Here the symbol asterisk denotes complex-conjugate.

References

Abramowitz M, Stegun IA (1964) Handbook of mathematical functions. National Bureau of Standards, Applied Mathematics Series, vol. 55. Government Printing Office, Washington, DC

Ballarotti MG, Saba MMF, Pinto O Jr (2005) High-speed camera observations of negative ground flashes on a millisecondscale. Geophys Res Lett 32:L23802. doi:10.1029/2005GL023889

Balser M, Wagner CA (1960) Observation of Earth – ionosphere cavity resonances. Nature 188:638–641

Balser M, Wagner CA (1962) Diurnal power variations of the Earth – ionosphere cavity modes and their relationships to worldwide thunderstorm activity. J Geophys Res 67(2):619–625

Belyaev PP, Polyakov SV, Rapoport VO, Trakhtengertz VY (1989) The theory of formation of the resonance spectrum structure of the atmospheric electromagnetic noise background in the range of short-period geomagnetic pulsations. Radiophys Quantum Electron (Izvestiya Vuzov. Radiofizika) 32:663–672

Bliokh PV, Nickolaenko AP, Filippov YuF (1980) In: Jones DLl (ed) Schumann resonances in the Earth – Ionosphere cavity. Peter Peregrinus, Oxford

Brook M, Nakano M, Krehbiel P, Takeuti T (1982) The electrical structure of the Hokuriku winter thunderstorm. J Geophys Res 87:1207–1215

Budden KG (1962) The waveguide mode theory of wave propagation. Prentice-Hall, Englewood Cliffs

Clayton M, Polk C (1974) Diurnal variations and absolute intensity of world-wide lightning activity, September 1970 to May 1971. In: Proceedings of the conference electrical processes in atmosphere, Garmisch-Partenkirchen

Füllekrug M (2004) The contribution of intense lightning discharges to the global atmospheric electric circuit during April 1998. J Atmos Solar-Terr Phys 66:1115–1119

Füllekrug M, Price C, Yair Y, Williams ER (2002) Oceanic lightning. Ann Geophys 20:133–137

Galejs J (1961) Terrestrial extremely low frequency noise spectrum in the presence of exponential ionospheric conductivity profiles. J Geophys Res 66:2789–2793

Galejs J (1965) Schumann resonances. J Res NBS Radio Sci 69D(8):1043–1055

Galejs J (1972) Terrestrial propagation of long electromagnetic waves. Pergamon Press, New York

Hayakawa M, Ohta K, Baba K (1994) Wave characteristic of tweek atmospherics deduced from the direction finding measurement and theoretical interpretations. J Geophys Res 99:10733–10743

Hayakawa M, Nickolaenko AP, Shvets AV, Hobara Y (2011) Recent studies of Schumann resonance and ELF transients. In: Wood MD (ed) Lightning: properties, formation and types, Chapter 3. Nova Science Publishers, Hauppauge, pp 39–71

Holtham PM, McAskill BJ (1988) The spatial coherence of Schumann activity in the polar casp. J Atmos Terr Phys 50:83–92

Jones DLl (1964) The calculation of the Q-factors and frequencies of earth-ionosphere cavity resonances for a two-layer ionospheric model. J Geophys Res 69:4037–4046

Mitra AP (1974) Ionospheric effects of solar flares. D. Reidel Pub. Comp., Dordrecht, 294 p

Nickolaenko AP (1981) One-dimensional distribution function of the vertical electric component in the ELF terrestrial radio noise. Radiophys Quantum Electron (Izvestiya Vuzov, Radiofizika) 24:34–42

Nickolaenko AP, Hayakawa M (2002) Resonances in the Earth-Ionosphere cavity. Kluwer Acad. Pub., Dordrecht

Nickolaenko AP, Hayakawa M, Hobara Y (2010) Q-bursts: natural ELF radio transients. Surv Geophys 31(4):409–425. doi:10.1007/s10712-010-9096-9

Ogawa T, Fraser-Smith AC, Gendrin R, Tanaka Y, Yasuhara M (1967) Worldwide simultaneity of occurrence of a Q-type ELF burst in the Schumann resonance frequency range. J Geomagn Geoelectr 19:377

Ogawa T, Murakami Y (1973) Schumann resonances frequencies and the conductivity profiles in the atmosphere. Contr Geophys Inst Kyoto Univ 13:13–20

Outsu J (1960) Numerical study of tweeks based on waveguide mode theory. Proc Res Inst Atmospherics Nagoya Univ 7:58–71

Polk C (1969) Relation of ELF noise and Schumann resonances to thunderstorm activity. In: Coronoti SC, Hughes J (eds) Planetary electrodynamics. Gordon and Breach, New York, pp 55–83

Raemer HR (1961a) On the spectrum of terrestrial radio noise at ELF. J Res NBS Radio Sci 65D:582

Raemer HR (1961b) On extremely low frequency spectrum of the Earth-Ionosphere cavity response to electrical storms J Geophys Res 66:1580–1584

Rakov VA (2003) A review of positive and bipolar lightning discharges. Bull Am Meteorol Soc 84:767–775

Rakov VA, Uman MA (2003) Lightning: physics and effects. Cambridge University Press, Cambridge, 687 pp

Rodger CJ (1999) Red sprites, upward lightning, and VLF perturbations. Rev Geophys 37:317–336

Rytov SM, Kravtsov YaA, Tatarsky VI (1978) Introduction to statistical radiophysics, P. 2, random fields. Nauka, Moscow (in Russian)

Schumann WO (1952a) On the free oscillations of a conducting sphere which is surrounded by an air layer and an ionospheric shell. Z Naturforschaftung 7a:149–154 (in German)

Schumann WO (1952b) On the damping of electromagnetic self-oscillations in the system earth-air-ionosphere. Z Naturforschaftung 7a:250–252 (in German)

Schumann WO (1957) Electrical self-oscillations of the cavity earth-air-ionosphere. Z Angew Phys 9:373–378 (in German)

Sentman DD (1987) Magnetic elliptical polarization of Schumann resonances. Radio Sci 22:595–606

Sentman DD (1989) Detection of elliptical polarization and mode splitting in discrete Schumann resonance excitation. J Atmos Terr Phys 51:507–519

Sentman DD (1995) Schumann resonances. In: Volland H (cd) Handbook of atmospheric electrodynamics, vol 1. CRC Press, Boca Raton, pp 267–310

Sentman DD, Fraser BJ (1991) Simultaneous observations of Schumann resonances in California and Australia: evidence for intensity modulation by the local height of the D region. J Geophys Res 96:15973–15984

Shalimov S, Bösinger T (2008) On distant excitation of the ionospheric Alfvén resonator by positive cloud-to-ground lightning discharges. J Geophys Res 113:A02303. doi:10.1029/2007JA012614

Shvets AV, Hayakawa M (1998) Polarization effects for tweek propagation. J Atmos Solar-Terr Phys 60:461–469

Simoes FA, Pfaff RF, Freudenreich HT (2011) Satellite observations of Schumann resonances in the Earth's ionosphere. Geophys Res Lett 38. doi:10.1029/2011GL049668

Surkov VV, Hayakawa M (2010) Schumann resonances excitation due to positive and negative cloud-to-ground lightning. J Geophys Res 115:D04101. doi:10.1029/2009JD012539

Surkov VV, Molchanov OA, Hayakawa M, Fedorov EN (2005) Excitation of the ionospheric resonance cavity by thunderstorms. J Geophys Res 110:A04308. doi:10.1029/2004JA010850

Surkov VV, Hayakawa M, Schekotov AY, Fedorov EN, Molchanov OA (2006) Ionospheric Alfvén resonator excitation due to nearby thunderstorms. J Geophys Res 111:A01303. doi:10.1029/2005JA011320

Surkov VV, Nosikova NS, Plyasov AA, Pilipenko VA, Ignatov VN (2013) Penetration of Schumann resonances into the upper ionosphere. J Atmos Solar-Terr Phys 97:65–74. doi:10.1016/j.jastp.2013.02.015

Sweeney JJ (1989) An investigation of the usefulness of extremely low-frequency electromagnetic measurements for treaty verification. Report No. UCRL-53899, Lawrence Livermore Laboratory, Livermore

Uman MA (1987) The lightning discharge. Academic, New York

Uman MA, Krider EP (1982) A review of natural lightnings: experimental data and modeling. IEEE Trans. Electromagn Compat 24:79–112

Wait JR (1960) On the propagation of ELF radio waves and the influence of a non-homogeneous ionosphere. J Geophys Res 65:597–600

Wait JR (1962) On the propagation of VLF and ELF radio waves when the ionosphere is not sharply bounded. J Res Nat Bur Stand Sect D 66:53–61

Wait JR (1972) Electromagnetic waves in the stratified media, 2nd edn. Pergamon Press, New York

Wait JR (1974) Guest editorial: historical background and introduction to the special issue on ELF communications. IEEE Trans Commun COM-22:353–354

Weissman MB (1988) $1/f$ noise and other slow, nonexponential kinetics in condensed matter. Rev Modern Phys 60:537–571

Williams E, Downes E, Boldi R, Lyons W, Heckman S (2007) Polarity asymmetry of sprite-producing lightning: a paradox? Radio Sci 42:RS2S17. doi:10.1029/2006RS003488

Chapter 5
Ionospheric Alfvén Resonator (IAR)

Abstract The topic of this chapter is the ionospheric Alfvén resonator (IAR) which has been the subject of a great deal of research during recent years. The IAR resonance cavity occupies a space between the conducting E layer and the topside ionosphere where there occurs the strong gradient of Alfvén velocity. The IAR accumulates the Alfvén wave energy in the ULF/ELF frequency range, typically between 0.5 and 7 Hz. In this chapter, the structure, models, and possible physical mechanisms for the IAR excitation are studied. Dispersion relation and the IAR resonance spectra at night and daytime conditions are calculated.

Keywords Dispersion relation • E layer • Exosphere • Hall and Pedersen conductivities • Ionospheric Alfvén resonator (IAR)

5.1 Structure and Models of IAR

Analysis of the plasma wave propagation in the low frequency range made in Sect. I has shown that there are two kinds of plasma waves: the shear Alfvén wave and the compressional wave which has been also termed the fast magnetosonic (FMS) wave. According to Eqs. (1.59) and (1.61) the phase velocity of the shear Alfvén waves in a homogeneous plasma depends on the angle included between the velocity vector and the Earth magnetic field, whereas the group velocity, V_A, is a constant and parallel to the ambient magnetic field $\mathbf{B}_0$. By contrast, the compressional wave propagates at constant velocity V_A in all directions. This means that the shear Alfvén wave predominantly propagates along the geomagnetic field lines, whereas the compressional mode propagates isotropically. As the shear Alfvén wave propagates along the geomagnetic field lines, both the wave modes have the same velocity, V_A, the so-called Alfvén speed, which depends on $B_0 = |\mathbf{B}_0|$ and on the plasma density ρ_0, according to Eq. (1.60).

Typical Alfvén speed profiles are sketched in Figs. 5.1 and 5.2, as a function of altitude for nighttime and daytime conditions and for different sunspot activity

V. Surkov and M. Hayakawa, *Ultra and Extremely Low Frequency Electromagnetic Fields*, Springer Geophysics, DOI 10.1007/978-4-431-54367-1__5,
© Springer Japan 2014

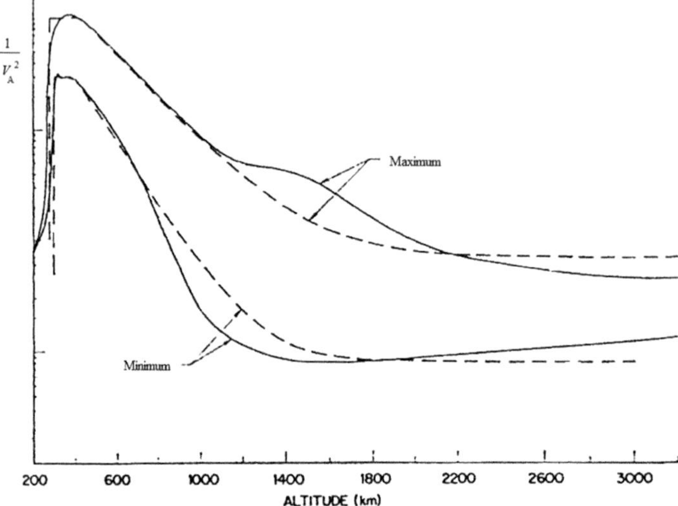

Fig. 5.1 A schematic daytime height profiles of the Alfvén speed for (1) maximum and (2) minimum of sunspot activity. Taken from Greifinger and Greifinger (1968)

(Greifinger and Greifinger 1968). It is clear from these figures that there is a minimum in the Alfvén speed at an altitude of about 300–400 km both at the daytime and nighttime conditions and that this minimum is located near the $F2$ ionization peak. In the exosphere, the region above the minimum, the altitude profile of the Alfvén speed shown in Fig. 5.1 is qualitatively similar to that shown in Fig. 5.2. In the region below the minimum, the variation of the Alfvén speed with altitude is dependent on ionospheric conditions.

In the first place an enhancement of the Alfvén speed results from the ionospheric plasma density fall off with height. The strong variation in the Alfvén speed in the topside ionosphere and an increase of the plasma conductivity in the gyrotropic E-layer that is in the bottom of the ionosphere, results in a strong variation of the Alfvén and FMS wave reflection indices from below and from above that gives rise to the formation of resonance cavity. The lower boundary of this resonator cavity is the conducting E-layer with enhanced Pedersen and Hall conductivities, that is, the altitude range 100–130 km, where nearly all plasma parameters undergo a strong impact. The semi-transparent IAR upper boundary is located at 600–1,200 km altitudes from the Earth's surface. The physical reason for the upper boundary occurrence is due to an exponential decrease of the background plasma density at this altitude range, which in turn results in an increase of the Alfvén speed by (1.60), thereby producing a partial reflection of Alfvén waves from the steep gradient of the wave velocity/plasma density.

This resonance cavity can serve as the ionospheric waveguide for the FMS mode, which can propagate along the waveguide walls around the Earth for a long distance. As the ionospheric inhomogeneities are ignored, the spherical form of

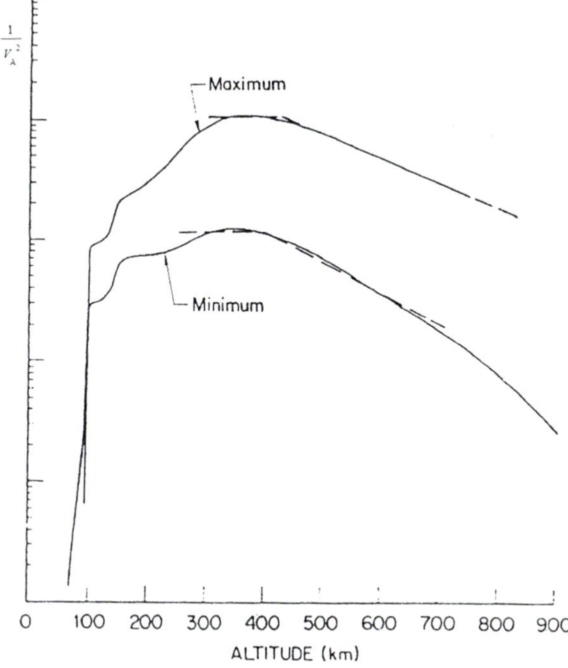

Fig. 5.2 A schematic nighttime height profiles of the Alfvén speed for (1) maximum and (2) minimum of sunspot activity. Taken from Greifinger and Greifinger (1968)

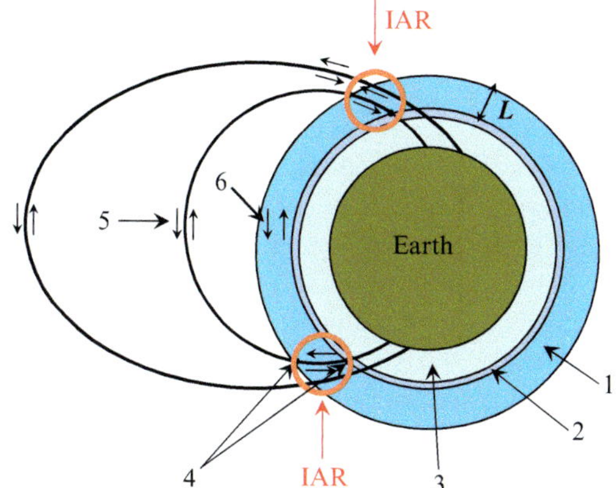

Fig. 5.3 A schematic drawing of the ionospheric Alfvén resonators (IAR). (1) the ionospheric waveguide; (2) E layer of the ionosphere; (3) the neutral atmosphere; (4) IAR; (5) magnetospheric MHD resonator; (6) resonant FMS waves propagating in the resonator. This simple model also explains the mechanism of the magnetospheric MHD resonances

the ionospheric waveguide, shown with blue area 1 in Fig. 5.3, is similar to the that of the Earth-Ionosphere waveguide shown with area 3. In particular the global standing FMS waves can be excited inside the resonance cavity in a close analogy to Schumann resonances. Notice that in reality a very powerful source such as a nuclear explosion is needed to excite the low-frequency resonant FMS modes. The properties of the ionospheric waveguide have been extensively investigated by a number of researchers (e.g., see Greifinger and Greifinger 1968) in connection with the problem of high-altitude large explosions.

The same cavity can serve as a resonator for the shear Alfvén waves, which propagate along the geomagnetic field lines. The ionospheric Alfvén resonator (IAR) arises due to the Alfvén wave reflection at the point of intersection between the magnetic field lines and the boundaries of the resonator as sketched in Fig. 5.3. The energy of the shear Alfvén wave can get trapped inside the IAR thereby exciting standing waves into the resonator. To summarize we note that the IAR is referred to as the class of field line resonances for shear Alfvén waves.

The idea of IAR was originally suggested by Polyakov (1976) and has been extensively studied by a number of authors (see Polyakov and Rapoport 1981; Belyaev et al. 1987, 1990; Lysak 1991; Trakhtengertz and Feldstein 1991). The existence of the IAR was well documented by ground-based observations in low latitudes (Hickey et al. 1996; Bösinger et al. 2002), in middle latitudes (Polyakov and Rapoport 1981; Belyaev et al. 1987, 1990; Hickey et al. 1996; Bösinger et al. 2002, 2004; Molchanov et al. 2004; Hebden et al. 2005), and even high latitudes (Belyaev et al. 1999; Demekhov et al. 2000; Yahnin et al. 2003; Semenova and Yahnin 2008). Based on Freja and FAST satellite onboard observations the IAR occurrence was also identified in space (e.g., Grzesiak 2000; Chaston et al. 1999, 2002, 2003).

To study the structure and mechanisms of the IAR excitation in more detail, we need to construct a suitably idealized model of the medium that is a reasonable approximation to the altitude variations of the plasma conductivity and the Alfvén speed. The ionospheric resonance cavity is localized at altitudes below 1–2 the Earth radius. When considering Alfvén waves propagating between the IAR walls at such distances, the magnetic field line curvature is of little importance and thus can be neglected. On the contrary, the plasma number density, the collision frequencies, the plasma conductivity, and other ionospheric parameters exhibit strong variations inside the IAR. On account of local character of the examined effect we shall consider a simplified plane-stratified model of the system Earth–atmosphere– ionosphere–magnetosphere, widely used in the studies of electromagnetic coupling between geospheres (e.g., Fujita and Tamo 1988; Pokhotelov et al. 2000; Surkov et al. 2004). For the sake of simplicity we adopt the model of the vertical external magnetic field in order to avoid the complexities connected with magnetic field inclination. A schematic drawing of our model is shown in Fig. 5.4. In this model, the conductive Earth, neutral atmosphere, E- and F-layers of the ionosphere, and magnetosphere, are assumed to be plane-stratified slabs of a constant thickness. The origin of the coordinate system is at the bottom of the ionosphere and z-axis is vertically upward. The gyrotropic E-layer of the ionosphere is shown with shaded area in Fig. 5.4. Furthermore, the plasma is assumed to be uniform within each layer in the direction perpendicular to the external magnetic field.

5.1.1 Model of Alfvén Speed Height Profile in the Exosphere

We start with the region above the Alfvén speed minimum, that has been termed the exosphere. In this region, the Alfvén speed height profile based on tabulated

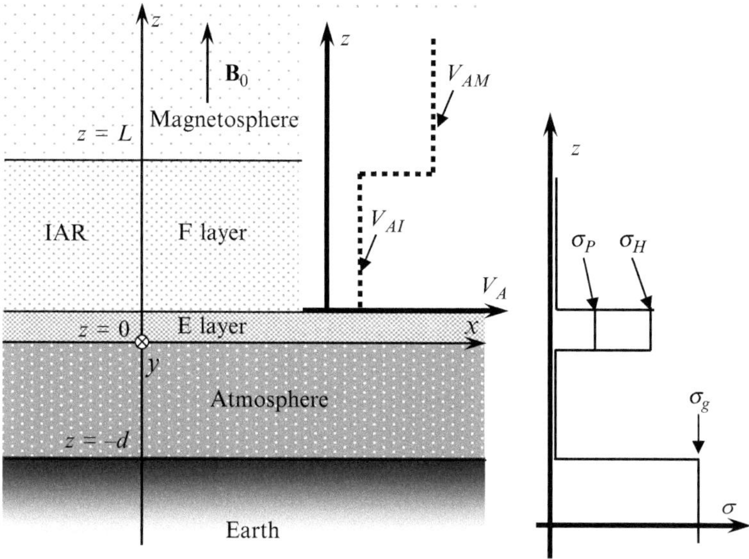

Fig. 5.4 A schematic drawing of a stratified medium model. The plots of the Alfvén velocity and the ionosphere/ground conductivities are shown in the *right panel*

Table 5.1 Values of ionospheric parameters used by Greifinger and Greifinger (1968) in fitting Eq. (5.1) to the profiles of Sims and Bostick (1963)

Local time and sunspot conditions	L, km	V_{A0}, km/s	ϵ^2
Night, sunspot maximum	300	720	0.0027
Night, sunspot minimum	450	395	0.0063
Day, sunspot maximum	350	375	0
Day, sunspot minimum	435	195	0

ionospheric numerical profiles of Sims and Bostick (1963) which may be considered as representative is quite well approximated by a simple analytical exospheric profile suggested by Greifinger and Greifinger (1968)

$$V_A^2 = \frac{V_{A0}^2}{\epsilon^2 + \exp\left[-2\left(z - z_0\right)/L\right]}. \quad (z > z_0) \qquad (5.1)$$

Here we have taken the positive z direction to be vertically upward, where z_0 denotes the coordinate of the Alfvén speed minimum. According to Greifinger and Greifinger (1968) the values of the parameters entering Eq. (5.1) for a variety of conditions are tabulated in Table 5.1.

Since $\epsilon^2 \ll 1$, this represents that the Alfvén speed increases exponentially from its minimum value V_{A0} at lower boundary z_0 of the exosphere to the constant value $V_{A0}\epsilon^{-1}$ in the outer magnetosphere when z tends to infinity. At nighttime condition this constant value varies within an interval $(5–14) \times 10^3$ km/s.

The use of such an analytical approximation makes it possible to express the IAR eigenfunctions in terms of Bessel functions and to obtain simple analytical expressions for the eigenfrequencies and the IAR damping/growth rates. In order to make our consideration as transparent as possible we, however, choose a simplified approximation (e.g., Pokhotelov et al. 2001), which describes the Alfvén velocity in terms of a piece-wise function, so that $V_A = V_{AI} \approx V_{A0}$ within the IAR ($0 < z < L$) and $V_A = V_{AM} \approx V_{A0}\epsilon^{-1}$ in the outer magnetosphere ($z > L$), where V_{AI} and V_{AM} are constant quantities and L denotes the characteristic width of the resonance cavity (IAR). Notice that V_{AI} is typically much smaller than Alfvén speed V_{AM} in outer magnetosphere. The model altitude profile of the Alfvén velocity used in this study is shown in Fig. 5.4 with dotted line.

The vertical z axis is positive parallel to $\mathbf{B}_0$, while the x and y axes are parallel to the plane boundaries of ionospheric layers. The region above the E-layer is supposed to be the area consisting solely of cold collisionless plasma. The plasma dielectric permittivity tensor, $\hat{\boldsymbol{\varepsilon}}$, in this region is assumed to be diagonal with components given by Eqs. (2.18) and (2.19).

5.1.2   Fourier Transform of Maxwell Equations

In what follows all perturbed quantities are considered to vary as $\exp(-i\omega t)$, where ω is the frequency. Let $\delta\mathbf{B}$ be the small perturbation of the geomagnetic field $\mathbf{B}_0$, so that $\delta B \ll B_0$. The Maxwell equation (2.17) is reduced then

$$\nabla \times \delta\mathbf{B} = -\frac{i\omega}{c^2}\hat{\boldsymbol{\varepsilon}}\,\boldsymbol{E}. \qquad (5.2)$$

We are thus left with the set of Maxwell equations (5.2) and (1.2), where $\mathbf{B}$ should be replaced by $\delta\mathbf{B}$. This set should be supplemented by Eq. (2.18) for the tensor of dielectric permittivity of plasma.

In what follows the z axis is positive parallel to the constant and homogeneous magnetic field $\mathbf{B}_0$. Since the medium is assumed to be uniform and infinite in the direction perpendicular to the unperturbed magnetic field $\mathbf{B}_0$, it is customary to seek for the solution of Maxwell equations in the form of Fourier transform over perpendicular coordinates x and y, for example,

$$\delta\mathbf{B}\,(\omega, \boldsymbol{\rho}, z) = \int\limits_{-\infty}^{\infty}\int\limits_{-\infty}^{\infty} \exp(ik_x x + ik_y y)\mathbf{b}\,(\omega, \mathbf{k}_\perp, z)\,dk_x dk_y, \qquad (5.3)$$

where $\boldsymbol{\rho} = (x, y)$ and $\mathbf{k}_\perp = \left(k_x, k_y\right)$ are the position vector and the wave vector, correspondingly, both perpendicular to the unperturbed magnetic field $\mathbf{B}_0$. The inverse Fourier transform is given by

$$\mathbf{b}\left(\omega,\mathbf{k}_\perp,z\right)=\frac{1}{(2\pi)^2}\int\limits_{-\infty}^{\infty}\int\limits_{-\infty}^{\infty}\exp(-ik_x x-ik_y y)\delta\mathbf{B}\left(\omega,\boldsymbol{\rho},z\right)dxdy.\qquad(5.4)$$

The same representation is true for the electric field $\mathbf{E}$. In the text we use the big letters to represent the functions of spatial variable $\boldsymbol{\rho}$, while the small letters are used to represent the functions of the perpendicular wave vector $\mathbf{k}_\perp$, that is, the Fourier transform of the electromagnetic field. Applying a Fourier transform (5.4) to Maxwell equations (5.2) and (1.2) yields

$$i\left(\mathbf{k}_\perp\times\mathbf{b}\right)+\hat{\mathbf{z}}\times\partial_z\mathbf{b}=-\frac{i\omega}{c^2}\hat{\boldsymbol{\varepsilon}}e,\qquad(5.5)$$

$$i\left(\mathbf{k}_\perp\times\mathbf{e}\right)+\hat{\mathbf{z}}\times\partial_z\mathbf{e}=i\omega\mathbf{b},\qquad(5.6)$$

where $\mathbf{b}\left(\omega,\mathbf{k}_\perp,z\right)$ and $\mathbf{e}\left(\omega,\mathbf{k}_\perp,z\right)$ denote the Fourier transforms of the electromagnetic variations and $\hat{\mathbf{z}}=\mathbf{B}_0/B_0$ is the unit vector parallel to $\mathbf{B}_0$.

5.1.3   Three- and Two-Potential Representation of Plasma Waves

As it follows from the analysis in Sect. 2.2, the electromagnetic perturbation in plasma can be split into the shear Alfvén and compressional/FMS modes. As shown in Appendix C, the electromagnetic field can be presented by scalar, Φ, and vector, $\mathbf{A}$, potentials or by three scalar potentials Φ, A, and Ψ. Particularly applying Fourier transforms to Eqs. (5.78) and (5.79) we come to the following field representation through the potentials

$$\mathbf{b}=i\mathbf{k}_\perp\partial_z\Psi+i\left(\mathbf{k}_\perp\times\hat{\mathbf{z}}\right)A+\hat{\mathbf{z}}k_\perp^2\Psi,\qquad(5.7)$$

and

$$\mathbf{e}=-i\mathbf{k}_\perp\Phi-\left(\mathbf{k}_\perp\times\hat{\mathbf{z}}\right)\omega\Psi+\hat{\mathbf{z}}\left(i\omega A-\partial_z\Phi\right).\qquad(5.8)$$

In the ULF/ELF range when $\omega\ll\omega_{pe}$, the absolute value of the parallel component (2.19) of the plasma dielectric permittivity is much greater than unity and thus can be assumed to be infinite whereas the total field-aligned Alfvén current which includes the conduction and displacement currents, must be finite as it follows from Eqs. (5.82) and (5.86). On account of equation $j_z=-i\omega\varepsilon_0\varepsilon_\parallel e_z$, one can conclude that the field-aligned electric component e_z tends to zero in this frequency range. Hence it follows that the potential A is coupled to Φ via $i\omega A=\partial_z\Phi$ and we come to two potential representation (Φ, Ψ) of the shear Alfvén and compressional waves given by Eqs. (5.84) and (5.85). In this case these equations can be split into two independent modes. The shear Alfvén mode is described through the potential Φ

$$\mathbf{b}_A = \frac{(\mathbf{k}_\perp \times \hat{\mathbf{z}})}{\omega} \partial_z \Phi, \quad \mathbf{e}_A = -i \mathbf{k}_\perp \Phi. \tag{5.9}$$

This equation demonstrates the basic properties of the Alfvén mode (for example, see Fig. 1.15) such that the magnetic perturbation is perpendicular to both the unperturbed magnetic field $\mathbf{B}_0$ and the wave vector $\mathbf{k}_\perp$, while the electric field is co-directed with $\mathbf{k}_\perp$.

The FMS/compressional mode can be represented via the potential Ψ

$$\mathbf{b}_F = i \mathbf{k}_\perp \partial_z \Psi + k_\perp^2 \Psi \hat{\mathbf{z}}, \quad \mathbf{e}_F = -(\mathbf{k}_\perp \times \hat{\mathbf{z}}) \omega \Psi. \tag{5.10}$$

Similarly, Eq. (5.10) demonstrates the basic properties of the FMS mode (for example, see Fig. 1.16), for which the electric perturbation is perpendicular to $\mathbf{B}_0$ and the wave vector $\mathbf{k}_\perp$.

In order to derive the equation for plasma waves in the magnetosphere one should substitute Eqs. (5.84) and (5.85) for $\mathbf{b}$ and $\mathbf{e}$ into Maxwell equations (5.5) and (5.6). First of all we note that the substituting of $\mathbf{b}$ and $\mathbf{e}$ into Eq. (5.6) gives identity. The formal proof of this statement is found in Appendix C (see Eqs. (5.87)–(5.91)). Substituting $\mathbf{b}$ into Eq. (5.5), taking into account of Eqs. (5.87) and (5.88) and rearranging, yields

$$\mathbf{k}_\perp \frac{\partial_z^2 \Phi}{\omega} - i(\mathbf{k}_\perp \times \hat{\mathbf{z}})\left(\partial_z^2 \Psi - k_\perp^2 \Psi\right) - \hat{\mathbf{z}}\frac{k_\perp^2 \partial_z \Phi}{\omega} = -\frac{i\omega}{c^2}\hat{\varepsilon}e \tag{5.11}$$

Then one should substitute Eqs. (5.85) and (2.18) for $\mathbf{e}$ and $\hat{\varepsilon}$ into Eq. (5.11). Working on the right-hand side of this equation and combining the terms, which include the factor $\mathbf{k}_\perp$, we come to the equation for shear Alfvén waves in the magnetospheric plasma

$$\partial_z^2 \Phi + \frac{\omega^2}{V_A^2} \Phi = 0. \tag{5.12}$$

Similarly, combining the terms which include the factor $\mathbf{k}_\perp \times \hat{\mathbf{z}}$, we come to the equation for FMS waves in the magnetospheric plasma

$$\partial_z^2 \Psi + \left(\frac{\omega^2}{V_A^2} - k_\perp^2\right) \Psi = 0. \tag{5.13}$$

5.1.4 Solution of Wave Equations in the Magnetosphere

We recall that in the framework of our model shown in Fig. 5.4 the Alfvén velocity is the constant value, $V_A = V_{AI}$, within the resonance cavity ($0 < z < L$), while $V_A = V_{AM}$ in outer space. In the region $z > L$ one should seek for the solution of Eq. (5.12) in the form of upward propagating Alfvén wave, that is

$$\Phi = C_1 \exp\left(\frac{i\omega z}{V_{AM}}\right), \tag{5.14}$$

where C_1 is undetermined coefficient. Indeed, on account of the factor $\exp(-i\omega t)$ we can reduce Eq. (5.14) to the following form

$$\Phi = C_1 \exp\left(i\omega\left\{\frac{z}{V_{AM}} - t\right\}\right), \tag{5.15}$$

which describes the Alfvén wave propagating upward at the velocity V_{AM}.

The implication of solution (5.15) is that the IAR upper boundary is transparent, in part, for the shear Alfvén waves that causes the leakage of the wave energy from the resonant cavity into the magnetosphere. This effect along with the energy loss due to Joule heating predominantly in the conducting E-layer results in the energy dissipation of the waves trapped in the resonance cavity, so that only several first IAR resonances can be detectable on the ground.

Actually the magnetospheric Alfvén waves can propagate along the magnetic field lines in both directions due to the wave reflection from the conjugate hemispheres (see Fig. 5.3) thereby exciting the field-line Alfvén resonances in the Earth's magnetosphere. In the next section we shall consider these resonances in more detail along with other kind of fundamental oscillations in the magnetosphere.

As before we seek for the solution of Eq. (5.13) in the form of upward propagating FMS wave ($z > L$), that is

$$\Psi = C_2 \exp\left(\frac{\lambda_M z}{L}\right), \tag{5.16}$$

where C_2 is undetermined coefficient. For convenience we have introduced the dimensionless function λ_M and dimensionless frequency x_0 via

$$\lambda_M^2 = k_\perp^2 L^2 - \epsilon^2 x_0^2, \quad x_0 = \frac{\omega L}{V_{AI}}. \tag{5.17}$$

The function λ_M has two bifurcation points, $\omega = \pm k_\perp V_{AM}$, in the complex plane of ω. The sign of λ_M in Eq. (5.17) should be chosen in such a way to satisfy the wave radiation condition at the magnetospheric end, whence it follows that the imaginary part of λ_M must be positive when $\omega > |k_\perp| V_{AM}$ (for the real $k_\perp$ and ω). In the low frequency limit, when $\omega < |k_\perp| V_{AM}$, the real part of the function λ_M becomes negative that means that the FMS mode exponentially decreases with altitude in the region $z > L$.

Inside the resonance cavity $(0 < z < L)$ the solutions of the wave equations (5.12) and (5.13) must include both the upward and downward propagating waves, which arise due to the wave reflection from the lower $(z = 0)$ and upper boundaries $(z = L)$ of the IAR. These solutions should be matched at this boundary under the requirement of continuity of the magnetic field components and transverse electric field $\mathbf{e}_\perp$ at $z = L$. Whence there follows the requirement of continuity of the potentials Φ and Ψ and their derivatives, $\partial_z \Phi$ and $\partial_z \Psi$, at $z = L$. The solution of the problem derived in Appendix D can be written as

$$\Phi = \Phi(0) \left[\cos \frac{\omega z}{V_{AI}} + i\beta_1 \sin \frac{\omega z}{V_{AI}} \right]. \tag{5.18}$$

$$\Psi = \Psi(0) \left[\cosh \frac{\lambda_I z}{L} + \beta_2 \sinh \frac{\lambda_I z}{L} \right]. \tag{5.19}$$

Here we made use of the following abbreviations

$$\lambda_I^2 = k_\perp^2 L^2 - x_0^2 \tag{5.20}$$

$$\beta_1 = \frac{1 + \epsilon - (1 - \epsilon) \exp(2ix_0)}{1 + \epsilon + (1 - \epsilon) \exp(2ix_0)}, \tag{5.21}$$

$$\beta_2 = \frac{\lambda_I + \lambda_M - (\lambda_I - \lambda_M) \exp(2\lambda_I)}{\lambda_I + \lambda_M + (\lambda_I - \lambda_M) \exp(2\lambda_I)}. \tag{5.22}$$

where $\epsilon = V_{AI}/V_{AM} \ll 1$ is the ratio of the Alfvén velocity in the IAR to that in the magnetosphere.

5.1.5 Boundary Conditions at the E Layer of the Ionosphere

At altitudes of the E-layer the electron gyrofrequency, ω_H, is much greater than the sum, ν_e, of electron–ion and electron–neutral collision frequencies. In contrast, the ion gyrofrequency, Ω_H, is much smaller than the sum, ν_i, of ion–electron and ion–neutral collision frequencies. This means that inside the E-layer the collisions play a crucial role in the behavior of ions whereas the electrons are guided by the Earth's magnetic field lines since they are magnetized. The tensor of plasma conductivity by Eq. (2.5) describes this kind of the anisotropy in the plasma conductivity which occurs in the E-layer of the ionosphere. Furthermore, in the ULF range the tensor components are nearly independent of the wave frequency, ω, because the value of ω is much smaller than all the collision frequencies and gyrofrequencies entering the tensor. In this case the components of the tensor can be expressed through Pedersen,

Hall and field-aligned conductivities given by Eqs. (2.13)–(2.15). Notice that all the plasma conductivities are highly altitude dependent, even though the plasma density itself is nearly uniform with height. Plots of the parallel, Pedersen and Hall conductivities for mid-latitude ionosphere and for $\omega = 0$ are shown in Fig. 2.5.

The plasma conductivity profiles undergo the diurnal variations due to changing solar radiation activity. In daytime, the UV radiation in the solar spectrum is incident on the neutral atmosphere and ionosphere, that results in increasing of the photoionization followed by the enhancement of plasma conductivity. Turning to the nighttime the conductivity profile is reduced in magnitude. However, the Hall and Pedersen conductivities keep a peak value within altitudes of the E-layer as displayed in Fig. 2.5.

As before we consider the plane-stratified model with vertical magnetic field $\mathbf{B}_0$, so that the z axis is positive parallel to $\mathbf{B}_0$ and is perpendicular to all the plane boundaries. This approach permits the use of Fourier transform (5.4) not only for the magnetosphere but also everywhere, that in turn makes it possible to simplify the boundary conditions at the E-layer.

In the frequency range of interest here the conduction current in the E-layer is much greater than the displacement one. As the displacement current is negligible, the Maxwell equation takes the form of Ampére's law given by Eq. (1.5). Applying a spatial Fourier transform (5.4) to this equation we come to the following equation:

$$i\left(\mathbf{k}_\perp \times \mathbf{b}\right) + \left(\hat{\mathbf{z}} \times \partial_z \mathbf{b}_\perp\right) = \mu_0 \mathbf{j}, \tag{5.23}$$

where $\mathbf{j}\left(\omega, \mathbf{k}_\perp, z\right)$ denotes the spatial Fourier transform of the current density $\mathbf{j}\left(\omega, \rho, z\right)$. Notice that this equation differs from Eq. (5.5) only on the right-hand side.

The Ohm's law for the ionospheric plasma is given by Eq. (2.6). Applying a Fourier transform to this equation yields

$$\mathbf{j} = \sigma_\| \mathbf{e}_\| + \sigma_P \mathbf{e}'_\perp + \sigma_H \left(\hat{\mathbf{z}} \times \mathbf{e}'_\perp\right) \tag{5.24}$$

where $\mathbf{e}$ is the electric field in a reference frame fixed at the Earth while $\mathbf{e}' = \mathbf{e} + \mathbf{v} \times \mathbf{B}_0$ is the electric field in a reference frame moving together with the neutral gas flow, $\mathbf{v}$ is a spacing Fourier transform of the mass velocity $\mathbf{V}$ of the neutral gas flow, σ_P and σ_H are the Pedersen and Hall conductivities, respectively. As is seen from Fig. 2.5 the parallel plasma conductivity, $\sigma_\|$, is much greater than the Pedersen and Hall conductivities at the altitudes above 100 km. The cause of this effect is the high mobility of the electrons along the magnetic field lines. Assuming that $\sigma_\| \to \infty$, the parallel electric field thus becomes infinitesimal, i.e., $\mathbf{e}_\| \to 0$.

Within the altitudes of the E-layer the ratio of plasma to neutral number densities is $\sim 10^{-7}$–10^{-9} for the day- and night-time conditions, respectively. Based on this value one can find that the influence of the motions of the electrons and ions on the neutrals motion is negligible. Therefore, we can consider the neutral wind velocity $\mathbf{v}$ in the E-layer as a given function, which can serve as a source for the electromagnetic perturbations (Surkov et al. 2004). For clarity we consider that the stationary convective electric field is absent, and the electromagnetic perturbations are solely due to the neutral wind and external sources in the magnetosphere or atmosphere.

Let l be a typical thickness of the conductive E layer of the ionosphere. Taking a maximum of σ_P in the ionosphere as a representative conductivity of the plasma, the skin-depth in the E layer of the ionosphere is estimated as $l_s \sim (\mu_0 \sigma_P \omega)^{-1/2}$. If $l \ll l_s$ or $\omega \ll (\mu_0 \sigma_P l^2)^{-1}$, the electromagnetic fields are slowly varying functions of height inside the E layer and thus "thin" ionosphere approximation can be applied (Lysak 1991; Pokhotelov et al. 2000). Using the following numerical parameters $l = 30\,\mathrm{km}$, $\sigma_P = 10^{-4}\,\mathrm{S/m}$, we find that the above approach is valid if $\omega \ll 10\,\mathrm{Hz}$.

In what follows we consider the E layer in the "thin" ionosphere approximation, that is $l \ll l_s$, because the IAR eigenfrequencies typically lie in the range of 0.5–3 Hz. In this notation we shall use the so-called height-integrated Pedersen and Hall conductivities

$$\Sigma_{P,H} = \int_0^l \sigma_{P,H}(z)\, dz, \tag{5.25}$$

which are measured in Ω^{-1}. Certainly, the boundaries of the actual E-layer cannot be determined exactly and the peaks of the Hall and Pedersen conductivities are situated at different altitudes due to the inhomogeneity of the actual ionosphere. This inaccuracy, however, is of no practical importance on account of exponential fall off of both conductivities with distance from the conducting E-layer. It is customary to use the normalized parameters $\alpha_P = \Sigma_P/\Sigma_w$ and $\alpha_H = \Sigma_H/\Sigma_w$, which are the ratios of the height-integrated Pedersen and Hall conductivities to the so-called wave conductivity $\Sigma_w = (\mu_0 V_{AI})^{-1}$, where V_{AI} is the Alfvén speed inside the IAR.

Substituting Eq. (5.24) for $\mathbf{j}$ into Eq. (5.23) we obtain the equation which can be integrated across the E-layer with respect to z from $z = 0$ to $z = l$. Taking only the projection of $\mathbf{j}_\perp$ perpendicular to $\mathbf{B}_0$ and making formally $l \to 0$ gives the boundary conditions at $z = 0$

$$V_{AI}\left(\hat{\mathbf{z}} \times [\mathbf{b}_\perp]\right) = \alpha_P \mathbf{e}'_\perp + \alpha_H \left(\hat{\mathbf{z}} \times \mathbf{e}'_\perp\right). \tag{5.26}$$

Here the square brackets denote the jump of magnetic field $\mathbf{b}_\perp$ across the E-layer, that is $[\mathbf{b}_\perp] = \mathbf{b}_\perp(0+) - \mathbf{b}_\perp(0-)$. In the thin slab approximation the electric field $\mathbf{e}' = \mathbf{e} + \mathbf{v} \times \mathbf{B}_0$ in Eq. (5.26) should be taken at $z = 0$. Besides, for the sake of simplicity, the Fourier transform of the wind velocity is assumed to be independent of the z-coordinate, that is $\mathbf{v} = \mathbf{v}(\omega, \mathbf{k}_\perp)$.

The appearance of the term $[\mathbf{b}_\perp]$ in Eq. (5.26) follows the general laws of the electrodynamics according to which the horizontal component of magnetic field must be discontinuous across the current flowing on an infinitely thin sheet (e.g., see monograph by Jackson (2001)). In our approach the jump of the horizontal magnetic field, i.e., the term $\hat{\mathbf{z}} \times [\mathbf{b}_\perp]$, arises just due to the presence of the surface Pedersen and Hall currents flowing in the thin E-layer.

5.1.6 Electromagnetic Field at the Atmosphere and in the Ground

The spectrum of IAR normal modes is its inner property, which is independent of source of the IAR excitation. Since the general IAR dispersion relation is of main interest here, the source of the IAR excitation is of minor importance at this point. In the next section some of the sources will be treated in detail and others in a more sketchy fashion. For the present we assume that the IAR excitation is due to neutral wind at the height of E-layer or due to the MHD waves coming from outer space, since it seems conceptually simpler than the lightning discharge.

Although the conductivity of the atmosphere increases with altitude approximately as an exponential function with the characteristic scale $z_a \approx 4$–7 km (see, for example, Eq. (3.1)), the plane slab model can be applied if the typical lateral sizes of the source are much greater than both z_a and l, i.e., if $k_\perp l \ll 1$ and $k_\perp z_a \ll 1$. In this approach the atmosphere and the ionosphere are considered as uniform media in lateral dimension in such a way that the atmospheric conductivity can be characterized via height-integrated conductivity $\Sigma_a \sim 10^{-3}\,\Omega^{-1}$ (e.g., Cole and Pierce 1965) similar to the height-integrated Pedersen and Hall conductivities (5.25) of the ionosphere. In practice $\Sigma_a \ll \Sigma_P$ and $\Sigma_a \ll \Sigma_H$ so that in the first approximation the neutral atmosphere $(-d < z < 0)$ can be considered as an insulator. Neglecting the displacement current $-i\omega\varepsilon_0\mathbf{e}$ as well, the ULF electromagnetic perturbations in the atmosphere obey Laplace equation

$$\nabla^2\mathbf{b} = 0. \tag{5.27}$$

The solid Earth $(z < -d)$ is supposed to be a uniform conductor with a constant conductivity σ_g. The conduction current $\sigma_g\mathbf{e}$ in the ground is much greater than the displacement one for all frequencies of interest here. This implies that the electromagnetic perturbations in the conducting ground are described by quasisteady Maxwell equation (1.15), which can be rewritten as follows:

$$-i\omega\mathbf{b} = v_m\nabla^2\mathbf{b}, \tag{5.28}$$

where $v_m = \left(\mu_0\sigma_g\right)^{-1}$ is the magnetic diffusion coefficient in the ground. The electric field in the ground is described by an equation analogous to Eq. (5.28).

Since the atmosphere is considered as an insulator and there are no sources of electromagnetic perturbations, the vertical electric current j_z flowing from the ionosphere into the atmosphere vanishes at the boundary between the ionosphere and atmosphere. As shown in Appendix D, this leads to the condition that $A = 0$ everywhere in the atmosphere. In this special case the magnetic variations in the atmosphere is dependent only on the potential Ψ. Although the TM mode can be excited in the atmosphere by the other sources such as lightning discharges and in this case the TM mode is described by both the potentials A and Φ. We shall return to this point later in the end of this section.

As before we may apply a Fourier transform to Eqs. (5.27) and (5.28). Taking into account Eqs. (5.7) and (5.8) we arrive at equations for the potential function Ψ. These equations should be supplemented by the proper boundary conditions at $z = -d$ that follow from the Ampére's law. On account of the continuity of the magnetic and horizontal electric fields at the boundary between the atmosphere and the ground we obtain that the scalar potential Ψ and its derivative $\partial_z \Psi$ to be continuous at this boundary. In greater detail the solution of this problem is examined in Appendix D. Combining Eqs. (5.115) and (5.116) we obtain the solution of problem in the atmosphere $(-d < z < 0)$

$$\Psi = \frac{\Psi(0)}{2\beta_3} \left\{ \left(1 + \frac{\xi}{k_\perp} \right) \exp\left[k_\perp (z + d) \right] \right.$$
$$\left. + \left(1 - \frac{\xi}{k_\perp} \right) \exp\left[-k_\perp (z + d) \right] \right\}, \tag{5.29}$$

where the parameter

$$\xi = \left(k_\perp^2 - i\mu_0 \sigma_g \omega \right)^{1/2} \tag{5.30}$$

plays a role of the propagation factor/"wave number" in the ground and

$$\beta_3 = \cosh(k_\perp d) + \frac{\xi}{k_\perp} \sinh(k_\perp d). \tag{5.31}$$

Taking the electromagnetic field representation (5.7)–(5.8) through the scalar potentials (A, Φ, Ψ), and using the solutions (5.18), (5.19), and (5.29) for the potentials, we can find the fields **b** and **e** both in the ionosphere and atmosphere. Substituting these fields into the boundary condition (5.26) gives the set of Eqs. (5.128) and (5.129) for the undetermined constants $\Psi(0)$ and $\Phi(0)$. The potential Ψ on the ground is derivable from $\Psi(0)$ through Eq. (5.29) as follows: $\Psi(-d) = \Psi(0)/\beta_{3.}$. The solution of these equations is found in Appendix D. Finally, we obtain

$$\Psi(-d) = \frac{B_0 F_0}{k_\perp^2 q \beta_{3.}}, \tag{5.32}$$

where the following abbreviations are introduced

$$F_0 = L V_{AI}^{-1} \left[(\mathbf{k}_\perp \cdot \mathbf{v}) \left(\alpha_H^2 + \alpha_P^2 + \beta_1 \alpha_P \right) - (\mathbf{k}_\perp \times \mathbf{v})_z \beta_1 \alpha_H \right], \tag{5.33}$$

$$q = (is + x_0 \alpha_P)(\beta_1 + \alpha_P) + x_0 \alpha_H^2, \tag{5.34}$$

and

$$s = (k_\perp L) \frac{\xi + k_\perp \tanh(k_\perp d)}{k_\perp + \xi \tanh(k_\perp d)} - \lambda_I \beta_2. \qquad (5.35)$$

It is clear that all the components of the electromagnetic perturbations at the ground surface are proportional to the factor $\Psi(-d)$ given by Eq. (5.32). In this notation, the zeros of the factor q in the denominator of Eq. (5.32) define the resonance structure of the IAR spectrum. We return to this point later.

Depending on the neutral wind velocity, $\mathbf{v}$, the function F_0 plays a role of forcing functions/sources for the IAR excitation. In contrast to Eqs. (5.14) and (5.16) where the shear Alfvén and compressional modes are uncoupled, Eq. (5.32) describes the interference of these two modes by virtue of the Hall conductivity σ_H. Indeed, only if $\alpha_H = 0$, the set of boundary equations (5.120) and (5.121) is split into two independent equations for the shear Alfvén (A, Φ) and compressional (Ψ) modes. As we shall see, this mode coupling plays a significant role in the IAR excitation and thus cannot be neglected. Furthermore, it is worth mentioning that Eqs. (5.120) and (5.121) describe the coupling of the ionospheric MHD modes with neutral wind motions in the ionosphere. To understand the forcing function F_0 in more detail, however, we need to consider the role of other sources of the IAR excitation such as the thunderstorm activity. We return to this matter in Sect. 5.3.

The denominator in Eq. (5.32) contains the factor β_3 (5.31) which increases strongly if $k_\perp d \gg 1$ because of the presence of the function $\exp(k_\perp d)$, so that the signals become practically undetectable owing to their smallness on the ground. This means that only large-scale perturbations with typical horizontal sizes $\sim k_\perp^{-1} \gg d \sim 100\,\mathrm{km}$ make a main contribution to the IAR spectrum observed on the ground. Before discussing this problem in any detail, we need to understand a little about the IAR dispersion relation and the IAR eigenfrequencies.

5.2 IAR Eigenfrequencies

5.2.1 Dispersion Relation of the IAR

The IAR is referred to as a class of magnetic field resonances for shear Alfvén waves. The resonance frequencies of the IAR are determined by the length, L, of the segment of magnetic field line, which is bounded from above and from below by the resonator sides. Based on the plane-stratified model of the IAR (Plyasov et al. 2012) have shown that the vertical electromagnetic structure inside IAR has a form of the standing mode: electric components have their anti-nodes in the vicinity of the magnetic nodes. Assuming that the integer number of the Alfvén half-wavelength keeps within L, the characteristic IAR eigenfrequencies can be estimated as $f_n \sim V_A n/(2L)$, where V_A is the Alfvén wave speed and n is integer value. Taking the following numerical values $V_A \sim (0.5\text{--}1) \times 10^3$ km/s and $L \sim 10^3$ km we obtain the rough estimate $f_n \sim (0.5\text{--}0.25)\,n$ (in Hz), where $n = 1, 2, \dots$. Thus, the IAR highlights eigenfrequencies of several Hz that lie below the Schumann resonances.

To study with rigorous formulation of the problem of eigen oscillations one should examine the dispersion relation of the IAR, that is the dependence of $\omega = \omega\,(k_\perp)$. In the framework of the model developed above, the IAR eigenfrequencies are defined by the zeros of the factor q in the denominator of Eq. (5.32). Hence taking $q = 0$, we get

$$(i s + x_0 \alpha_P)\,(\beta_1 + \alpha_P) + x_0 \alpha_H^2 = 0. \tag{5.36}$$

This equation defines an implicit dependence of eigenfrequencies on $k_\perp$.

5.2.2 Shear Alfvén and FMS Modes for the Case of Zero Hall Conductance

We start our analysis with the simplified case of small Hall conductance Σ_H. In the first approximation we omit the terms proportional to α_H^2 in Eq. (5.33) for F_0 and Eq. (5.34) for q whereas the terms linear in α_H can be kept (Pokhotelov et al. 2001). The general dispersion relation (5.36) in this case decouples into two branches/modes. The first one corresponds to the roots of the following equation:

$$\beta_1 + \alpha_P = 0. \tag{5.37}$$

The dispersion relation (5.37) does not depend on $k_\perp$ and correspond to the shear Alfvén mode. Indeed, on account of Eq. (5.21) for β_1, Eq. (5.37) can be rewritten as

$$\exp\,(2 i x_0) = \left(\frac{1 + \epsilon}{1 - \epsilon}\right) \frac{(1 + \alpha_P)}{(1 - \alpha_P)}. \tag{5.38}$$

Decomposing the dimensionless frequency x_0 in Eq. (5.38) into its real and imaginary parts, i.e., $x_0 = \eta + i\gamma$, one finds

$$\exp\,(2 i \eta - 2\gamma) = \left(\frac{1 + \epsilon}{1 - \epsilon}\right) \left|\frac{1 + \alpha_P}{1 - \alpha_P}\right| \exp\,(2 i \vartheta_\alpha + 2 i \pi n), \tag{5.39}$$

where n is an integer, $n = 1, 2, \dots$ and

$$\vartheta_\alpha = \frac{1}{2}\,\arg\,\frac{(1 + \alpha_P)}{(1 - \alpha_P)}. \tag{5.40}$$

At the nighttime conditions the Pedersen conductivity is small so that $\alpha_P < 1$. It follows from Eq. (5.40) that $\vartheta_\alpha = 0$ when $0 \le \alpha_P < 1$ and $\vartheta_\alpha = \pi/2$ when $\alpha_P > 1$.

The latter condition is usually valid at the daytime. The solution of Eq. (5.39) can be written in the form $\eta_n = \pi n$ if $0 \leq \alpha_P < 1$ and $\eta_n = \pi \left(n - \frac{1}{2}\right)$, if $\alpha_P > 1$, while the wave damping factor is

$$\gamma = -\epsilon - \frac{1}{2} \ln \left| \frac{1 + \alpha_P}{1 - \alpha_P} \right|. \tag{5.41}$$

The set of values η_n define dimensionless eigenfrequencies of the normal modes of the IAR when the Hall conductivities is neglected. As we shall see, the IAR excitation seems to be the most effective at the nighttime, when the height-integrated Pedersen conductivity is small compared to wave conductivity Σ_w, so that $\alpha_P < 1$. On account of Eq. (5.17) for x_0 the dimensional values of the nighttime IAR eigenfrequencies are

$$f_n = \frac{\omega_n}{2\pi} = \frac{V_{AI} n}{2L}, \quad n = 1, 2, \ldots, \tag{5.42}$$

which coincide with the rough estimate made in the beginning of this section. A simple interpretation of Eq. (5.42) is that the integer number of half-wavelength keeps within the IAR length; that is $L = \lambda_n n / 2$ where the wavelength $\lambda_n = V_{AI} / f_n$.

At the daytime condition when the Pedersen conductivity is large enough to satisfy the inequality $\alpha_P > 1$, the eigenfrequencies can be rewritten as

$$f_n = \frac{V_{AI}}{2L} \left(n - \frac{1}{2}\right), \quad n = 1, 2, \ldots \tag{5.43}$$

It should be noted that the former relation can be considered in terms of the fact that the integer number of half-wavelength and one fourth of the wavelength keep within the IAR length, that is $L = (\lambda_n / 2)(n + 1/2)$.

In the remainder of this subsection we focus attention on the attenuation of the normal modes. The first term on the right-hand side of Eq. (5.41) corresponds to the IAR damping factor due to the wave energy leakage through the resonator upper wall, whereas the second term describes the losses due to the ionosphere Joule dissipation. It should be noted that in the case $\alpha_P \to 1$ the damping factor γ in Eq. (5.41) tends to infinity. This only means that here the dispersion due to the Hall conductivity that has been neglected in Eq. (5.37) must play a major role.

Another family of roots in Eq. (5.36) corresponds to the FMS/compressional mode which is described by

$$i s + x_0 \alpha_P = 0, \tag{5.44}$$

where s is given by Eq. (5.35). By analogy with Eq. (5.38) we can rewrite Eq. (5.44) as

$$\exp\left(2\lambda_I\right) = \frac{\left(\lambda_M + \lambda_I\right)}{\left(\lambda_M - \lambda_I\right)} \frac{\left(\alpha_P + \alpha_g - i\lambda_I/x_0\right)}{\left(\alpha_P + \alpha_g + i\lambda_I/x_0\right)}, \qquad (5.45)$$

where

$$\alpha_g = \left(\frac{ik_\perp L}{x_0}\right) \frac{\xi + k_\perp \tanh\left(k_\perp d\right)}{k_\perp + \xi \tanh\left(k_\perp d\right)}. \qquad (5.46)$$

The functions λ_M, λ_I, ξ and α_g given by Eqs. (5.17), (5.20), (5.30), and (5.46) depend on $k_\perp$. This means that Eq. (5.45) defines an inexplicit dependence of dimensionless frequency on perpendicular wave number, i.e., $x_0 = x_0\left(k_\perp\right)$.

The range of very small values of $k_\perp$, or $k_\perp L \ll 1$, is not of great importance for practical applications since this limit corresponds to the large-scaled perturbations $\sim k_\perp^{-1} \gg L \sim 10^3$ km. Furthermore, these scale sizes are of the order of the Earth radius and thus cannot be considered in the framework of the plane-stratified model developed in this section. So, we restrict our analysis to the opposite case of $k_\perp L \gg 1$. This implies that the typical size of FMS waves propagating inside the IAR is much smaller than the resonator scale L. Not surprisingly, these waves have a dispersion relation as though they were in an infinite space. In other words, one may expect that the dispersion relation of the FMS wave has to be insensitive to the boundary condition at the resonator walls. Indeed, noticing that formally the value $\lambda_I = 0$ satisfies Eq. (5.45) because in this case both parts of Eq. (5.45) are equal to unity, we come to the equation

$$x_0 = k_\perp L \qquad \text{or} \qquad \omega = k_\perp V_{AI}, \qquad (5.47)$$

which corresponds to the ordinary dispersion relation (2.26) for the FMS mode in a homogeneous plasma. Nevertheless, this relation can be considered only as a first approximation because it does not satisfy the original equation (5.44). One can expand Eq. (5.45) in a power series of small parameter $\left(k_\perp L\right)^{-1}$ to find the function $\omega = \omega\left(k_\perp\right)$ in the next approximation (Surkov et al. 2004).

5.2.3   Mode Coupling: The Role of the Ionospheric Hall Conductivity

In the above consideration the shear and FMS modes are uncoupled. It is a conventionally idealized model which can be relevant for the nighttime conditions. However, during daytime conditions, the situation may become more complex since the effects due to finite Hall conductivity start to play an important role. Analysis of the general dispersion relation equation (5.36) has shown that despite the strong mode coupling equation (5.36) consists of two branches as before (Surkov et al. 2004). Each branch includes a discrete set of normal modes. In Fig. 5.5 we plot

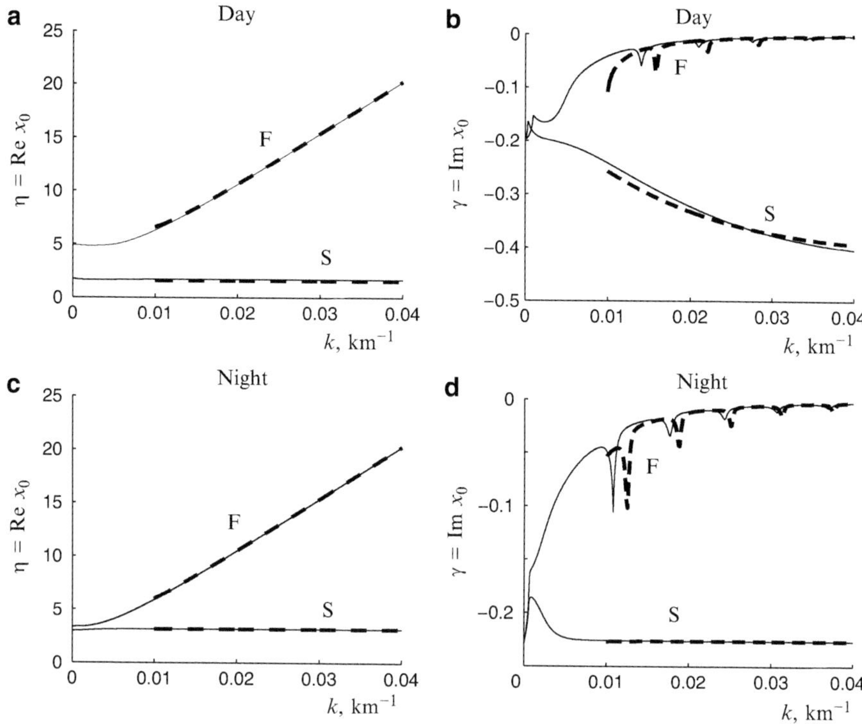

Fig. 5.5 The plots of the real (**a,c**) and imaginary (**b,d**) parts of the dimensionless frequency x_0 as a function of perpendicular wave number k for the fundamental mode. S and F denote the shear Alfvén and fast magnetosonic modes, respectively. The *dashed lines* correspond to the approximate analytical formulae. Taken from Surkov et al. (2004)

the results of numerical analysis of Eq. (5.36) for the fundamental mode ($n = 1$). Figure 5.5a,b show the real and imaginary parts of dimensionless frequency x_0 as a function of $k_\perp$ (in inverse kilometers) for the daytime conditions. A similar plot for the nighttime conditions is depicted in Fig. 5.5c,d. The numerical values for the various magnetospheric, ionospheric, and other parameters are: $V_{AI} = 500\,\mathrm{km/s}$, $V_{AM} = 5 \times 10^3\,\mathrm{km/s}$, $L = 500\,\mathrm{km}$, $d = 100\,\mathrm{km}$, and $\sigma_g = 2 \times 10^{-3}\,\mathrm{S/m}$. For the daytime ionosphere (Fig. 5.5a,b) the height-integrated conductivities are $\Sigma_P = 5\,\mathrm{Ohm}^{-1}$ and $\Sigma_H = 7.5\,\mathrm{Ohm}^{-1}$, respectively, so that $\alpha_P = 3.14$ and $\alpha_H = 4.71$. The nighttime parameters of the ionosphere are as follows: $\Sigma_P = 0.2\,\mathrm{Ohm}^{-1}$ and $\Sigma_H = 0.3\,\mathrm{Ohm}^{-1}$ (Fig. 5.5c,d), so that $\alpha_P = 0.126$ and $\alpha_H = 0.188$.

The real part of x_0, which is denoted by η, defines the eigenfrequency of the fundamental mode. As is seen from Fig. 5.5a,c, the fundamental eigenfrequency η of the shear Alfvén mode, shown with solid lines S, practically does not depend on $k_\perp$. At the same time, the FMS mode shown in Fig. 5.5a,c with solid lines F exhibits approximately linear response to $k_\perp$ if $k_\perp > 0.01\,\mathrm{km}^{-1}$. As we have noted above, this mode tends asymptotically to the dependence $x_0 = k_\perp L$ or $\omega = k_\perp v_{AI}$, which is typical for the FMS mode in a uniform plasma.

The imaginary part of x_0 or dimensionless attenuation coefficient γ is displayed in Fig. 5.5b,d with solid lines S and F for the shear Alfvén and FMS modes, respectively. It is obvious from this figure that both modes strongly depend on $k_\perp$. Interestingly enough the absolute value of attenuation factor of the shear Alfvén mode is much stronger than that of the FMS mode. The behavior of FMS mode is also distinguished from that of shear Alfvén mode by some oscillations, as seen in these figures. The small peaks in γ shown in Fig. 5.5b,d can be due to interference of the shear and compression modes because they disappear when $\alpha_H = 0$, that is, if the modes are decoupled.

To clarify the mode properties, an analytical solution of the dispersion relation equation (5.36) is required. The interested reader is referred to the paper by Surkov et al. (2004) for details on the approximate analytical solutions shown in Fig. 5.5 with dotted lines.

It follows from the numerical and analytical studies that the shear Alfvén mode attenuation is much greater than that of the fast mode especially as $k_\perp > 0.01\,\mathrm{km}^{-1}$. It means that the excitation of the FMS mode in the IAR is quite possible and it can play an important role in the formation of the IAR spectrum. It should be noted nevertheless that the eigenfrequencies and attenuation factors of the FMS mode become close to those for the shear Alfvén mode as $k_\perp$ approaches to zero and the difference between them disappears as the ground conductivity is neglected.

5.3   Sources of IAR Excitation

5.3.1   Possible Physical Mechanisms for IAR Excitation

Special credit has been paid in the past to the study of the physical mechanisms for the IAR excitation. It is generally believed that the main mechanism of the IAR excitation in low latitudes is due to the global thunderstorm activity (Polyakov and Rapoport 1981; Belyaev et al. 1987, 1990; Bösinger et al. 2002). From this viewpoint, the electromagnetic noise energy stemming from the thunderstorms can excite both the Schumann and IAR resonances. The global thunderstorm activity includes about 2,000 thunderstorms operating in the atmosphere at the same time. As we have noted above, the most intensive of them are located in Central Africa, South America, and South-East Asia. The CG lightning discharges result in the electromagnetic emission propagating in the Earth-Ionosphere waveguide predominantly in the form of TM mode. This mode can convert into TE mode, in part, due to the interaction with gyrotropic E-region of the ionosphere. The coupling of these two modes caused by Hall conductivity results in the formation of both shear Alfvén and FMS waves in the bottom E-region ionosphere. The Alfvén wave energy can get trapped in the F-region ionosphere thereby exciting the IAR and producing the geomagnetic perturbations with the spectral resonance structure (SRS) signature. As it was shown by Belyaev et al. (1987, 1990), the shear Alfvén mode connected with the TE mode in the atmosphere can exhibit SRS observed on the ground far from the thunderstorm center.

Recently Surkov et al. (2005a,b, 2006) and Fedorov et al. (2006) have reported that the calculated IAR spectra due to the tropic thunderstorm activity are on one or two order of magnitude lower than that observed at middle latitudes. The model calculations of the power spectra are in favor of the nearby thunderstorms as a possible cause for the IAR excitation at middle latitudes. Perhaps, an impulsive magnetic background from regional thunderstorms makes a significant contribution to the low-frequency part of SRS (Fedorov et al. 2006; Schekotov et al. 2011; Pilipenko 2011). The upper atmospheric discharges, associated with TLEs, can be even more effective in the IAR excitation (Sukhorukov and Stubbe 1997). We note that at middle latitudes there is a natural source of free energy stored in the ionospheric neutral wind motions which can excite the IAR similar to the operation of a police whistle (Surkov et al. 2004; Molchanov et al. 2004).

On the contrary, at high latitudes other sources of the free energy can come into play. The basic mechanism of the IAR excitation at high latitudes usually refers to the resonant energy transfer from the magnetospheric convective flow to the IAR eigenmodes (Trakhtengertz and Feldstein 1981, 1984, 1987, 1991) and development of the fast feedback instability induced by the large-scale ionospheric shear flows (Lysak 1991; Trakhtengertz and Feldstein 1991; Pokhotelov et al. 2000, 2001). Lysak (1991, 1993, 1999) has included a self-consistent analysis of the "fast feedback instability" due to the modification of the ionospheric conductivity by precipitating energetic electrons that may increase the rate of energy transfer from the convective flow to the IAR. The term "fast" was used in order to distinguish this type of instability from the slow feedback instability of the global ionosphere-magnetosphere resonator studied by Atkinson (1970), Sato and Holzer (1973), and Sato (1978). The energetics of the feedback instability was reviewed by Lysak and Song (2002). It should be noted that the feedback instability can serve as a basic mechanism of the IAR excitation only at high latitudes where the convection electric fields can reach quite strong values. In some cases the IAR manifests itself as the anomalous ULF transients and can be observed in the upper ionosphere on board the low-orbiting satellites above strong atmospheric weather systems (Fraser-Smith 1993). In due time Sukhorukov and Stubbe (1997) considered the nonlinear conversion of the lightning discharges energy into the IAR eigenmodes. The nonlinear theory of the IAR was recently developed by Pokhotelov et al. (2003, 2004) and Onishchenko et al. (2004).

To summarize, we note that the spectral structure and attenuation factors of IAR have been successfully modeled by analytical (e.g., see Belyaev et al. 1989; Lysak 1991; Surkov et al. 2004) and numerical (e.g., see Pokhotelov et al. 2001; Polyakov et al. 2003) models, while the excitation mechanisms have not been firmly established yet.

5.3.2 Observations of the IAR Spectra

As we have noted above, the SRS of IAR is mainly evident during the night time irrespective of season, at low (e.g., see Bösinger et al. 2002), middle (e.g., see Hebden et al. 2005), and even high (e.g., see Semenova and Yahnin 2008) latitudes. Now we first consider the IAR signature detected at the mid-latitudes at a remote site near Karimshino station (52.94°N, 158.25°E, $L = 2.1$) in Kamchatka peninsula (Fedorov et al. 2006; Surkov et al. 2006). The reader is referred to the work by Uyeda et al. (2002) for details about the equipment of the Russian–Japanese geophysical observatory in Kamchatka. The ground-based observations at this point have shown that the IAR signature predominantly occurred at local nighttime. Analysis of data obtained at Karimshino station has demonstrated that some spectrograms should be interpreted as impulse IAR excitation rather than permanent one. To illustrate this, we have chosen a representative time interval from 21 h till 22 h (local time) on 13 September 2000. During this interval, half-minute samplings were used to analyze the spectra of the ULF geomagnetic variations. Typical half-minute recordings, power spectrum densities, and dynamic spectrograms of the magnetic variations are shown in Figs. 5.6, 5.7, 5.8. The upper panels marked H SR and D SR display the time-dependence of D (the magnetic declination) and H (the horizontal component) components of magnetic variations in pT in the frequency ranges of 6–20 Hz. In the middle row of the panels, the data displayed in the top row are low pass filtered 0.25–4.0 Hz. The time in seconds is referenced to 21 h. For example, the 30 s interval shown in Fig. 5.6 begins at 21:31:30. The function $\log 10\,[p\,(B)]$ (B in pT) is shown in Figs. 5.6, 5.7, 5.8 in the second and fourth panels in the first two rows, where $p\,(B)$ is the amplitude probability distribution of B. The power spectrum densities of time-derivative of the magnetic variations are shown in the second and fourth panels in the third row.

Morlet wavelet decomposition (Mallat 1999) was used in order to obtain the normalized dynamic spectrograms of both components that are marked H and D at the lower row of the panels. The vertical axis on these panels corresponds to the wavelet center frequency. What attracts our first attention is that the enhancement of the dynamic spectrograms occurs just at the moment of the sharp impulses, which occasionally happen at the middle panels. For example, it is obvious from Fig. 5.6 that the impulse occurrence at the low-frequency channel (0.25–4 Hz) at the moment $t \approx 1{,}914$ s is accompanied by an enhancement of the dynamic spectrograms of both components. It should be noted that the SRS builds up as a result of this impulse and exhibits typical frequencies close to the IAR eigenfrequencies. Similarly, two impulses seen on the second line in Fig. 5.7 cause the distinct SRSs of the dynamic spectrograms. It is worth mentioning that the pattern of the dynamic spectrograms are consistent with the half-minute power spectra displayed in Figs. 5.6, 5.7, 5.8 at the lower row. During a quiet period shown in Fig. 5.8 the SRS is not so distinct except for the low-frequency domain of the spectrograms whereas the SRS is clearly seen in the power spectra. The same resonance structure with distinct IAR signature

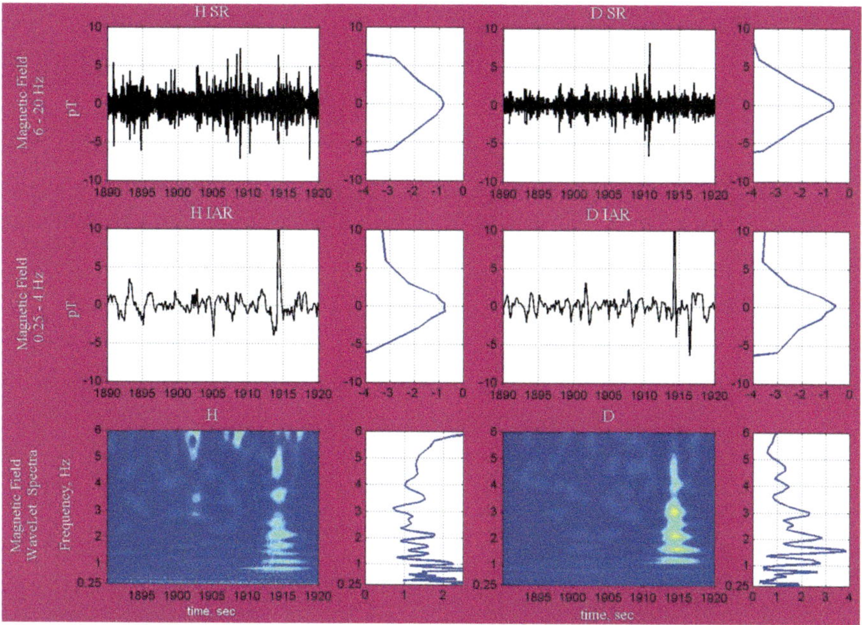

Fig. 5.6 Typical temporal dependences (*upper and middle lines* of panels), dynamic spectrograms, and power spectra (*lower line* of panels) of magnetic background noise observed at Karimshino station, Kamchatka peninsula on 13 September 2000. The time in seconds is counted from the starting moment 2,100 LT. For details see the text. Taken from Surkov et al. (2006)

can be seen in Fig. 5.9. In making this plot we have used the whole interval from 21 h till 22 h and D component recordings related to only the low-frequency channel.

Some understanding of the experimental data alluded to above is necessary for adequate interpretation of the phenomena. An electromagnetic wave originating from a lightning discharge undergoes dispersion and dissipation in the Earth-Ionosphere cavity in such a way that the low-frequency part of the wave spectrum falls off more rapidly with distance than does the high-frequency part. This is due to that the low frequencies determine the lightning near field which depends on distance r as r^{-3} whereas the high frequencies make a main contribution to the far-field spectrum which tends to decrease as r^{-1}. On the other hand, the quasistatic field falls off faster with distance than the wave field. This means that in the ULF frequency range the IAR spectrum from nearby lightning can be more intense than that of more distant discharges.

As is seen from the upper row of the panels in Figs. 5.6, 5.7, and 5.8, there are many impulses that can be associated with thunderstorm activity and only some of these impulses are accompanied by the sharp impulses in the frequency range of 0.25–4 Hz. In the above notation, these latter impulses are assumed to be a result from the nearly lightning discharges. In order to estimate the number of lightning discharges per unit time we chose the signal discrimination level as

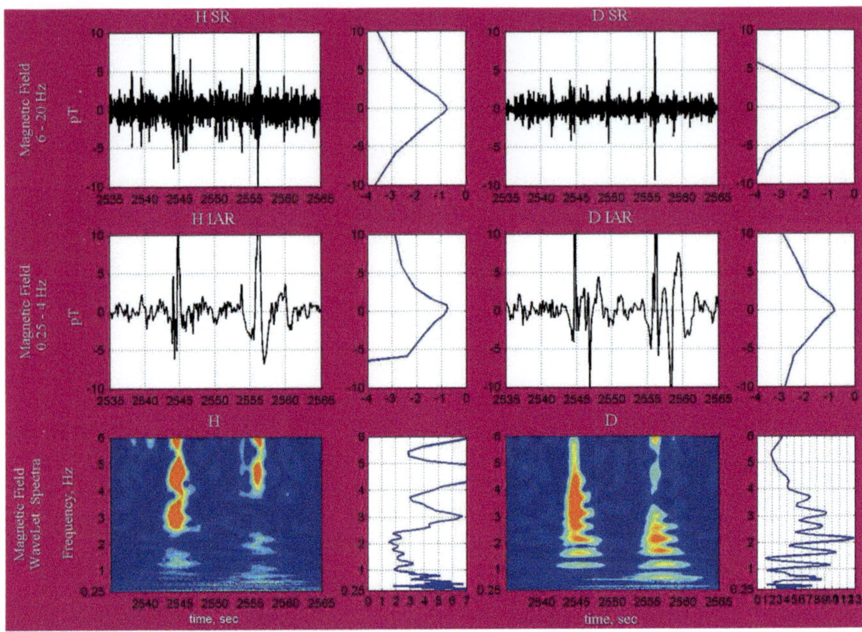

Fig. 5.7 Same as in Fig. 5.6, but for other temporal sampling. Taken from Surkov et al. (2006)

5 pT. Total number of the impulses, ΔN, with amplitude, which is greater than this level, increases with time as shown in Fig. 5.10. The lines 1 and 2 correspond to the frequency filters 6–20 and 0.25–4 Hz, respectively. Averaging over interval of 1 h results in a mean rate number of the impulses of about $\nu_1 = 0.144\,\mathrm{s}^{-1}$ and $\nu_2 = 0.023\,\mathrm{s}^{-1}$, which is typical for the nighttime conditions at Karimshino station. A thunderstorm typically produces the rate number of about 0.03–$0.05\,\mathrm{s}^{-1}$, that is close to ν_2. From here we may assume that one nearby thunderstorm and 3–6 remote ones make a major contribution to the rate number shown in Fig. 5.10. Concerning the signals of remote thunderstorms, it should be noted that the intensity of the H component is larger than that of the D component. The impulses of the H component (6–20 Hz) that are displayed in Fig. 5.10 with dashed lines 3 occur more frequently than in the D component shown with dashed line 4. On the other hand, in the frequency range of 0.25–4 Hz the rate numbers of both components are very close to each other.

Most of intense signals, which can be associated with nearby thunderstorm, have a bipolar structure as is seen from Figs. 5.6 and 5.7. It appears that the first impulse in the signals is due to the primary wave radiated by the return stroke. Since the interval between positive and negative impulses is typically ∼2 s, one may assume that the second impulse results from Alfvén wave reflection from the Alfvén velocity gradient at the upper boundary of the resonance cavity. If the typical size of the resonance cavity is 500–1,000 km, the arrival time of the reflected Alfvén wave

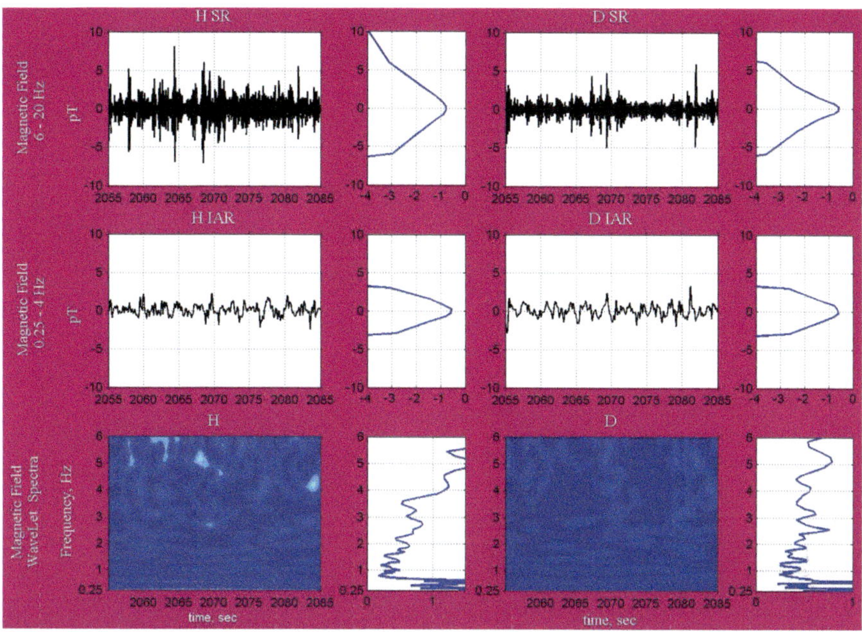

Fig. 5.8 Same as in Figs. 5.6 and 5.7, but for other temporal sampling. Taken from Surkov et al. (2006)

Fig. 5.9 Nighttime power spectra of 1-h temporal sequence (from 2,100 till 2,200 LT) observed at Karimshino station on 13 September 2000. Taken from Surkov et al. (2006)

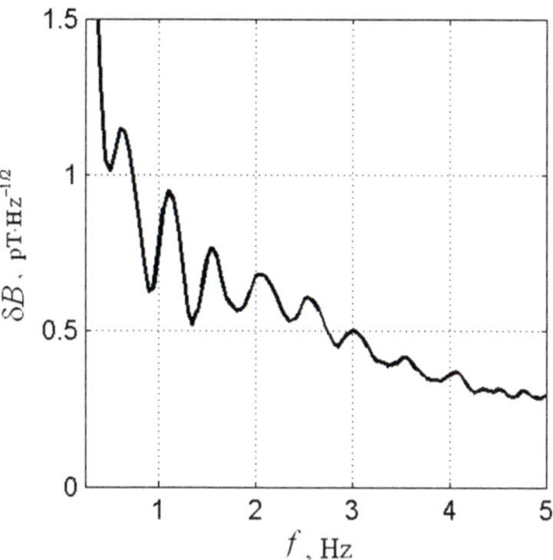

is estimated as 2 s, which is close to the signal duration. As is seen from Fig. 5.7 (the second line panels, D-component), the signal occasionally contains three distinct impulses at least. This implies a possibility for multiple wave reflections from the IAR upper and lower boundaries.

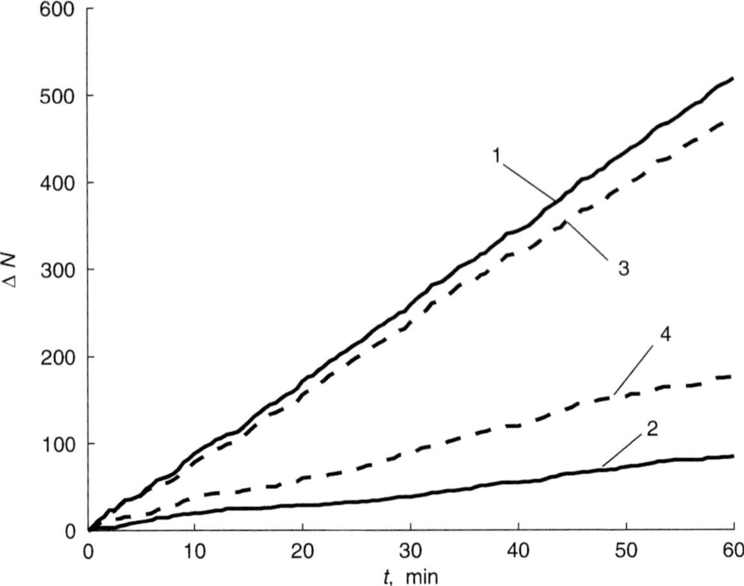

Fig. 5.10 Time variations of the magnetic impulse sum. The threshold level for the impulse amplitude is 5 pT. The time interval is same as in Fig. 5.9. The lines 1 and 2 correspond to the frequency channels 6–20 and 0.25–4 Hz, accordingly. The impulses of solely H component (6–20 Hz) is shown with *dash lines 3* while D component is shown with *dash line 4*. Taken from Surkov et al. (2006)

Overall the SRS signature recorded at low-latitude station (Bösinger et al. 2002, 2004) located at a remote site in the island of Crete, Greece (35.15°N, 25.20°E, $L = 1.3$) is consistent with the morphology of the resonance spectra detected at mid ($L = 2.1$, Fedorov et al. 2006) and high latitudes ($L = 5.2$, Yahnin et al. 2003). The IAR excitation is usually observed at nighttime condition but practically not at all during daytime. In contrast to mid and high latitudes, the resonance spectra of the low-latitude IAR have been detected every night during a half year of operation of the station at Crete (Bösinger et al. 2002). Taking into account that the island of Crete is near the African thunderstorm center, we can assume that the lightning activity in this region is capable of sustaining daily generation of the SRS at low latitudes. One more features of the low-latitude IAR is that the average frequency difference between two adjacent spectral maxima is very small (0.2 Hz) and does not exhibit local time variations from evening to nighttime. The seasonal dependence of the resonance spectra was found to be very weak, but distinct.

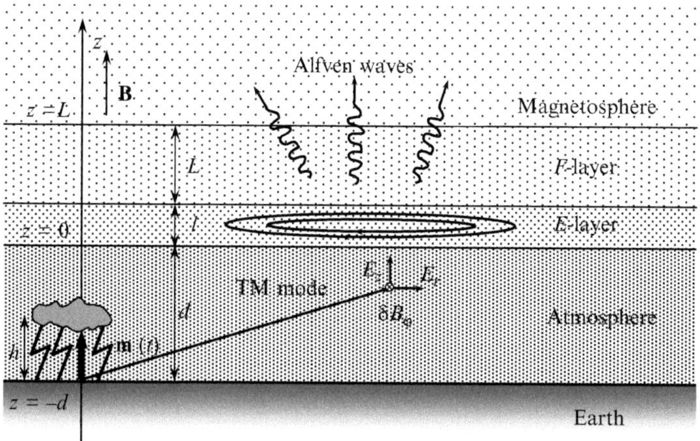

Fig. 5.11 Schematic illustration of a stratified medium model and a cloud-to-ground (CG) lightning discharge

5.3.3   IAR Excitation Due to a Solitary CG Lightning Discharge

We previously discussed the effect of the neutral wind in the ionosphere as a possible source of the generation of normal mode in the resonance cavity. In what follows we first study the lightning discharges as a source for the IAR excitation. As we shall see, the electromagnetic waves caused by the discharge can penetrates into the top ionosphere thereby exciting the IAR resonator. Suppose now that a vertical CG lightning discharge has appeared in the atmosphere at the height h above the ground. The discharge is considered as a lumped vertical electric current moment $\mathbf{m}(t)$ located in the atmosphere in the z-axis so that its coordinate is $z = h - d$ (Fig. 5.11). We recall that the current moment equals to the current produced by the medial return stroke multiplied by the lightning channel length. As before the atmosphere is supposed to be an insulator. So far as the geomagnetic field $\mathbf{B}_0$ is assumed to be pointed vertically upward, the problem becomes axially symmetrical and thus one should use the cylindrical coordinates z, r, φ. In this case all the components of the electromagnetic perturbations are independent of φ, so that $\partial/\partial\varphi = 0$. A primary field excited by the vertical current moment in the atmosphere consists of only three components, E_r, E_z, and δB_φ, which is termed the TM mode. In what follows we show that the components of the TE mode, i.e., E_φ, δB_r, and δB_z, can also arise in the atmosphere due to the mode coupling through Hall conductivity in the ionosphere. The Hall currents in the gyrotropic E-layer of the ionospheres thus play a role of the secondary source exciting the TE mode in the atmosphere.

As before all the perturbed quantities are considered to vary as $\exp(-i\omega t)$, so the Maxwell equations (4.1), (4.2) for the TM mode in the neutral atmosphere $(-d < z < 0)$ can be written as

$$\partial_z \delta B_\varphi = \frac{i\omega}{c^2} E_r,$$ (5.48)

$$\frac{1}{r}\partial_r \left(r\delta B_\varphi\right) = -\frac{i\omega}{c^2} E_z + \frac{\mu_0 m\,(\omega)}{2\pi r}\delta\,(z+d-h)\,\delta\,(r),$$ (5.49)

$$\partial_z E_r - \partial_r E_z = i\omega\delta B_\varphi.$$ (5.50)

where $m\,(\omega)$ is the Fourier transform of the lightning current moment and $\delta\,(x)$ denotes Dirac's delta-function. The second term on the right-hand side of Eq. (5.49) describes the current density due to the lightning discharge.

As before the ground is assumed to be a uniform conductor with constant conductivity σ_g. In the case of the axially symmetrical problem Eq. (5.28) for the TM mode in the ground can be written as

$$\partial_z \delta B_\varphi = -\mu_0 \sigma_g E_r,$$ (5.51)

$$\frac{1}{r}\partial_r \left(r\delta B_\varphi\right) = \mu_0 \sigma_g E_z.$$ (5.52)

We recall that the TM mode components can be expressed through the scalar potentials Φ and A via Eqs. (5.96), (5.98), and (5.100). It is customary to seek for the solution of the axially symmetrical problem in the form of Bessel transform. Substituting the potentials Φ and A into the Maxwell's equations for the neutral atmosphere and for the ground and applying the Bessel transform to those equations, we obtain a set of equations for the functions $\Phi\,(k, z, \omega)$ and $A\,(k, z, \omega)$, where k is the parameter of the Bessel transform given by Eq. (5.134). These equations should be supplemented by the proper boundary conditions at $z = 0$. A detailed solution of this problem is found in Appendix E. In the atmosphere, altitude range $0 > z > h - d$, the solution (5.145) is reduced to the following:

$$A = A\,(-d)\cosh\{k\,(d+z)\} - \frac{\mu_0 m\,(\omega)}{2\pi k}\sinh\{k\,(z+d-h)\},$$ (5.53)

where $A\,(-d)$ is the value of the potential A at the ground surface while the potential Φ is derivable from A through Eq. (5.132).

As we have noted above, the TE mode can also arise in the atmosphere due to the mode coupling via the Hall conductivity in the ionosphere. A detailed analysis of Maxwell's equation for the TE mode in the neutral atmosphere and in the ground is given in Appendix E. The TE mode components, δB_r, δB_z, and E_φ, can be expressed through the potential function Ψ via Eq. (5.131). Applying a Bessel transform to Maxwell's equation for the atmosphere and conducting ground with the proper boundary condition at $z = -d$ gives a set of equations for the potential $\Psi(k, z, \omega)$. It is interesting to note also that these equations completely coincide with Eqs. (5.112) and (5.113) for the atmosphere and ground derived for the "plane symmetry" problem. The only difference is that the "wave number" $k_\perp$ in Eqs. (5.112) and (5.113) should be replaced by the factor k. This means that the solution of the axially symmetrical problem for $\Psi(k, z, \omega)$ is the same as that given by Eqs. (5.29)–(5.31). This analogy can be extended to the ionosphere and magnetosphere equations.

Before discussing the wave equation for the plasma, it is useful to give some concerns about the axially symmetrical problem. In our model the Maxwell equations (4.2) and (5.2) for the magnetosphere are split into two independent set of equations in analogy to the plane symmetry problem. Depending on the components E_r, E_z and δB_φ, the first set of equations describes the shear Alfvén wave and thus is related to the TM mode in the atmosphere. The second equation set, which contains the components δB_r, δB_z and E_φ, corresponds to the FMS wave and is related to the TE mode in the atmosphere.

As before we assume that the parallel plasma conductivity is infinite so that the electromagnetic field can be presented via two scalar potential, Ψ and Φ. In greater details this problem is examined in Appendix E where the Bessel transform (5.134) is applied to the Maxwell equations to yield the equations for scalar potentials. These equations can be reduced to the form that completely coincide with Eqs. (5.12) and (5.13) for the shear Alfvén and FMS waves in the plane problem. In other words, after applying the Fourier and Bessel transforms to the Maxwell equations, both the problems, plane and axially symmetrical, lead to the same wave equations for the potentials Φ and Ψ, and thus both the problems can be studied jointly. This conclusion holds true for all the problems we are going to study in this section.

Some complication appears in the E layer studies due to the importance of mode coupling to the shear Alfvén and FMS modes. For simplicity, the neutral wind velocity is assumed to be independent of azimuthal angle φ in the ionosphere, so that we have an axially symmetrical problem. In the thin E layer approximation, which is valid as $l \ll l_s$, where l is the thickness of the E layer and l_s is the skindepth, we obtain the boundary conditions (5.167) and (5.168) at $z = 0$ for the jump of the potentials and their derivatives across E layer.

The solution of the problem for the neutral atmosphere and the ground is treated in any detail in Appendix E. The potential functions in the ionosphere and the atmosphere is matched by virtue boundary conditions (5.167)–(5.168) to yield the general solution of the problem. To an accuracy of the factor $(V_{AI}/c)^2 \approx 10^{-6} \ll 1$, the electromagnetic field at the ground is given by Eqs. (5.181)–(5.182). Taking the inverse Bessel transform of the functions b_r and b_φ, these equations are reduced to the following:

$$\delta B_r(\omega, r, -d) = M g_r(\omega, r) + \frac{i L B_0}{V_{AI}} w_r(\omega, r), \qquad (5.54)$$

$$\delta B_\varphi(\omega, r, -d) = M g_\varphi(\omega, r). \qquad (5.55)$$

The spectrum of lightning discharge is given by $m(\omega) = MF(\omega)$, where M is the "magnitude" of the current moment and the function $F(\omega)$ describes a shape of the spectrum. Here we have made the following abbreviations

$$g_r = -\frac{i\mu_0\alpha_H LF(\omega)}{2\pi}\int_0^\infty \frac{\cosh(kh)}{\sinh(kd)}\frac{\xi k J_1(kr)}{q\beta_3}\,dk, \qquad (5.56)$$

$$g_\varphi = \frac{\mu_0 F(\omega)}{2\pi}\int_0^\infty \frac{\cosh\{k(d-h)\}}{\sinh(kd)}k J_1(kr)\,dk \qquad (5.57)$$

$$w_r = \int_0^\infty \left[\beta_1\alpha_H v_\varphi - \left(\alpha_H^2 + \alpha_P^2 + \beta_1\alpha_P\right)v_r\right]\frac{\xi k J_1(kr)}{q\beta_3}\,dk \qquad (5.58)$$

where the functions ξ, β_3, and q are given in Eqs. (5.30), (5.31), and (5.34), respectively, $J_1(kr)$ is the Bessel function of the first kind and the first order. As is seen from Eqs. (5.54)–(5.58), only the radial component δB_r includes the resonance factor q, while the perpendicular component δB_φ does not exhibit the resonance properties. The resonance factor q in turn depends on the "wave numbers" $\xi^2 = k^2 - i\mu_0\omega\sigma_g$, $\lambda_I^2 = k^2 L^2 - x_0^2$, and $\lambda_M^2 = k^2 L^2 - \epsilon^2 x_0^2$ for the ground, ionosphere, and magnetosphere, respectively. The factor β_3 in denominators of the functions g_r and w_r describes the field damping with distance due to the presence of exponential functions. The latter factor is a function of d and ξ; that is, a function of the atmospheric and ground parameters.

The radial component given by Eq. (5.54) consists of two terms. The first one describes the IAR excitation due to the CG lightning discharge whereas the second term is caused by the neutral gas flow inside the E-layer. Alternatively, the perpendicular component (5.55) can be excited by only the lightning discharges. However, the last statement holds true only in the case of axial-symmetrical distribution of the neutral wind velocities in the ionosphere.

5.3.4 A Model Calculation of the IAR Spectra

To study the contribution of a solitary lightning discharge to the SRS of the IAR, we now shall ignore the neutral gas flow in the E layer, so that the neutral wind velocity is assumed to be zero. Figure 5.12 shows a model calculation of the lightning-generated spectra of the resonance component δB_r for the typical nighttime parameters of the mid-latitude ionosphere. The lightning flash contains three CG return strokes. The discharge shape and the spectrum, $F(\omega)$, are approximated by Eqs. (3.7) and (3.8). The numerical parameters of the lightning discharge used in making this plot are given in Chap. 3. The numerical values for the various magnetospheric, ionospheric, and other parameters are: $V_{AI} = 500\,\text{km/s}$, $V_{AM} = 5 \times 10^3\,\text{km/s}$, $L = 500\,\text{km}$, $d = 100\,\text{km}$, $z = -d$, $h = 0$, $\sigma_g = 2 \times 10^{-3}\,\text{S/m}$,

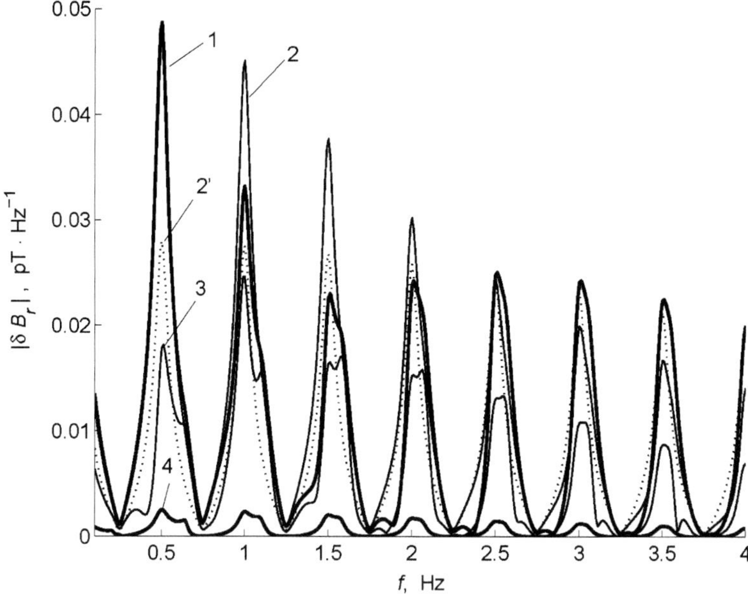

Fig. 5.12 A model calculation of the nighttime IAR spectra excited by a solitary CG lightning discharge. The radial/resonant component δB_r on the ground is shown with lines 1–4, which correspond to the distances $r = 100$, 300, 1,000, and 10,000 km, respectively. The approximate analytical solution (Eqs. (5.59) and (5.60)) at distance $r = 300$ km is shown with *dotted line 2′*. Taken from Surkov et al. (2006)

$\Sigma_P = 0.2 \ \Omega^{-1}$ and $\Sigma_H = 0.3 \ \Omega^{-1}$ (nighttime conditions). In this figure the lines 1–4 correspond to the distances $r = 100$, 300, 1,000, and 10,000 km, respectively. It is obvious from Fig. 5.12 that the spectra exhibit distinct resonance structure in such a way that the resonance frequencies are close to the IAR eigenfrequencies. By symmetry of the problem the radial component of the magnetic perturbation must tend to zero when $r \to 0$. The calculations have shown that the spectrum magnitude reaches a peak at the distance of about 300 km.

If the thunderstorm activity occurs at the distance $r \geq 10^3$ km, that is far away from the ground-recording station, then Eqs. (5.54)–(5.57) are simplified since the integrands include the rapidly oscillating function $J_1 (kr)$ with the short period $k \sim r^{-1} \leq 10^{-3}$ km^{-1}. The other slowly varying functions under the integral sign may be moved through the integral at $k = 0$ to yield

$$\delta B_r (\omega, r, -d) = M g_r \approx \frac{\mu_0 \alpha_H L M F (\omega) \xi_1 (\omega)}{2\pi r d \left[1 + \xi_1 (\omega) d \right] q_1 (\omega)}, \qquad (5.59)$$

where $\xi_1 = (-i \mu_0 \sigma \omega)^{1/2}$, and

$$q_1 = \left(\frac{i \xi_1 (\omega) L}{1 + \xi_1 (\omega) d} + x_0 \{ \beta_1 (\omega) + \alpha_P \} \right) \{ \beta_1 (\omega) + \alpha_P \} + x_0 \alpha_H^2 . \qquad (5.60)$$

In a similar fashion we get

$$\delta B_\varphi (\omega, r, -d) = M g_\varphi \approx \frac{\mu_0 M F (\omega)}{2\pi r d}. \tag{5.61}$$

Here we have set $h = 0$. To illustrate the approximate solution given by Eqs. (5.59) and (5.60), a plot of the solution is shown in Fig. 5.12 with dotted line $2'$ that corresponds to the distance $r = 300$ km. This dependence approximates the numerical solution shown with line 2 to an accuracy of tens percentages.

Some understanding of the above approximation can be achieved by comparing Eq. (5.61) with Eq. (4.40) for quasi-steady magnetic field of CG lightning occurring between the perfectly conducting ionosphere and the Earth. In the case of small distances, if $r \ll R_e$, we can simplify Eq. (4.40) with taking into account that $R_e / \cot (\theta/2) \approx R_e \theta/2 \approx r/2$. In such a case Eq. (4.40) becomes identical to Eq. (5.61), that is, the solution of the spherical–symmetric problem transforms to the solution of plane problem. This implies that the approximate solution (5.61) for the component δB_φ can be derivable from the simple model of the perfectly conducting ionosphere and the Earth.

In conclusion we discuss the assumption of vertical geomagnetic field used in this study. At first we note that the primary electromagnetic field, that is, the TM mode excited by a CG discharge is practically independent of both the state of the ionosphere and the dip angle of the geomagnetic field. For example, when deriving the perpendicular component δB_φ given by Eq. (5.61), the ionosphere in the first approximation can be considered as a perfect conductor, so that the above assumption is of minor importance.

On the other hand, the transformation of the TM-polarization into the TE-polarization is caused by the mode coupling through the Hall currents in the bottom ionosphere. The basic equations describing this coupling depend on inclination of the geomagnetic field and local time may greatly affect the ionospheric parameters. This means that the magnitude and resonant frequencies of the TE mode are sensitive to the local dip angle of the geomagnetic field in contrast to the TM mode. Since the IAR is referred to as the class of field-line resonances, the IAR eigenfrequencies depend on the length L of the field line segment confined by the IAR boundaries. Taking Eq. (5.42) for the rough estimate of the nighttime resonant frequencies, that is, $f_n = V_{AI} n / (2L)$, where n is an integer value, we can conclude that f_n can increase with magnetic latitude because of the decrease in L. Despite we have ignored the actual variations of V_{AI} with altitude, the above conclusion is in accordance with observations. For example, Bösinger et al. (2002) have reported that the IAR resonant frequencies at low-latitude station are on average smaller by a factor of 2–3 than those at high latitude. In particular, the fundamental resonant frequency varies from 0.4 to 1.0 Hz with increase in latitude.

The amplitudes of the resonances seem to be not so sensitive to the local dip angle of the geomagnetic field. In this notation, the most important factors are the intensity of local lightning activity and the distance from the observation site to the World thunderstorm centers. However, it appears that the primary TM mode excited

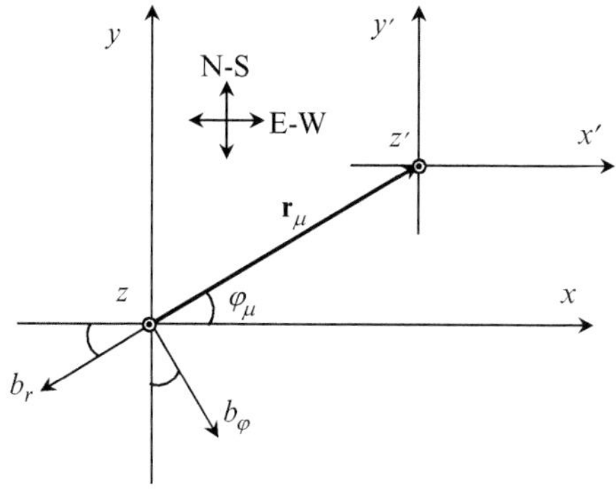

Fig. 5.13 A schematic drawing of reference and local coordinate systems. Here z and z' axes are "out of paper", b_r and b_φ are the component of magnetic variations caused by a thunderstorm located at $\mathbf{r} = \mathbf{r}_\mu$

by the stroke in the atmosphere keeps on one order of magnitude greater than the TE mode irrespective of the ionospheric parameters and the magnetic field inclination. For example, according to Belyaev et al. (1990) the ratio $\delta B_r / \delta B_\varphi$ lies in the range of 0.08–0.5 depending on the ionosphere parameters.

5.3.5   IAR Excitation Due to Random Lightning Process

We now focus attention on ULF background electromagnetic noise originated from lightning activity as a possible cause for the IAR excitation. On the basis of the approach developed in Sect. 4.3 we shall treat the lightning activity as a stochastic process. Let N be the number of thunderstorm centers simultaneously operating around the ground-based recording station. As before we use a local coordinate system, which has the x axis east-ward, the y axis to the north, and z axis vertically upward. At first we are interested in solely nearby thunderstorms, so a plane-stratified model of the medium is used. Let r_μ and φ_μ be the polar coordinates of the thunderstorms epicenters, where $\mu = 1, 2, \ldots N$, as shown in Fig. 5.13. A typical size of the thunderstorm is assumed to be smaller than the distance from the recording station, so that we ignore the lightning discharge distribution inside a thunderstorm area.

Now we also introduce a reference frame x', y' and z' fixed to the thunderstorm with number μ. In our model $\mathbf{b}\left(\mathbf{r}_\mu, t - t_{n_\mu}\right)$ denotes the random variations of the magnetic field caused by a single lightning discharge happened at the accidental moment t_{n_μ}, where $n_\mu = 1, 2, \ldots$ is a number of the lightning discharge. Considering a vertical CG discharge, we use a cylindrical coordinate system in which the lightning discharge is in the direction of the polar z' axis. In this case $\mathbf{b}$ is independent of azimuthal angle φ. As has already been stated, only radial component, b_r, among two horizontal ones contains the resonance factor, which dominates the IAR resonance properties. In the Cartesian reference frame fixed to the ground-recording station, the horizontal magnetic field can be expressed through the radial and azimuthal components as follows:

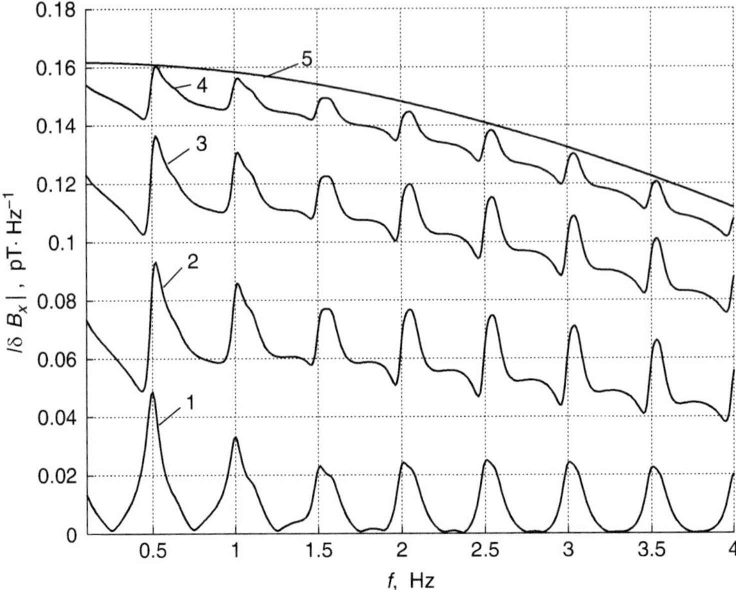

Fig. 5.14 A model calculation of the nighttime IAR spectra due to a solitary lightning discharge at distance $r = 300$ km. The component b_x is shown with lines 1–5, which correspond to the angles $\varphi = 0, \pi/8, \pi/4, 3\pi/8$ and $\pi/2$, respectively. Taken from Surkov et al. (2006)

$$b_x = b_\varphi \sin \varphi_\mu - b_r \cos \varphi_\mu, \quad \text{and}$$

$$b_y = -b_\varphi \cos \varphi_\mu - b_r \sin \varphi_\mu. \tag{5.62}$$

On the ground surface at $z = z' = -d$ the components b_r and b_φ are random values, which depend on polar radius r_μ and the time interval $t - t_{n_\mu}$.

Experimental recordings of the horizontal field contain a certain mixture of both resonant, b_r, and non-resonant, b_φ, components. The observations depend on the angle φ between the direction to lightning flash and to the x axis (Fig. 5.13). To illustrate this, we have calculated the spectra of a single CG discharge at fixed distance $r = 300$ km and different angles. The lines 1–5 in Fig. 5.14 correspond to the angles $\varphi = 0, \pi/8, \pi/4, 3\pi/8$, and $\pi/2$, respectively. Not surprisingly, the most distinct signature of the SRS is expected for the angle $\varphi = 0$ when the signal is dependent on only the radial field in contrast to the case of $\varphi = \pi/2$ when the signal contains only perpendicular, b_φ, component.

As is seen from Eqs. (5.54) and (5.55), the magnetic field of the single lightning discharge is proportional to the current moment magnitude, M_{n_μ}, and thus can be written similarly to Eq. (4.42), i.e., $\mathbf{b}\left(\mathbf{r}_\mu, t - t_{n_\mu}\right) = M_{n_\mu} \mathbf{G}\left(\mathbf{r}_\mu, t - t_{n_\mu}\right)$, where M_{n_μ} is a random value whereas $\mathbf{G} = \left(G_r, G_\varphi, G_z\right)$ are deterministic functions, describing the shape of single lightning discharge. The Fourier transform of these functions is defined by the functions g_r and g_φ given by Eqs. (5.56) and (5.57). In the reference frame fixed to the ground-recording station the Cartesian components of the field (b_x, b_y, b_z) are related to the cylindrical components (b_r, b_φ, b_z) by virtue Eq. (5.62). The net magnetic field is a sum of random fields, $\mathbf{b}_{n_\mu}$, originated from the lightning totality.

Consider now a number of thunderstorm centers distributed around the ground-recording station. Since the lightning activity is treated as a random process, the net magnetic field variation, $\mathbf{B}(t)$, is a random quantity as well. The necessary summation of the random magnetic fields over all the lightning discharges is found in Sect. 4.3 and the result of this summation is given by Eq. (4.43). As has already been stated, the mean value of the net magnetic field is close to zero and thus is not interesting for practise. More usually we measure the power spectrum or spectral density of correlation matrix of the random process. A general definition of this matrix for a steady stochastic process is given by Eq. (4.45) while the basic properties of the matrix are examined in Appendix B in greater detail. In what follows we consider only the correlation between the horizontal component of magnetic field.

As before, the lightning appearance is assumed to be a Poisson random process. In such a case the spectral density of the correlation matrix is described by equations analogous to Eqs. (4.47) and (4.48). Only difference is that the functions g_θ and g_φ in these equations should be replaced by the function g_r and g_φ, given by Eqs. (5.56) and (5.57).

It follows from these equations that both g_r and g_φ are proportional to the function $F(\omega)$ while the spectral density of correlation matrix is proportional to $|F(\omega)|^2$. In the low-frequency limit the spectral density $m(\omega) = MF(\omega)$ of the current moment generated by $-CG$ return stroke is given by Eq. (4.50). The IAR eigenfrequencies lie in the ULF region where $\omega \ll \omega_4$ so that Eq. (4.50) is simplified to

$$MF(\omega) \approx I\left(\frac{I_3}{\omega_3} + \frac{I_4}{\omega_4}\right). \tag{5.63}$$

The empirical parameters I_3, I_4, ω_3 and ω_4 determine, in fact, the magnitude and duration of CC which follows the return stroke. As is seen from Eq. (5.63), the spectrum of single lightning discharge is practically constant in the ULF region.

To study the random magnetic variations in a little more detail, we approximate the actual thunderstorm distribution around the recording station with the idealized configuration that is displayed in Fig. 5.15 (Surkov et al. 2006). The thunderstorm sites shown in Fig. 5.15 with little circles are randomly distributed along the circumferences in such a way that the mean number density of the thunderstorms or thunderstorm number per unit square, $\Delta N/\Delta S$, is approximately constant. In the model the circumference radii, r_n, and the thunderstorm numbers, N_n, on the circumference are chosen in the following way $r_n = r_1 n$ and $N_n = 4n$, where $n = 1, 2, 3, \dots$ If $n \to \infty$, the summation of the thunderstorm numbers over the

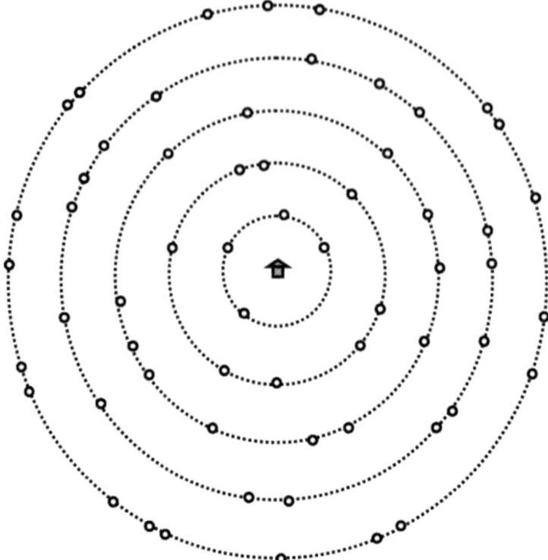

Fig. 5.15 Sketch of model thunderstorm distribution around the ground-recording station, with the location of the thunderstorm sites (*little circles*) and the station (the "house" in the center of the picture) indicated. The thunderstorm sites are randomly distributed along the "equidistant" circumferences so that the circumference radii are $r_n = r_1 n$, where $n = 1, 2, 3, \ldots$, and the mean number density of the sites is approximately constant. Taken from Surkov et al. (2006)

area with radius r_n gives $\Delta N / \Delta S \approx 2/\left(\pi r_1^2\right)$, which is a constant value as we have assumed above. Hence r_1 can be expressed through the mean distance, $\langle r \rangle$, between the thunderstorms $r_1 = \langle r \rangle (2/\pi)^{1/2}$.

In Figs. 5.16 and 5.17 we plot the spectra of the correlation function ψ_{xx} versus frequency $f = \omega/(2\pi)$. We have chosen the above parameters for the lightning discharges as well as the ionospheric parameters for the nighttime conditions. The spectrum due to four thunderstorms located on the first circumference with radius $r_1 = 550\,\text{km}$ is shown in Fig. 5.16 with lines 1. The same but for $r_1 = 400\,\text{km}$ is shown in Fig. 5.17 with lines 1. The mean number rate of lightning flashes is chosen to be $\nu = 0.05\,\text{s}^{-1}$ for each thunderstorm, that is a typical value for the single thunderstorm center. The thunderstorm sites were randomly distributed along the circumference with the help of a random-number generator. In making the plots in Figs. 5.16 and 5.17 we have used two different random samplings of the thunderstorm site distribution. In the next place we take into account the total contribution of all the thunderstorms located on the first and second circumferences. The areas occupied by these thunderstorms have the radii $r_2 = 1,100$ and $800\,\text{km}$, respectively. The spectra due to these 12 thunderstorms are shown in Figs. 5.16 and 5.17 with lines 2. Moreover, each spectrum shown in these figures is related to a certain area occupied by the thunderstorms, which make a contribution to the spectrum. The spectra shown with lines 1–7 correspond to the area radii $r = 0.55$, 1.1, 1.65, 2.2, 3.3, 4.4, and 5.5 thousands km (Fig. 5.16) and $r = 0.4$, 0.8, 1.2, 1.6, 2.4, 3.2, and 4.0 thousands km (Fig. 5.17). Total number of the thunderstorms restricted by the latest radius (line 7) is equal to 220.

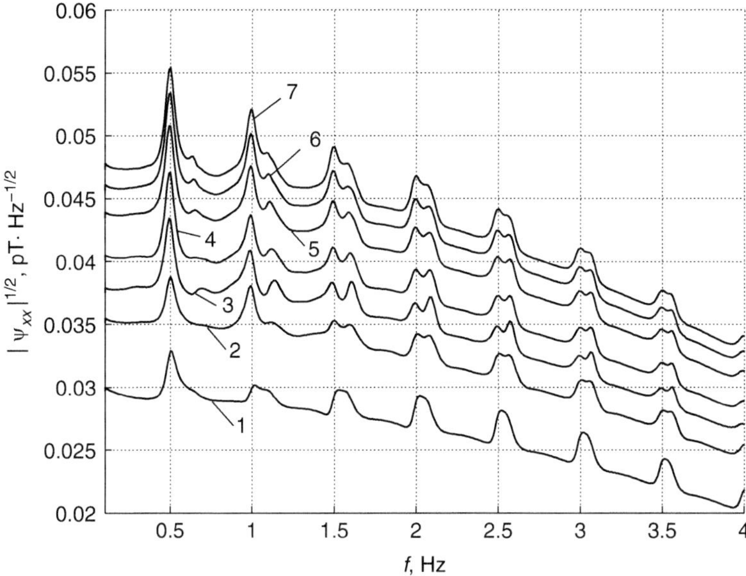

Fig. 5.16 Square root of the spectra of correlation function, $(\psi_{xx}/2\pi)^{1/2}$, on the ground for nighttime conditions. The thunderstorm sites were randomly distributed along the circumferences as shown in Fig. 5.15. In the model the spectra shown with lines 1–7 correspond to different number of the thunderstorms, which make a contribution to the spectra. These numbers are defined by the radius of area occupied by the thunderstorm sites. In making the plots of ψ_{xx} shown with lines 1–7 we have used $r = 0.55, 1.1, 1.65, 2.2, 3.3, 4.4$, and 5.5 thousands km. Taken from Surkov et al. (2006)

It is obvious from Fig. 5.16 that all the spectra exhibit distinct SRS and the resonant frequencies are close to the IAR eigenfrequencies. Starting with distance of $\sim$1–2 thousand kilometers the relative peaks weakly depend on the thunderstorm number. This means that only a few nearby thunderstorms dominate in the SRS signature which can be detected on the ground. The remote thunderstorms are of little importance since the relative amplitudes of peaks are practically independent of distance as $r > (1.5 - 2) \times 10^3$ km. The plots shown in Fig. 5.17 partly contradict to this notation since the first two spectra do not exhibit any distinct SRS. By chance, in this case the nearby thunderstorm distribution was so symmetric that all the random angles φ_μ were close to $\pi/2$ and $3\pi/2$. In such a case the resonant component g_r makes a little contribution to the correlation function ψ_{xx}.

Model calculations of the spectrum peaks versus distance at fixed frequencies are displayed in Fig. 5.18. The fixed frequencies $f = 0.5, 1.0$, and 1.5 Hz correspond to the first nighttime IAR eigenmodes. As before the distance, r, denotes the radius of the area, S, which makes a contribution to the spectra. In making the plot of ψ_{xx} we have used the model sketched in Fig. 5.15 and $r_1 = 400$ km (lines 1–3) and 550 km (lines 4–6).

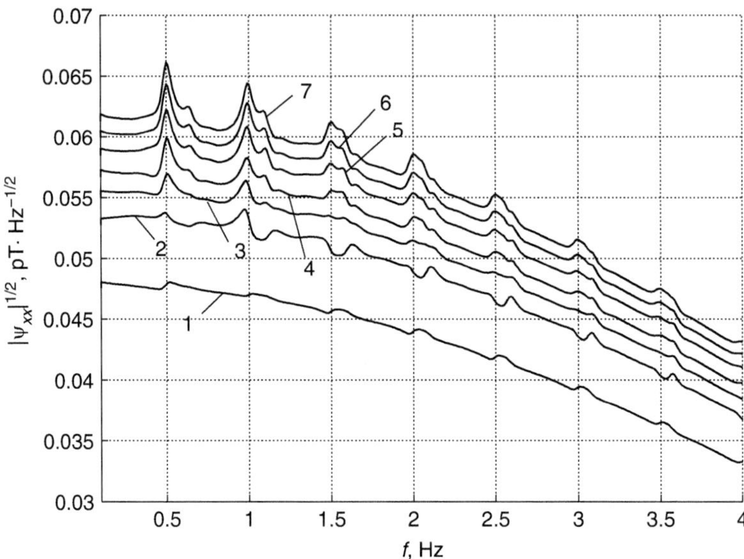

Fig. 5.17 Same as in Fig. 5.16, but for $r = 0.4, 0.8, 1.2, 1.6, 2.4, 3.2$, and 4.0 thousands km. Taken from Surkov et al. (2006)

To explain this dependence, we note that the correlation function ψ_{xx} can be represented as a sum of correlation functions produced by individual thunderstorms since they are statistically independent of each other. This means that the amplitude of the correlation function can be roughly estimated as an integral over the area S

$$\psi_{xx} \sim \int_S \psi_1 \frac{\Delta N}{\Delta S} dS, \tag{5.64}$$

where ψ_1 denotes the contribution of a single thunderstorm to the correlation function and $\Delta N/\Delta S$ is the thunderstorm number per unit square. It follows from Eqs. (5.59)–(5.61) that $\psi_1 \sim r^{-2}$. In our model the mean value of $\Delta N/\Delta S$ is assumed to be constant so that we can move it outside the integral. Taking the notice of $dS = 2\pi r dr$, and performing integration in Eq. (5.64), we obtain that $\psi_{xx} \sim \ln r$, as $r \to \infty$. This logarithmic dependence of the correlation function versus distance is compatible with that shown in Fig. 5.18. It should be noted that Fig. 5.18 only illustrates an increase of the absolute peak, whereas the relative peak value (with deduction of background noise) does not increase with distance.

As we have noted above, it is generally believed that the main excitation of the IAR at low latitude is due to the global thunderstorm activity (e.g., see Bösinger et al. 2004). This is due to the fact that the World thunderstorm centers are mainly concentrated in the vicinity of tropics. In order to examine the possibility for the same mechanism of the IAR excitation at middle and high latitudes, we consider one

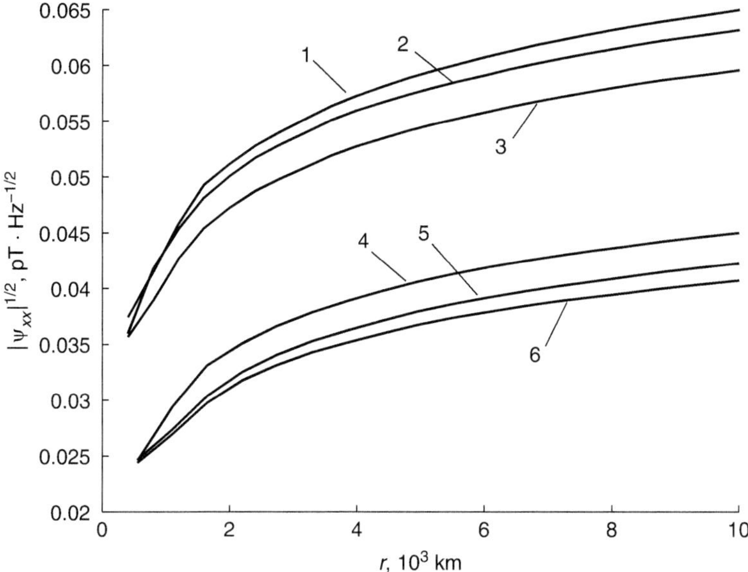

Fig. 5.18 Spectra of correlation function $(\psi_{xx}/2\pi)^{1/2}$ as a function of distance r at fixed frequencies. The distance r denotes the radius which confines the area occupied by randomly distributed thunderstorms that makes a contribution to the spectrum. The lines 1–3 correspond to the fixed resonant frequencies $f = 0.5$, 1.0, and $1.5\,\text{Hz}$, respectively. In making these plots we have used the distance $r_1 = 400\,\text{km}$ between circumferences shown in Fig. 5.15. The lines 4–6 correspond to the same frequencies and distance $r_1 = 550\,\text{km}$. Taken from Surkov et al. (2006)

World thunderstorm center located far from the observation cite. In such a case we can neglect the flash distribution inside this thunderstorm center. Thus, the spectral density of the correlation matrix is simplified to

$$\psi_{rr} = 2\pi\nu\left\langle M^2\right\rangle|g_r\,(r,\omega)|^2 \quad \text{and} \quad \psi_{\varphi\varphi} = 2\pi\nu\left\langle M^2\right\rangle|g_\varphi\,(r,\omega)|^2, \qquad (5.65)$$

where ν is the average rate of the lightning discharges in the area occupied by the World thunderstorm center. Substituting approximating Eqs. (5.59) and (5.61) for the functions g_r and g_φ into Eq. (5.65) yields

$$[\psi_{rr}\,(\omega)]^{1/2} = \frac{\mu_0\alpha_H L}{rd}\left|\frac{F\,(\omega)\,\xi_1\,(\omega)}{[1+\xi_1\,(\omega)\,d]\,q_1\,(\omega)}\right|\left(\frac{\nu\left\langle M^2\right\rangle}{2\pi}\right)^{1/2}, \qquad (5.66)$$

$$[\psi_{\varphi\varphi}\,(\omega)]^{1/2} = \frac{\mu_0\,|F\,(\omega)|}{rd}\left(\frac{\nu\left\langle M^2\right\rangle}{2\pi}\right)^{1/2}. \qquad (5.67)$$

The numerical calculations have shown that this estimate is consistent in magnitude with the IAR observations if the distance r is not far as 1,000 km from the thunderstorm center (Surkov et al. 2005a). This implies that this mechanism can explain, in principle, the IAR observations at low latitudes. On the other hand, this estimate disagrees sharply with the IAR spectrum magnitude recorded far from the thunderstorm center. For example, substituting the parameter $r = 10^4$ km into Eqs. (5.66) and (5.67) gives the value 0.01–0.02 pT/Hz$^{1/2}$, that is one or two order of magnitude smaller than the power spectrum observed at middle latitudes. For example, one may compare this value with data gathered at Karimshino station in Kamchatka peninsula (Molchanov et al. 2004), also see Figs. 5.6, 5.7, 5.8, 5.9.

Finally we arrive at the following conclusions. The stochastic model of lightning activity predicts that the global thunderstorm can make a main contribution to SRS of IAR at low latitudes whereas the nearby thunderstorms are the most appropriate candidate to explain the IAR observations at middle latitudes.

5.3.6   IAR Excitation Due to Ionospheric Neutral Wind

In the remainder of this section we focus our attention on the neutral winds in the bottom ionosphere as a possible origin of the IAR excitation in the mid- and high-latitudes. In this important altitude range, the electric fields are basically generated by neutral winds in the E-layer of the ionosphere (Kelley 1989). As for the acoustic energy transfer from the neutral gas flow into the energy of electromagnetic vibrations, we note that this mechanism is similar to the acoustic autovibration in such a system as "a police whistle" (Surkov et al. 2005b). Indeed, let us imagine a vertical cylindrical case/shell bounded from one end and opened from another. It is known that an aerial flux externally tangent to the open end of the shell results in excitation of the vertical gas vibrations inside the shell that in turn gives rise to amplification of aerial column eigenmodes. In such a case the energy flux coming from the external source is governed solely by the aerial column itself. The fluctuations of the tangent aerial flux whose frequencies are close to the aerial column eigenfrequencies in the shell can give rise to enhancement of the eigenmode magnitudes.

Our analysis shows that the similar scenario may operate in the ionospheric resonance cavity. In this case the neutral wind in the lower ionosphere can serve as an energy source for excitation of the shear Alfvén and FMS waves in the IAR. First, a part of the gas kinetic energy is transferred into the energy of the electric current in the conductive ionospheric slab, which is then converted into the energy of the shear and fast modes. Some of this energy is lost/dissipated due to the ionospheric Joule heating and wave energy leakage through the upper boundary of IAR into the magnetosphere.

Consider again an idealized slab geometry that ignores the magnetic field dip angle but includes conductivity variations with height along the magnetic field direction. The IAR excitation is supposed to be only due to the ionospheric wind system. The solution of this problem given by Eqs. (5.29) and (5.32) describes the potential function Ψ in the atmosphere. Substituting this potential function into Eqs. (5.111) and rearranging, we obtain

$$\mathbf{b} = \frac{B_0 F_0 \left(\hat{\mathbf{z}} + i \mathbf{k}_\perp \xi / k_\perp^2 \right)}{q \beta_3}, \tag{5.68}$$

where the functions β_3, F_0, and q are given by Eqs. (5.31), (5.33) and (5.34).

Now we suppose that the acoustic perturbations in neutral gas propagate horizontally with constant speed $\mathbf{U}$ parallel to the wave vector $\mathbf{k}_0$. The value of U can be close to the velocity of IGW at the altitudes of E-layer. For the sake of simplicity, we assume that the mass velocity of the neutral gas flow has the following form:

$$\mathbf{V} = \mathbf{V}_\mathrm{m} \exp\left[i \mathbf{k}_0 \cdot (\mathbf{r} - \mathbf{U}t) \right], \tag{5.69}$$

where $\mathbf{V}_\mathrm{m}$ is the amplitude of the neutral gas variations and $\mathbf{r}$ is a position vector perpendicular to the vertical z axis. For a large-scale flow pattern the neutral gas should be considered practically incompressible, i.e., $\nabla \cdot \mathbf{V} = 0$ (Kelley 1989). Hence, $\mathbf{k}_0 \cdot \mathbf{V} = 0$ so that $\mathbf{k}_0$ is orthogonal to $\mathbf{V}$.

A temporal Fourier transform of Eq. (5.69) is given by

$$\mathbf{v}\left(\omega, \mathbf{r}\right) = \frac{\mathbf{V}_\mathrm{m} \exp\left(i \mathbf{k}_0 \cdot \mathbf{r} \right)}{i \left(\omega - \mathbf{k}_0 \cdot \mathbf{U} \right)}. \tag{5.70}$$

A spatial Fourier transform of Eq. (5.70), $\mathbf{v}\left(\omega, \mathbf{k}_\perp\right)$, has the pole which corresponds to $\mathbf{k}_\perp = \mathbf{k}_0$. The magnetic field $\mathbf{b}\left(\omega, \mathbf{k}_\perp\right)$ is proportional to $\mathbf{v}\left(\omega, \mathbf{k}_\perp\right)$ and thus Eq. (5.68) contains the same pole. Performing an integration of Eq. (5.68) over $\mathbf{k}_\perp$ gives the inverse Fourier transform of the magnetic field, that is $\mathbf{b}\left(\omega, \mathbf{r}\right)$. To study the contribution of the pole $\mathbf{k}_\perp = \mathbf{k}_0$ into this integral, one should take a residue of Eq. (5.68) at this point. Then the function F_0 is reduced to the form

$$F_0 = \frac{i L (\mathbf{k}_0 \times \mathbf{V}_\mathrm{m})_z \beta_1 \alpha_P}{V_{AI} \left(\omega - \mathbf{k}_0 \cdot \mathbf{U} \right)}. \tag{5.71}$$

Substituting Eq. (5.71) into Eq. (5.68) one can estimate the spectrum peaks caused by the neutral gas flow at E-layer. The calculations of this spectrum at various ionospheric and atmospheric parameters have shown a distinct SRS of the electromagnetic variations on the ground surface (Surkov et al. 2004). In making these calculations the numerical values of the gas flow parameters were as follows: $V_\mathrm{m} = 50\,\mathrm{m/s}$, $U = 500\,\mathrm{m/s}$, and $k_0 = 0.01\text{–}0.02\,\mathrm{km}^{-1}$. For the nighttime conditions the magnitude of the first spikes was of the order of $30\text{–}100\,\mathrm{pT/Hz}^{1/2}$. This result should be considered as only a rough estimate of the magnitude because the latter strongly depends on k_0, the source spectrum and on other parameters. For example, the increase in k_0 results in the decrease of the power spectra and the magnitude of the spikes. This follows from the fact that the denominator in Eq. (5.68) contains β_3 (5.31) which includes the exponential functions of $k_0 d$, so that if $k_0 d \gg 1$, the signals become practically undetectable on the ground.

The fluctuations of the neutral wind can lead to the resonant amplification of the energy flux flowing from the wind into the resonance cavity if the fluctuation frequency range is close to the IAR eigenfrequencies. Such fluctuations caused by the gas turbulence are usually observed in the vicinity of turbopause and we assume that they can occur in the E-layer. The gas flow pattern is characterized by the Reynolds number $\mathrm{Re} = \Delta V \lambda \rho / \eta$, where ρ is the mass density of the neutral gas, η is the coefficient of gas viscosity due to molecular collisions, ΔV denotes the variation of the mean gas velocity, and λ is the characteristic scale of the variations. In order to find the frequency range typical for the turbulent pulsations we need to estimate the Reynolds number. The value of η can roughly be estimated as $\eta \sim (k_B T \overline{m}_n)^{1/2} / \sigma_c$, where k_B is Boltzmann constant, T is the gas temperature, $\overline{m}_n$ denotes the mean molecule mass, and σ_c is the collisional cross-section of the neutral particles. At the altitudes range of 100–130 km the nitrogen molecules are predominant. So, at the E-layer the average molecule mass $\overline{m}_n$ is equal to $\approx$27–28 units of proton mass that approximately corresponds to nitrogen molecules while the typical collisional cross-section $\sigma_c \sim 0.8 \times 10^{-18}$ m^2. Using these parameters one can find the rough estimate $\eta \sim 1.6 \times 10^{-5}$ Pa s. It should be noted that the effective gas viscosity can be much larger than the molecular viscosity calculated above, for example, due to the interaction between eddies in the gas flow located below turbopause (Kelley 1989). Our estimation is rather relevant to the E-layer where the turbulent mixing gradually decreases.

The mass density and the neutrals number density fall off approximately exponentially with altitude in the atmosphere. Inside the ionospheric E-layer $n_m \sim \left(7 \times 10^{17} – 10^{19}\right)$ m^{-3} depending on the altitude. Choosing $n_m = 2 \times 10^{18}$ m^{-3} as an average value one obtains $\rho = n_m \overline{m}_n \approx 9 \times 10^{-8}$ kg/m^3. The altitude profile of the wind velocity is subject to diurnal and seasonal variations. Typically, the diurnal wind variations increase with altitude from 10–30 m/s at 95 km up to 100–150 m/s at 200 km. So, the value $\Delta V \sim (10 - 100)$ m/s seems to be a relevant estimate for the wind velocity fluctuations at the altitudes of 100–130 km. Taking $\lambda \sim (1 - 10)$ km as a characteristic spatial scale of such fluctuations, we finally obtain $\mathrm{Re} \sim 60 - 6 \times 10^3$.

In this picture, one can assume that such a great value of the Reynolds number exceeds the critical value which is necessary for transition from the laminar to turbulent gas flow. To study the possible effect of the gas turbulence in a little more detail we consider a scaling law for turbulent gas flow. The Kolmogorov theory (e.g., see Landau and Lifshits 1986) assumes that if a neutral flow is stirred at some wavelength λ, the certain structures will be formed in a so-called inertial subrange in k space, where the energy will cascade to larger and larger values of k, i.e., from the large- to the small scales. The cascade is bounded from below by the value λ^{-1} and from above by the so-called Kolmogorov dissipative scale $k_{\mathrm{m}} = \lambda^{-1} \mathrm{Re}^{3/4}$, where the influence of the molecular viscosity becomes significant and the energy dissipation occurs. So within the interval $\lambda^{-1} \ll k \ll \lambda^{-1} \mathrm{Re}^{3/4}$ the energy is transferred from eddy to eddy with no net energy gain or loss. In an isotropic homogeneous medium the omnidirectional spectral density of the mechanical energy of the turbulent flow has a power law spectrum $\propto k^{-5/3}$. The typical frequencies of turbulent pulsations are evaluated as $\omega \sim kV$, where V is the smoothed mean velocity that slowly varies along the flow. Hence we find that the Kolmogorov spectrum is localized in the frequency range given by

$$\frac{V}{\lambda} \ll \omega \ll \frac{V}{\lambda}\mathrm{Re}^{3/4}, \tag{5.72}$$

where the lower margin corresponds to large-scale pulsations whereas the upper one stands for the dissipative turbulence scale, i.e., for the smallest pulsations in the turbulent flux. For instance, taking the parameters $V = 100\,\mathrm{m/s}$, $\Delta V = 50\,\mathrm{m/s}$ and $\lambda = 10\,\mathrm{km}$ one obtains $0.01 \ll \omega \ll 4\,\mathrm{Hz}$. These estimates show that the typical frequency band of the gas flow turbulence can be close to the eigenfrequencies of the ionospheric resonance cavity that results in the most effective transfer of the gas kinetic energy into the Alfvén and FMS wave energy.

The analysis has demonstrated that, in principle, the wind-driven ionospheric currents are capable of producing IAR excitation and observable perturbations at the ground level. One test of this theory can use the fact that the SRS signature should be sensitive to the magnitude of the wind velocity at the ionospheric altitudes. A standard technique for measuring the wind velocity is based on the Doppler shift of electromagnetic waves reflected from the ionosphere. When this book is being prepared, these tests have not been carried out.

In summary, the principal results of this section are as follows:

1. The IAR dispersion relation is split into two coupled modes, the shear Alfvén mode and the FMS mode. The eigenfrequencies of the Alfvén mode practically do not depend on the perpendicular wave number $k_\perp$, whereas the eigenfrequencies of the FMS mode approximately follow the linear dependence on $k_\perp$. The FMS mode damping rate decreases with the increase in $k_\perp$ while the Alfvén mode exhibits the opposite tendency.
2. It follows from the theory that the IAR power spectra exhibit the SRS only during the nighttime conditions. This conclusion agrees with the observations both at middle and low latitudes. The nighttime conditions are thus more preferable for the IAR spectrum observation.
3. Overall, the predicted IAR spectra are consistent in magnitude and resonant frequencies with observations. The typical IAR eigenfrequencies lie in the range of 0.5–5 Hz depending on the magnetic latitude and the ionospheric parameters. The average frequency difference Δf between two adjacent peaks in the spectrum is about 0.2–0.5 Hz.
4. The predicted shape of the IAR spectrum is practically independent of the shape of single lightning spectrum while the IAR spectrum magnitude depends on both the mean number of lightning discharges per second and the magnitude of low-frequency part of the lightning spectrum. This implies that the CC of return strokes can greatly affect the magnitude of IAR spectra.

5. At low latitudes the basic source of IAR excitation is believed to be the ULF electromagnetic noise produced by random lightning discharges due to the global thunderstorm activity. However, at middle latitudes far from tropical thunderstorm centers, the regional thunderstorms can make a main contribution to the observed IAR power spectra. The solitary CG lightning discharge in the vicinity of ground-recording station may result in the impulse IAR excitation which is capable of producing observable SRS signature on the ground. Additionally, the IAR excitation at mid-latitudes can be associated with the turbulent motions of the neutral winds. At high latitudes other mechanisms can play a key role in the generation of IAR eigenmodes. Among them are the magnetospheric convective flow and the fast feedback instability induced by the precipitating energetic electrons.

Appendix C: Vector and Scalar Potentials of Electromagnetic Field

General Description

In this section we introduce the standard vector and scalar potentials of the electromagnetic field in a conducting medium immersed in the external magnetic field $\mathbf{B}_0$. To treat the electric and magnetic fields, we need Maxwell's equations, which, in their full form, are given by Eqs. (1.1)–(1.4). If we seek for the solution of these equations in the form $\mathbf{B} = \mathbf{B}_0 + \delta\mathbf{B}$, where $\delta\mathbf{B}$ is the small variation of $\mathbf{B}_0$, the electromagnetic field can be represented through the vector potential, $\mathbf{A}$, and scalar potential, Φ, as follows (Jackson 2001)

$$\delta\mathbf{B} = \nabla \times \mathbf{A}, \tag{5.73}$$

$$\mathbf{E} = -\nabla\Phi - \partial_t\mathbf{A}. \tag{5.74}$$

Considering two important cases when the external field $\mathbf{B}_0$ is a constant value or when $\mathbf{B}_0$ denotes the Earth's magnetic field in the dipole approximation given by Eq. (1.32), we have the condition $\nabla \times \mathbf{B}_0 = 0$. Taking the notice of this condition and substituting the field presentation given by Eqs. (5.73) and (5.74) into Maxwell equations (1.2) and (1.3) converts these equations into identities.

Let z axis be positive parallel to the external/unperturbed magnetic field $\mathbf{B}_0$ and $\hat{\mathbf{z}} = \mathbf{B}_0/B_0$ be a unit vector parallel to $\mathbf{B}_0$. In this notation the total vector potential can be written as $\mathbf{A} = A\hat{\mathbf{z}} + \mathbf{A}_\perp$, where the second term represents the perpendicular component of the vector potential. We choose the calibration equation for the vector potential in the form

$$\nabla_\perp \cdot \mathbf{A}_\perp = 0, \tag{5.75}$$

where $\nabla_\perp$ denotes the perpendicular component of the gradient, that is $\nabla_\perp = \left(\partial_x, \partial_y\right)$, where the symbols $\partial_x = \partial/\partial x$ and $\partial_y = \partial/\partial y$ denote the partial derivatives with respect to x and y, respectively. It follows from Eq. (5.75) that the vector $\mathbf{A}_\perp$ can be written in the form

$$\mathbf{A}_\perp = \nabla_\perp \times (\Psi\hat{\mathbf{z}}), \tag{5.76}$$

where Ψ is the second scalar potential. Indeed, substituting Eq. (5.76) for $\mathbf{A}_\perp$ into Eq. (5.75) gives an identity. Hence we get

$$\mathbf{A} = A\hat{\mathbf{z}} + \nabla_\perp \times (\Psi\hat{\mathbf{z}}). \tag{5.77}$$

Subsisting Eq. (5.76) for $\mathbf{A}_\perp$ into Eqs. (5.73) and (5.74) and rearranging yields

$$\delta\mathbf{B} = (\nabla_\perp A) \times \hat{\mathbf{z}} + \nabla_\perp \partial_z \Psi - \hat{\mathbf{z}}\nabla_\perp^2 \Psi, \tag{5.78}$$

and

$$\mathbf{E} = -\nabla_\perp \Phi - \nabla_\perp \times (\hat{\mathbf{z}}\partial_t \Psi) - \hat{\mathbf{z}} (\partial_z \Phi + \partial_t A). \tag{5.79}$$

Potentials of Shear Alfvén and Compressional Waves in Plasma

The representation of the electromagnetic field via potentials is of frequent use in plasma waves physics. In specific cases the general wave equations can be split into two independent sets of equations in such a way that the scalar potentials Φ and A describe the shear Alfvén mode while the potential Ψ corresponds to the compressional mode.

As the plasma is immersed in the external magnetic field, the plasma conductivity exhibits anisotropy, which can be described by the tensor of the plasma conductivity (2.5) or by the tensor of dielectric permittivity (2.18). As the field-aligned plasma permittivity $\varepsilon_\parallel$, that is, the tensor component parallel to the magnetic field $\mathbf{B}_0$ tends to infinity, the parallel electric field becomes infinitesimal, that is $E_z = 0$. This implies that $\partial_z \Phi + \partial_t A = 0$, so that the component A can be expressed through Φ. The same is true if the field-aligned plasma conductivity $\sigma_\parallel \to \infty$. In particular, if all perturbed quantities are considered to vary as $\exp(-i\omega t)$, then

$$i\omega A = \partial_z \Phi. \tag{5.80}$$

In fact this means that the shear Alfvén and compressional modes can be described through two scalar potentials, say Φ and Ψ, instead of three potentials. For example, the shear Alfvén mode can be represented via only the potential Φ

$$\delta\mathbf{B}_A = \frac{(\nabla_\perp \partial_z \Phi) \times \hat{\mathbf{z}}}{i\omega}, \quad \mathbf{E}_A = -\nabla_\perp \Phi. \tag{5.81}$$

As can be seen from Eq. (5.81), the magnetic and electric fields of the shear Alfvén mode are both perpendicular to the external magnetic field $\mathbf{B}_0$. This conclusion is consistent with the analysis made in Chap. 3 and is illustrated in Fig. 1.15. Nevertheless, the total field-aligned Alfvén current , j_{z_A}, including the conduction and displacement currents, is nonzero. Substituting Eq. (5.81) for $\delta\mathbf{B}_A$ into Eq. (1.1), yields

$$j_{z_A} = \frac{i\nabla_\perp^2 \partial_z \Phi}{\mu_0 \omega}. \tag{5.82}$$

The FMS/compressional mode can be expressed by the potential Ψ as follows:

$$\delta\mathbf{B}_F = \nabla_\perp \partial_z \Psi - \hat{\mathbf{z}}\nabla_\perp^2 \Psi, \quad \mathbf{E}_F = i\omega\nabla_\perp \times (\hat{\mathbf{z}}\Psi). \tag{5.83}$$

It follows from Eq. (5.83) that the electrical field of the compressional mode is perpendicular to the external magnetic field as shown in Fig. 1.16, while the parallel current density $j_{z_C} = 0$.

Fourier Transform over Space

As before, we assume that a local coordinate system has the z axis positive parallel to the magnetic field $\mathbf{B}_0$. The direct and inverse Fourier transforms of the electromagnetic perturbations over the coordinates x and y perpendicular to $\mathbf{B}_0$ are given by Eqs. (5.3) and (5.4). Applying the same Fourier transform to Eqs. (5.78) and (5.79) gives the relationships (5.7) and (5.8) between the components, $\mathbf{b}$ and $\mathbf{e}$, of electromagnetic field and potential functions, A, Φ, and Ψ in the $(\omega, \mathbf{k}_\perp)$ space, where ω is the frequency and $\mathbf{k}_\perp = (k_x, k_y)$ stands for the perpendicular wave vector. In the magnetosphere and ionosphere the potentials A and Φ are related through Eq. (5.80). Combining this equation and Eqs. (5.7) and (5.8), we come to the two potential field representation

$$\mathbf{b} = i\mathbf{k}_\perp \partial_z \Psi + \frac{(\mathbf{k}_\perp \times \hat{\mathbf{z}})}{\omega}\partial_z \Phi + k_\perp^2 \Psi\hat{\mathbf{z}}, \tag{5.84}$$

and

$$\mathbf{e} = -i\mathbf{k}_\perp \Phi - (\mathbf{k}_\perp \times \hat{\mathbf{z}})\omega\Psi. \tag{5.85}$$

Here the potential Φ describes the shear Alfvén while the potential Ψ corresponds to the FMS mode.

In a similar fashion we may obtain the Fourier transform of the parallel electric current density produced by the shear Alfvén wave

$$j_{z_A} = -\frac{i k_\perp^2 \partial_z \Phi}{\mu_0 \omega}. \tag{5.86}$$

As we have noted above, the field representation through the vector and scalar potentials satisfies the Faraday law given by Eq. (1.2). It is useful to demonstrate, additionally, that Eqs. (5.84) and (5.85) satisfy a Fourier transform of the Faraday equation given by Eq. (5.6). In other words, we now show that substituting of Eqs. (5.7) and (5.8) for $\mathbf{b}$ and $\mathbf{e}$ into Eq. (5.6) gives an identity. To verify this statement one should take into account that

$$\mathbf{k}_\perp \times (\hat{\mathbf{z}} \times \mathbf{k}_\perp) = k_\perp^2 \hat{\mathbf{z}}, \tag{5.87}$$

and

$$\hat{\mathbf{z}} \times (\mathbf{k}_\perp \times \hat{\mathbf{z}}) = \mathbf{k}_\perp. \tag{5.88}$$

In this notation the first term on the right-hand side of Eq. (5.6) is reduced to

$$i\,(\mathbf{k}_\perp \times \mathbf{e}) = i\omega k_\perp^2 \Psi \hat{\mathbf{z}} + i\,(i\omega A - \partial_z \Phi)\,(\mathbf{k}_\perp \times \hat{\mathbf{z}}). \tag{5.89}$$

The second term of Eq. (5.6) can be converted to

$$\hat{\mathbf{z}} \times \partial_z \mathbf{e} = i\,\partial_z \Phi\,(\mathbf{k}_\perp \times \hat{\mathbf{z}}) - \omega \partial_z \Psi \mathbf{k}_\perp. \tag{5.90}$$

Combining Eqs. (5.89) and (5.90) and rearranging we come to the following equation

$$i\,(\mathbf{k}_\perp \times \mathbf{e}) + \hat{\mathbf{z}} \times \partial_z \mathbf{e} = -\omega A\,(\mathbf{k}_\perp \times \hat{\mathbf{z}}) - \omega \partial_z \Psi \mathbf{k}_\perp + i\omega k_\perp^2 \Psi \hat{\mathbf{z}} = i\omega \mathbf{b}, \tag{5.91}$$

that coincides with Eq. (5.6), which is required to be proved.

Cylindrical Coordinates

In the course of the main text, some of the phenomena are considered in the cylindrical coordinates r, φ, and z. On account of the representation of the perpendicular divergence operator in the cylindrical coordinates, the calibration equation (5.75) reduces to

$$\frac{1}{r}\partial_r\,(rA_r) + \frac{1}{r}\partial_\varphi A_\varphi = 0. \tag{5.92}$$

This equation holds true if

$$A_r = \frac{1}{r}\partial_\varphi\Psi, \quad \text{and} \quad A_\varphi = -\partial_r\Psi. \tag{5.93}$$

Finally we arrive at the following representation

$$\mathbf{A} = \frac{\hat{\mathbf{r}}}{r}\partial_\varphi\Psi - \hat{\boldsymbol{\varphi}}\partial_r\Psi + \hat{\mathbf{z}}A, \tag{5.94}$$

where $\hat{\mathbf{r}}$, $\hat{\boldsymbol{\varphi}}$, and $\hat{\mathbf{z}}$ stand for the unit vectors. Substituting Eq. (5.94) for $\mathbf{A}$ into Eq. (5.73) yields

$$\delta B_r = \frac{1}{r}\partial_\varphi A_z - \partial_z A_\varphi = \frac{1}{r}\partial_\varphi A + \partial_{rz}^2\Psi, \tag{5.95}$$

$$\delta B_\varphi = \partial_z A_r - \partial_r A_z = \frac{1}{r}\partial_{z\varphi}^2\Psi - \partial_r A, \tag{5.96}$$

$$\delta B_z = \frac{1}{r}\partial_r\left(rA_\varphi\right) - \frac{1}{r}\partial_\varphi A_r = -\frac{1}{r}\partial_r\left(r\partial_r\Psi\right) - \frac{1}{r^2}\partial_{\varphi\varphi}^2\Psi. \tag{5.97}$$

Similarly, substituting Eq. (5.94) for $\mathbf{A}$ into Eq. (5.74) yields

$$E_r = -\partial_r\Phi + \frac{i\omega}{r}\partial_\varphi\Psi, \tag{5.98}$$

$$E_\varphi = -\frac{1}{r}\partial_\varphi\Phi - i\omega\partial_r\Psi, \tag{5.99}$$

$$E_z = -\partial_z\Phi + i\omega A. \tag{5.100}$$

Here, as we have noted above, the terms depending on the potentials Φ and A describe the shear Alfvén mode, whereas the terms depending on the potential Ψ correspond to the compressional mode.

Appendix D: Solutions of the Boundary Problems

Solution of the Problem Associated with IAR

In the magnetosphere ($z > L$) the solution of wave equations for the potentials Φ and Ψ is given by Eqs. (5.15) and (5.16). Inside the IAR region ($0 < z < L$) the solution of Eq. (5.12) describing Alfvén waves can be written as

$$\Phi = \Phi\left(0\right)\cos\frac{\omega z}{V_{AI}} + C_3\sin\frac{\omega z}{V_{AI}}, \tag{5.101}$$

where C_3 and $\Phi(0)$ are undetermined coefficients. In order to match the solutions (5.15) and (5.101) at the boundary $z = L$, one should take into account a requirement of continuity of the potential Φ and its derivative $\partial_z \Phi$. Whence we get

$$C_3 \sin x_0 + \Phi(0) \cos x_0 = C_1 \exp(i x_0), \tag{5.102}$$

$$C_3 \cos x_0 - \Phi(0) \sin x_0 = i \epsilon C_1 \exp(i x_0). \tag{5.103}$$

where $x_0 = \omega L / V_{AI}$ denotes the dimensionless frequency and $\epsilon = V_{AI}/V_{AM}$. The set of Eqs. (5.102)–(5.103) can be solved for C_3 to yield

$$C_3 = i \Phi(0) \left[\frac{1 + \epsilon - (1 - \epsilon) \exp(2i x_0)}{1 + \epsilon + (1 - \epsilon) \exp(2i x_0)} \right]. \tag{5.104}$$

Substituting Eq. (5.104) for C_3 into Eq. (5.101), we come to Eq. (5.18), which describes the potential Φ inside the IAR region.

Similarly, the solution of Eq. (5.13) describing FMS waves in the region $0 < z < L$ can be written as

$$\Psi = \Psi(0) \cosh \frac{\lambda_I z}{L} + C_4 \sinh \frac{\lambda_I z}{L}, \tag{5.105}$$

where the function λ_I is given by Eq. (5.20). As before C_4 and $\Psi(0)$ denote undetermined coefficients. On account of the continuity of the potential Ψ and its derivative $\partial_z \Psi$ at the boundary $z = L$ we get

$$\Psi(0) \cosh \lambda_I + C_4 \sinh \lambda_I = C_2 \exp \lambda_M, \tag{5.106}$$

$$\Psi(0) \lambda_I \sinh \lambda_I + C_4 \lambda_I \cosh \lambda_I = C_2 \lambda_M \exp \lambda_M, \tag{5.107}$$

where the function λ_M is given by Eq. (5.17). The set of Eqs. (5.106)–(5.107) can be solved for C_4 to yield

$$C_4 = \Psi(0) \left[\frac{\lambda_I + \lambda_M - (\lambda_I - \lambda_M) \exp(2\lambda_I)}{\lambda_I + \lambda_M + (\lambda_I - \lambda_M) \exp(2\lambda_I)} \right]. \tag{5.108}$$

Substituting Eq. (5.108) for C_4 into Eq. (5.105), we come to Eq. (5.19), which describes the potential Ψ inside the IAR.

Magnetic Field Perturbations in the Atmosphere and in the Solid Earth

Since there are no sources in the neutral atmosphere $(-d < z < 0)$, the ULF electromagnetic perturbations excited by the ionospheric current in the atmosphere are described by Laplace equation (5.27). A spatial Fourier transform of this equation is given by

$$\partial_{zz}^2 \mathbf{b} - k_\perp^2 \mathbf{b} = 0. \tag{5.109}$$

A vertical electric current j_z flowing from the conducting ionosphere must be zero at the boundary $z = 0$ and everywhere in the layer $-d < z < 0$ because the atmosphere is an insulator. Taking Ampere's law, applying a Fourier transform to this equation, using Eq. (5.87) and the representation (5.7) of the magnetic field via the potentials, we obtain

$$\mu_0 j_z = (\nabla \times \mathbf{b})_z = (\mathbf{k}_\perp \times \mathbf{b}_\perp)_z = k_\perp^2 A = 0, \tag{5.110}$$

whence it follows that $A = 0$ in the atmosphere including the upper boundary $z = 0$. Thus the magnetic field in the atmosphere is derivable from only the potential Ψ

$$\mathbf{b} = \left(k_\perp^2 \hat{\mathbf{z}} + i \mathbf{k}_\perp \partial_z \right) \Psi. \tag{5.111}$$

Substituting Eq. (5.111) for $\mathbf{b}$ into Eq. (5.109) yields

$$\partial_{zz}^2 \Psi - k_\perp^2 \Psi = 0. \tag{5.112}$$

The solid Earth $(z < -d)$ is supposed to be a uniform conductor with a constant conductivity σ_g. The low frequency electromagnetic field in the solid Earth is described by the quasisteady Maxwell equation (5.28). Applying a Fourier transform to this equation, using Eq. (5.7), and rearranging, we obtain

$$\partial_{zz}^2 \Psi - \xi^2 \Psi = 0, \tag{5.113}$$

where $\xi^2 = k_\perp^2 - i\mu_0\sigma_g\omega$ is the squared "wave" number/propagation factor in the ground.

Now we need to solve Eq. (5.112) and (5.113) for the atmosphere and for the solid Earth, respectively, and then match the solutions at the boundary $z = -d$. The solution of Eq. (5.113) decays at infinity $(z \to -\infty)$ and is

$$\Psi = \Psi(-d) \exp\left[\xi(z+d)\right], \quad (\text{Re}\,\xi > 0). \tag{5.114}$$

The solution of Eq. (5.112) can be written as

$$\Psi = C_+ \exp(-k_\perp z) + C_- \exp(k_\perp z), \tag{5.115}$$

where the constants C_+ and C_- can be expressed through the constant $\Psi(-d)$ making allowance for the continuity of Ψ and $\partial_z \Psi$ at the boundary $z = -d$. This yields

$$C_\pm = \frac{1}{2}\Psi(-d)\left(1 \pm \frac{\xi}{k_\perp}\right). \tag{5.116}$$

Substituting Eq. (5.116) for $C_\pm$ into Eq. (5.115) gives the solution of the problem in the region $-d < z < 0$.

Boundary Conditions at the E-Layer of the Ionosphere

In Sect. 5.1 we have derived the boundary condition (5.26) at $z = 0$ in the approximation of an infinitely thin conducting E-layer. Now we shall express this boundary condition through scalar potentials of the electromagnetic field. In deriving this condition we shall take into account the following properties of triple vector and scalar products

$$\hat{\mathbf{z}} \times (\mathbf{v} \times \hat{\mathbf{z}}) = \mathbf{v}_\perp, \tag{5.117}$$

and

$$\mathbf{k}_\perp \cdot (\mathbf{v} \times \hat{\mathbf{z}}) = \hat{\mathbf{z}} \cdot (\mathbf{k}_\perp \times \mathbf{v}) = (\mathbf{k}_\perp \times \mathbf{v})_z. \tag{5.118}$$

Eq. (5.26) contains the jump of perpendicular magnetic field across the conducting E-layer. Taking the magnetic field representation (5.7) through the scalar potentials (A, Ψ) we obtain that $[\mathbf{b}_\perp] = i\mathbf{k}_\perp [\partial_z \Psi] + i(\mathbf{k}_\perp \times \hat{\mathbf{z}})[A]$, where the square brackets denote the jump of the functions across the E layer, for example, $[A] = A(0+) - A(0-)$. Substituting $[\mathbf{b}_\perp]$ into the boundary condition (5.26), taking into account the continuity of $\mathbf{e}_\perp$ at $z = 0$, using the potentials (A, Φ, Ψ) according to Eqs. (5.7)–(5.8)), and combining these equations with Eqs. (5.87)–(5.88) and (5.117) we find that

$$iV_{AI}\left\{[A]\mathbf{k}_\perp + [\partial_z \Psi](\hat{\mathbf{z}} \times \mathbf{k}_\perp)\right\}$$
$$= -\mathbf{k}_\perp\left\{\alpha_H \omega \Psi + i\alpha_P \Phi\right\} + (\hat{\mathbf{z}} \times \mathbf{k}_\perp)\left\{\alpha_P \omega \Psi - i\alpha_H \Phi\right\}$$
$$+ B_0\left\{\alpha_H \mathbf{v}_\perp + \alpha_P(\mathbf{v} \times \hat{\mathbf{z}})\right\}. \tag{5.119}$$

Here we have just used the identity $\mathbf{v} \times \mathbf{B}_0 = B_0(\mathbf{v} \times \hat{\mathbf{z}})$.

Taking the scalar and cross product of Eq. (5.119) with $\mathbf{k}_\perp$, taking into account Eq. (5.118) and rearranging, we get

$$[A(0)] = \frac{i\alpha_H x_0}{L}\Psi(0) - \frac{\alpha_P}{V_{AI}}\Phi(0) - f_1, \tag{5.120}$$

$$[\partial_z\Psi(0)] = -\frac{i\alpha_P x_0}{L}\Psi(0) - \frac{\alpha_H}{V_{AI}}\Phi(0) + f_2, \tag{5.121}$$

where x_0 is a dimensionless frequency defined in Eq. (5.17). Here we made use of the following abbreviations

$$f_1 = \frac{iB}{V_{AI}k_\perp^2}\left\{\alpha_P(\mathbf{k}_\perp \times \mathbf{v})_z + \alpha_H(\mathbf{k}_\perp \cdot \mathbf{v})\right\}, \tag{5.122}$$

$$f_2 = \frac{iB}{V_{AI}k_\perp^2}\left\{\alpha_P(\mathbf{k}_\perp \cdot \mathbf{v}) - \alpha_H(\mathbf{k}_\perp \times \mathbf{v})_z\right\} \tag{5.123}$$

Depending on the neutral wind velocity, $\mathbf{v}$, the functions f_1 and f_2 play a role of forcing functions/sources for the IAR excitation.

Now we use Eq. (5.29) for Ψ in the atmosphere to connect the potential Ψ and its derivative at the interface $z = 0$ between the atmosphere and the ionosphere

$$\partial_z\Psi(0-) = k_\perp\Psi(0)\frac{\xi + k_\perp\tanh(k_\perp d)}{k_\perp + \xi\tanh(k_\perp d)}, \tag{5.124}$$

where ξ is given by Eq. (5.30). Here minus in the argument of the function Ψ in Eq. (5.124) denotes that the derivative should be taken just below the E-layer.

As is seen from Eqs. (5.120) and (5.121), the boundary conditions at $z = 0$ relate the jump of values of $\partial_z\Psi$ and A just above and below the E layer of the ionosphere. It follows from Eq. (5.19) that just above E-layer the function $\partial_z\Psi$ is

$$\partial_z\Psi(0+) = \Psi(0)\frac{\beta_2\lambda_I}{L}, \tag{5.125}$$

where the function β_2 is given in Eq. (5.22). Notice that the values of $\Psi(0)$ are the same in both Eq. (5.124) and Eq. (5.125) because the function Ψ must be continuous at $z = 0$. Subtracting Eq. (5.125) from Eq. (5.124) brings about the jump of derivative $\partial_z\Psi$ across the E layer

$$[\partial_z\Psi(0)] = \Psi(0)\left(\frac{\lambda_I\beta_2}{L} - k_\perp\frac{\xi + k_\perp\tanh(k_\perp d)}{k_\perp + \xi\tanh(k_\perp d)}\right). \tag{5.126}$$

As we have noted above, the potential $A = 0$ in the atmosphere, so that the jump of function A across the E layer equals the value of A just above E-layer, that is $[A(0)] = A(0+)$. According to Eqs. (5.80) and (5.18), the jump of A is given by

$$[A(0)] = \frac{[\partial_z\Phi(0+)]}{i\omega} = \frac{\beta_1\Phi(0)}{V_{AI}}, \tag{5.127}$$

Substituting Eqs. (5.126) and (5.127) for the jump of functions A and $\partial_z \Psi$ into boundary conditions (5.120) and (5.121), we are thus left with the set

$$\frac{i x_0 \alpha_H}{L} \Psi(0) - \frac{(\beta_1 + \alpha_P)}{V_{AI}} \Phi(0) = f_1, \tag{5.128}$$

$$\frac{\alpha_H}{V_{AI}} \Phi(0) + \frac{(i x_0 \alpha_P - s)}{L} \Psi(0) = f_2, \tag{5.129}$$

where the dimensionless frequency x_0 is again defined in Eq. (5.17), and the functions f_1 and f_2 are given by Eqs. (5.122) and (5.123). Equations (5.128) and (5.129) can be solved for $\Psi(0)$.

Appendix E: Solutions of the Axially Symmetrical Problem

TM Mode in the Neutral Atmosphere and in the Ground

In Sect. 5.3 we study the electromagnetic field excited by the vertical CG lightning discharge which is located on the vertical z axis in the neutral atmosphere. The problem is axially symmetrical since the geomagnetic field $\mathbf{B}_0$ is assumed to be directed vertically upward. The components of the electromagnetic perturbations can be expressed through potential functions A, Φ and Ψ in cylindrical coordinates z, r, φ via Eqs. (5.95)–(5.100). For the axially symmetrical problem these equations are simplified to

$$\delta B_\varphi = -\partial_r A, \quad E_r = -\partial_r \Phi, \quad E_z = -\partial_z \Phi + i \omega A, \tag{5.130}$$

and

$$E_\varphi = -i \omega \partial_r \Psi, \quad \delta B_r = \partial_{rz}^2 \Psi, \quad \delta B_z = -\frac{1}{r} \partial_r (r \partial_r \Psi). \tag{5.131}$$

To treat the TM mode generated by the vertical CG discharge in the atmosphere, Maxwell's equations are required, which are given by the set of Eqs. (5.48)–(5.50). As is seen from Eq. (5.130) the TM mode components δB_φ, E_r, and E_z are represented by the potentials Φ and A and do not depend on Ψ. Substituting these components into Eqs. (5.48)–(5.50) and rearranging, we obtain that Eq. (5.50) is reduced to identity, while Eqs. (5.48) and (5.49) take the forms

$$\partial_z A = \frac{i \omega}{c^2} \Phi, \tag{5.132}$$

$$\partial_{zz}^2 A + \frac{1}{r} \partial_r (r \partial_r A) + \frac{\omega^2}{c^2} A = -\frac{\mu_0 m(\omega)}{2 \pi r} \delta(z + d - h) \delta(r), \tag{5.133}$$

where $m(\omega)$ stands for Fourier transform of the lightning current moment and $-d < z < 0$.

We shall seek for the solution of Eqs. (5.132) and (5.133) in the form of Bessel transform, for example,

$$A(k, z, \omega) = \int_0^\infty A(r, z, \omega) J_0(kr) r\, dr, \qquad (5.134)$$

where $J_0(kr)$ is the Bessel function of the first kind of zero order and the parameter k of the Bessel transform plays a role of the perpendicular wave number. For brevity, here we made use of the same designation for the potential $A(r, z, \omega)$ and its Bessel transform $A(k, z, \omega)$. The same representation is true for the potentials Φ and Ψ.

Applying the Bessel transform to Eq. (5.133) we obtain

$$\partial_z^2 A - k_a^2 A = -\frac{\mu_0 m(\omega)}{2\pi} \delta(z + d - h), \qquad (5.135)$$

where $k_a^2 = k^2 - \omega^2/c^2$, and $A = A(k, z, \omega)$. We shall restrict our study to the low-frequency limit when $k_a \approx k$. Integrating Eq. (5.135) from $z = h - d - \varepsilon$ to $z = h - d + \varepsilon$ and then formally taking $\varepsilon \to 0$, we come to the following condition at $z = h - d$

$$[\partial_z A] = -\frac{\mu_0 m(\omega)}{2\pi}, \qquad (5.136)$$

where the square brackets denote the jump of z-derivative of A at $z = h - d$. The solution of Eq. (5.135) in both regions, $z < h - d$ and $z > h - d$, should be matched via the condition (5.136). The solution of Eq. (5.135) under the requirement that A is continuous at that boundary has the form

$$A = C_1 \exp(kz) + C_2 \exp(-kz) - \frac{\mu_0 m(\omega) \eta(z')}{2\pi k} \sinh(kz') \qquad (5.137)$$

where C_1 and C_2 are arbitrary constants, $z' = z + d - h$, and $\eta(z')$ is the step-function; i.e., $\eta = 1$ if $z' > 0$ and $\eta = 0$ if $z' < 0$.

Now we shall treat electromagnetic fields in the ground, which is considered as a uniform conducting half-space ($z < -d$) with constant conductivity σ_g. For the axially symmetrical problem Maxwell's equations for the ground are given by Eqs. (5.51) and (5.52). Substituting Eq. (5.130) and for the TM mode into those equations, yields

$$\partial_z A = -\mu_0 \sigma_g \Phi, \qquad (5.138)$$

$$\partial_{zz}^2 A + \frac{1}{r} \partial_r (r \partial_r A) + i\omega \mu_0 \sigma_g A = 0. \qquad (5.139)$$

Taking the Bessel transform of Eq. (5.139) we get

$$\partial_z^2 A - \xi^2 A = 0, \tag{5.140}$$

where $A = A(k, z, \omega)$ and as before $\xi^2 = k^2 - i\mu_0\omega\sigma_g$ stands for the propagation factor in the ground. We mention in passing that Eq. (5.140) is analogous to Eq. (5.113) for the scalar potential Ψ in the ground. Taking into account that A should tend to zero when $z \to -\infty$ we chose the solution of Eq. (5.140) in the form

$$A = A(-d) \exp\left[\xi(z + d)\right], \tag{5.141}$$

where $\mathrm{Re}\,\xi > 0$ and $A(-d)$ denotes the value of potential A on the ground surface.

The components δB_φ and E_r must be continuous at $z = -d$ whence it follows that A and Φ must be continuous at $z = -d$. Taking into account Eqs. (5.132) and (5.138) for A and Φ, the boundary condition on the ground surface takes the following form

$$\partial_z A(-d + 0) = -\frac{i\omega\varepsilon_0}{\sigma_g}\partial_z A(-d - 0). \tag{5.142}$$

Taking into account the boundary conditions at $z = -d$ and combining Eqs. (5.137) and (5.141) gives a set of algebraic equations for undefined constants. These equations can be solved for C_1 and C_2 to yield

$$C_1 = \frac{A(-d)}{2}\left(1 - \frac{i\omega\varepsilon_0\xi}{k\sigma_g}\right)\exp(kd), \tag{5.143}$$

$$C_2 = \frac{A(-d)}{2}\left(1 + \frac{i\omega\varepsilon_0\xi}{k\sigma_g}\right)\exp(-kd). \tag{5.144}$$

Substituting Eqs. (5.143) and (5.144) into Eq. (5.137) yields the solution of the problem in the neutral atmosphere

$$A = A(-d)\left\{\cosh\left[k(d + z)\right] - \chi\sinh\left[k(d + z)\right]\right\}$$
$$- \frac{\mu_0 m(\omega)\eta(z')}{2\pi k}\sinh(kz'). \tag{5.145}$$

Here we introduce the dimensionless parameter $\chi = i\omega\varepsilon_0\xi/(\sigma_g k)$. Substituting Eq. (5.145) for A into Eq. (5.132), we obtain

$$\Phi = \frac{kc^2}{i\omega}\left[A(-d)\left\{\sinh\left[k(d + z)\right] - \chi\cosh\left[k(d + z)\right]\right\}\right.$$
$$\left. - \frac{\mu_0 m(\omega)\eta(z')}{2\pi k}\cosh(kz')\right]. \tag{5.146}$$

Choosing the typical parameters $\sigma_g = 10^{-3}\,\text{S/m}$, $\omega = 2\pi\,\text{Hz}$, $k = 10^{-2}$–$10^{-4}\,\text{km}^{-1}$, one can find that $\chi \approx 10^{-5}$–10^{-7}, i.e., this value can be neglected. This means that the TM mode in the atmosphere is practically independent of the ground conductivity, but that is not the case for the TE mode treated below in any details.

As is seen from Eq. (5.146), there is a discontinuity in the potential Φ at $z = h - d$ due to the presence of a step function $\eta(z')$ in the last term. This leads to an interesting question of how this discontinuity may influence the electric and magnetic fields in Eqs. (5.130) and (5.131). In other words, this gives rise to the question of whether the spatial field representation will save the continuity despite the presence of the discontinuity in the potential Φ. Some insight into this problem can be achieved by taking into account that this discontinuity results from the use of the point dipole approach for the lightning discharge. In order to clarify the problem, consider for simplicity a point electric dipole immersed in an infinite space. In such a case the solution of the problem can be written as

$$b_\varphi = \frac{m(\omega)}{4\pi} \exp\left(-k\left|z'\right|\right), \tag{5.147}$$

and

$$e_r = \pm \frac{i k m(\omega)}{4\pi\varepsilon_0\omega} \exp\left(-k\left|z'\right|\right), \tag{5.148}$$

where the sign plus in Eq. (5.148) corresponds to $z\prime > 0$ whereas the sign minus corresponds to $z\prime < 0$. Applying the inverse Bessel transform to Eq. (5.148) we get

$$e_r = \pm \frac{i m(\omega)}{4\pi\varepsilon_0\omega} \int_0^\infty k^2 \exp\left(-k\left|z'\right|\right) J_1(kr)\,dk = \frac{i m(\omega)\,r z'}{4\pi\varepsilon_0\omega\,(z'^2 + r^2)^{5/2}}. \tag{5.149}$$

As is seen from Eq. (5.149), we have obtained the function, which is continuous everywhere except for the point $z\prime = r = 0$. This means that despite the discontinuity in Eq. (5.148) at $z\prime = 0$ the inverse Bessel transform gives the continuous radial electric field. It follows from this example that the Bessel transform of the field of the point source may be represented by the discontinuous function. So one may expect that the inverse Bessel transform of Eq. (5.146) will result in a continuous function describing a spatial field representation.

TE Mode in the Neutral Atmosphere and in the Ground

As we have noted in Chap. 5, the TE mode in the atmosphere may occur due to the excitation of secondary sources such as the Hall current in the ionosphere.

Since there are no sources of the TE mode in the atmosphere the Maxwell equations (4.1) and (4.2) can be written as

$$\partial_r \delta B_z - \partial_z \delta B_r = \frac{i\omega}{c^2} E_\varphi, \tag{5.150}$$

$$\partial_z E_\varphi = -i\omega \delta B_r, \tag{5.151}$$

$$\partial_r E_\varphi = i\omega \delta B_z. \tag{5.152}$$

Substituting Eqs. (5.130) and (5.131) into Eqs. (5.150)–(5.152), one can express the TE mode components through the potential Ψ. As a result we come to the single wave equation for the potential Ψ

$$r^{-1}\partial_r \left(r \partial_r \Psi \right) + \partial^2_{zz} \Psi = -\frac{\omega^2}{c^2} \Psi. \tag{5.153}$$

Applying Bessel transform to this equation we get

$$\partial^2_{zz} \Psi - k_a^2 \Psi = 0, \qquad (-d < z < 0), \tag{5.154}$$

where $k_a^2 = k^2 - \omega^2/c^2 \approx k^2$. It should be noted that if one changes the parameter k by the perpendicular "wave number" $k_\perp$, Eq. (5.154) coincides with Eq. (5.112) for the case of "plane" atmosphere. Similarly, one can derive an equation for the ground that is completely coincident with Eq. (5.113). This means that the solution of the plane problem given by Eqs. (5.115) and (5.116) holds true in the axially symmetrical case.

The Ionosphere and Magnetosphere

In the model the space $z > 0$ consists of a solely cold collisionless plasma, which is described by Maxwell equations (4.2) and (5.2) and the plasma dielectric permittivity tensor (2.18). When the cylindrical coordinates are applied, the Maxwell equations are split into two independent sets of equations. The first set includes the components of the shear Alfvén waves, i.e., E_r, E_z and δB_φ

$$\partial_z \delta B_\varphi = \frac{i\omega}{V_A^2} E_r, \tag{5.155}$$

$$\frac{1}{r}\partial_r \left(r \delta B_\varphi \right) = -\frac{i\omega \varepsilon_\parallel}{c^2} E_z, \tag{5.156}$$

$$\partial_z E_r - \partial_r E_z = i\omega \delta B_\varphi. \tag{5.157}$$

The second one is for the three components of the FMS wave, i.e., E_φ, δB_r, and δB_z

$$\partial_r \delta B_z - \partial_z \delta B_r = \frac{i\omega}{V_A^2} E_\varphi, \tag{5.158}$$

$$\partial_z E_\varphi = -i\omega \delta B_r, \tag{5.159}$$

$$\partial_r E_\varphi = i\omega \delta B_z. \tag{5.160}$$

Since $\varepsilon_\parallel$ has been assumed to be infinite, the parallel electric field in the magnetosphere becomes infinitesimal, i.e., $E_z \to 0$. As before, the electromagnetic field is derivable by the scalar potentials A, Φ, and Ψ through Eqs. (5.130), (5.131), and (5.80). Substituting these equations into the set of Eqs. (5.155)–(5.160), and rearranging under the requirement that $i\omega A = \partial_z \Phi$, yields

$$\partial_z^2 \Phi + \frac{\omega^2}{V_A^2} \Phi = 0, \tag{5.161}$$

$$\partial_{zz}^2 \Psi + \frac{1}{r}\partial_r \left(r\partial_r \Psi\right) = -\frac{\omega^2}{V_A^2}\Psi. \tag{5.162}$$

Applying Bessel transforms to Eqs. (5.161) and (5.162), we come to the equations that are completely similar to Eqs. (5.12) and (5.13) for the shear Alfvén and compressional waves in the plane problem.

E Layer of the Ionosphere

In order to obtain the boundary conditions at the bottom of the ionosphere we now consider the conductive E layer of the ionosphere. In the framework of the axially symmetrical problem the neutral wind velocity is assumed to be independent of azimuthal angle φ in the ionosphere. Substituting Eq. (5.24) for the current density into Ampere's equation (1.5), taking the notice of $\mathbf{B} = \delta\mathbf{B} + \mathbf{B}_0$, and using cylindrical coordinates, we obtain

$$\mu_0^{-1}\partial_z \delta B_\varphi = \sigma_H \left(E_\varphi - V_r B_0\right) - \sigma_P \left(E_r + V_\varphi B_0\right), \tag{5.163}$$

$$\mu_0^{-1} \left(\partial_z \delta B_r - \partial_r \delta B_z\right) = \sigma_P \left(E_\varphi - V_r B_0\right) + \sigma_H \left(E_r + V_\varphi B_0\right), \tag{5.164}$$

where σ_P and σ_H are the Pedersen and Hall conductivities, and V_r and V_φ are the components of the neutral flow velocity. In what follows we use a thin E layer approximation, which is valid if a typical thickness of the E layer, l, is much smaller than the skin-depth in the ionosphere $l_s \sim \left(\mu_0 \sigma_P \omega\right)^{-1/2}$. Integrating Eqs. (5.163) and (5.164) with respect to z across the E layer from $z = 0$ to $z = l$ and making formally $l \to 0$, gives the boundary conditions at $z = 0$

$$\mu_0^{-1} \left[\delta B_\varphi \right] = \Sigma_H E_\varphi - \Sigma_P E_r - B_0 \left(\Sigma_H V_r + \Sigma_P V_\varphi \right), \tag{5.165}$$

$$\mu_0^{-1} \left[\delta B_r \right] = \Sigma_P E_\varphi + \Sigma_H E_r + B_0 \left(\Sigma_H V_\varphi - \Sigma_P V_r \right), \tag{5.166}$$

where the square brackets denote the jump of magnetic field across the E-layer, and the height-integrated Pedersen and Hall conductivities, Σ_P and Σ_H, are given by Eq. (5.25). As before the wind velocity $\mathbf{V}$ is assumed to be independent of the z-coordinate.

Substituting Eqs. (5.130) and (5.131) for the potentials into Eqs. (5.165) and (5.166) and applying a Bessel transform to these equations, we get

$$[A(0)] = \frac{i \alpha_H x_0}{L} \Psi(0) - \frac{\alpha_P}{V_{AI}} \Phi(0) - F_1, \tag{5.167}$$

$$[\partial_z \Psi(0)] = -\frac{i \alpha_P x_0}{L} \Psi(0) - \frac{\alpha_H}{V_{AI}} \Phi(0) + F_2, \tag{5.168}$$

where the Bessel transform of the potentials A, Φ and Ψ are given by Eq. (5.134). Here as before $\alpha_P = \Sigma_P / \Sigma_w$ and $\alpha_H = \Sigma_H / \Sigma_w$ are the ratios of the height-integrated Pedersen and Hall conductivities to the wave conductivity $\Sigma_w = (\mu_0 V_{AI})^{-1}$, and the dimensionless frequency x_0 is again defined in Eq. (5.16). Additionally we made use of the following abbreviations:

$$F_1 = \frac{B_0}{k V_{AI}} \left(\alpha_H v_r + \alpha_P v_\varphi \right), \tag{5.169}$$

$$F_2 = \frac{B_0}{k V_{AI}} \left(\alpha_P v_r - \alpha_H v_\varphi \right), \tag{5.170}$$

where v_r and v_φ are the Bessel transform of the radial and azimuthal components of the wind velocity

$$v_{r,\varphi}(\omega, k) = \int_0^\infty V_{r,\varphi}(\omega, r) \, r J_1(kr) \, dr. \tag{5.171}$$

It is interesting to note that Eqs. (5.167)–(5.170) are identical to Eqs. (5.120)–(5.123) if the parameter $\mathbf{k}_\perp$ is replaced by the factor $\mathbf{k} = i k \hat{\mathbf{r}}$ where $\hat{\mathbf{r}} = \mathbf{r}/r$ is a unite vector directed along the vector $\mathbf{r}$.

Electromagnetic Perturbations at the Ground Surface

We start with calculation of the jump of the potential A across the E layer. Taking into account that the value $A\,(0+)$ can be found from Eq. (5.127) and combining this equation with Eq. (5.53) we obtain that

$$[A\,(0)] = \frac{\beta_1}{V_{AI}}\Phi\,(0) - A\,(-d)\cosh\,(kd) + \frac{\mu_0 m\,(\omega)}{2\pi k}\sinh\,\{k\,(d-h)\}\,. \quad (5.172)$$

Substituting Eq. (5.172) into Eq. (5.167) and rearranging leads to

$$\frac{ix_0\alpha_H}{L}\Psi\,(0) - \frac{(\beta_1 + \alpha_P)}{V_{AI}}\Phi\,(0) = f, \quad (5.173)$$

where

$$f = \frac{B_0}{kV_{AI}}\left(\alpha_H v_r + \alpha_P v_\varphi\right) - A\,(-d)\cosh\,(kd) + \frac{\mu_0 m\,(\omega)}{2\pi k}\sinh\,\{k\,(d-h)\}\,. \quad (5.174)$$

Here one can see an analogy between Eqs. (5.173) and (5.128), which was derived for the plane problem. These two equations differ only in the source functions which stay on the right-hand sides of these equations.

As has already been intimated, considering an analogy between the plane and cylindrical problems, Eq. (5.168) can be reduced to the equation analogous to Eq. (5.129), i.e.

$$\frac{\alpha_H}{V_{AI}}\Phi\,(0) + \frac{(ix_0\alpha_P - s)}{L}\Psi\,(0) = \frac{B_0}{kV_{AI}}\left(\alpha_P v_r - \alpha_H v_\varphi\right), \quad (5.175)$$

Finally, one should take into account the continuity of the potential Φ at $z = 0$. As it follows from Eq. (5.132)

$$\Phi\,(0) = \frac{kc^2}{i\omega}\left\{A\,(-d)\sinh\,(kd) - \frac{\mu_0 m\,(\omega)}{2\pi k}\cosh\,[k\,(d-h)]\right\}\,. \quad (5.176)$$

The set of Eqs. (5.173), (5.175), and (5.176) can be solved for $A\,(-d)$, $\Phi\,(0)$ and $\Psi\,(0)$ to yield

$$A\,(-d) \approx \frac{\mu_0 m\,(\omega)\cosh\,[k\,(d-h)]}{2\pi k\sinh\,(kd)}, \quad (5.177)$$

$$\Psi\,(0) \approx \frac{iL}{kq}\left\{\frac{\mu_0\alpha_H m\,(\omega)\cosh\,(kh)}{2\pi\sinh\,(kd)}\right.$$

$$\left. + \frac{B_0}{V_{AI}}\left[(\alpha_H^2 + \alpha_P^2 + \beta_1\alpha_P)\,v_r - \beta_1\alpha_H v_\varphi\right]\right\}, \quad (5.178)$$

where β_1 and q are given by Eqs. (5.21) and (5.34). In deriving Eqs. (5.177) and (5.178) we have neglected the terms which contain the factor $(V_{AI}/c)^2 \approx 10^{-6} \ll 1$. The potential Φ is derivable from Eq. (5.177) through

$$\Phi(-d) = -\frac{i\xi A(-d)}{\mu_0 \sigma_g}. \tag{5.179}$$

Combining Eqs. (5.29) and (5.178) we obtain

$$\Psi(-d) = \frac{iL}{kq\beta_3} \left\{ \frac{\mu_0 \alpha_H m(\omega) \cosh(kh)}{2\pi \sinh(kd)} \right.$$
$$\left. + \frac{B_0}{V_{AI}} \left[\left(\alpha_H^2 + \alpha_P^2 + \beta_1 \alpha_P\right) v_r - \beta_1 \alpha_H v_\varphi \right] \right\}, \tag{5.180}$$

where the function β_3 is given by Eq. (5.31).

Applying a Bessel transform to Eqs. (5.130) and (5.131) one can express the components of electromagnetic field through the potentials. The derivatives $\partial_z \Psi$ and $\partial_z \Phi$ in Eqs. (5.130) and (5.131) are derivable from Eqs. (5.114) and (5.146) via $\partial_z \Psi(-d) = \xi \Psi(-d)$ and $\partial_z \Phi(-d) = -ik^2 c^2 A(-d)/\omega$. On account of these expressions we get

$$b_r(-d) = -k\xi \Psi(-d), \quad b_\varphi(-d) = kA(-d), \quad b_z(-d) = k^2 \Psi(-d), \tag{5.181}$$

$$e_r(-d) = k\Phi(-d), \quad e_\varphi(-d) = i\omega k \Psi(-d), \quad e_z(-d) \approx ik^2 c^2 A(-d)/\omega. \tag{5.182}$$

References

Atkinson G (1970) Auroral arcs: results of the interaction of a dynamic magnetosphere with the ionosphere. J Geophys Res 75:4746–4755

Belyaev PP, Polyakov SV, Rapoport VO, Trakhtengertz VY (1987) Discovery of the resonance spectrum structure of atmospheric EM noise background in the range of short-period geomagnetic pulsations. Reports of the USSR Academy of Sciences (Dokl. Akad. Nauk SSSR), vol 297, pp 840–843 (English Translation)

Belyaev PP, Polyakov SV, Rapoport VO, Trakhtengertz VY (1990) The ionospheric Alfvén resonator. J Atmos Terr Phys 52:781–787

Belyaev PP, Bösinger T, Isaev SV, Kangas J (1999) First evidence at high latitudes for the ionospheric Alfvén resonator. J Geophys Res 104:4305–4317

Bösinger T, Haldoupis C, Belyaev PP, Yakunin MN, Semenova NV, Demekhov AG, Angelopoulos V (2002) Spectral properties of the ionospheric Alfvén resonator observed at a low-latitude station ($L = 1.3$). J Geophys Res 107:1281. doi:10.1029/2001JA005076

Bösinger T, Demekhov AG, Trakhtengertz VY (2004) Fine structure in ionospheric Alfvén resonator spectra observed at low latitude. Geophys Res Lett 31:L13802

Chaston CC, Carlson CW, Peria WJ, Ergun RE, McFadden JP (1999) Fast observations of the inertial Alfvén waves in the dayside aurora. Geophys Res Lett 26:647–650

Chaston CC, Bonnell JW, Carlson CW, Berthomier M, Peticolas LM, Roth I, McFadden JP, Ergun RE, Strangeway RJ (2002) Electron acceleration in the ionospheric Alfvén resonator. J Geophys Res 107:1413. doi:10.1029/2002JA009272

Chaston CC, Bonnell JW, Carlson CW, McFadden JP, Ergun RE, Strangeway RJ (2003) Properties of small-scale Alfvén waves and accelerated electrons from FAST. J Geophys Res 108:8003. doi:10.1029/2002JA009420

Cole RK, Pierce ET (1965) Electrification in the earths atmosphere. J Geophys Res 70:2735–2749

Demekhov AG, Trakhtengertz VY, Bösinger T (2000) Pc 1 waves and ionospheric Alfvén resonator: generation or filtration? Geophys Res Lett 27:3805–3808

Fedorov E, Schekotov AJ, Molchanov OA, Hayakawa M, Surkov VV, Gladichev VA (2006) An energy source for the mid-latitude IAR: world thunderstorm centers, nearby discharges or neutral wind fluctuations? Phys Chem Earth 31:462–468

Fraser-Smith AC (1993) ULF magnetic fields generated by electrical storms and their significance to geomagnetic pulsation generation. Geophys Res Lett 20:467–470

Fujita S, Tamao T (1988) Duct propagation of hydromagnetic waves in the upper ionosphere, 1, electromagnetic field disturbances in high latitudes associated with localized incidence of a shear Alfvén wave. J Geophys Res 93. doi: 10.1029/88JA03338

Greifinger C, Greifinger S (1968) Theory of hydromagnetic propagation in the ionospheric waveguide. J Geophys Res 76:7473–7490

Grzesiak M (2000) Ionospheric Alfvén resonator as seen by Freja satellite. Geophys Res Lett 27:923–926

Hebden SR, Robinson TR, Wright DM, Yeoman T, Raita T, Bösinger TA (2005) Quantitative analysis of the diurnal evolution of ionospheric Alfvén resonator magnetic resonance features and calculation of changing IAR parameters. Ann Geophys 23:1711–1721

Hickey K, Sentman DD, Heavner MJ (1996) Ground-based observations of ionospheric Alfvén resonator bands. EOS Trans AGU 77(46, Fall Meeting Supplementary):F92

Jackson JD (2001) Classical electrodynamics, 3rd edn. Wiley, New York

Kelley MC (1989) The earth's ionosphere. Academic Press, New York

Landau LD, Lifshits EM (1986) Hydrodynamics. Nauka, Moscow

Lysak RL (1991) Feedback instability of the ionospheric resonator cavity. J Geophys Res 96:1553–1568

Lysak RL (1993) Generalized model of the ionospheric Alfvén resonator. In: Lysak RL (ed) Auroral plasma dynamics. Geophysical monograph series, vol 80. American Geophysical Union, Washington, DC, p 121

Lysak RL (1999) Propagation of Alfvén waves through the ionosphere: dependence on ionospheric parameters. J Geophys Res 104:10017–10030

Lysak RL, Song Y (2002) Energetics of the ionospheric feedback interaction. J Geophys Res 107:1160. doi:10.1029/2001JA00308

Mallat S (1999) A wavelet tour of signal processing. Academic Press, Paris/New York

Molchanov OA, Schekotov AY, Fedorov EN, Hayakawa M (2004) Ionospheric Alfvén resonance at middle latitudes: results of observations at Kamchatka. Phys Chem Earth Parts A/B/C 29:649–655

Onishchenko OG, Pokhotelov OA, Sagdeev RZ, Treumann RA, Balikhin MA (2004) Generation of convective cells by kinetic Alfvén waves in the upper ionosphere. J Geophys Res 109:A03306. doi:10.1029/2003JA010248

Pilipenko VA (2011) Impulsive coupling between the atmosphere and ionosphere/magnetosphere. Space Sci Rev. doi:10.1007/s11214-011-9859-8

Plyasov AA, Surkov VV, Pilipenko VA, Fedorov EN, Ignatov VN (2012) Spatial structure of the electromagnetic field inside the ionospheric Alfvén resonator excited by atmospheric lightning activity. J Geophys Res Space Phys 117:A09306. doi:10.1029/2012JA017577

Pokhotelov OA, Pokhotelov D, Streltsov A, Khruschev V, Parrot M (2000) Dispersive ionospheric Alfvén resonator. J Geophys Res 105:7737–7746

Pokhotelov OA, Khruschev V, Parrot M, Senchenkov S, Pavlenko VP (2001) Ionospheric Alfvén resonator revisited: feedback instability. J Geophys Res 106:25813–25824

Pokhotelov OA, Onishchenko OG, Sagdeev RZ, Treumann RA (2003) Nonlinear dynamics of the inertial Alfvén waves in the upper ionosphere: parametric generation of electrostatic convective cells. J Geophys Res 108:1291. doi:10.1029/2003JA009888

Pokhotelov OA, Onishchenko OG, Sagdeev RZ, Balikhin MA, Stenflo L (2004) Parametric interaction of kinetic Alfvén waves with convective cells. J Geophys Res 109:A03305. doi:10.1029/2003JA010185

Polyakov SV (1976) On the properties of the ionospheric Alfvén resonator. In: KAPG symposium on solar-terrestrial physics, vol 3. Nauka, Moscow, pp 72–73

Polyakov SV, Rapoport VO (1981) The ionospheric Alfvén resonator. Geomagn Aeron (English Translation) 21:610–614

Polyakov SV, Ermakova EN, Polyakov AS, Yakunin MN (2003) Formation of the spectra and polarization of background ULF electromagnetic noise at the earth's surface. Geomagn Aeron 42:240–248

Sato T (1978) A theory of quiet auroral arcs. J Geophys Res 83:1042–1048

Sato T, Holzer TE (1973) Quiet auroral arcs and electrodynamic coupling between the ionosphere and the magnetosphere. J Geophys Res 78:7314–7329

Schekotov A, Pilipenko V, Shiokawa K, Fedorov E (2011) ULF impulsive magnetic response at mid-latitudes to lightning activity. Earth Planets Space 63:119–128

Semenova NV, Yahnin AG (2008) Diurnal behavior of the ionospheric Alfvén resonator signatures as observed at high latitude observatory Barentsburg. Ann Geophys 26:2245–2251

Sims WE, Bostick FX (1963) Atmospheric Parameters for Four Quiescent Earth Conditions, Report No. 132, Electrical Engineering Research Laboratory, University of Texas, Sept 1

Sukhorukov AI, Stubbe P (1997) Excitation of the ionospheric resonator by strong lightning discharges. Geophys Res Lett 24:829–832

Surkov VV, Pokhotelov OA, Parrot M, Fedorov EN, Hayakawa M (2004) Excitation of the ionospheric resonance cavity by neutral winds at middle latitudes. Ann Geophys 22:2877–2889

Surkov VV, Molchanov OA, Hayakawa M, Fedorov EN (2005a) Excitation of the ionospheric resonance cavity by thunderstorms. J Geophys Res 110:A04308. doi:10.1029/2004JA010850

Surkov VV, Pokhotelov OA, Fedorov EN, Onishchenko OG (2005b) Police whistle type excitation of the ionospheric Alfvén resonator at middle latitudes. In: Physics of Auroral Phenomena, Proceedings of XXVIII Annual Seminar, Apatity, pp 108–114

Surkov VV, Hayakawa M, Schekotov AY, Fedorov EN, Molchanov OA (2006) Ionospheric Alfvén resonator excitation due to nearby thunderstorms. J Geophys Res 111:A01303. doi:10.1029/2005JA011320

Trakhtengertz VY, Feldstein AY (1981) Effect of the nonuniform Alfvén velocity profile on stratification of magnetospheric convection. Geomagn Aeron (English Translation) 21:711

Trakhtengertz VY, Feldstein AY (1984) Quiet auroral arcs: ionospheric effect of magnetospheric convection stratification. Planet Space Sci 32:127–134

Trakhtengertz VY, Feldstein AY (1987) About excitation of small-scale electromagnetic perturbations in ionospheric Alfvén resonator. Geomagn Aeron (English Translation) 27:315

Trakhtengertz VY, Feldstein AY (1991) Turbulent Alfvén boundary layer in the polar ionosphere, 1, excitation conditions and energetics. J Geophys Res 96:19,363–19,374

Uyeda S, Nagao T, Hattori K, Noda Y, Hayakawa M, Miyaki K, Molchanov O, Gladychev V, Baransky L, Schekotov A, Belyaev G, Fedorov E, Pokhotelov O, Andreevsky S, Rozhnoi A, Khabazin Y, Gorbatikov A, Gordeev E, Chebrov V, Lutikov A, Yunga S, Kosarev G, Surkov V (2002) Russian-Japanese complex geophysical observatory in Kamchatka for monitoring of phenomena connected with seismic activity. In: Hayakawa M, Molchanov OA (eds) Seismo electromagnetics: lithosphere-atmosphere-ionosphere coupling. Terrapub, Tokyo, pp 413–419

Yahnin AG, Semenova NV, Ostapenko AA, Kangas J, Manninen J, Turunen T (2003) Morphology of the spectral resonance structure of the electromagnetic background noise in the range of 0.1 − 4 Hz at $L = 5.2$. Ann Geophys 21:779–786

Chapter 6
Magnetospheric MHD Resonances and ULF Pulsations

Abstract This chapter deals with the low-frequency MHD oscillation of the whole magnetosphere and ULF pulsations including their origins and magnetospheric plasma instabilities. We discuss briefly magnetospheric models and the generation of field-line resonances (FLRs) and cavity modes. Properties of MHD waves propagating in the solar wind are covered. In the remainder of this chapter we examine source mechanisms of natural electromagnetic ULF noises.

Keywords Cavity mode • Field-line resonances (FLRs) • Plasma instabilities • Space weather • ULF electromagnetic noises

6.1 Structure of Global Magnetospheric Oscillations

6.1.1 An Axisymmetric Magnetosphere Model

It is customary to believe that the MHD oscillation of the whole magnetosphere was originally studied by Dungey (1954) who derived equations for the eigen oscillation of an axisymmetric magnetosphere. The normal magnetospheric MHD oscillation, which is independent of azimuthal angle φ, can be split into two practically uncoupled modes: toroidal and poloidal modes depending on their polarization. In the poloidal modes, the electric field oscillates in the azimuthal direction, while the magnetic field and plasma velocity pulsate across magnetic shells. By contrast, in the case of the toroidal modes, the electric field is in the meridional plane, while the magnetic field and plasma velocity oscillate in the azimuthal direction. These types of the magnetospheric MHD oscillation have been studied intensively for several decades and much is now known of their properties (Radoski 1967a,b; Radoski and Carovillano 1969; Cummings et al. 1969; Krylov and Lifshitz 1984).

V. Surkov and M. Hayakawa, *Ultra and Extremely Low Frequency Electromagnetic Fields*, Springer Geophysics, DOI 10.1007/978-4-431-54367-1_6,
© Springer Japan 2014

The neutral gas is so rarefied at magnetospheric heights that the number density of ions is much greater than that of neutrals, so that the plasma can be considered fully ionized in this region. Furthermore, the plasma itself is so tenuous that $v_{ei} \ll \Omega_i$ and thus the medium can be treated as a collisionless magnetized plasma. In such a case we can use the MHD approach, in which, as we have noted in Sect. 1.2.2, the plasma is considered as a single fluid having infinite conductivity. This means that in a reference frame moving at plasma velocity $\mathbf{V}$ the electrical field $\mathbf{E}' = \mathbf{E} + \mathbf{V} \times \mathbf{B}$ vanishes in both directions parallel and perpendicular to $\mathbf{B}$, whence it follows that in a reference frame fixed to the Earth $\mathbf{E} = \mathbf{B} \times \mathbf{V}$.

The magnetospheric plasma dynamics is described by Eq. (1.13), which relates the plasma velocity to the forces acting on the plasma. This equation in its general form contains the pressure gradient, the terms describing the gravitational and "viscous" forces, and the magnetic/Ampere's force given by $\mathbf{j} \times \mathbf{B}$. In the reference frame fixed to the Earth this equation includes all the inertial forces acting on the plasma due to the Earth spin. The pressure gradient and the magnetic force dominate if the typical frequencies are smaller than 0.1 Hz. In this frequency range the gravity, viscosity, and inertial terms in Eq. (1.13) can be neglected.

Following Dungey (1954, 1963) we first assume that the geomagnetic field and electric currents in the magnetosphere are large enough so that the magnetic force in Eq. (1.13) is much greater than the pressure gradient. It should be noted that the plasma motion parallel to the magnetic field lines must be due to only the pressure gradient ∇P since the magnetic force $\mathbf{j} \times \mathbf{B}$ is always perpendicular to $\mathbf{B}$. On the other hand the plasma motion parallel to $\mathbf{B}$ does not greatly affect the magnetic field. In this picture the cancel of ∇P in Eq. (1.13) is not so a burdensome condition. Finally, we have

$$\rho d\mathbf{V}/dt = \mathbf{j} \times \mathbf{B}, \qquad (6.1)$$

where ρ is the plasma mass density, and the total time-derivative d/dt is given by Eq. (1.12). To treat the plasma dynamics, Maxwell's equations are required, whose full forms are given by Eqs. (1.1)–(1.4). Since the vacuum displacement current $\partial_t \mathbf{D}$ can be ignored due to the high plasma conductivity, Eq. (1.1) is simplified to the form in which the curl of magnetic field is related to the conduction current $\mathbf{j}$ through Eq. (1.5).

Substituting $\mathbf{E} = \mathbf{B} \times \mathbf{V}$ into Eq. (1.2) gives Eq. (1.18). The meaning of this equation is that the magnetic field is frozen to the conducting plasma and thus can be considered to move with the plasma. The concept of "frozen-in" magnetic field lines has been discussed in more detail in Sect. 1.2.1.

Let $\delta \mathbf{B}$ be the small perturbation of the ambient/geomagnetic field $\mathbf{B}_0$, so that $\mathbf{B} = \mathbf{B}_0 + \delta \mathbf{B}$, and $|\delta \mathbf{B}| \ll |\mathbf{B}_0|$. The unperturbed geomagnetic field $\mathbf{B}_0$ is not a function of time. In the first approximation, one can replace the term $\mathbf{V} \times \mathbf{B}$ in Eq. (1.18) with $\mathbf{V} \times \mathbf{B}_0$. After these simplifications we come to the following equation:

$$\partial_t \delta \mathbf{B} = \nabla \times (\mathbf{V} \times \mathbf{B}_0). \qquad (6.2)$$

The same approximations can be applied to Eq. (1.35) to yield

$$\mathbf{E} = \mathbf{B}_0 \times \mathbf{V}. \tag{6.3}$$

Here $\mathbf{E}$ denotes the perturbation of the electric field since a constant electric field is assumed to be absent. Returning to the equation of motion (6.1) and substituting Eq. (1.5) for $\mathbf{j}$ into this equation, we have

$$\mu_0 \rho d\mathbf{V}/dt = (\nabla \times \mathbf{B}) \times \mathbf{B}. \tag{6.4}$$

In what follows we restrict our analysis to the case of a dipole approximation of the geomagnetic field $\mathbf{B}_0$, given by Eq. (1.30). Substituting $\mathbf{B} = \mathbf{B}_0 + \delta\mathbf{B}$ into Eq. (6.4), taking into account that $\nabla \times \mathbf{B}_0 = 0$, and considering the small amplitude waves, so that $d\mathbf{V}/dt \approx \partial_t \mathbf{V}$, the equation of plasma motion is reduced to

$$\mu_0 \rho \partial_t \mathbf{V} = (\nabla \times \delta\mathbf{B}) \times \mathbf{B}_0. \tag{6.5}$$

Taking the cross product of both sides of Eq. (6.5) with $\mathbf{B}_0$ and substituting Eq. (6.3) for $\mathbf{B}_0 \times \mathbf{V}$ into this equation yields

$$\mu_0 \rho \partial_t \mathbf{E} = \mathbf{B}_0 \times \left[(\nabla \times \delta\mathbf{B}) \times \mathbf{B}_0 \right]. \tag{6.6}$$

Now we should use Eq. (1.55) for triple cross product with $\mathbf{A}_1 = \mathbf{A}_3 = \mathbf{B}_0$ and $\mathbf{A}_2 = \delta\mathbf{B}$. Applying this equation to the right-hand side of Eq. (6.6) yields

$$\mu_0 \rho \partial_t \mathbf{E} = B_0^2 (\nabla \times \delta\mathbf{B}) - \mathbf{B}_0 \left[\mathbf{B}_0 \cdot (\nabla \times \delta\mathbf{B}) \right]. \tag{6.7}$$

The set of Eqs. (6.2), (6.3), and (6.7) constitutes the suitable single-fluid description of dynamics of a magnetized plasma. A general analytical solution of the plasma dynamics problem is not yet at hand although the numerical solutions, which can be applied to the actual magnetosphere, have been studied in detail (e.g., see Lee and Lysak 1989, 1990; Alperovich and Fedorov 2007).

As we have noted above, to the first order the Earth magnetic field is described through the dipole approximation. If the polar z axis is positive parallel to the Earth's magnetic moment $\mathbf{M}_e$ and the origin of the coordinate system is in the Earth center, then the Earth's dipole magnetic field is the axially symmetrical one and has only the components B_r and B_θ given by Eq. (1.33) while $B_\varphi = 0$.

In this approximation we consider the axially symmetrical problem, in which all the values are independent on φ. As we shall see, in this case the equation set is split into two independent parts: the first one contains the components of electromagnetic perturbations δB_φ, E_r, E_θ and azimuthal velocity V_φ, and the second one contains the components δB_r, δB_θ, E_φ, V_r, and V_θ. The first mode is referred to as the shear Alfvén wave and the next one is the FMS/compressional wave. According to geophysical terminology, the standing quasi-Alfvén wave which contains the azimuthal magnetic field δB_φ is termed the toroidal mode, while the standing compressional wave $(\delta B_r, \delta B_\theta)$ is referred to as the poloidal mode.

So, the polarization of the Alfvén oscillations differs from that of the poloidal mode. The electric field of the poloidal mode has only an azimuthal component, while the magnetic field perturbations and plasma velocity are directed across the magnetic shell.

6.1.2 Toroidal Mode

The shear Alfvén mode could have an important role to play in the generation of field line resonances (FLRs hereafter). These kinds of standing waves in the magnetosphere have frequently been observed onboard the satellites (e.g., see Glassmeier 1995) so that a spatial structure and the source mechanisms of these waves are of a special interest in geophysical studies.

Using the spherical coordinates, the azimuthal components of Eq. (6.2) can be written as

$$\partial_t \delta B_\varphi = r^{-1} \left\{ \partial_r \left[r \left(\mathbf{V} \times \mathbf{B}_0 \right)_\theta \right] - \partial_\theta \left[\left(\mathbf{V} \times \mathbf{B}_0 \right)_r \right] \right\}. \tag{6.8}$$

Substituting the angular and radial components of $\mathbf{V} \times \mathbf{B}_0$ into Eq. (6.8) yields

$$\partial_t \delta B_\varphi = r^{-1} \left\{ \partial_r \left(r V_\varphi B_r \right) - \partial_\theta \left(V_\varphi B_\theta \right) \right\}. \tag{6.9}$$

The components B_r and B_θ of undisturbed Earth's magnetic field are given by Eq. (1.33). Substituting these components into Eq. (6.9) and rearranging this equation yields

$$\partial_t \delta B_\varphi = r \sin \theta \left(\mathbf{B}_0 \cdot \nabla \right) \left(\frac{V_\varphi}{r \sin \theta} \right), \tag{6.10}$$

where we have introduced the differential operator acting on the functions of variables r and θ:

$$\mathbf{B}_0 \cdot \nabla = \frac{\mu_0 M_e}{4 \pi r^3} \left(2 \cos \theta \partial_r + \frac{\sin \theta}{r} \partial_\theta \right). \tag{6.11}$$

Here M_e denotes the Earth's magnetic dipole moment.

The azimuthal component of the equation of motion (6.5) has the form

$$\rho \partial_t V_\varphi = \frac{M_e}{4 \pi r^4} \left\{ \partial_\theta \left(\delta B_\varphi \sin \theta \right) + 2 \cos \theta \partial_r \left(r \delta B_\varphi \right) \right\}. \tag{6.12}$$

Combining this equation with Eq. (1.33) for the components B_r and B_θ and rearranging, we find that

$$\mu_0 \rho r \sin \theta \partial_t V_\varphi = \left(\mathbf{B}_0 \cdot \nabla \right) \left(r \sin \theta \delta B_\varphi \right). \tag{6.13}$$

The electric field components can be found from Eq. (6.3)

$$E_r = B_\theta V_\varphi, \quad E_\theta = -B_r V_\varphi. \qquad (6.14)$$

It is interesting to note that the scalar product of the electric field (6.14) with the vector $\mathbf{B}_0 = (B_r, B_\theta, 0)$ equals zero, that is, the vector $\mathbf{E}$ is perpendicular to the Earth's dipole magnetic field $\mathbf{B}_0$. This implies that the electric field of the toroidal mode is perpendicular to the Earth magnetic field shells.

Considering the toroidal mode, we are thus left with the set of Eqs. (6.10), (6.13), and (6.14) for the functions δB_φ, V_φ, E_r, and E_θ. To study the standing waves, in effect finding the normal modes of the axisymmetric magnetosphere, all perturbed quantities are considered to vary as $\exp(-i\omega t)$, where ω is the frequency. Replacing the time-derivatives ∂_t with the factor $-i\omega$ and eliminating δB_φ from Eqs. (6.10) and (6.13) yields

$$\mu_0 \rho r \omega^2 \sin\theta V_\varphi + (\mathbf{B}_0 \cdot \nabla)\left[r^2 \sin^2\theta \, (\mathbf{B}_0 \cdot \nabla)\left(\frac{V_\varphi}{r\sin\theta}\right)\right] = 0. \qquad (6.15)$$

The only differential operator occurred at this equation is the operator $(\mathbf{B}_0 \cdot \nabla)$, which defines in fact the directional derivative. In other words, this equation contains only derivative along the magnetic field lines. This means that Eq. (6.15) describes oscillations of the azimuthal velocity V_φ and magnetic shells that originate from the rotation of the field lines about the symmetry axis. Each magnetic shell can vibrate independently of each other. All the field lines belonging to the same magnetic shell must vibrate synchronously, that is, the magnetic shell vibrates as a whole. To study eigen oscillations of the shell, therefore, it is sufficient to consider the oscillations of one of the magnetic field lines.

Equation (6.15) should be supplemented by the proper boundary conditions at the ends of the magnetic field lines, that is at the points where the field lines intersect the high conducting ionospheric E layer and the Earth's surface. The skin-depth in the ionosphere at frequencies $f \lesssim 0.1\,\mathrm{Hz}$ exceeds the thickness of the ionospheric conductive layer. In this notation, the E layer is usually treated in a "thin" ionosphere approximation, while the Earth can be considered as a perfect conductor, which reflects the electromagnetic waves totally. More usually we use the boundary conditions of the impedance type in which the electric and magnetic field components tangential to the ionosphere are related in a linear fashion.

In order to make our consideration as transparent as possible we, however, choose a simplified approximation, considering the wave reflection off a perfect conductor surface. In such a case, the boundary condition at the end of field line is $\mathbf{E} = 0$. On account of Eq. (6.14) one can derive the boundary condition $V_\varphi = 0$ at the ends of the magnetic field line. It should be noted that these relations can serve as proper boundary conditions rather for the sunlit ionosphere because of the high ionospheric conductivity at daytime.

The field line shape of the dipole magnetic field is described by Eq. (1.34). This equation relates the polar radius, r, to the magnetic latitude λ. If polar angle θ is expressed through the magnetic latitude λ (northern hemisphere) via $\lambda = \theta - \pi/2$, the equation for magnetic field lines can be written as $r = LR_e \sin^2 \theta$, where R_e is the mean Earth radius, L is the McIllwain parameter. In this notation the operator $\mathbf{B}_0 \cdot \nabla$ taken along the field line can thus be rewritten as

$$\mathbf{B}_0 \cdot \nabla = \frac{\mu_0 M_e}{4\pi L^4 R_e^4 \sin^7 \theta} \frac{d}{d\theta}. \tag{6.16}$$

Substituting Eq. (6.16) for $\mathbf{B}_0 \cdot \nabla$ into Eq. (6.15) and rearranging yields (Dungey 1954, 1963; Cummings et al. 1969)

$$\frac{d}{d\theta}\left(\frac{1}{\sin\theta} \frac{d}{d\theta} \frac{V_\varphi}{\sin^3\theta} \right) + \frac{\rho \left(4\pi R_e^4\right)^2 L^8 \omega^2 \sin^{10}\theta}{\mu_0 M_e^2} V_\varphi = 0. \tag{6.17}$$

We recall that the plasma velocity V_φ in Eq. (6.17) must be equal to zero at two intersection points where the corresponding field line crosses the bottom of the ionosphere. Substituting $r = R_e$ into Eq. (1.34) one can find the angles corresponding to these intersection points. We are thus left with the equation $\sin^2\theta_0 = \cos^2\lambda_0 = L^{-1}$ for the sought polar angles θ_0 and magnetic latitude λ_0. Finally, the proper boundary conditions for Eq. (6.17) take the form $V_\varphi(\theta_0) = V_\varphi(\pi - \theta_0) = 0$, where $\theta_0 = \arcsin L^{-1/2}$ and $\pi - \theta_0$ is the angle corresponding to the conjugate intersection point. The atmospheric depth is disregarded here.

Owing to the complexity of Eq. (6.17) the general analysis of this differential equation encounters some difficulty. As would be expected, considering the finite length of the field line segment bounded these two interception points, the given boundary problem has periodic solutions. The fundamental toroidal mode of Eq. (6.17) has numerically been studied by Dungey (1954, 1963). For example, according to this calculation made at the plasma density $\rho = 10^{-18}\,\mathrm{kg/m^3}$, the period of the fundamental mode can be approximated by the formula

$$T \approx \frac{0.6}{\sin^8 \theta_0}, \quad \text{(in second)}. \tag{6.18}$$

Taking the numerical values of the magnetic latitude $\lambda_0 = 45°$, $55°$, $65°$ and $70°$, the typical periods of the fundamental mode are as follows: $T = 10\,\mathrm{s}$, $54\,\mathrm{s}$, $11\,\mathrm{min}$ and $55\,\mathrm{min}$, correspondingly, while the corresponding eigenfrequencies lie much below the IAR and Schumann resonances.

Thus, Eq. (6.17) describes the toroidal field oscillation in the magnetosphere or the standing shear quasi-Alfvén waves in the dipole approximation of the geomagnetic field. In this case the plasma velocity has only an azimuthal component and the "frozen in" magnetic field lines therefore vibrate within the resonance shell. The toroidal (twisting) oscillations manifest themselves through the azimuthal magnetic component and through the electric component orthogonal to the magnetic shell. Such quasi-Alfvén modes are referred to as the class of the FLRs.

6.1.3   Poloidal Mode

Now we consider another important mode of the magnetospheric oscillations, that is, the poloidal (breathing) mode, which contains the set of field components δB_r, δB_θ, V_r, V_θ, and E_φ. In contrast to the toroidal mode, which is mainly due to the shear Alfvén waves the poloidal mode is caused by the compressional waves, which propagate isotropically. To treat the basic properties of this mode we first take the azimuthal component of Eq. (6.7)

$$-i\mu_0\rho\omega E_\varphi = r^{-1}B_0^2\left[\partial_r\left(r\delta B_\theta\right) - \partial_\theta\delta B_r\right].\tag{6.19}$$

As before all the functions are assumed to vary as $\exp\left(-i\omega t\right)$.

Maxwell equation $\nabla\times\mathbf{E} = i\omega\delta\mathbf{B}$ in spherical coordinates now is reduced to

$$\partial_\theta\left(\sin\theta E_\varphi\right) = i\omega\delta B_r r\sin\theta,\tag{6.20}$$

$$r^{-1}\partial_r\left(rE_\varphi\right) = -i\omega\delta B_\theta.\tag{6.21}$$

The set of Eqs. (6.19)–(6.21) can be solved for E_φ to yield the equation for poloidal mode

$$\frac{\omega^2 r}{V_A^2}E_\varphi + \partial_r^2\left(rE_\varphi\right) + \frac{1}{r}\partial_\theta\left[\frac{1}{\sin\theta}\partial_\theta\left(\sin\theta E_\varphi\right)\right] = 0.\tag{6.22}$$

Moreover the azimuthal plasma velocity $V_\varphi = 0$ while the components V_r and V_θ can be expressed through E_φ as follows:

$$V_r = -\frac{B_\theta}{B_0^2}E_\varphi,\quad V_\theta = \frac{B_r}{B_0^2}E_\varphi.\tag{6.23}$$

It follows from Eq. (6.23) that the scalar product of the poloidal mode velocity $\mathbf{V}_p = (V_r, V_\theta, 0)$ with $\mathbf{B}_0$ is equal to zero so that the plasma velocity $\mathbf{V}_p$ is perpendicular to the magnetic shell in contrast to the quasi-Alfvén oscillations. The poloidal mode is not guided by the field lines and can cover the whole magnetosphere or the large part of that. These modes are referred to as the class of cavity modes, which can propagate via FMS/compressional waves. In a homogeneous plasma, the phase velocity of the compressional waves is independent of the angle included between the plasma velocity vector and the geomagnetic field. Not surprisingly, the cavity oscillations due to compressional waves can fill the whole magnetosphere and the cavity mode spectrum is dependent on conditions at the outer boundary of the magnetosphere, that is, at the magnetopause.

Notice that the FMS/poloidal mode results in considerable variations of the field-aligned magnetic field, whereas the field-aligned electric current is small. On the contrary, the field-aligned current of the toroidal quasi-Alfvén mode has a finite value while the longitudinal magnetic field is small.

6.1.4 Azimuthal Harmonics

From the above analysis it is clear that in the axisymmetric magnetosphere model the basic equations can be split into two independent sets for the shear Alfvén and compressional waves, which can propagate independently of each other. According to the FLR theory, the magnitude of the standing shear Alfvén wave can reach a peak value in the vicinity of resonance magnetic shells under certain resonant conditions, whereas the standing compressional wave is associated with variations of the magnetic field perpendicular to the magnetic shells (Radoski 1967a; Radoski and Carovillano 1969; Southwood 1974; Chen and Hasegawa 1974; Krylov and Fedorov 1976; Krylov and Lifshitz 1984). In a general way, that is, in an arbitrary ambient magnetic field, the MHD-wave equations for these modes can be coupled. For example, if φ dependence of the normal modes is taken into account, all the functions should be expanded in a series of azimuthal harmonics $\exp(im\varphi)$, where $m = 0, 1, 2, \ldots$ is azimuthal wave number. Ignoring the φ dependence, we thereby have chosen $m = 0$ in the above equations. Furthermore, the shear and compressional Alfvén waves in the magnetospheric plasma can be coupled through the boundary conditions at the conducting E-layer of the ionosphere due to both the tensor character of the ionospheric plasma conductivity and finite value of Hall and Pedersen conductivities.

If $m \neq 0$ and $m \neq \infty$, the set of MHD equations for magnetospheric oscillations does not reduce to independent equations for the toroidal and poloidal modes. Nevertheless, away from the resonance magnetic shells the coupling between these modes is weak under the requirement that $m \sim 1$, and in the first approximation they can be considered as independent modes. The interactions between the toroidal and poloidal modes become significant only in a narrow region in the vicinity of the resonance shell (Leonovich and Mazur 1993; Leonovich 2000).

In the case of $m \gg 1$ the coupling between the shear Alfvén and compressional modes is so strong that they cannot be divided into two individual modes. Both of these modes manifest themselves as a single MHD mode, which is more likely to be similar to the Alfvén wave rather than to compressional one (Leonovich and Mazur 1993; Leonovich 2000). Nevertheless, such a quasi-Alfvén wave combines the properties of both modes, i.e., strong localization across the magnetic shell, that is typical for the shear Alfvén wave, and the presence of a considerable constituent of the field-aligned magnetic field variations, that is typical for the compressional wave.

If $m \to \infty$, the azimuthal scale across the main magnetic field tends to zero. This implies that the derivatives over φ in the MHD equations become much greater than those with respect to other variables. For this special case the MHD-wave equations can be split into two independent groups in analogy to the case of $m = 0$. As would be expected, considering the small value of the transverse scale, the plasma velocity V_φ is small and the plasma movement is mainly concentrated within the meridional plane (Dungey 1954, 1963).

6.2   Field-Line Resonance (FLR)

6.2.1   MHD Box Model

We now take up one more viewpoint on the magnetospheric resonances. To study the MHD-waves coupling in a little more detail, however, we need to consider a simple approximate model of the magnetosphere sketched in Fig. 6.1. In the model the dipole geomagnetic field lines are replaced by straightened field lines in such a way that the area marked in Fig. 6.1 with M is transformed into the parallelepiped/box magnetosphere shown in the bottom of Fig. 6.1. The y axis which was originally in the west–east direction is now transformed into a straight line that is infinite in length. In this picture the y-coordinate in the box model corresponds to the azimuthal direction/coordinate φ in the reference frame fixed to the Earth spin axis.

The box contains a cold magnetized plasma immersed in a straight magnetic field, $\mathbf{B}_0 = B_0(x)\,\hat{\mathbf{z}}$, which is a function of x. Both the plasma mass density, ρ, and the Alfvén velocity, V_A, also depend on only x, which plays a role of radial coordinate in the equatorial plane. The magnetic field lines are finite in length in the z direction and there are boundary conditions at the ends of lines. The box surfaces $z = 0$ and $z = l_1$ correspond to the southern and northern ionospheres. The box surface $x = 0$ represents the equatorial region of the ionosphere while the plane $x = l_2$ corresponds to the outer boundary of the magnetosphere. This model was originally suggested by Radoski (1966, 1967a,b) and has been termed the MHD box.

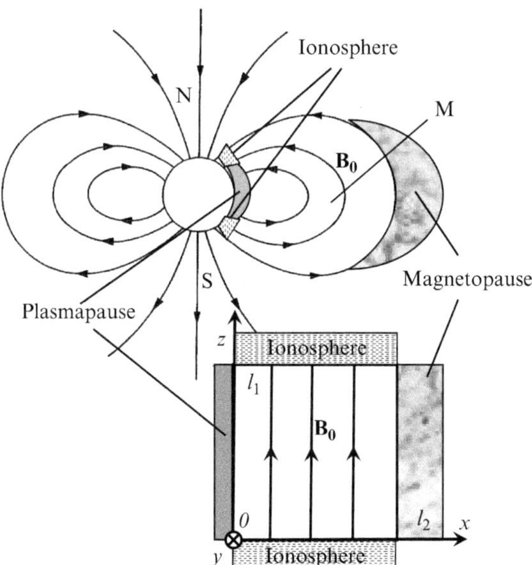

Fig. 6.1 Sketch of MHD-box model of the magnetosphere. The figure is partly adapted from Southwood and Kivelson (1982)

Plasma oscillations can be excited by sources situated both inside and outside the box. In order to take into account the internal source of excitation of the normal modes we now add the driven current, $\mathbf{J}_d$, to the conduction current on the right-hand side of Maxwell equation (1.5). The driven current is assumed to be a given function. Hence, Eq. (6.5) for the plasma motion is reduced to

$$\mu_0 \rho \partial_t \mathbf{V} = (\nabla \times \delta \mathbf{B}) \times \mathbf{B}_0 - \mu_0 (\mathbf{J}_d \times \mathbf{B}_0). \tag{6.24}$$

As is seen from Eqs. (6.3) and (6.24) the electric field $\mathbf{E}$ and the plasma acceleration $\partial_t \mathbf{V}$ are perpendicular to the unperturbed magnetic field $\mathbf{B}_0$. So we seek for the solution of the problem in the form $\mathbf{E} = (E_x, E_y, 0)$ and $\mathbf{V} = (V_x, V_y, 0)$. All perturbed quantities are considered to vary as $\exp(ik_y y)$, where k_y is the perpendicular wave number. Thus Eq. (6.24) is reduced to

$$\mu_0 \rho \partial_t V_x = (\partial_z \delta B_x - \partial_x \delta B_z - \mu_0 J_y)\, B_0, \tag{6.25}$$

$$\mu_0 \rho \partial_t V_y = (\partial_z \delta B_y - i k_y \delta B_z + \mu_0 J_x)\, B_0, \tag{6.26}$$

where J_x and J_y are the projections of the driven current $\mathbf{J}_d$.

In this approach the Faraday's law (1.2) reads

$$\partial_z E_y = \partial_t \delta B_x, \tag{6.27}$$

$$\partial_z E_x = -\partial_t \delta B_y, \tag{6.28}$$

$$i k_y E_x - \partial_x E_y = \partial_t \delta B_z. \tag{6.29}$$

The plasma velocity is related to the electric field through Eq. (6.3), that is

$$V_y = -E_x / B_0, \quad V_x = E_y / B_0. \tag{6.30}$$

Substituting V_x and V_y into Eqs. (6.25) and (6.26) we come to the set of equations for the electromagnetic fields. If $k_y = 0$, this set is split into two uncoupled sets of equation describing the shear Alfvén $(E_x,\ \delta B_y,\ V_y)$ and FMS waves $(E_y,\ \delta B_x,\ \delta B_z,\ V_x)$. A close analogy exists with axisymmetric magnetic field, in which, as we have noted, both the modes are uncoupled in an extreme case of azimuthal harmonics with $m = 0$. It should be noted that the first mode (E_x) corresponds to the toroidal field in the axisymmetric magnetosphere, whereas the second mode (E_y) corresponds to the poloidal field.

Assuming that all the perturbed quantities vary in time as $\exp(-i\omega t)$ and solving this set of equations for E_x and E_y we come to the following wave equations:

$$(\partial_z^2 + \omega^2 / V_A^2)\, E_x = -i\mu_0 \omega J_x, \tag{6.31}$$

$$(\partial_x^2 + \partial_z^2 + \omega^2 / V_A^2)\, E_y = -i\mu_0 \omega J_y, \tag{6.32}$$

where

$$V_A(x) = \frac{B_0(x)}{[\mu_0 \rho(x)]^{1/2}}, \tag{6.33}$$

is the Alfvén velocity. Despite the fact that the wave equations (6.31) and (6.32) are independent, both the MHD modes can be coupled through the boundary conditions at the E region of the ionosphere. So we need to consider the effect of the boundary conditions on the spectrum of normal oscillations.

6.2.2   FLR Eigenfrequencies

For all frequencies of interest here the E region of the ionosphere can be considered as a thin conductive layer with integral Pedersen and Hall conductivities. The boundary condition (5.26) at the E layer relates the jump of horizontal magnetic field across this layer to the horizontal electric field. We recall that the interaction between the magnetospheric MHD waves and the ionosphere depends on value of the dimensionless parameters α_P and α_H, which are equal to the ratio of height-integrated Pedersen Σ_P and Hall Σ_H conductivities to the Alfvén wave parallel conductance of the magnetosphere $\Sigma_w = (\mu_0 V_A)^{-1}$. As would be expected, considering the FLRs due to the shear Alfvén waves propagation, the ionospheric Pedersen conductivity plays an important role in closing of the field-aligned Alfvén currents in the ionosphere, that is in the closing of the field lines perpendicular to **B**. In this picture the Hall conductivity in the ionosphere is of minor importance and in the first approximation it can be ignored (e.g., Krylov and Fedorov 1976; Krylov and Lifshitz 1984). In this approach the wave perturbations coming from the magnetosphere cannot penetrate through the conducting ionosphere so that we can neglect the variations of the magnetic field below the ionosphere. In this way the boundary condition (5.26) at the ionosphere reduces to

$$\delta B_y = \pm \mu_0 \Sigma_P^{\pm} E_x, \quad \text{and} \quad -\delta B_x = \pm \mu_0 \Sigma_P^{\pm} E_y, \tag{6.34}$$

where the sign plus on the right-hand side of Eq. (6.34) corresponds to the northern ionosphere $(z = l_1)$ and the sign minus corresponds to the southern ionosphere $(z = 0)$. Furthermore, Σ_P^{+} stands for the height-integrated Pedersen conductivity of the northern ionosphere while Σ_P^{-} denotes the same value for the southern ionosphere. Combining Eqs. (6.27), (6.28), and (6.34) we finally obtain

$$\partial_z \mathbf{E}_\perp = \pm i \omega \mu_0 \Sigma_P^{\pm} \mathbf{E}_\perp, \tag{6.35}$$

where $\mathbf{E}_\perp = (E_x, E_y)$ and $z = 0$ or l_1.

We choose first to study the free Alfvén oscillations at $J_x = 0$. In such a case the solution of Eq. (6.31) can be written

$$E_x = C_1 \exp(i\omega z / V_A) + C_2 \exp(-i\omega z / V_A), \tag{6.36}$$

where C_1 and C_2 are undetermined constants. Substituting Eq. (6.36) for E_x into boundary conditions (6.35) we come to an algebraic set of equations for constants C_1 and C_2. This set of equations has a nontrivial solution under the requirement that the system determinant equals to zero, whence it follows

$$\exp\left(-\frac{2i\omega l_1}{V_A}\right) = R_+ R_-,\tag{6.37}$$

where

$$R_+ = \frac{1 - \alpha_P^+}{1 + \alpha_P^+} \quad \text{and} \quad R_- = \frac{1 - \alpha_P^-}{1 + \alpha_P^-}\tag{6.38}$$

denote the reflection coefficients for the northern and southern ionospheres, respectively. These coefficients vary from 1 to -1 with changing $\alpha_P^\pm$ from zero to infinity.

Decomposing the frequency in Eq. (6.37) into its real and imaginary parts, $\omega = \omega' + i\omega''$, one finds the solution of Eq. (6.37) in the form

$$\omega_n'(x) = \pi n V_A(x) / l_1,\tag{6.39}$$

when $R_+ R_- > 0$ and

$$\omega_n'(x) = \{\pi V_A(x) / l_1\}(n - 1/2),\tag{6.40}$$

when the inverse inequality, $R_+ R_- < 0$, is valid. Here n is integer, $n = 1, 2, \dots$ In both of these cases the imaginary part of the frequency is given by

$$\omega''(x) = \frac{V_A(x)}{2l_1} \ln |R_+ R_-|.\tag{6.41}$$

If $k_y \neq 0$, the general solution and eigenfunctions of the problem are found in Appendix F. In this case the normal modes are coupled through the boundary conditions at the conjugate ionospheres. The sole exception corresponds to two opposite extreme cases of zeroth and infinite Pedersen conductivities when the shear Alfvén and FMS modes become independent. In these extreme cases the wave vector $k_n = \omega_n'/V_A$ in Eq. (6.39) coincides with that given by Eq. (6.125).

The general solution of the problem can be expanded in a series of the orthonormal eigenfunctions, $q_n(z)$, given by Eq. (6.126). Arbitrary perturbations of E_x and δB_y appear as a sum of modes, each of which changes harmonically in time. Considering the amplitudes E_{xn} and δB_{yn} of the normal oscillation with frequency $\omega = \omega_n(x)$ and rearranging Eq. (6.126) we get

$$E_{xn} \propto q_n = \frac{\sin k_n z}{k_n} + \frac{i \cos k_n z}{k_n \alpha_P^-}, \quad \delta B_{yn} \propto \frac{dq_n}{dz},\tag{6.42}$$

where $k_n = \omega_n / V_A = \left(\omega_n' + i\omega'' \right) / V_A$ is not a function of x and the real and imaginary parts of ω are given by Eqs. (6.39)–(6.41).

As long as x is fixed, Eqs. (6.39)–(6.41) describe the spectrum of normal Alfvén oscillations. It should be emphasized that this spectrum depends on x and thus is continuous. The real part of the frequency $\omega_n'(x)$ represent the eigenfrequency of the n-th harmonic. Notice that all the eigenfrequencies are equidistant, whereas the damping factor $\omega''(x)$ is independent of n. From these equations it is clear that the magnetic shell which has a constant value of x will vibrate as a whole. In this picture all the segments of the field lines at constant x will covibrate so that it is sufficient to study the normal oscillations of one of these field lines. The net Alfvén perturbations can be thus considered as a superposition of the independent normal oscillations of the field lines/magnetic shells.

It is clear that Eq. (6.39) describes the frequencies of the half-wave mode in such a way that an integer number of half-waves lies on the field line while Eq. (6.40) corresponds to the quarter-wave mode. In this case the field line length equals to $1/4, 3/4, 5/4\ldots$ of the wavelength.

This kind of field line oscillations has a close analogy with normal oscillations of a taut elastic string. The sense of the magnetic force is similar to that of elastic forces due to tension in a stretched string since the restoring force in Alfvén oscillations arises due to tension in the magnetic field line. In addition, the shear Alfvén waves play a role of the transverse elastic waves propagating along the string.

In the framework of the "MHD-box" model, we note that the energy loss is mostly due to the Joule dissipation at the ends of field lines, that is in the ionosphere. The interpretation we make is that the Joule dissipation results from the Pedersen conductivity, which is subject to diurnal variations. It is usually the case that at the nighttime ionosphere Σ_P is much smaller than Σ_w so that the parameter α_P is small. On the contrary, at the dayside ionosphere the plasma conductivity is so high that α_P is greater than unity.

Consider first the extreme case of a small Pedersen conductivity in the conjugate ionospheres when α_P tends to zero at both ends of the field line, that means that $R_+ R_- > 0$. From here it follows that the set of eigenfrequencies, $\omega_n'(x)$, is described by Eq. (6.39), while the damping factor, $\omega''(x)$, in Eq. (6.41) tends to zero. As it is seen from Eq. (6.42), in this case $\partial_z E_x = 0$ at the ends of field line. This means that the transverse electric field E_x has a node in the equatorial plane and it has the antinodes at the ends of the field line. The same is true for the plasma velocity V_y, which is related to E_x through Eq. (6.30). On the contrary, it follows from Eq. (6.42) that the transverse magnetic field δB_y has the nodes at the ends of the field line. To illustrate this, the symmetrical profiles of the first and second harmonics of δB_y are shown in Fig. 6.2 with solid ($n = 1$) and dotted ($n = 2$) lines.

Before leaving this case it is useful to return to the analogy between the field lines and elastic strings. Considering the taut elastic string dead at its two ends and replacing V_A by the velocity of the elastic wave, we note that Eq. (6.39) can describe the eigenfrequencies of the elastic string with the length l_1.

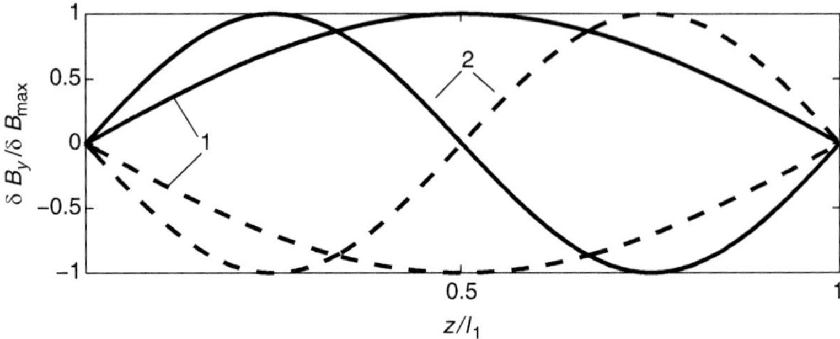

Fig. 6.2 Profiles of the standing Alfvén waves at the magnetospheric shell in the extreme cases $\alpha_P^{\pm} \to 0$. The first and second harmonics of the transverse magnetic field δB_y are shown with *solid* ($n = 1$) and *dotted* ($n = 2$) *lines*

In the inverse case of $\alpha_P^{\pm} \to \infty$, the set of eigenfrequencies is defined by Eq. (6.39) as before. Moreover, the damping factor in Eq. (6.41) is equal to zero as well. In this case $E_x = 0$ at the ends of field line. The interpretation we make is that the Joule dissipation of energy in the ionosphere can be neglected since the electromagnetic field cannot penetrate into the ionosphere due to its infinite conductivity. The components E_x and V_y have an antinode in the equatorial plane and the nodes at the ends of the field line, whereas δB_y has the antinodes at the ends of the field line. Notice that the analogous eigenfrequencies have the elastic string dead in its middle.

If $\alpha_P^{\pm}$ is finite and nonzero, Eqs. (6.39)–(6.41) generally describe the spectrum of damped Alfvén oscillations, which are similar to oscillations of the stretched elastic string with energy losses at the claimed end points.

If a homogeneous confined space is studied, it is usually the case that the spectrum of normal field oscillations is discrete. On the basis of the "MHD-box" model, we have found, however, that the spectrum of the Alfvén oscillations is continuous. It is not surprising that there is one-dimensional (1D) inhomogeneity across straight field lines. In some sense, the actual Earth magnetic field is inhomogeneous across the magnetic shells. This implies that the spectrum of the FLR of the Earth magnetic field depends on the magnetic shell, which is a function of the McIllwain parameter L. Under nominal magnetospheric conditions one may expect an increase of the oscillation period with L or with radial distance, at least at auroral latitude. Below we show that this conclusion is consistent with the observations. It can be shown that in a curvilinear magnetic field the major features of the FLR are the same except for the effect of polarization splitting of the FLR-spectrum. This effect is due to the difference of the convergency/divergency rate of the magnetic field lines within the meridional and equatorial plains. The interested reader is referred to the text by Leonovich and Mazur (1993) and Leonovich (2000) for details about the dependence of the FLR-resonance frequencies on polarization.

6.2.3   Cavity Mode

In this section we consider as before $k_y = 0$ in order to treat each wave mode separately. In a cold homogeneous plasma the phase velocity of the compressional/FMS wave is independent of the angle between the wave vector $\mathbf{k}$ and the external magnetic field $\mathbf{B}_0$. In the MHD box model the properties of this mode are defined by Eq. (6.32), which is an ordinary 2D (two dimensional) wave equation for an inhomogeneous plasma. This means that in contrast to the Alfvén waves which are guided by the field lines, the compressional waves can propagate in all directions and fill the whole resonance cavity. In the magnetosphere these waves are therefore referred to as the class of cavity modes.

We will seek for the solution of Eq. (6.32) in terms of the series

$$E_y = \sum_n A_n(x) q_n(z), \tag{6.43}$$

where the orthonormal eigenfunctions $q_n(z)$ of the problem are given by Eq. (6.126). These eigenfunctions satisfy the boundary conditions (6.34) at the southern $(z = 0)$ and at the northern ionospheres $(z = l_1)$. Substituting Eq. (6.43) for E_y and Eq. (6.126) for q_n into Eq. (6.32) yields

$$\sum_{m=1}^{\infty} \left\{ A_m'' + \left(k_A^2 - k_m^2 \right) A_m \right\} q_m = -i\mu_0 \omega J_y, \tag{6.44}$$

where $k_A(x) = \omega / V_A(x)$, and the prime denotes derivative with respect to x. The eigenvalues $k_n(\omega)$ are the roots of Eq. (6.123). In two extreme cases of the non-conducting ionosphere $\left(\Sigma_P^{\pm} = 0 \right)$ and of the perfect conducting ionosphere $\left(\Sigma_P^{\pm} \to \infty \right)$ there are only real roots

$$k_n = \pi n / l_1, \quad \text{where } n = 1, 2, 3 \ldots, \tag{6.45}$$

which are independent of the frequency ω. Overall, if the Pedersen conductivities $\Sigma_P^{\pm}$ are finite and nonzero values, the eigenvalues are complex.

Consider first the problem of free oscillations assuming for the moment that $J_y = 0$. Multiplying both sides of Eq. (6.44) by $q_n(z)$, integrating these equations over z from 0 to l_1 and using the condition (6.127) of orthogonality of the eigenfunctions $q_n(z)$, we come to

$$A_n'' + \kappa^2 A_n = 0, \tag{6.46}$$

where

$$\kappa^2(x) = k_A^2(x) - k_n^2(\omega). \tag{6.47}$$

The plasma velocity normal to the boundaries of equatorial ionosphere and magnetopause is assumed to be equal to zero, that is $V_x = 0$ at $x = 0$ and $x = l_2$. From Eq. (6.30) it follows that E_y and A_n are both zero at $x = 0$ and $x = l_2$. Equation (6.46) under this homogeneous boundary condition is a special case of the so-called Sturm–Liouville problem for determination of the resonance frequencies of the cavity mode. To estimate the fundamental frequency of the normal oscillations, consider the case of constant Alfvén speed when the parameter κ in Eq. (6.47) is independent of x. The solution of Eq. (6.46) can be thus written as $A_n = C_n \sin \kappa x$, where C_n is the undetermined constant. Under boundary conditions alluded to above the parameter κ is given by an equation similar to Eq. (6.45), that is

$$\kappa_m = \pi m / l_2, \tag{6.48}$$

where m is integer. For simplicity, the eigenvalues $k_n\,(\omega)$ is assumed to be given by Eq. (6.45). Combining this equation with Eqs. (6.47) and (6.48) we obtain the set of resonance frequencies

$$\omega_{n,m} = V_A \left(\frac{n^2}{l_1^2} + \frac{m^2}{l_2^2} \right)^{1/2}. \tag{6.49}$$

In the framework of the MHD box model the curvature of Earth magnetic field is ignored. To give a numerical estimate of the fundamental eigenfrequency ($n = m = 1$), it is necessary at this point to find a suitable estimate of the parameters appearing in Eq. (6.49). We recall that the x axis approximates the radial direction. If the outer boundary of the magnetosphere $l_2 = L R_e$ corresponds to the McIllwain parameter $L \approx 5$, the corresponding length of the field line $l_1 \approx 7.7 R_e$. Substituting the Earth radius $R_e = 6.4 \times 10^3$ km and the Alfvén speed $V_A = 10^3$ km/s into Eq. (6.49), we get $f_{11} = \omega_{11} / (2\pi) \approx 0.02$ Hz. The cavity resonance period of the fundamental harmonic is about $T_{11} = f_{11}^{-1} \approx 50$ s. It should be noted that we have obtained only the rough estimate of the period and frequency of the fundamental harmonic.

As would be expected, an FMS-wave in the magnetosphere may increase in amplitude as the wave frequency is close to the frequencies of the global resonances. This kind of oscillations can cover a significant part of the magnetosphere. In the framework of the MHD box model the properties of the cavity mode are similar to that of the TE mode excited in the inhomogeneous resonator. Since the y direction corresponds to the azimuthal coordinate of the magnetosphere, the transverse electric field E_y corresponds to the azimuthal component E_ϕ. In some sense, the cavity mode is identical in its properties to the poloidal mode in the curved magnetic field. Some concerns about the mode coupling and the energy dissipation are found in the next sections.

6.2.4   The Mode Coupling

In this section we consider the field variations excited by the plasma perturbations coming from the outer space into the magnetosphere. Such perturbations can be resulted from the interaction between the solar wind and the Earth's magnetic field at the magnetopause. The curvature of the magnetospheric boundary is ignored since the perturbations wavelength is assumed to be much smaller than the size of the magnetospheric cavity. The MHD box model of the medium is a reasonable approximation at this point to proceed analytically and to treat the FLR structure.

The inner region of the magnetosphere is assumed to be free of the driving/external current so that $J_y = 0$. To specify the problem, we assume that E_y is a given function at the magnetospheric boundary $x = l_2$. As one example, let $E_y = E_0(z) \exp\left(-i\omega t + i k_y y\right)$ at the box surface $x = l_2$ while $E_y = 0$ at $x = 0$. This function can describe the perturbation coming from outer space into the magnetosphere or the surface wave generated at the magnetopause. Here ω is the frequency of this wave. (In general ω is a complex value.) The inhomogeneous boundary conditions at $x = l_2$ are important in the sense that they play a role of source of field variations in the magnetosphere.

If $k_y = 0$, then the shear Alfvén and compressional waves are described by independent Eqs. (6.31)–(6.32). As before we seek for the solution for E_y in terms of the series (6.43) in eigenfunctions $q_n(z)$ where the expansion coefficients $A_n(x)$ satisfy the differential equation (6.46). If the boundary function $E_0(z)$ can be expressed as a series of eigenfunctions $q_n(z)$, that is

$$E_0(z) = \sum_n d_n q_n(z), \tag{6.50}$$

then the expansion coefficients are given by

$$d_n = \int_0^{l_1} E_0(z)\, q_n(z)\, dz. \tag{6.51}$$

Whence it follows that the boundary conditions reduce to $A_n(0) = 0$ and $A_n(l_1) = d_n$.

We will study Eq. (6.46) by using a qualitative method since the explicit form of the function $V_A(x)$ is unknown. As is seen from Eq. (6.33), the Alfvén speed depends on both the Earth magnetic field $\mathbf{B}_0$ and the plasma density. The plasma density falls off more rapidly with distance x than $\mathbf{B}_0$ does, and hence the Alfvén speed generally increases with distance. At the outer boundary of the magnetosphere ($x = l_1$) the Alfvén speed is on one or two order of magnitude greater than that at the conducting layer of the ionosphere ($x = 0$). Furthermore, if near the boundary $x = l_1$ the wave frequency is so small that the inequality $\omega / V_A(x) \ll |k_n(\omega)|$ takes place, then the parameter κ in Eq. (6.46) is approximately constant. This means that the exponential functions can fit the approximate solutions of Eq. (6.46) in this region. For example, considering for the moment that V_A is a constant, we get

$$A_n = d_n \frac{\sinh \lambda_n x}{\sinh \lambda_n l_1}, \quad \text{where} \quad \lambda_n = \left(k_n^2 - k_A^2\right)^{1/2}. \tag{6.52}$$

It follows from this equation that the boundary perturbations E_y decay in amplitude with decreasing the distance x due to the exponential fall off of all the coefficients A_n. The interpretation we make is that the FMS waves must attenuate when propagating from the outer boundary to the inner region of the magnetosphere. This tendency is valid for the case of space-varying Alfvén speed except for the region where the FLR occurs.

In the resonance region the coupling of the shear Alfvén and the FMS waves cannot be ignored. The mode coupling is studied in more detail in Appendix F. As we have noted above, if $k_y \neq 0$, then Eqs. (6.25)–(6.29) cannot be split into two independent sets of equations. In this case the FMS wave can excite the shear Alfvén wave and vice versa. The interaction between these two modes may greatly affect the field amplitude as the resonance condition

$$k_A^2(x) = k_n^2 \tag{6.53}$$

holds true. If $x = \xi$ is a root of Eq. (6.53), then at this point the wave frequency ω equals to one of the Alfvén resonance frequencies $\omega_n(x)$ that are given by Eqs. (6.39)–(6.41).

As has already been stated in Appendix F, the amplitude the Alfvén mode which includes the components δB_y, E_x, and V_y, has a peak of Lorentz form near the FLR position $x = \xi$. The schematic representation of the amplitude of δB_y as a function of x/ξ is displayed in Fig. 6.3 with solid line 1. According to Eqs. (6.139) and (6.141), the amplitude of the components δB_x, E_y, and V_x has a maximum at the same resonance point but this maximum is not so distinct as is shown in Fig. 6.3 with dashed line 2. It is worth mentioning that, as shown in Appendix F, the phase of the resonance components E_x and δB_y, changes by π when crossing the maximum.

The next singular point $x = \eta$, can be found from the following equation

$$k_A^2(x) = k_n^2 + k_y^2. \tag{6.54}$$

The implication here is that the roots of this equation correspond to turning points $x = \eta$ where solutions change from being oscillatory in nature to characteristically growing or decaying with coordinate x. At the turning point the wave reflection occurs. It should be noted that if $V_A(x)$ is not a monotonic function there may be more turning and resonance points.

The MHD box model is based on an idealized field geometry that ignores the magnetic field line curvature and dip angle but includes the field variations with radial distance and boundary conditions at the ionosphere and magnetopause. The MHD box model provides us with a qualitative theory of the FLR in the

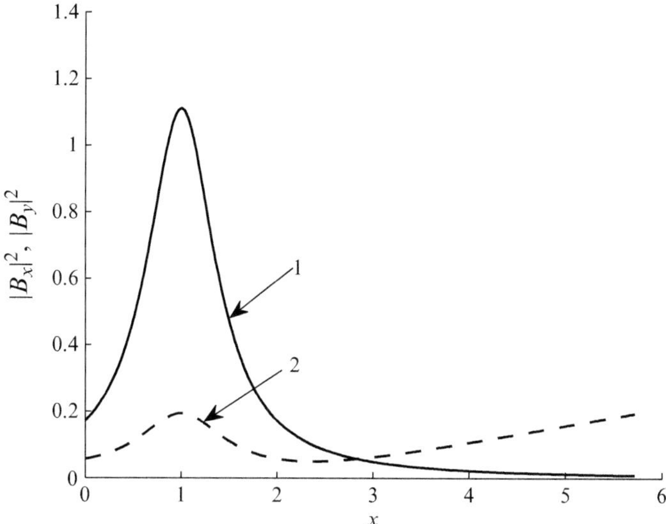

Fig. 6.3 The perpendicular magnetic field components in the vicinity of field line resonance position versus normalized variable x/ξ_n. The components $\left|\delta B_y\right|^2$ and $\left|\delta B_x\right|^2$ are shown with *solid line 1* and *dashed line 2*, respectively

magnetosphere. We recall that in the model the coordinate x plays a role of the radial distance in the actual magnetosphere. In this picture the resonance components, δB_y and E_x, are analog of the azimuthal magnetic field variations and the radial electric field variations, respectively. A schematic plot of the amplitude variations as a function of L-shell is shown Fig. 6.4.

As one example, consider the MHD wave excited due to, say, plasma instabilities at the magnetopause. A scheme of penetration of MHD wave from the magneto-spheric boundary into its inner region can be summarized as follows. At first the initial perturbations propagate as an FMS-wave from the magnetospheric boundary to the turning point where wave reflection occurs. The electromagnetic field in this region may be oscillatory in character. Once the turning point has been passed, the amplitude of the FMS-wave falls off exponentially with distance up to the region where the FLR conditions will occur. This implies that in this region the wave frequency becomes close to the Alfvén resonance frequency of the magnetic shell. In the vicinity of the resonance shell the energy of the FMS-wave is transferred in part into the energy of the Alfvén oscillations by virtue of the mode coupling to the shear Alfvén and FMS modes. The shear Alfvén wave can get trapped in this region thereby exciting the FLR. At the resonance point a phase shift of π between the toroidal field components (δB_y and E_x in the box model) on both sides of the resonance is apparent. Some complication arises in this scenario as there are several turning points or resonance shells.

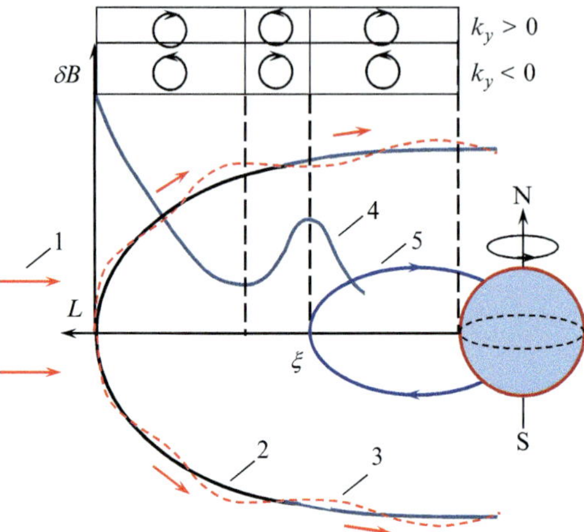

Fig. 6.4 A schematic plot of amplitude variations of FMS wave excited at the magnetopause and sense of the wave polarization as a function of L-shell in the magnetosphere. *1*—solar wind, *2*—magnetopause, *3*—boundary surface wave caused by Kelvin–Helmholz instability, *4*—plot of wave amplitude, *5*—resonance field line

Attenuation of the resonance oscillation is basically due to the Joule dissipation caused by the Pedersen conductivity in the ionosphere. In addition, azimuthal propagation of the waves leads to the energy losses in the magnetotail.

The Hall conductivity scarcely affects the FLR but it may play a crucial role in occurrence of the magnetic perturbations under the ionosphere, that is in the atmosphere and on the ground surface. As the Hall conductivity is ignored, the incident shear Alfvén wave cannot excite the field perturbation in the atmosphere. If only the Hall conductivity is finite, the shear Alfvén wave can be transformed in the ionosphere into both the reflected and transmitted field of the FMS wave. In other words, the FLR-related field observed on the ground builds up as a result of the mode coupling in the ionosphere via the Hall conductivity followed by the penetration of the FMS mode through the atmosphere towards the ground.

A variety of mechanisms of coupling between the shear Alfvén and FMS waves in a realistic magnetospheric environment have been studied in numerous papers. With some care the terms "shear Alfvén wave" and "FMS" are applicable to the actual MHD waves propagating in the magnetosphere since in most cases these pure eigenmodes do not exist. However these two terms are extremely important for understanding of wave processes in the planetary magnetosphere. The interested reader is referred to the text by Glassmeier (1995) for a more complete review on mode coupling in actual magnetosphere.

6.2.5 Wave Polarization

As we have noted above, near the resonance the phase of the resonance components changes by π abruptly, so one may expect a corresponding change in the wave polarization in the vicinity of the resonance point. From Eq. (6.131) it follows that the amplitudes of E_x and E_y are related through

$$\frac{E_x}{E_y} \sim \frac{ik_y}{\left(k_A^2 - k_n^2 - k_y^2\right)} \frac{dE_y}{dx} \frac{1}{E_y}. \tag{6.55}$$

In the region $x < \eta$ the sense of polarization depends on only the signs of k_y and dE_y/dx. In other words, the sense of polarization is a function of the direction of propagation in y, and it depends on whether the amplitude increases or decreases with the radial distance (Southwood 1974).

In the vicinity of the resonance point $x = \xi$ Eq. (6.55) is reduced to

$$\frac{E_x}{E_y} \sim -\frac{i}{k_y} \frac{dE_y}{dx} \frac{1}{E_y}. \tag{6.56}$$

This implies that the sense of polarization switches on each side of a maximum and or minimum in amplitude. For example, consider an eastward propagating MHD wave, which corresponds to $k_y > 0$. In the region where $dE_y/dx < 0$ the wave is the clockwise-polarized and vice versa. We recall that the y axis corresponds to the west–east direction, while x axis is in radial direction. In this picture the expected polarization and the wave amplitude as a function of L for magnetic equator plane is schematically shown at the upper panel of Fig. 6.4. For the resonance shell the polarization tends to be linear. In the case of a westward propagating wave $\left(k_y < 0\right)$ the sense of polarization is reversed. Looking down on the Earth from above in the northern hemisphere, this wave would have clockwise polarization south of the resonant site ξ and anticlockwise polarization north of the resonance. These properties of the polarization would thus be expected to be valid on the ground despite the influence of the conducting E layer of the ionosphere.

On the basis of data recorded at a chain of stations at geomagnetic latitudes between 59°N and 77°N within 2° of longitude 302°E Samson et al. (1971) have studied the diurnal and latitude variations of the amplitude and polarization of the long-period pulsation. Their basic results for fixed frequency of 5 mHz are schematically shown in Fig. 6.5. The pulsation amplitude reaches a peak value at the line which belongs to auroral zone. Across this line the rotation sense of the horizontal polarization changes from counterclockwise to clockwise or vice versa at midday.

It is generally believed that the ULF pulsations in the frequency range 10^{-2}–10^{-3} Hz originate from the interaction between the solar wind and planetary magnetosphere. In this picture an FMS wave propagating in the magnetosphere can build up as a result of Kelvin–Helmholtz instability at the magnetopause.

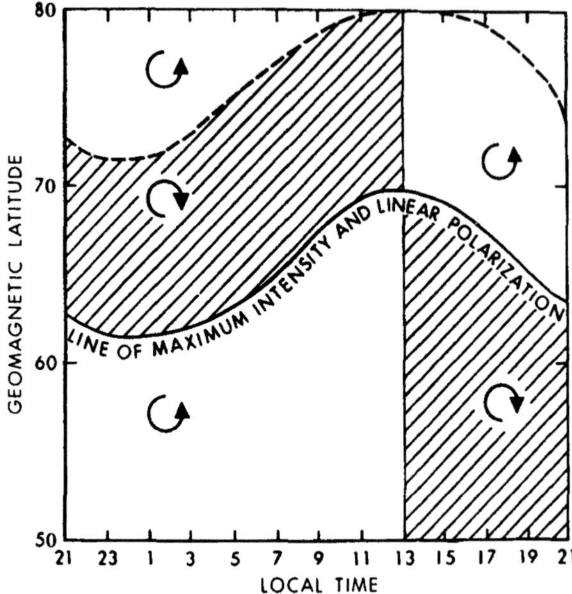

Fig. 6.5 A schematic plot of the diurnal and latitudinal variations of the amplitude and sense of rotation of the horizontal component for 5 mHz pulsations as observed by Samson et al. (1971). The amplitude of the pulsations reaches a peak value at auroral zone, with latitude changing in local time as shown with line of maximal intensity. The horizontal polarization switches sense across this line. Taken from Glassmeier (1995)

When propagating from the magnetopause into the magnetosphere the FMS wave decays in amplitude until the resonance point will occur, as shown in Fig. 6.4. The switch in polarization at this point is consistent with waves propagating eastwards $(k_y < 0)$ in the afternoon and westwards $(k_y > 0)$ in the morning. On account of the fact that the radial derivative of the azimuthal electric component $(dE_y/dx$ in the MHD-box model) changes sign across the resonant point, a four-quadrant pattern arises due to the FLR phenomenon much as observed by Samson et al. (1971) (Fig. 6.5).

6.2.6   *Effect of the Ionosphere on Ground-Based Observation*

As has already been stated, the amplitude of the FMS-waves falls off exponentially as they propagate towards the ionosphere and their amplitude becomes smaller than that of Alfvén waves. This means that the polarization of the MHD waves incident to the ionosphere can be considered to be basically corresponding to that of Alfvén mode.

The ionospheric plasma may greatly affect the ground-based observation of the ULF pulsations. The dominant effect is the attenuation of MHD waves and rotation

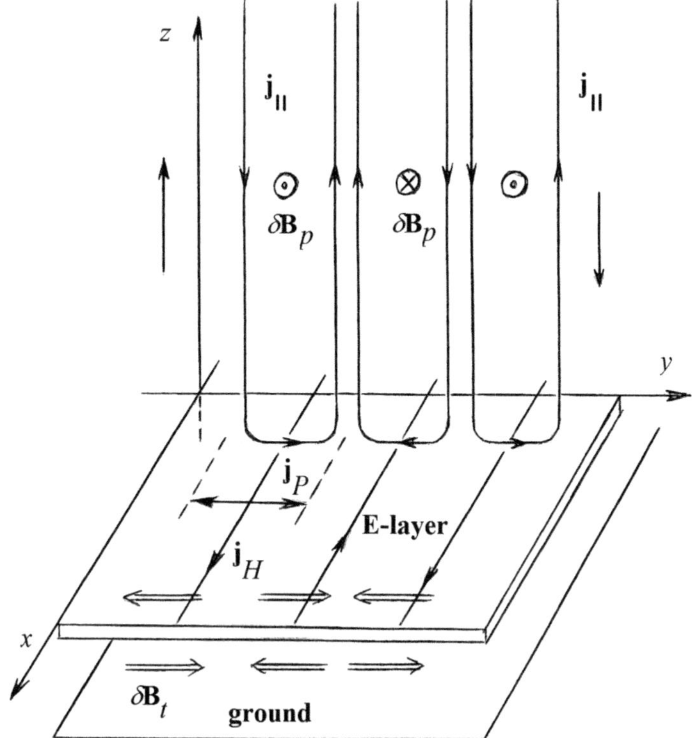

Fig. 6.6 A schematic illustration of the electric current system and magnetic perturbations resulted from Alfvén wave interaction with the ionosphere

of polarization mainly due to the conducting E layer of the ionosphere. The high frequency range of the signal spectrum undergoes strong attenuation since the skin length of the conducting ionosphere is inversely proportional to the square root of frequency. In other words, the ionosphere acts as a spatial low-pass filter for the signals observed on the ground.

To illustrate the rotation of polarization, we will consider the effect of an incident Alfvén wave on the high-latitude ionosphere. To make our consideration as transparent as possible, the Earth magnetic field is assumed to be homogenous and positively parallel to vertical z axis. A plane harmonic Alfvén wave propagates along the magnetic field perpendicular to the ionosphere that are in the plane x, y. All perturbed values are assumed to vary as $\exp(iky - i\omega t)$. In the magnetosphere the Alfvén wave carries transverse polarization of electromagnetic perturbations and field-aligned current, $\mathbf{j}_\parallel$, as schematically shown in Fig. 6.6. The conducting E layer of the ionosphere shorts out the field-aligned current thereby exciting the sheet current system. As the magnetic $\delta\mathbf{B}$ and the electric $\mathbf{E}$ perturbations in the magnetosphere are directed along the x and y axes, respectively, the Pedersen current $\mathbf{j}_P = \sigma_P \mathbf{E}$ in the E layer of the ionosphere is parallel to y axis while the Hall current $\mathbf{j}_H = \sigma_H (\hat{\mathbf{z}} \times \mathbf{E})$ is parallel to x axis. The Pedersen and the Hall conductivities of the E layer are assumed to be constant.

Thus the field-aligned currents of the incident and reflected Alfvén waves and the ionospheric sheet system of currents, shown in Fig. 6.6, can be split into poloidal and toroidal current systems. The first one includes the field-aligned currents and the ionospheric Pedersen currents. In such a case the magnetic perturbation, $\delta\mathbf{B}_p$, are concentrated inside the poloidal current system so that the magnetic effect is undetectable below the ionosphere. In general, formal proof of this assertion can be found in McHenry and Clauer (1987). The toroidal system builds up as a result of the Hall currents flowing in the E layer of the ionosphere. In our model these currents are closed in the infinity ($x \to \pm\infty$). The magnetic field of the toroidal currents, $\delta\mathbf{B}_t$, is perpendicular to $\delta\mathbf{B}_p$. This means that the magnetic perturbations in the atmosphere are perpendicular to the magnetospheric field of the Alfvén wave. The ionosphere therefore changes the wave polarization by $90°$. In our model we have ignored the field line curvature and the inhomogeneous distribution of the Pedersen and Hall conductivities in the ionosphere. Actually the ionosphere produces a rotation of the polarization plane in the angle range from $0°$ to $90°$. The detailed calculations of this problem are found in numerous papers (e.g., see review by Glassmeier 1995 for details). Notice that the latitude variations of the ULF pulsation period, as observed in space and on the ground, are in favor of the ionospheric rotation effect. In particular the latitude dependence of the Alfvén resonance oscillations in space is detected in azimuthal component (D component), whereas the ground-based observation exhibits the same dependence in meridional field (H component) that is consistent with the rotation of components by the angle $90°$.

6.3   Sources of ULF Pulsations

6.3.1   Observations of ULF Pulsations

Observations and study of ULF MHD waves is certainly necessary as they transmit energy, momentum, and most importantly they provide us with information about magnetospheric dynamics. A variety of these waves occurring in the magnetosphere and ionosphere result in the generation of ULF geomagnetic pulsations that have been identified in both ground-based and satellite observations. Periods and frequencies of the ULF pulsation vary from 0.2 to 600 s, and from several milliHertz to several Hertz, respectively. Below is the frequency range of magnetic storms. The amplitudes of the ULF pulsation typically change from 0.1 to 50 nT.

In standard geophysical practice the ULF pulsations are classified according to their period. They can be also divided into two classes depending on whether the pulsation accompany substorms or not (e.g., see Jacobs 1970; Nishida 1978). The latter class includes the regular quasiharmonic oscillations, which are termed Pc oscillations (Pulsations continuous). This class of the ULF pulsations can be in turn

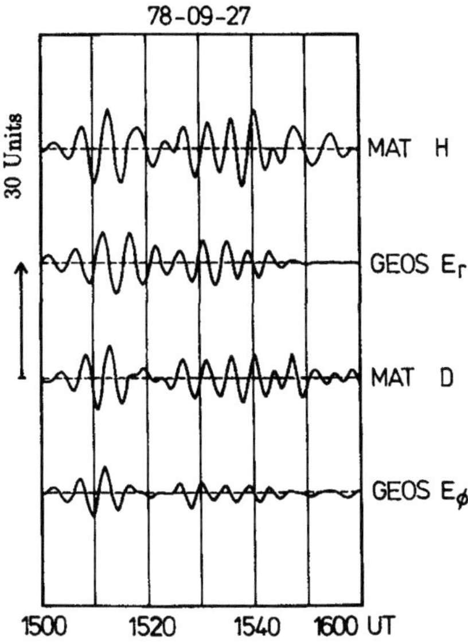

Fig. 6.7 A Pc5 pulsation event recorded at the ground-based observatory of the Scandinavian Magnetometer Array in the Finmark (shown with MAT) and in the magnetosphere by the geostationary satellite GEOS 2 (shown with GEOS). The H and D components of the magnetic variations have been recorded at the ground-based station while the components of electric variations, E_r and E_ϕ, were measured on board the satellite. Units of the magnetic and electric variations are nT and 0.1 mV/m, respectively. Taken from Glassmeier (1995)

split into five spectral subclasses Pc1 (period 0.2–5 s), Pc2 (5–10 s), Pc3 (10–45 s), Pc4 (45–150 s), and Pc5 (150–600 s). The period of the Pc oscillations are controlled by both the parameters of interplanetary space and the resonance properties of the Earth magnetosphere. The class of irregular pulsations, which have been termed Pi pulsations (Pulsations irregular), consists of two subclasses Pi1 (1–40 s) and Pi2 (40–150 s). These pulsations are a signature of onset of magnetospheric substorms which build up as a result of the plasma and solar energy penetration into the magnetosphere from the interplanetary space during magnetic storms and active processes on the Sun. Certainly, this classification is fairly relative. There are complicated and unusual pulsations, in which the regular and irregular oscillations are mixed.

The typical amplitude of Pc5 pulsations, about 10–50 nT, is the largest among the ULF pulsation. An example of Pc5 pulsations simultaneously observed at the ground-based station and in the magnetosphere by the geostationary satellite GEOS 2 is shown in Fig. 6.7 (Glassmeier 1995). This event is in favor of the magnetospheric origin of the Pc5 pulsations. The Pc5 pulsations are latitude-dependent and are frequently localized within narrow regions extended along geomagnetic parallels. It is usually the case that the period of the Pc5 pulsation falls off with decreasing latitude of the sighting point (Ohl 1962, 1963; Annexstad and

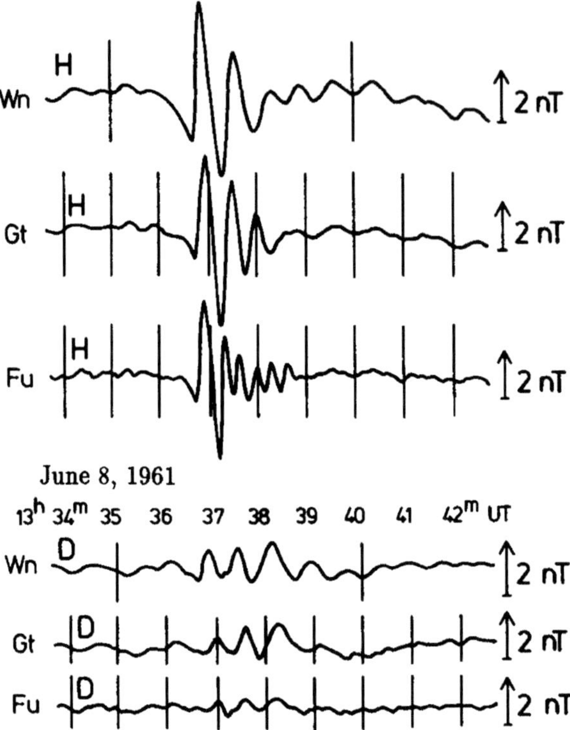

Fig. 6.8 Large-scale ULF pulsations observed at the geomagnetic observatories Wingst (Wn), Göttingen (Gt), and Fürstenfeldbruck (Fu). H and D components of the magnetic variations are displayed in the upper and bottom panels, respectively. Taken from Voelker (1962), and Glassmeier (1995)

Wilson 1968). This property keeps partly for the case of Pc4 pulsation. It should be noted that at the latitudes below $50° - 60°$ these pulsations can be masked by the global magnetic variations with time-independent period. The fundamental Pc5 pulsations occur primary in the sunlit hemisphere. This asymmetry probably arises from magnetospheric structure asymmetries, which result from the existence of the magnetotail and the plasmospheric convexity in the nightside magnetosphere.

The damped-type Pc3–Pc4 oscillations of global extension are illustrated in Fig. 6.8 (Voelker 1962). These oscillations with latitude-dependent period have been recorded at three ground-based stations located at Wingst (the northernmost station), Göttingen, and Fürstenfeldbruck (the southernmost station). The increase of the oscillation period with L is compatible with the above analysis. The typical amplitude of the Pc4 pulsations varies within 5–20 nT while the Pc3 amplitude is smaller than 10 nT.

Other kind of the damped-type oscillations is shown in Fig. 6.9 (Glassmeier 1995). The H-component of the geomagnetic variations caused by a magnetospheric substorm is displayed at the upper panel. The sharp decrease of the magnetic field at

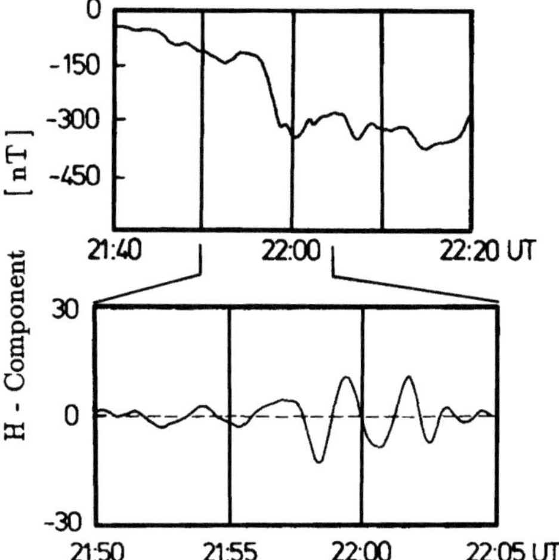

Fig. 6.9 An example of magnetic field variations during an isolated magnetospheric substorm (*upper panel*) and a Pi2 pulsation (*bottom panel*). The data displayed in the *bottom panel* are high pass filtered time series of the upper record. Taken from Glassmeier (1995)

21:57 UT is an evident signature of the substorm onset. The data at the bottom panel is high pass filtered to yield a magnetogram of the Pi2 pulsation with period about 150 s and amplitude about 10 nT. The trains of Pi2 pulsations are usually observed at the nightside of the Earth. Their amplitude about 1–50 nT tends to maximize in the vicinity of aurora region at midnight.

The long-period pulsation trains, such as Pc3, Pc4, Pc5, and Pi2 pulsations, are believed to be due to eigenoscillations of the Earth magnetosphere (Kato 1962; Zibyn and Yu 1965). The Alfvén oscillations can explain the basic properties of the Pc5 pulsations and, in some cases, the patterns of the Pc4 pulsations. The observed dependence of pulsation period on the latitude is consistent with that predicted by the FLR theory we have treated in the previous sections (Guglielmi and Troitskaya 1973). Likewise, a number of pulsation events exhibit a rather localized wavefield of extension 100–200 km in north–south direction and about 500–1,000 km in east–west direction (Glassmeier 1980). The additional argument that is in favor of the resonance origin of the Pc4, Pc5 pulsations is that these pulsations are well correlated at the magneto-conjugate points (Guglielmi and Troitskaya 1973). Lanzerotti and Fukunishi (1974) have found that in the ground magnetic observation the odd mode of the Alfvén oscillations is prevailed so that the amplitude of the oscillation reaches a peak value at the equator. On the other hand the amplitude of these pulsations grows with increase of the latitude (Ziesolleck et al. 1993). This suggests that the source of the pulsations is at the periphery of the magnetosphere. The fundamental mode of the FLR oscillations manifests itself through Pc5 pulsations as observed on board the satellites OGO 5 and GEOS 2 (Singer and Kivelson 1979; Junginger et al. 1984). A signature of the fundamental mode together with higher harmonic of the same field-line shell has been detected by Singer et al. (1979), Baumjohann and Glassmeier (1984) and Engebretson et al. (1986).

A detailed review of observations of the ULF pulsation is outside the scope of this section, and the interested reader is referred to the excellent tutorial review by Glassmeier (1995) for detail about the horizontal polarization, wave propagation across the ambient magnetic field and other properties of the ULF fields.

Only a few observations may have interpreted as global poloidal eigenoscillations that may be associated with the cavity mode (Higbie et al. 1982; Kivelson et al. 1984). An observational hint toward the existence of cavity mode has been reported by Crowley et al. (1987) on the basis of measurements of the ionospheric Pedersen conductivity and damping rates of the ULF pulsations

The short-period pulsations such as Pi1, Pc1, Pc2 contain a wide variety of shape compared to the long-period pulsations. To describe this diversity of the short-period pulsations, we use the additional nomenclature including "pearl-type micropulsations," "interval of pulsations of diminishing periods" (IPDP), "hydromagnetic whistler," "pulsation burst" (Guglielmi and Troitskaya 1973), "continuous emissions", and so on. A typical amplitude of the short-period pulsations is smaller than 1 nT, and these pulsations cover the frequency range from 0.025 to 5 Hz. It is generally believed that the main excitation of the short-period pulsations is due to kinetic plasma instabilities (Trakhtengerts and Rycroft 2008) resulted in the generation of MHD and ion-cyclotron waves in the frequency range of Pi1, Pc1, and Pc2.

6.3.2 Kelvin–Helmholtz Instability at the Magnetopause

The most prominent mechanism for Pc5 pulsations is thought to be the Kelvin–Helmholtz instability at the Earth's magnetopause (Dungey 1954). This effect is believed to be due to surface waves propagating at the flanks of the magnetopause. The surface wave may be in turn excited due to the interaction between the solar wind and the planetary magnetic field as illustrated in Fig. 6.4 (Atkinson and Watanabe 1966; Kivelson and Southwood 1985). This kind of instability may arise in a fluid flow at the boundary between two regions which are separated by a tangential discontinuity of the fluid velocity. This means that the fluid flow velocities are both parallel to the boundary and have a jump across the boundary whereas the fluid pressure is kept continuous. Consider a small random variation of the equilibrium position of the boundary. This variation, shown in Fig. 6.10 with a bulge of the boundary surface, results in restriction of the effective cross section of the flow in the upper region. From the principle of the fluid flux conversation it follows that the fluid velocity V must increase in this region. According to Bernoulli's law for an inviscid fluid

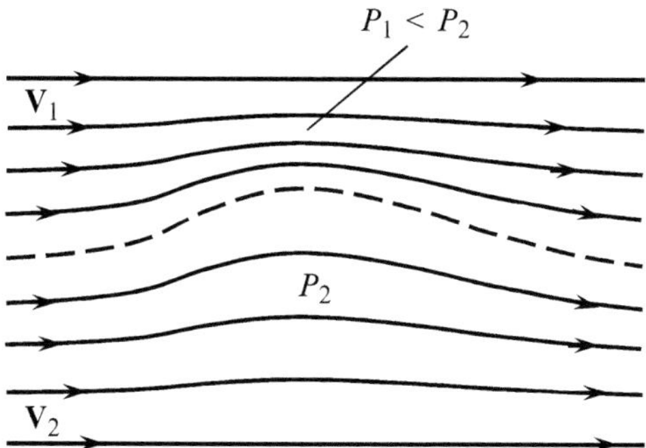

Fig. 6.10 A schematic illustration of the mechanism of Kelvin–Helmholtz instability in an inviscid fluid. The tangential discontinuity of a fluid flow is shown with *dotted line*. $\mathbf{V}_1$ and $\mathbf{V}_2$ denote the unperturbed velocities of the fluid flow on both sides of the tangential discontinuity. The fluid pressure, P_1, just over the bulge of the boundary is smaller than that under the bulge

$$\frac{\rho V^2}{2} + P = \text{const,} \tag{6.57}$$

a flow velocity increase is accompanied by the fluid pressure decrease over the bulge shown in Fig. 6.10. The pressure difference between two regions leads to further enhancement of the instability and the initial perturbation of the boundary position that result in the generation of surface waves.

In familiar hydrodynamics the tangential discontinuities are always unstable with respect to small perturbations that result in their fast turbulization. Magnetic field stabilizes the flow of conducting fluid in such a way that the tangential discontinuities in the fluid may be stable. This is due to the fact that the fluid velocity perturbations across the ambient magnetic field give rise to extension of the field lines frozen in the conducting fluid that in turn results in the generation of forces aiming to restore the unperturbed fluid flow.

The condition of instability of the tangential discontinuity in a conducting fluid/plasma is the following (e.g., see the text by Landau and Lifshitz 1982 for details)

$$\frac{B_1^2 + B_2^2}{\mu_0} < \frac{\rho_1 \rho_2}{\rho_1 + \rho_2} (V_1 - V_2)^2, \tag{6.58}$$

where B_1 and B_2 are magnetic fields on both sides of the boundary, ρ_1 and ρ_2 are the corresponding fluid/plasma densities, and $V_1 - V_2$ stands for the relative flow velocity.

This equation can be applied to the Kelvin–Helmholtz instability arising in the magnetospheric plasma at the magnetopause. In such a case ρ_1, V_1, and B_1 may denote the plasma and field parameters of solar wind in the magnetosheath while the same values with inferior index 2 describe the magnetospheric plasma. To estimate this effect we suppose that $\rho_1 \approx \rho_2$, and $B_1 \approx B_2$. On account of the relation $V_1 \gg V_2$ we also assume that there are no plasma motion in the magnetosphere, i.e., $V_2 \approx 0$. Substituting these values into Eq. (6.58) yields

$$V_1 > 4V_A. \qquad (6.59)$$

The implication here is that if only the plasma flow in the magnetosheath is super-Alfvénic, then the Kelvin–Helmholtz instability can develop. As alluded to earlier in Sect. 1.2, the solar wind is supersonic near the Earth orbit. Across the bow shock shown in Fig. 1.8, the solar wind density and temperature increase abruptly whereas the wind velocity decreases, allowing for the presence of subsonic flow around the Earth magnetosphere. In other words, in the vicinity of local magnetic noon the solar wind flow stalls and becomes sub-Alfvénic. Toward the flanks of the magnetosphere the stream accelerates in a such way that the flow velocity becomes super-Alfvénic again. This implies that the Kelvin–Helmholtz instability may occur at the flanks of the magnetosphere, i. e., around dusk and dawn.

A number of experimental data is consistent with the solar-wind-related mechanism for excitation of the Kelvin–Helmholtz instability followed by the long-period ULF pulsations (e. g., see the text by Glassmeier 1995). First, it is usually the case that the Pc5 pulsations are observed at the flanks of the magnetosphere, especially at the downside, that are in favor of the mechanism of the Kelvin–Helmholtz instability. Moreover, analysis of the observation shows that the vector of the phase wave velocity is directed toward the tail of the magnetosphere. Second, the ULF pulsation activity and polarization characteristics are clearly controlled by the solar wind.

As discussed above, the instability of the tangential discontinuity may result in the turbulization of plasma flow followed by generation of a rather broad spectrum of perturbations. This mechanism is capable of exciting different FLRs, which may therefore form a continuous spectrum of the resonant field. This conclusion contradicts with the observations since there usually occurs only one resonance. To explain this contradiction Kivelson and Southwood (1985) have suggested that the Kelvin–Helmholtz instability caused by surface waves first results in excitation of the fundamental and higher harmonics cavity modes, that is the global poloidal eigenoscillations of the magnetosphere. As has already been stated, the spectrum of cavity modes is discrete. When these modes are excited they produce a frequency filter for wideband spectrum of the initial perturbations. At this point the FLRs can be excited by virtue of the shear Alfvén mode coupling to the resonant cavity modes. In other words, the energy of unstable surface waves may transform into the energy of poloidal oscillations which in turn can propagate across field lines up to the resonance magnetic shell thereby producing the FLR due to the mode coupling.

6.3.3   Magnetospheric Plasma Instabilities

Instabilities of the internal magnetospheric plasma distributions can be another mechanism for generation of the ULF pulsations. For one example, consider now kinetic instabilities which result in the generation of MHD and ion-cyclotron waves in the frequency range of Pc1, Pc2, and Pi1 pulsations. In a linear approximation the ion-cyclotron instability arises from the energy exchange between the wave field and charged particles. This interaction becomes the most effective under the resonance condition (e.g., see the texts by Ginzburg (1970), and Trakhtengerts and Rycroft (2008) for detail)

$$\omega - k_z V_z = n\Omega_i, \tag{6.60}$$

where ω is the wave frequency, V_z is projection of the thermal velocity of particles on ambient magnetic field, k_z is the same projection of wave vector, Ω_i is gyrofrequency of the ions, and n is integer, that is $n = 0, \pm 1, \pm 2, \ldots$ The kinematic meaning of this condition can be understood in a local reference frame fixed at the Larmor center of the particle. In this reference frame the wave frequency $\omega' = \omega - k_z V_z$ either equals zero ($n = 0$) or a multiple of the ion gyrofrequency ($n \neq 0$). Depending on the plasma particle distribution of the velocities the interaction between the MHD wave and the resonant particles may result in either enhancement or damping of the magnitude of oscillation.

This kind of the plasma instability can be due to the energetic protons (~ 10–$100\,$keV) of the ring current region ($L \sim 3$–6) because of anisotropy of the proton distributions with respect to the proton velocities (Cornwall 1965). The kinetic energy of the plasma particles trapped in the ring current region can thus serve as a source for energy transfer towards the ULF pulsations due to either ion-cyclotron instability mechanism or collisionless Landau damping. Notice that despite small concentration the helium ions present in plasma of the radiation ring may greatly affect the ion-cyclotron instability in the frequency range of Pc1 (Dowden 1966).

The particle bounce motion between the mirror points above the northern and southern ionospheres may cause the resonance interaction between the bounce motion and the MHD waves. The bounce motion is accompanied by large-scale drift of the plasma particles approximately perpendicular to the Earth magnetic field lines. This drift caused by the gradient and curvature of the Earth magnetic field take the particles entirely around the Earth as shown in Fig. 1.10. The MHD wave field can resonate with the bounce and drift plasma motions as the wave frequency is related to the bounce and drift frequencies through special resonance condition (Karpman et al. 1977; Southwood 1980). Moreover, the deviation of the particle distribution function from the equilibrium function is necessary to provide the drift-bounce instability. It is believed that this mechanism is capable of explaining the origin of small-scale (wave number $m \sim 50$–100) azimuthal poloidal Pc4 pulsations and giant pulsations Pg (Takahashi 1988).

Available candidates for a source of the ULF pulsations are the so-called firehose and drift mirror plasma instabilities, which are driven by anisotropies of the plasma pressure. We cannot come close to exploring these topics in any detail, but the interested reader is referred to the text by Glassmeier (1995) for a more complete treatise on magnetospheric plasma instabilities.

6.3.4 MHD Waves Propagating in Solar Wind

An indirect hint toward the existence of a variety of MHD waves, which can propagate in the solar wind, has been provided by numerous observations and has been supported by a number of theoretical studies (e.g., Kwok and Lee 1984; Takahashi et al. 1984; Yumoto 1984; Engebretson et al. 1987). A wide variety of the solar-wind-generated waves cover a wideband frequency range including ULF pulsations region. It appears that such waves can cross the Earth's bow shock, magnetosheath, magnetopause and then penetrate deep into the magnetosphere and plasmasphere. There may also be an indirect way for the energy transfer from the interplanetary space to the magnetosphere. For example, the Alfvén and FMS wave energy can be transferred into the particle kinetic energy and back to the wave energy via ionospheric interactions.

6.3.5 Reconstruction of the Magnetospheric Configuration

All the excitation mechanisms alluded to above share a common trait since they are generated from the different kinds of plasma instabilities that can arise inside the magnetosphere, at the magnetopause, or outside the magnetosphere in the solar wind. In some sense, these mechanisms can serve as more or less permanent sources of the ULF pulsations. In the next subsection we consider more impulsive sources such as SSC and magnetic storm associated Pc5 pulsations and fast transients. The SSC is due to the sudden changes of the solar wind flow followed by variations of the dynamic pressure from the solar wind on the Earth's magnetosphere, that in turn may be sufficient in order to change position of the dayside magnetopause. The large-scale reconstruction of the magnetopause and the whole magnetosphere results in the generation of ULF pulsations, which is believed to be almost axisymmetric. The azimuthal wave numbers m associated with these pulsations seem to be close to zero. This means that both main modes of the magnetospheric eigenoscillations, i.e., toroidal and poloidal modes, can propagate through the magnetosphere independently of each other. It appears that a number of ULF pulsations are related to magnetospheric substorms in the magnetotail (Baumjohann and Glassmeier 1984).

6.4   ULF Electromagnetic Noises

6.4.1   Main Sources of the ULF Noises

The global electromagnetic resonances which have been considered in this section, cover a wide frequency range from several mHz to 30–35 Hz. The first Schumann resonances cover the overall range from 7–8 to 30–35 Hz, which are in the ELF frequency band. The IAR eigenfrequencies lie in the range from 0.5–0.25 to 3–5 Hz. The FLRs and the cavity mode eigenfrequencies are below this range since they typically cover the interval 10^{-2}–10^{-3} Hz. In what follows we focus on the natural ULF noise, that is at the frequencies which are even smaller than above resonant frequencies covering the range of the global electromagnetic resonances.

The Earth electromagnetic field is subject to a variety of random forces such as the variations of solar radiations, incident MHD waves and global magnetospheric resonances, fluctuation of the ionospheric currents, changes in the world thunder-storm activity, and so on. A variety of magnetospheric MHD waves acting on the Earth ionosphere give rise to a wideband spectrum of electromagnetic perturbations, which can be detected on the ground surface. Throughout the frequency range from HF to ULF the flux density of natural magnetic variations increase with a decrease in frequency in such a way that the amplitude of the spectral density varies from 10^{-21}–10^{-24} W/(m^2 Hz) at frequency 10^9–10^{10} Hz up to 10^{-3}–10^{-1} W/(m^2 Hz) at frequency $\sim 10^{-3}$ Hz (Lanzerotti 1978). Figure 6.11 taken from Lanzerotti et al. (1990) shows spectra of background magnetic variations measured in the wideband frequency range, which cover the ten-decades from 10^{-5} to 10^5 Hz. Interestingly enough the noise in the ELF/VLF range is an overall approximate inverse relation between the noise amplitude and frequency (Lanzerotti et al. 1990; Fraser-Smith 1995). This implies that there is an overall approximate inverse relation between the noise power amplitude and frequency. Notice that a power law spectrum of noise, which is referred to as the class of $1/f$ noise, or flicker noise, is usually observed in all electric devices over a very broad frequency range (e.g., see Rytov et al. 1978; Weissman 1988). There exists other tendency in the frequency range from 10^{-5} to 10^{-1} Hz where in the first approximation the noise amplitude is inversely proportional to $f^{-1.5}$. As indicated in Fig. 6.11, the spectrum of the noise amplitude in the intermediate interval can be approximated by a power law proportional to f^{-n} with the exponent n laying in the range 1.0–1.5. However, the value of n appears to vary considerably, depending on the case study, measurement technique and on the instruments arranged at the ground-recording station.

Knowledge of these tendencies for the natural low-frequency noise is of special interest in geophysical studies, since it gives information about spatiotemporal variations of the natural ULF electromagnetic noise and their source mechanism. This knowledge is also important from a scientific point of view, because, as pointed out by Fraser-Smith (1995), it is not understood at present why there exists such a relation between the ULF noise amplitude and frequency.

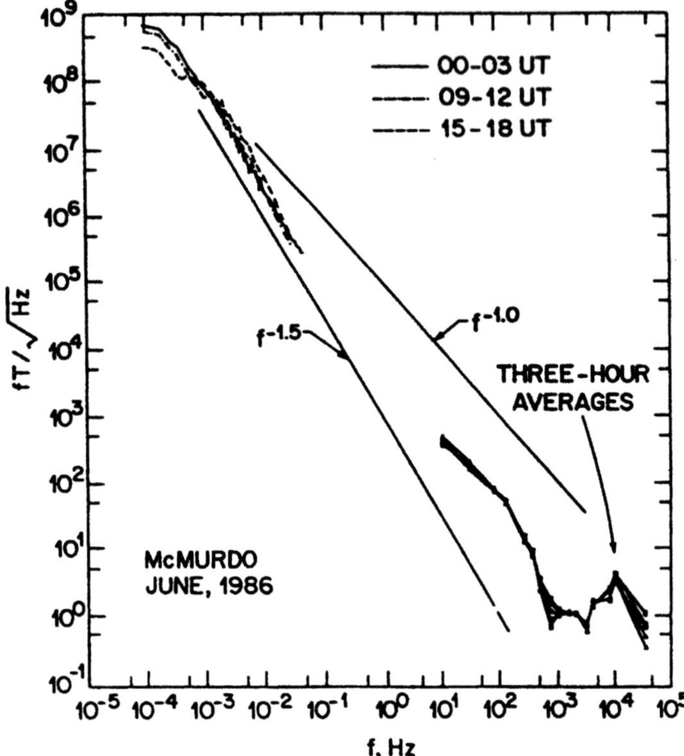

Fig. 6.11 Measured monthly average 3-h power spectra for magnetic field variations. The data were gathered at Arrival Heights, Antarctica near McMurdo Station, during June 1986. Taken from Lanzerotti et al. (1990)

To a certain extent the sources of the natural ULF variations can be divided into two general classes, depending on whether they are external or internal to the magnetosphere. It is now generally accepted that the external sources are mainly due to the interaction of the Earth's magnetosphere with solar wind and with MHD waves coming from outer space. Under certain orientation of the interplanetary magnetic field (IMF) and the Earth's magnetic field lines, when partial reconnection of the field lines occurs, the small quasiperiodic variations of the IMF ($\sim 10\,\text{nT}$) may result in generation of the ground-based variations with amplitude of about several hundred Tesla (Pilipenko et al. 2000). The energy of turbulent noise generated in the magnetosheath can penetrate through the magnetopause and thus can get trapped in the magnetosphere thereby exciting ULF noise and MHD waves.

An important example of internal sources is the global lightning activity, considering that there is about 2×10^3 thunderstorm in progress around the world at any time.

The sources of natural ULF noise covering the frequency range 10^{-4}–10^{-2} Hz have not yet been adequately explored. It is customary to conjecture that the MHD waves traveling through the magnetosphere can transfer a variety of electromagnetic noises from the outer regions of the magnetosphere towards the Earth. The high frequency region of the noise spectrum is lost in the conducting E layer of the ionosphere. In this picture the E layer plays a major role in formation of the ULF noise in the neutral atmosphere. Furthermore, the ionospheric current variations due to fluctuations of the ionospheric plasma conductivity and of neutral wind velocity can produce an additional random perturbation in the ULF region.

6.4.2   Model and Basic Equations

In what follows we focus our attention on the two possible sources, which are the incident MHD waves and the ionospheric current fluctuations originated from variation of the neutral gas flow in the altitude range of the ionospheric E layer. The field fluctuations in the magnetosphere and ionosphere can excite a random electromagnetic field in the atmosphere and on the ground surface. At first we consider the fields of the MHD wave and of the wind-driven currents in the ionosphere as given deterministic functions, which play a role of forcing functions. Solving this problem we can find the transfer matrices, which relate the fields in the ionosphere and magnetosphere with the fields in neutral atmosphere. Since the characteristic spatial size of the ULF variations is supposed to be smaller than the Earth radius, the curvature of the magnetic field lines is disregarded. This implies that the undisturbed geomagnetic field is considered as a homogeneous one.

To approximate the actual variation of medium parameters with altitude, we consider a plane-stratified medium model, which consists of the magnetosphere, conducting ionosphere, neutral atmosphere and conducting earth, as shown in Fig. 6.12. Consider first the conducting E layer of the ionosphere. We use a traditional coordinate system in which the y axis is directed westward, the x axis to the north, and z axis vertically upward. The origin of the local coordinate system is situated on the boundary between the bottom of the ionosphere and the neutral atmosphere. The vector of the Earth magnetic field is situated at the meridional x, z plane and makes an angle γ with respect to the horizontal axis x. The inclination angle is chosen in such a way that γ is positive for the northern hemisphere.

Let $\delta \mathbf{B}$ be a small perturbation of the geomagnetic field $\mathbf{B}_0$, i.e., $\delta B \ll B_0$. In the frequency range of interest the conduction current is much greater than the displacement one so the Ampere's law (1.5) holds at the E-layer. The Ohm's law for the ionospheric plasma of the E-layer is given by Eq. (2.6). Combining these equations we get

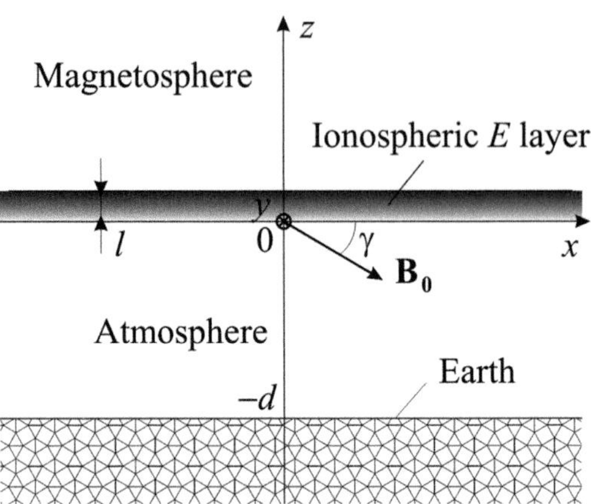

Fig. 6.12 Schematic illustration of a stratified medium model

$$\partial_y \delta B_z - \partial_z \delta B_y = \mu_0 \left\{ \sigma_\parallel E_\parallel \cos\gamma + J_\perp \sin\gamma + J_x^{(w)} \right\}, \tag{6.61}$$

$$\partial_z \delta B_x - \partial_x \delta B_z = \mu_0 \left\{ \sigma_P E_y - \sigma_H \left(E_x \sin\gamma + E_z \cos\gamma \right) + J_y^{(w)} \right\}, \tag{6.62}$$

$$\partial_x \delta B_y - \partial_y \delta B_x = \mu_0 \left\{ -\sigma_\parallel E_\parallel \sin\gamma + J_\perp \cos\gamma + J_z^{(w)} \right\}, \tag{6.63}$$

where as before $\sigma_\parallel$ denotes the field-aligned plasma conductivity, σ_H and σ_P are the Hall and Pedersen conductivities. Here we made use of the following abbreviations:

$$E_\parallel = E_x \cos\gamma - E_z \sin\gamma, \tag{6.64}$$

$$J_\perp = \sigma_P \left(E_x \sin\gamma + E_z \cos\gamma \right) + \sigma_H E_y. \tag{6.65}$$

The wind-driven current density is given by

$$J_x^{(w)} = B_0 \left(\sigma_H V_\perp - \sigma_P V_y \right) \sin\gamma, \quad J_z^{(w)} = J_x^{(w)} \cot\gamma,$$

$$J_y^{(w)} = B_0 \left(\sigma_H V_\perp + \sigma_P V_y \right), \tag{6.66}$$

where V_x, V_y and V_z are the component of the mass velocity of the neutral wind, and $V_\perp = V_x \sin\gamma + V_z \cos\gamma$. Notice that the neutral gas dominates below 130 km in such a way that the charged particles cannot greatly affect the neutral gas flow. This implies that the mass gas velocity can be considered as a given/forcing function which affects the electromagnetic fields and conduction currents inside the conducting E layer of the ionosphere. Furthermore, the parallel plasma conductivity in this region is much greater than the Hall and Pedersen ones. Assuming that $\sigma_\parallel \to \infty$, the parallel electric field $\mathbf{E}_\parallel$ thus becomes zero, i.e.,

$$E_z \sin\gamma = E_x \cos\gamma. \tag{6.67}$$

In order to eliminate the nonzero parallel current $\sigma_\parallel E_\parallel$ from Eqs. (6.61) and (6.63) one should slightly rearrange these equations. Equation (6.61) multiplied by $\sin\gamma$ plus equation (6.63) multiplied by $\cos\gamma$ gives

$$\left(\partial_y\delta B_z - \partial_z\delta B_y\right)\sin\gamma + \left(\partial_x\delta B_y - \partial_y\delta B_x\right)\cos\gamma = \mu_0\left(J_\perp + J_x^{(w)}\sin\gamma + J_z^{(w)}\cos\gamma\right).$$

$$(6.68)$$

In the frequency range $f < 0.1\,\text{Hz}$ the thickness, l, of the E layer is much smaller than the skin-depth in the ionosphere. In this notation the "thin" layer approximation can be used in order to derive the boundary conditions at the E layer of the ionosphere. This approximation is described in more detail in Sect. 5. Integrating of Eq. (6.68) with respect to z across the E-layer, making formally $l \to 0$, and taking into account Eq. (6.67), gives the boundary conditions at $z = 0$

$$-\sin\gamma\left[\delta B_y\right] = \mu_0\left(\Sigma_P E_x / \sin\gamma + \Sigma_H E_y + I^{(w)}\right),\qquad(6.69)$$

where the square brackets denote the jump of magnetic field across the E-layer, Σ_P and Σ_H are the height-integrated Pedersen and Hall conductivities given by Eq. (5.25). Here $I^{(w)} = I_x^{(w)}\sin\gamma + I_z^{(w)}\cos\gamma$ stands for the height-integrated wind-driven currents, i.e.

$$I_x^{(w)} = \int_0^l J_x^{(w)}dz \quad\text{and}\quad I_z^{(w)} = \int_0^l J_z^{(w)}dz.\qquad(6.70)$$

Similarly, integrating of Eq. (6.62) with respect to z across the E-layer yields

$$[\delta B_x] = \mu_0\left(\Sigma_P E_y - \Sigma_H E_x / \sin\gamma + I_y^{(w)}\right),\qquad(6.71)$$

where $I_y^{(w)}$ is another component of the height-integrated wind-driven current, i.e.

$$I_y^{(w)} = \int_0^l J_y^{(w)}dz.\qquad(6.72)$$

In the framework of our model the region above the E-layer is supposed to be the area consisting solely of a cold collisionless plasma, which is described by Eq. (5.2). In the ULF frequency range the absolute value of parallel components of the plasma dielectric permittivity, $\varepsilon_\parallel$, is much greater than perpendicular ones and thus can be assumed to be infinite. This means that the parallel electric field $\mathbf{E}_\parallel$ equals approximately zero, and we come to Eq. (6.67). Thus, we can eliminate the parallel current from Eq. (5.2) in analogy to the procedure used for the derivation of Eq. (6.68). Whence, we get

$$\left(\partial_y \delta B_z - \partial_z \delta B_y\right) \sin \gamma + \left(\partial_x \delta B_y - \partial_y \delta B_x\right) \cos \gamma = -\frac{i\omega}{V_A^2} \left(E_x \sin \gamma + E_z \cos \gamma\right),$$

$$\tag{6.73}$$

$$\partial_z \delta B_x - \partial_x \delta B_z = -\frac{i\omega}{V_A^2} E_y, \tag{6.74}$$

where $V_A = c/\varepsilon_\perp^{1/2}$ is the Alfvén velocity. These equations should be supplemented by the Faraday's law given by Eq. (4.2), where $\mathbf{B}$ should be replaced by $\delta \mathbf{B}$. The neutral atmosphere $(-d < z < 0)$ is considered as an insulator, and the solid Earth $(z < -d)$ as a uniform conductor with a constant conductivity σ_g. If the displacement current in both media is disregarded, the electromagnetic perturbations are described by Eqs. (5.27) and (5.28).

6.4.3 Transfer Matrices

Solution of the above problem with proper boundary conditions relates the magnetic perturbations on the ground surface with the forcing functions, i.e., the amplitudes of the MHD waves and of the wind-driven ionospheric current. We seek for the solution of the problem in the form of spatiotemporal Fourier transform. This implies that all quantities vary as $\exp(i\mathbf{k}\cdot\mathbf{R} - i\omega t)$, where $\mathbf{k}$ is horizontal wave vector, and $\mathbf{R} = (x, y)$. Let $\delta\mathbf{b}\,(\mathbf{k}, \omega, z)$ and $\delta\mathbf{e}\,(\mathbf{k}, \omega, z)$ be Fourier transforms of the magnetic and electric field variations, respectively. Let $\delta\mathbf{b}^{(m)}\,(\mathbf{k}, \omega)$ be the spectral amplitude of the incident Alfvén and FMS waves in the ionosphere, while $\delta\mathbf{I}^{(w)}\,(\mathbf{k}, \omega)$ stands for the height-integrated wind-driven ionospheric current. This latter value denotes a Fourier transform of the functions $I_x^{(w)}$, $I_y^{(w)}$ and $I_z^{(w)}$ given by Eqs. (6.70) and (6.72), respectively. In consequence of linearity of both Maxwell equations and boundary conditions, the spectral densities of the ionospheric and atmospheric fields are coupled in a linear fashion through the transfer matrices $\hat{\mathbf{M}}^{(w)}\,(\mathbf{k}, \omega, z)$ and $\hat{\mathbf{M}}^{(m)}\,(\mathbf{k}, \omega, z)$

$$\delta\mathbf{b}\,(\mathbf{k}, \omega, z) = \hat{\mathbf{M}}^{(w)}\,(\mathbf{k}, \omega, z) \cdot \delta\mathbf{I}^{(w)}\,(\mathbf{k}, \omega) + \hat{\mathbf{M}}^{(m)}\,(\mathbf{k}, \omega, z) \cdot \delta\mathbf{b}^{(m)}\,(\mathbf{k}, \omega). \tag{6.75}$$

We now omit the detailed derivation of the transfer matrices. The interested reader is referred to the paper by Surkov and Hayakawa (2007, 2008) for details. If the vertical ambient magnetic field is assumed then an analytical solution of the problem can be found for arbitrary value of $\mathbf{k}$. As the magnetic field $\mathbf{B}_0$ is vertically downward one should therefore substitute $\gamma = \pi/2$ in the basic equations. In such a case the set of Eqs. (6.73), (6.74), and (4.2) can be split into two independent sets, which describe the shear Alfvén and FMS waves propagating in the magnetospheric plasma. Similarly, the set of Eqs. (5.27) and (5.28) for the atmosphere and the ground can be split into two independent sets, which describe the TM and TE modes in the atmosphere. These two modes are coupled through boundary conditions at the ionosphere, that is via Eqs. (6.69), (6.71) and via the continuity condition for $\delta\mathbf{b}$ and for the horizontal components of $\delta\mathbf{e}$. In this notation the wavefield of the incident Alfvén and FMS waves in the magnetosphere is assumed to be given functions whereas the waves reflected from the ionosphere should be found to fit the solutions in the magnetosphere and the atmosphere. Considering the large-scale perturbations with the scale size $2\pi/k \sim 10^3$ km, we restrict our analysis on an extreme case of $\omega \ll kV_A \approx 30$ Hz ($f = \omega/2\pi \ll 5$ Hz) and even on the case of stronger inequality $\omega \ll k_x^2/\left(\mu_0\sigma_g\right) \approx 0.03$ Hz ($f \ll 0.005$ Hz). In such a case the matrix $\hat{\mathbf{M}}^{(w)}$ for the ground surface $z = -d$ can be simplified to

$$\hat{\mathbf{M}}^{(w)} \approx \frac{i\mu_0\exp\left(-kd\right)}{2g_3} \begin{pmatrix} ik_x f_-/k & ik_x f_+/k & 0 \\ ik_y f_-/k & ik_y f_+/k & 0 \\ f_- & f_+ & 0 \end{pmatrix}, \tag{6.76}$$

where $g_3 = 1 + \alpha_P$. Here we made use of the following abbreviations:

$$f_+ = \frac{k_y\alpha_H + k_x g_3}{k}, \quad f_- = \frac{k_x\alpha_H - k_y g_3}{k}, \quad . \tag{6.77}$$

The third column in the matrix consists of zeros because the wind velocity component parallel to the vertical magnetic field $\mathbf{B}_0$ cannot excite the magnetic perturbations. The similar expression for matrix $\hat{\mathbf{M}}^{(m)}$ can be found in the paper by Surkov and Hayakawa (2008).

Furthermore, the study is simplified if the components of the horizontal wave vector satisfy the requirement $k_x \gg k_y$. This implies that the azimuthal scale size of the perturbations is much greater than that of the meridional perturbations. In fact, we assume a 1D distribution of the height-integrated ionospheric current, $\delta\mathbf{I}^{(w)}\left(x,t\right)$, as a source of 2D random electromagnetic fields in the surroundings. The east–west neutral winds at the altitude range of the E layer can excite the ring current in the ionosphere thereby producing this type of perturbations. For one more example, it is worth mentioning that the Pc5 pulsations have a rather localized wavefield of extension 100–200 km in north–south direction.

On the basis of this simplifying assumption, the analytic form of the transfer matrices can be simplified. Considering the large-scale perturbations the matrix $\hat{\mathbf{M}}^{(w)}$ at $z = -d$ is given by

$$\hat{\mathbf{M}}^{(w)} \approx \frac{i\mu_0\exp\left(-kd\right)}{2g_2} \begin{pmatrix} i\alpha_H\sin\gamma & ig_2 & i\alpha_H\cos\gamma \\ 0 & 0 & 0 \\ k\alpha_H\sin\gamma/k_x & kg_2/k_x & k\alpha_H\cos\gamma/k_x \end{pmatrix}, \tag{6.78}$$

where $g_2 = \alpha_P + \sin\gamma$. The TM mode in the atmosphere contains the components δb_y, δe_x, and δe_z that are identical with those of the shear Alfvén wave in the magnetosphere. Both these modes ares coupled by virtue of boundary conditions at $z = 0$. The second strings in the matrices consist of zeros because $\delta b_y = 0$ everywhere under the ionosphere including at $z = -d$. This means that the magnetic field due to Alfvén mode cannot penetrate through the conducting ionosphere to generate the magnetic perturbations in the atmosphere on the ground surface. So, in this model only the TE mode which includes the components δb_x, δb_z, and δe_y can contribute to the magnetic variation in the atmosphere.

As is seen from these two expressions for $\hat{\mathbf{M}}^{(w)}$, the propagation of the ULF perturbations through the conducting ionosphere and the neutral atmosphere towards the ground is accompanied by the wavefield damping with the exponential factor $\exp(-kd)$. It should be noted that in the low-frequency limit the transfer matrix is not a function of frequency. One may suppose that these conclusions hold not only for the cases examined above but also for an arbitrary angle γ and wave vector $\mathbf{k}$.

6.4.4 Correlation Matrix of Random Fields

In this section the MHD waves propagating from the magnetosphere towards the ionosphere and the ionospheric wind-driven currents are treated as the random functions of coordinate and time. Let $\delta\mathbf{B}(\mathbf{r}, t)$ and $\delta\mathbf{E}(\mathbf{r}, t)$ be the random electromagnetic variations at the point $\mathbf{r} = (x, y, z)$ produced by the fluctuations of the random electromagnetic fields in the ionosphere. In practice, the mean value of the random magnetic variations is close to zero. Therefore, we will be interested in the correlation matrix/product moment, which has the form

$$\Psi^{(B)}_{nm}\left(\mathbf{r}, t, \mathbf{r}', t'\right) = \left\langle \delta B_n\left(\mathbf{r}, t\right) \delta B_m^*\left(\mathbf{r}', t'\right)\right\rangle, \tag{6.79}$$

where the brackets $\langle\rangle$ denotes the averaging over all available realizations of the random process, the symbol $*$ denotes a complex-conjugate value and the inferior indexes n and m are taken on the values x, y, and z. This correlation matrix describes the spatial and temporal correlation of the field components $\delta B_n(\mathbf{r}, t)$ and $\delta B_m^*(\mathbf{r}', t')$ taken at different points $\mathbf{r}$ and $\mathbf{r}'$, and at different time t and t'.

In a similar fashion we may introduce the correlation matrix, $\Psi^{(E)}_{nm}$, of the electric field fluctuations. Notice that Eq. (6.79) satisfies both real and complex random fields. In a similar fashion we may introduce the correlation matrix of the forcing function fluctuations, $\Psi^{(m)}_{nm}$ and $\Psi^{(w)}_{nm}$.

It is clear that the spectral amplitudes of the forcing functions, $\delta\mathbf{b}^{(m)}(\omega, \mathbf{k})$ and $\delta\mathbf{I}^{(w)}(\omega, \mathbf{k})$, and of the magnetic, $\delta\mathbf{b}(\mathbf{k}, \omega, z)$, and electric, $\delta\mathbf{e}(\mathbf{k}, \omega, z)$, field fluctuations are random functions as well. By contrast, the transfer matrices are considered to be deterministic/given functions. The spectra of random electromagnetic fluctuations on the ground surface are related to the spectral amplitudes, $\delta\mathbf{b}^{(m)}(\omega, \mathbf{k})$ and $\delta\mathbf{I}^{(w)}(\omega, \mathbf{k})$, through the linear equation (6.75).

Considering the ground-based observation we first study the spectral density of correlation matrix of the magnetic perturbations given by

$$\psi_{nm}^{(B)}\left(\omega,\mathbf{k},\omega',\mathbf{k}'\right) = \left\langle \delta b_n\left(\omega,\mathbf{k}\right)\delta b_m^*\left(\omega',\mathbf{k}'\right)\right\rangle, \tag{6.80}$$

where the symbols in the brackets denote the spectral amplitudes of the ground-based field fluctuations.

Substituting Eq. (6.75) for $\delta\mathbf{b}$ into Eq. (6.80) and rearranging yields

$$\psi_{nm}^{(B)} = \sum_{l=1}^{3}\sum_{p=1}^{3}\left(\hat{M}_{nl}^{(w)}\hat{M}_{mp}^{(w)*}\psi_{lp}^{(w)} + \hat{M}_{nl}^{(m)}\hat{M}_{mp}^{(m)*}\psi_{lp}^{(m)}\right), \tag{6.81}$$

where

$$\psi_{lp}^{(w)} = \left\langle \delta I_l^{(w)}\left(\omega,\mathbf{k}\right)\delta I_p^{(w)*}\left(\omega',\mathbf{k}'\right)\right\rangle, \tag{6.82}$$

$$\psi_{lp}^{(m)} = \left\langle \delta b_l^{(m)}\left(\omega,\mathbf{k}\right)\delta b_p^{(m)*}\left(\omega',\mathbf{k}'\right)\right\rangle. \tag{6.83}$$

Here we have assumed that the forcing functions, $\delta\mathbf{b}^{(m)}\left(\omega,\mathbf{k}\right)$ and $\delta\mathbf{I}^{(w)}\left(\omega,\mathbf{k}\right)$, are statistically independent of each other.

In what follows we focus on the correlation matrix, which describes the contribution of the ionospheric wind-driven currents to the natural electromagnetic noise observed on the ground surface. We choose first to study the case of vertical Earth's magnetic field. The fluctuations of height-integrated ionospheric current, $\delta\mathbf{I}^{(w)}\left(\mathbf{R},t\right)$, is considered as a 2D random field of $\mathbf{R} = (x,y)$. This random field is assumed to be uniform in time so that shift of the initial time has no effect on the random process. In this notation the spatiotemporal correlation functions, $\Psi_{lp}^{(w)}$, and their linear combination in Eq. (6.81) depend on the time difference $\tau = t - t'$. If the random field is uniform in space, the shift of the origin of coordinate system O is insignificant, so that the correlation functions must depend on only relative distance $L = |\mathbf{L}| = |\mathbf{R} - \mathbf{R}'|$. Since the plasma conductivity is anisotropic in the E layer, $\Psi_{lp}^{(w)}$ cannot depend only on relative distance. So we assume that $\Psi_{lp}^{(w)}$ is a function of both $L_x = |x - x'|$ and $L_y = |y - y'|$. In such a case the spectral density of this random process is delta-correlated both over k_x, k_y and ω

$$\psi_{lp}^{(w)}\left(\mathbf{k},\omega,\mathbf{k}',\omega'\right) = \delta\left(\omega-\omega'\right)\delta\left(k_x - k_x'\right)\delta\left(k_y - k_y'\right)G_{lp}\left(\omega,\mathbf{k}\right). \tag{6.84}$$

Here the function $G_{lp}\left(\omega,k\right)$ is derivable through the spatial distribution of the correlation function $\Psi_{lp}^{(w)}\left(\mathbf{L},\omega\right)$

$$G_{lp}\left(\omega,\mathbf{k}\right) = \frac{1}{4\pi^2}\int_{-\infty}^{\infty}\int_{-\infty}^{\infty}\Psi_{lp}^{(w)}\left(\mathbf{L},\omega\right)\exp\left(-i\mathbf{k}\cdot\mathbf{L}\right)dL_x dL_y. \tag{6.85}$$

Now we choose for study the Gaussian-shaped form of the correlation function. Since the random fields are anisotropically distributed on the ground surface, the function $\Psi_{lp}^{(w)}(\mathbf{L}, \omega)$ may depend on two correlation radii, so it can be chosen in the form

$$\Psi_{lp}^{(w)}(\mathbf{L}, \omega) = F_{lp}(\omega) \exp\left(-\frac{L_x^2}{\rho_x^2(\omega)} - \frac{L_y^2}{\rho_y^2(\omega)}\right). \tag{6.86}$$

The radii $\rho_x(\omega)$ and $\rho_y(\omega)$ characterize the correlations of the wind-current fluctuation in the x and y-directions, respectively. The functions $F_{lp}(\omega)$ depend on how the height-integrated current $I_l^{(w)}$ is correlated with the current $I_p^{(w)}$. In particular, if these currents are statistically independent of each other, then $F_{lp}(\omega) = 0$. In the subsequent discussion we define both the specific form of the correlation radius and the factor $F_{lp}(\omega)$.

Substituting Eq. (6.86) for $\Psi_{lp}^{(w)}$ into Eq. (6.85), performing integration over L_x and L_y, yields

$$G_{lp}(\omega, \mathbf{k}) = \frac{F_{lp}\rho_x\rho_y}{4\pi} \exp\left(-\frac{k_x^2\rho_x^2 + k_y^2\rho_y^2}{4}\right). \tag{6.87}$$

In situ measurements the horizontal magnetic field variations are greater than the vertical one. As one example, we consider now the correlation function $\Psi_{xx}^{(B)}(\mathbf{L}, \omega)$. In fact, this correlation function describes the spatial correlation of the spectral components $\delta B_x(\mathbf{r}, \omega)$ and $\delta B_x^*(\mathbf{r}', \omega)$ taken at different points $\mathbf{r}$ and $\mathbf{r}'$ at frequency ω. For practical purposes, it is interesting to study the spectral density/power spectrum, which is based on a single-stationed three-component magnetometer recording. If the data for the power spectra is gathered in a single point, only autocorrelation function is available. In this notation, substituting Eqs. (6.76), (6.84), (6.87) into Eq. (6.81), applying an inverse Bessel transform, and setting $\mathbf{L} = \mathbf{R} - \mathbf{R}' = 0$, yields

$$\Psi_{xx}^{(B)}(0, \omega) = \frac{\mu_0^2\rho_x\rho_y}{16\pi g_3} \int_{-\infty}^{\infty}\int_{-\infty}^{\infty} \left\{F_{xx}f_-^2 + (F_{xy} + F_{yx}) f_-f_+ + F_{yy}f_+^2\right\}$$

$$\times \exp\left(-\frac{k_x^2\rho_x^2 + k_y^2\rho_y^2}{4} - 2kd\right) \frac{k_x^2}{k^2} dk_x dk_y, \tag{6.88}$$

In the extreme case of small correlation radii, i.e., $2d \gg \rho_x, \rho_y$, one can find that

$$\Psi_{xx}^{(B)}(\omega) = \Psi_{xx}^{(B)}(0, \omega) \approx \frac{\mu_0^2\rho_x(\omega)\rho_y(\omega)\Theta(\omega)}{256d^2}, \tag{6.89}$$

where

$$\Theta\left(\omega\right) = \left(\frac{3\alpha_H^2}{g_3} + g_3\right) F_{xx}\left(\omega\right) + \left(\frac{\alpha_H^2}{g_3} + 3g_3\right) F_{yy}\left(\omega\right)$$

$$+2\alpha_H \left\{F_{xy}\left(\omega\right) + F_{yx}\left(\omega\right)\right\}. \tag{6.90}$$

Within the altitudes of the E-layer the ratio of plasma to neutrals number densities is 10^{-7}–10^{-9} for the day- and night-time conditions, respectively. This means that the motions of electrons and ions practically have no effect on the pattern of neutral has flow. In contrast, the moving neutrals drag the ions thereby exciting the wind-driven ionospheric currents. In our model we leave out of account the diurnal variations and fluctuation of the ionospheric plasma conductivity due to the variation of solar radiation and other causes. This implies that the spatiotemporal distribution of the wind-driven currents is basically governed by the hydrodynamic processes and fluctuations of the neutrals flow in the ionosphere. Such fluctuations may propagate with the velocities of acoustic and atmospheric gravity waves, which frequently occur at the altitudes of the E-layer. In this picture the correlation radius, $\rho_c\left(\omega\right)$, of the random fields can be roughly estimated as (Surkov and Hayakawa 2007)

$$\rho_c\left(\omega\right) \sim V_a\left(\omega\right) T = \frac{2\pi V_a\left(\omega\right)}{\omega}. \tag{6.91}$$

Here $V_a\left(\omega\right)$ denotes the acoustic wave velocity or the mass velocity of the neutrals and T stands for a typical period of ionospheric parameter variations.

Finally, using Eq. (6.91) to estimate the correlation radii, ρ_x and ρ_y, we obtain the following rough estimate of the power spectrum

$$\Psi_{xx}^{(B)}\left(\omega\right) \sim \frac{\mu_0^2 \pi^2 V_a^2\left(\omega\right) \Theta\left(\omega\right)}{64\omega^2 d^2}. \tag{6.92}$$

In a similar fashion we may examine the 2D field of the electromagnetic fluctuations that can be expressed via the transfer matrix (6.78). Using this line of reasoning, the spectral density of correlation matrix $\psi_{nm}^{(B)}$ is found to be given by Eq. (6.81) where the coefficients $\hat{M}_{nl}^{(w)}$ and $\hat{M}_{mp}^{(w)*}$ stand for the components of the matrix (6.81). The 1D random fields, $\delta\mathbf{I}^{(w)}\left(x,t\right)$, of the height-integrated currents is considered to be uniform in the ionosphere in such a way that the spatiotemporal correlation functions, $\Psi_{lp}^{(w)}$, and their linear combinations depend only on the time difference $\tau = t - t'$ and relative distance $\rho = |x - x'|$. As before we choose for study the Gaussian-shaped form of the correlation function of the ionospheric current fluctuations

$$\Psi_{lp}^{(w)}\left(\rho, \omega\right) = F_{lp}\left(\omega\right) \exp\left(-\frac{\rho^2}{\rho_c^2\left(\omega\right)}\right), \tag{6.93}$$

where $\rho_c(\omega)$ stands for the correlation radius. The spectral density of this random process is given by

$$\psi_{lp}^{(w)}\left(k_x, \omega, k_x', \omega'\right) = \delta\left(\omega - \omega'\right) \delta\left(k_x - k_x'\right) G_{lp}\left(k_x, \omega\right), \qquad (6.94)$$

where δ denotes Dirac's function and

$$G_{lp}\left(k_x, \omega\right) = \frac{F_{lp}\left(\omega\right) \rho_c\left(\omega\right)}{2\pi^{1/2}} \exp\left(-\frac{k_x^2 \rho_c^2\left(\omega\right)}{4}\right). \qquad (6.95)$$

Combining these equations, applying an inverse Fourier transform, and performing the integration over k_x' and k_x, leads to the spatial distribution of the spectral density of the horizontal magnetic field variations. Setting $\rho = x - x' = 0$ yields

$$\Psi_{xx}^{(B)}\left(\omega\right) = \frac{\mu_0^2 \Theta_1\left(\omega\right)}{4} \exp\left(\frac{4d^2}{\rho_c^2\left(\omega\right)}\right) \left\{1 - \mathrm{erf}\left(\frac{2d}{\rho_c\left(\omega\right)}\right)\right\}, \qquad (6.96)$$

where $\mathrm{erf}\left(x\right)$ denotes the error function. Here the function $\Theta_1\left(\omega\right)$ is given by

$$\Theta_1\left(\omega\right) = \sum_{l=1}^{3} \sum_{p=1}^{3} \hat{m}_{xl}^{(w)} \hat{m}_{xp}^{(w)*} F_{lp}\left(\omega\right), \qquad (6.97)$$

where $\hat{m}_{lp}^{(w)}$ stands for the matrix elements appearing in Eq. (6.78). When considering the extreme case $2d \gg \rho_c$, Eq. (6.96) is simplified to

$$\Psi_{xx}^{(B)}\left(\omega\right) = \frac{\mu_0^2 \Theta_1\left(\omega\right) \rho_c\left(\omega\right)}{8\pi^{1/2} d}. \qquad (6.98)$$

Substituting Eq. (6.91) for $\rho_c\left(\omega\right)$ into Eq. (6.98) gives the estimate of the power spectrum on the ground surface for the case of 1D distributions of the ionospheric current fluctuations

$$\Psi_{xx}^{(B)}\left(\omega\right) \sim \frac{\mu_0^2 \pi^{1/2} \Theta_1\left(\omega\right) V_a\left(\omega\right)}{4\omega d}. \qquad (6.99)$$

When this result is compared with Eq. (6.92), it is apparent that the 2D-case correlation function falls off more rapidly with frequency than does the 1D-case correlation function. In the analysis that follows, we show that Eq. (6.92) is better consistent in magnitude with the observations than does Eq. (6.99).

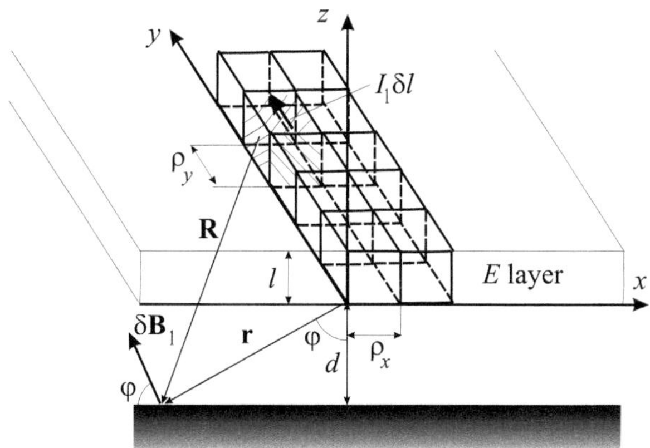

Fig. 6.13 A simplified model of random ionospheric currents that are used to gain better understanding of the solution with rigorous formulation of the problem. The current fluctuations are correlated inside each cell with sizes ρ_x and ρ_y but not correlated with respect to each other. **R** is the position vector drawn from the cell to the observation point, and $\delta\mathbf{B}_1$ is the magnetic variation caused by the current element $I_1\delta l$

6.4.5   Rough Estimate of Spectral Density

To gain better understanding of the results alluded to above, it is necessary to give a simple interpretation of these results on the basis of a simplified model of the medium. To be specific, we consider E region of the ionosphere as a thin isotropically conducting layer, and only the wind-driven current flowing in the y-direction is taken into account. First, we note that the fluctuations of this current can be considered as the correlated current fluctuations inside the region with horizontal sizes of the order of $\rho_x(\omega)$ and $\rho_y(\omega)$. Consider such a region as shown in Fig. 6.13 with the shaded area, as an elementary current element. The magnetic perturbations, δB_1, originated from a solitary current element on the ground surface can be estimated via Biot–Savart law

$$\delta B_1 = \frac{\mu_0 r I_1 \delta l}{4\pi \left(r^2 + y^2\right)^{3/2}}, \tag{6.100}$$

where $I_1\delta l$ denotes the current moment, $r = \left(x^2 + d^2\right)^{1/2}$ is the distance shown in Fig. 6.13, and d is thickness of the neutral atmosphere. The horizontal component is related to δB_1 through $\delta B_{1x} = \delta B_1 \cos\varphi = \delta B_1 d/r$. The effective length of the current element is estimated as follows: $\delta l \sim \rho_y$ while the current amplitude can be expressed through the height-integrated wind-driven current density, $I_y^{(w)}$, via $I_1 \sim I_y^{(w)}\rho_x$. Dividing the ionosphere into the "coherent" regions with sizes ρ_x and ρ_y as shown in Fig. 6.13, we obtain that the number of such "coherent" currents covering the area $dxdy$ is of the order of $dN \sim dxdy/\left(\rho_x\rho_y\right)$. Since these currents are uncorrelated, the net amplitude of the magnetic variations is close to zero whereas the sum of squared amplitudes is proportional to dN; that is, the contribution of the area $dxdy$ is of the order of $d\left(\delta B_x\right)^2 = \delta B_{1x}^2 dN$. Combining above relationships with Eq. (6.100) and integrating gives the amplitude of the net squared magnetic variations

$$\delta B_x^2 = \frac{\mu_0^2 \left(I_y^{(w)}\right)^2 \rho_x \rho_y}{16\pi^2} \int\limits_{-\infty}^{\infty} \int\limits_{-\infty}^{\infty} \frac{dxdy}{\left(x^2 + y^2 + d^2\right)^3}. \tag{6.101}$$

Performing integration over x and y, taking into account that the power spectrum of the magnetic noise is proportional to δB_x^2, and replacing $\left(I_y^{(w)}\right)^2$ by the spectral density of random current fluctuation $\Theta(\omega)$, we obtain

$$\Psi_{xx}^{(B)}(\omega) = \frac{\mu_0^2 \rho_x(\omega)\, \rho_y(\omega)\, \Theta(\omega)}{32\pi d^2}. \tag{6.102}$$

This rough estimate coincides with Eq. (6.89) to an accuracy of the numerical factor $\pi/8$. This detailed calculation made in previous sections is totally consistent with the simple model presented above.

6.4.6   Flicker-Noise of Ionospheric Currents

The ionospheric currents and conductivity are subject to violent changes from the action of many forces: variations of the solar radiation, MHD waves and particle precipitation from the magnetosphere, fluctuations of the plasma number density, turbulence occurring in the plasma and neutral gas flows, and etc. A close analogy exists with conductivity of the electric devices, in which the low-frequency current fluctuations are supposed to be due to slow fluctuations of both the medium resistance and the source emissivity, which are in turn provided by a superposition of a great number of random processes with different relaxation times. This kind of electromagnetic noise is termed flicker-noise or $1/f$ noise since overall the power spectrum of this noise, $F(f)$, tends to decrease inversely proportional to the frequency, i.e.,

$$F(f) = K \frac{\langle J \rangle^m}{f^n}, \tag{6.103}$$

where $\langle J \rangle$ is the mean current density, K, m, and n are the empirical constants, and $f = \omega/(2\pi)$ is frequency. The exponent n in Eq. (6.103) varies within the interval $0.8 < n < 1.2$, but in most cases n is close to unity while $m \approx 2$ (Rytov et al. 1978; Weissman 1988). This universal dependence has been observed in gas-discharge devices, electrolytes, granulated resistance, germanium and silicon diodes, photoelectric cells, contact resistances, thermistor, and etc. Surprisingly, our understanding of the flicker-noise is not so good as it should be, given its commonplace occurrence. This type of noise is supposed to be provided by a superposition of a large number of random processes with different relaxation times, including slow fluctuations of both the medium resistance and the source emissivity.

Here we assume the presence of the flicker-noise in the spectral density of the ionospheric wind-driven currents. In such a case the mean current density in Eq. (6.103) should be replaced by the mean height-integrated currents, $\left\langle I_l^{(w)} I_p^{(w)} \right\rangle^{1/2}$, in such a way that Eq. (6.103) is transformed to

$$F_{lp}(f) = K \frac{\left\langle I_l^{(w)} I_p^{(w)} \right\rangle^{m/2}}{f^n}, \tag{6.104}$$

Notice that if the currents $I_l^{(w)}$ and $I_p^{(w)}$ are uncorrelated, then the function $F_{lp}(f)$ vanishes.

Substituting Eq. (6.104) into Eq. (6.92) gives an order-of-magnitude estimate of power spectrum on the ground surface. For simplicity we choose the case $F_{xy} = F_{yx} = 0$ and $m = 2$ that gives

$$\Psi_{xx}^{(B)} = \frac{\mu_0^2 V_a^2 K}{256 d^2 f^{2+n}} \left\{ \left(\frac{3\alpha_H^2}{g_3} + g_3 \right) \left\langle I_x^{(w)} \right\rangle^2 + \left(\frac{\alpha_H^2}{g_3} + 3g_3 \right) \left\langle I_y^{(w)} \right\rangle^2 \right\} \tag{6.105}$$

where $\left\langle I^{(w)} \right\rangle$ stands for the mean amplitude of the height-integrated ionospheric current, which can be estimated as

$$\left\langle I_x^{(w)} \right\rangle = B_0 \left(\Sigma_H \left\langle V_\perp \right\rangle - \Sigma_P \left\langle V_y \right\rangle \right) \sin \gamma,$$

$$\left\langle I_y^{(w)} \right\rangle = B_0 \left(\Sigma_H \left\langle V_\perp \right\rangle + \Sigma_P \left\langle V_y \right\rangle \right), \tag{6.106}$$

If the dispersion of the acoustic wave velocity is neglected, that is, V_a is a constant value, then the spectral density $\Psi_{xx}^{(B)}(f)$ is inversely proportional to f^{n+2}. Recently Surkov and Hayakawa (2007) have found that the presence of flicker noise in the atmospheric background current, can provide the same dependence of the ULF power spectrum on frequency.

To make a theoretical plot of the spectral amplitude we use the numerical values of the ionospheric and atmospheric parameters alluded to above. Taking the notice that $V_A = 5 \times 10^2$ km/s is best suited in the altitude range of the E layer, the ionospheric parameters $\alpha_P = 3.14$ and $\alpha_H = 4.71$ are chosen for the daytime ionosphere while $\alpha_P = 0.126$ and $\alpha_H = 0.188$ for the nighttime conditions. The mass velocity of the neutral gas in the ionosphere is estimated as $\left\langle V_y \right\rangle \sim \left\langle V_\perp \right\rangle \sim V_a \sim 10^2$ m/s and the angle of magnetic field inclination $\gamma = \pi/2$.

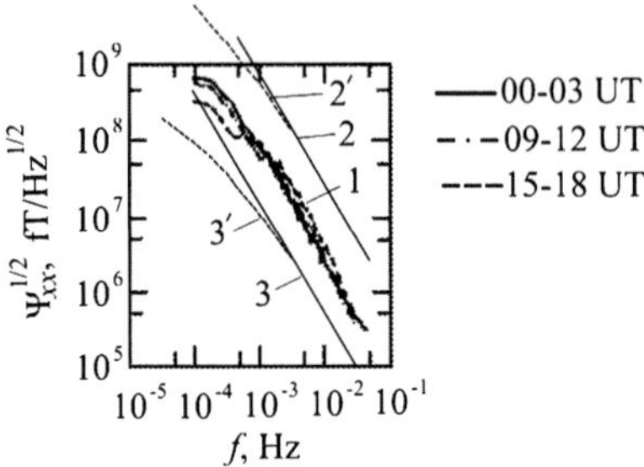

Fig. 6.14 Measured monthly average 3-h power spectra taken from Lanzerotti et al. (1990) (curve 1) and calculated power spectra of magnetic noise for the nighttime (2) and daytime (3) ionospheric parameters. The numerical calculations from improved equation (6.107) are shown with lines 2′ and 3′

The parameters appearing in empirical Eq. (6.103) is chosen as follows: $K = 1$ and $m = 2$. When Eq. (6.105) is compared with the evidence from ULF measurements (Lanzerotti et al. 1990), it is apparent that the spectral index $n = 1$ is a best fit value.

A model calculation of the square root of the spectral amplitude with a best fit value $n = 1$ and of the power spectrum recorded at Arrival Heights, Antarctica in June 1986 (Lanzerotti et al. 1990) are presented in Fig. 6.14 as a function of frequency f. The observational data taken from Lanzerotti et al. (1990) are shown with line 1 while our model calculations are plotted with line 2 (daytime conditions) and 3 (nighttime conditions). It is obvious from Fig. 6.14 that the observational data are sandwiched between the theoretical lines 1 and 2. It should be noted that there are some uncertainties in the ionospheric current parameters, for example, in the constant K in Eq. (6.104).

We recall that 1D distribution of the ionospheric wind-driven currents results in the 2D spectral amplitude in inverse proportion to the squared frequency, which in turn leads to a discrepancy between the predicted and measured spectra.

The observational data slightly deviate from the straight line as is seen in the upper corner of Fig. 6.14. In this frequency range the correlation radius may be greater than or equal to the distance between the Earth and the ionosphere. In such a case the approximate solution given by Eq. (6.89) should be replaced by the more accurate solution. To gain better understanding of this behavior of the observational data, consider the case $\rho_x = \rho_y = \rho_c(\omega)$. Substituting Eqs. (6.84), (6.87) into Eq. (6.81) and applying an inverse Bessel transform yields

$$\Psi_{xx}^{(B)}(\omega) = \frac{\mu_0^2 \Theta(\omega)}{32} \left[1 - \frac{2\pi^{1/2} d}{\rho_c(\omega)} \exp\left(\frac{4d^2}{\rho_c^2(\omega)}\right) \left\{ 1 - \mathrm{erf}\left(\frac{2d}{\rho_c(\omega)}\right) \right\} \right].$$

$$(6.107)$$

Given the above parameters and based on Eq. (6.107), the numerical calculations are shown in Fig. 6.14 with lines 2′ and 3′. In the low-frequency limit, when $\rho_c(\omega) \gg 2d$, the expression in square bracket tends to unity whence it follows that $\Psi_{xx}^{(B)}(\omega) \propto \Theta(\omega) \propto \omega^{-1}$. This means that the spectral index of the power spectrum must fall off with a decrease in frequency followed by the decrease in inclination angle of the lines 2′ and 3′ as shown in Fig. 6.14.

The lines 3 and 3′, which correspond to the nighttime parameters of the ionosphere, lie below the experimental data. To explain this discrepancy with observations, one may assume the presence of supplementary sources, which contribute the ULF noise at nighttime.

6.4.7   Neutral Gas Turbulence

Turbulence of neutral gas flow in the altitude range of the E layer can serve as an alternative excitation source of the ULF electromagnetic noise. As we have noted above, if a neutral gas flow is stirred in some region with size λ, turbulization of flow may occur in a so-called inertial subrange, $\lambda^{-1} \ll k \ll \lambda^{-1}\mathrm{Re}^{3/4}$, in k space. It is usually the case that the Reynolds number, Re, tends to maximize in the vicinity of turbopause and it can be large enough in the E-layer, that is about 10^2–10^4 as it follows from the assessment we made in Sect. 5.3.6. The Kolmogorov spectrum covers the frequency range given by Eq. (5.72). Assuming for the moment that the smoothed mean mass velocity of the gas flow is $V = 10^2$ m/s, and the typical scale of the turbulization of flow is $\lambda = 10^2$ km we get the estimate $1.6 \times 10^{-4} \ll f \ll (0.005$–$0.16)$ Hz. According to the Kolmogorov theory for an isotropic homogeneous medium, in this frequency region the mechanical energy of the turbulent flow has a power law spectrum $\propto k^{-5/3}$.

The correlation matrix of the ionospheric wind-driven current can be expressed through the spectral density of the mass velocity fluctuations $\left\langle \delta V_l \delta V_p^* \right\rangle$ which in turn is proportional to the spectral density of the mechanical energy. Since the typical frequencies of turbulent pulsations are evaluated as $\omega \sim kV$, we can thus assume that the functions $F_{lp}(\omega) \propto \omega^{-5/3}$. Considering 2D distribution of the height-integrated currents in the ionosphere we come to the dependence $\Psi_{xx}^{(B)}(\omega) \propto \omega^{-11/3}$. The spectral index 11/3 of the correlation function $\Psi_{xx}^{(B)}(\omega)$ slightly differs from the best fit value 3, which corresponds to the data shown in Fig. 6.14. The best hope for that is the case of 1D distribution of the wind-driven ionospheric currents when we obtain $\Psi_{xx}^{(B)}(\omega) \propto \omega^{-8/3}$.

6.4.8   Random Variations of Background Atmospheric
Current and Conductivity

The mean value of the background atmospheric current density due to atmospheric conductivity is about $(3.5$–$4) \times 10^{-12}$ A/m^2. Recently Davydenko et al. (2004) have studied the electric environment of a mesoscale convective system (MCS). The typical size of the MCS-trailing stratiform region was estimated to be about

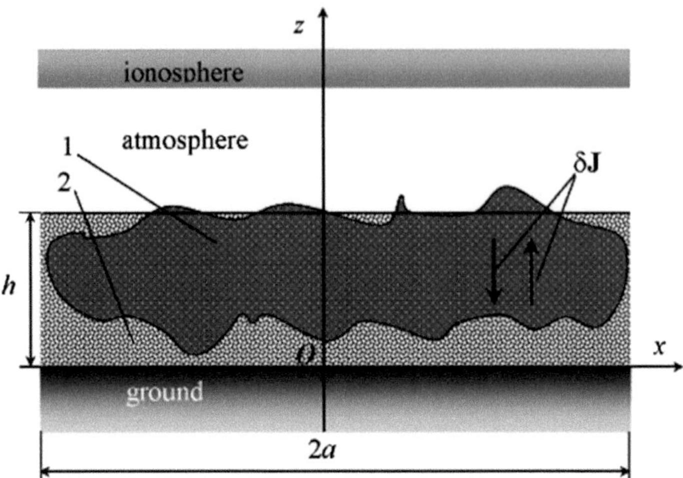

Fig. 6.15 A schematic drawing of current fluctuations in the vicinity of large-scale atmospheric inhomogeneities

200 km and the total vertical current in this region was 25 A, which is much greater than the contribution of an isolated thunderstorm ($\sim$ 0.4 A). In the region surrounding the thunderstorm, the atmospheric current density reached a peak value of about $(3.5–1.2) \times 10^{-9}$ A/m^2. Interestingly, both the atmospheric current in the MCS-trailing stratiform region and the mean current due to lightnings discharges were upward-directed. Considering the important role played by the atmospheric background current in formation of the global electric circuit, it is expected that the random current fluctuations may attribute to the electromagnetic noise.

In this notation the atmospheric current flowing in the vicinity of large-scale atmospheric inhomogeneities such as thunderstorms or hurricanes is treated as a stochastic process. We estimate the spectrum and amplitude of correlation function of the electromagnetic noise caused by the random current and conductivity fluctuations and discuss whether such a noise contributes to natural electromagnetic background in the range of $10^{-4}–10^{-2}$ Hz.

Following Surkov and Hayakawa (2007) we assume that the current variations are upward or downward inside the perturbed region and are zero outside, as shown in Fig. 6.15. To simplify the problem the actual inhomogeneous region labeled 1 is replaced by a cylindrical region with the radius a and the height h labeled 2. As before the random current fields are assumed to be steady, uniform, and isotropic inside the inhomogeneity, which, in turn, implies that the spectral density of the process is delta-correlated. To relate the electromagnetic spectra with current fields, a transfer matrix should be found, and then we can calculate the correlation matrix and power spectra of the random electromagnetic field.

In order to derive the main results we consider a simple way using the following line of reasoning. First, we note that the fluctuations of the atmospheric current

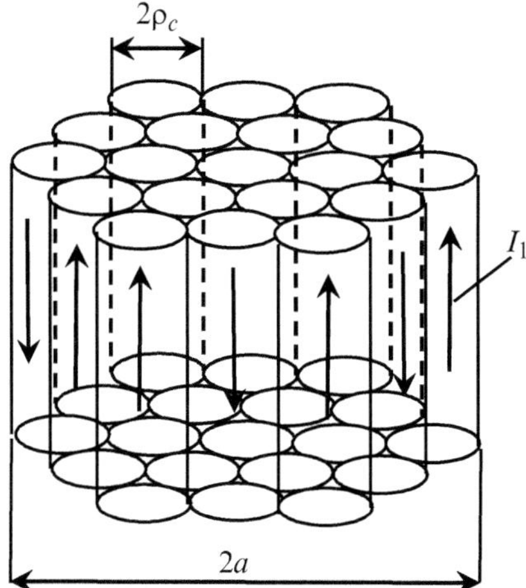

Fig. 6.16 A simplified model used to give simple interpretation and to indicate physical meaning of the results. The atmospheric current fluctuations are correlated inside each vertical cylinder but are not correlated with respect to each other

can be considered as the correlated current fluctuations inside the vertical cylinder with radius of the order of $\rho_c\,(\omega)$. The net current flowing through the cross-section of this cylinder is estimated as $I_1 = \pi\rho_c^2\,\langle\delta J\rangle$, where $\langle\delta J\rangle$ is the mean amplitude of the current density fluctuations. In the first approximation we neglect the coupling due to the magnetic field generated by each current cylinder. Dividing the whole perturbed region into parts/cylinders with radii ρ_c, as shown in Fig. 6.16, we obtain that the number of such "coherent currents" is of the order of $N \sim a^2/\rho_c^2$. Since these currents are uncorrelated, the net amplitude of the electric current variations is proportional to the square root of the current number, that is $I = I_1 N^{1/2} = \pi a \rho_c\,\langle\delta J\rangle$. At far distances the magnetic field of the vertical current can be expressed through the current moment Ih. Assuming for the moment that the non-conductive atmosphere with thickness d is "sandwiched" between two conductive plates, which approximate the ionosphere and the ground, the solution of the problem is given by Eq. (4.40). Replacing the current moment $M\,(\omega)$ by the value Ih and taking into account that $\cot(\theta/2) \approx 2/\theta \approx 2R_e/r$ yields

$$\delta B_\phi = \frac{\mu_0 I h}{2\pi r d},\qquad (6.108)$$

where δB_ϕ denotes the azimuthal magnetic field and r is radial distance to the vertical current moment. If the exponential atmospheric conductivity [Eq. (3.1)] is allowed for, the parameter d in Eq. (6.10) should be replaced by the vertical scale, H, of the conductivity variations with height. Such a characteristic scale can serve as an effective thickness of "insulator" layer. Substituting I and H into Eq. (6.108), we obtain

$$\delta B_\phi = \frac{\mu_0 a \rho_c h}{2rH} \langle \delta J \rangle .$$

(6.109)

Taking into account that the power spectrum of the magnetic noise is proportional to δB_ϕ^2 and replacing $\langle \delta J \rangle^2$ by the spectral density of random current fluctuation $F(f)$ yields

$$\Psi^{(B)}(r, f) \sim \left(\frac{\mu_0 a \rho_c h}{2rH} \right)^2 F(f),$$

(6.110)

where $f = \omega / (2\pi)$.

The background atmospheric currents mainly depend on the air conductivity which, in turn, is subject to violent changes from the action of the winds, precipitations, air humidity, pressure and temperature and etc. Thus, there may be many causes of the background conductivity and current fluctuations. This means that there may be the same mechanisms, which lead to the flicker-noise spectral density. Substituting Eq. (6.103) for $F(\omega)$ and Eq. (6.91) for $\rho_c(f)$ into Eq. (6.110), we finally obtain the rough estimate of the power spectrum

$$\Psi^{(B)}(r, f) \sim \frac{K \langle \delta J \rangle^m}{f^{n+2}} \left(\frac{\mu_0 a V_a h}{2rH} \right)^2 .$$

(6.111)

When $h \ll H$, the above equation coincides with the result obtained by Surkov and Hayakawa (2007) to an accuracy of factor 2.

On the basis of the assumption that the ULF atmospheric background current fluctuations exhibit a power law noise with the spectral index n we have found that the power spectra of magnetic noise must vary on overall inversely proportional to f^{n+2}. It is interesting to note that if n is close to unity, then the magnetic noise power spectra vary as f^{-3}, which is totally compatible with the measured dependence in the frequency range of 5×10^{-4} to 5×10^{-2} Hz (Lanzerotti et al. 1990).

The theoretical line calculated from Eq. (6.111) approximately coincides with line 3 shown in Fig. 6.14 that is lower, but nearly parallel to the experimental data shown with line 1. It should be noted that we have considered only a single atmospheric inhomogeneity as the source of the electromagnetic noise. Meanwhile, the worldwide thunderstorms, whirlwinds, or hurricanes, which are in progress around the world at any time, may contribute to the net electromagnetic nose.

Before leaving this section, it should be noted that the ULF electric field noise in the atmosphere due to background current variations is estimated to be of the order of 20–$0.7 \, \mu\text{V}/(\text{m Hz}^{1/2})$ (Surkov and Hayakawa 2007). Such electric variations are practically undetectable since its amplitude lies below the actual electric noise level. This leads us to the conclusion that the atmospheric electric noise must arise due to another causes.

It should be emphasized that the mechanism we have treated above is not unique because other mechanisms of $1/f$ noise such as MHD waves and solar activity might also be operative.

6.4.9 *Electric Field Pulsations at Fair-Weather Conditions*

In this section we consider reasonably steady status of the Earth's electric field. This implies that no processes of charge separation are taking place in the atmosphere (Chalmers 1967). Here we focus our attention alone on short-term ULF electric field pulsations at fair-weather conditions in lower atmosphere. The study of the pulsations of both the atmospheric electric field and charge density is indicative of the existence of certain relation between these pulsations and the turbulent stirring of charged particles in surface air as well as the drift of space charge (e.g., see the papers by Ogden and Hutchinson 1970; Yerg and Johnson 1974; Anderson 1982; Hoppel et al. 1986; Anisimov et al. 1999). It is believed that the charge density behaves like a passive air-entraining admixture and the electric field spectra are therefore controlled by the neutral-gas turbulence that can drag the aero-electric structures in the near-surface atmospheric layer. In contrast to the traditional problem of the atmospheric turbulence which deals with the fluctuations of mass velocity and gas temperature, the electric field fluctuations are nonlocal values since they depend on the spatial distribution of atmospheric charges around observation point. Thus, the electric field and charge density pulsations in the lower atmosphere are a significant indicator of atmospheric dynamics at fair-weather conditions.

Anisimov et al. (2002) have reported that at the frequencies of 0.01–0.1 Hz the spectral density, $\psi^{(E)}$, of the electric field pulsations in the surface atmospheric layer obeys the power law $\psi^{(E)} \propto f^{-n}$. Under the fair-weather and fog conditions, the spectral index n varies in the range of 1.23–3.36 with the most probable value from 2.25 to 3.0. The study of the temporal variations has shown that the structured pulsations alternate with unstructured variations of the electric field. The spectral index of the structured pulsation lies within interval 2.03–3.36 whereas the spectra of the unstructured variations is characterized by $n = 1.23$–2.89. Furthermore, these latter variations have small amplitude and energy.

The structured pulsations are thought to be due to the aero-electric structures flying at a low altitude, which is of the order of the structure size. As would be expected, the main energy of the aero-electric pulsation is concentrated in the near-surface atmospheric layer. The theory predicts the spectral density of such electric variations to be $\psi^{(E)} \propto f^{-11/3}$.

The unstructured variations can be resulted from the distant submesoscale aero-electric structures, which move along with mean atmospheric air flow. If the turbulent pulsations in these structures lie in the inertial subrange, the Kolmogorov theory predicts that the power spectrum of the pulsations is proportional to $f^{-5/3}$. The spatial horizontal size of such structures is estimated as 0.5–1 km and their lifetime is not less than 10–20 min.

Under fog conditions the amplitude of electric pulsation was found to increase more than one order of magnitude whereas the spectral index of the fog aeroelectric field pulsations does not differ drastically from the fair-weather spectrum index (Anisimov et al. 2002).

6.4.10 Monitoring of Near-Earth Plasma

The measurements of the global electromagnetic resonances and ULF fields is extremely important in the study of both the magnetospheric plasma dynamics and the Earth's magnetosphere status as a whole. There exists a close analogy with seismology, in which seismic waves are used to study the Earth's interior structure. The fundamental difference between the two areas is that the position and spectrum of the seismic sources are usually known with assurance whereas we have only a rough measure of the source properties of the MHD waves incident to the ionosphere. The monitoring of near-earth plasma density and the study of the ionosphere conductivity have their basis in separating the resonance effects from the ULF natural and man-made noises. The idea of hydromagnetic diagnostics of the magnetosphere based on the resonance spectrum of a field line was originally suggested by Obayashi (1958) and Dungey (1963). The problem of the diagnostics can be split into two basic tasks; that is, the measurement of the FLR-frequencies and solution of the inverse problems to determine plasma parameters in the magnetosphere (Guglielmi 1974, 1989; Baransky et al. 1985, 1990). Plausibility of this technique is restricted by an instability of the solution of the inverse problem since this solution is rather sensitive to small perturbations of the initial data. In practice, such perturbations are present not only due to the measurement inaccuracy but also because of the variability of the magnetosphere itself.

Much progress toward better understanding of the global ULF electromagnetic resonances and noises has been achieved in the past decades, that results in the appearance of "hydromagnetic seismology" of the near-earth space.

6.4.11 Space Weather

Overall the space weather describes today's status of the space environment including the conditions on the Sun and in the solar wind, magnetosphere, ionosphere, and thermosphere. When the space environment is disturbed by the variable output of particles and radiation from the Sun, it can influence the performance and reliability of space-borne equipment including computer memories. The geomagnetic storms, substorms, cosmic and solar rays give rise to degradation of spacecraft material, primarily solar battery. The interrelation between the fluxes of high energy particles and onboard anomalies has been well documented. The failure quota due to geophysical factors can reach about 60 % of total spacecraft failures (e.g., Pilipenko et al. 2006). Drastic deterioration of the space weather, that is fast increase in solar and geomagnetic activity, may greatly affect the ground-based technological systems and can endanger human life or health. At geostationary orbits the most dangerous effect is the influence of energetic particles on spacecraft performance. Depending on the particle energy, it can produce electrostatic charging followed by the faulty operation of electronics. The space-borne equipment errors have occurred during a magnetospheric storm just after a sharp enhancement of the relativistic electrons flux in the magnetosphere. The effect of these electrons is appearance of static negative charges, which can be irregularly distributed on the satellite surface because of different electric properties of surface elements. The potential drop between adjacent details of the satellite can reach tenth kilovolts that may result in dielectric breakdown and solar battery damage. Due to that the magnetospheric electrons with energy about 100 keV are termed "killer electrons." One of the best known events is a breakdown of American satellite TELSTAR during a magnetic storm in 1997, that resulted in the paging disconnection in considerable regions of the USA.

Sudden magnetospheric perturbations may greatly affect systems of communication and navigation including satellite navigation (GPS, GALLILEO and GLONASS) which in turn result in unforeseen contingencies (e.g., Yasuda et al. 2011; Takada et al. 2012). Air carriage comes up against such serious problems as a complete or particular loss of communication during the flight, delay of flight or changing of the flight routine, increase in fuel consumption, and fall off of gross weight. Moreover, the fluxes of high energy protons (with the energies greater than 100 MeV) of the solar flares can trigger health hazard for pilots and air travelers because of enhanced radiation background onboard. For example, the FAA (Federal Aviation Administration), which is primarily responsible for the safety and regulation of civil aviation in the USA, has reported that due to the strong solar flares on October 29 and 30, 2003 the global American system of precise GPS-positioning WAAS (Wide Area Augmentation System) was nonserviceable for aviation for 15 and 11 h, respectively. The intense radio bursts of clockwise-polarized waves in the frequency range of L1 and L2, which is usually used for satellite navigation, have been observed on December 2006. This results in the complete loss of GPS-signal for 10 min. With the current trend in miniaturization of electronic equipment, the impact of solar energetic particles greatly increases the risk of radiation damage of particular elements which may result in false operation and the generation of incorrect commands. For example, the failure probability of main memory module due to an impact of individual solar high energy particle was estimated to be one event per 200 h during the flight in polar region.

In the interest of air traffic, the space weather monitoring needs to include measurements of solar particle fluxes and Roentgen radiation, ejections of coronal plasma, and other characteristics of the solar activity provided by geostationary satellites as well as the observations of the particle flux variations in polar region and Earth radiation belts provided by polar satellites. Indices of solar and geomagnetic activities, measurements of absorption in D region of the ionosphere and polar cap, and other ground-based observations provide us with additional information (Burov et al. 2013).

For monitoring and study of the space weather the modern geophysics has a powerful tool such as space stations between the Sun and the Earth, flotilla of satellites in near-earth orbits, solar radio-telescopes, a network of ground-based radars and magnetometers. One of the challenges of magnetospheric research is to know enough about the solar activity and geomagnetic storms to make it possible for us to forecast the space weather. Much emphasis has been put on studies of this problem during the last few decades due to the increasing deployment of radiation- current- and field-sensitive systems in space and complex technological systems on the Earth. Despite much success in the study of this problem, the space weather forecasting has not become purely an engineering problem and it remains a formidable task to be accomplished.

Appendix F: FLR Structure

In Sect. 6.2 we study the FLR theory on the basis of the "MHD box" model. The plasma dynamics is described by Eqs. (6.25)–(6.30) where all the perturbed quantities are assumed to vary as $\exp\left(-i\omega t + ik_y y\right)$. Eliminating the plasma velocity from this set of equations we are thus left with the set

$$i\omega E_x / V_A^2 = \partial_z \delta B_y - ik_y \delta B_z + \mu_0 J_x, \tag{6.112}$$

$$-i\omega E_y / V_A^2 = \partial_z \delta B_x - \partial_x \delta B_z - \mu_0 J_y, \tag{6.113}$$

$$\partial_z E_y = -i\omega \delta B_x, \tag{6.114}$$

$$\partial_z E_x = i\omega \delta B_y, \tag{6.115}$$

$$\partial_x E_y - ik_y E_x = i\omega \delta B_z. \tag{6.116}$$

Eliminating the variations of magnetic fields δB_x, δB_y, and δB_z from Eqs. (6.112)–(6.116) we come to the set

$$\left(\partial_z^2 + k_A^2\right) E_x = ik_y \left(\partial_x E_y - ik_y E_x\right) - i\mu_0 \omega J_x, \tag{6.117}$$

$$\left(\partial_x^2 + \partial_z^2 + k_A^2\right) E_y = ik_y \partial_x E_x - i\mu_0 \omega J_y, \tag{6.118}$$

where $k_A^2 = \omega^2 / V_A^2$. Notice that if $k_y = 0$, Eqs. (6.117) and (6.118) are similar to Eqs. (6.31) and (6.32). The boundary conditions at the "MHD box" sides, which correspond to the northern ($z = l_1$) and the southern ionospheres ($z = 0$), are given by Eq. (6.35)

Using a variable separation method, we seek for the solution of Eqs. (6.117) and (6.118) in terms of the series

$$E_x = \sum_n a_n(x) q_n(z) \quad \text{and} \quad E_y = \sum_n b_n(x) q_n(z), \tag{6.119}$$

where the eigenfunctions, $q_n(z)$, of the problem must satisfy the following equation

$$\frac{d^2 q_n}{dz^2} + k_n^2 q_n = 0, \tag{6.120}$$

where k_n denotes the eigenvalues of the problem. The solution of Eq. (6.120) is given by

$$q_n = C_1 \sin k_n z + C_2 \cos k_n z. \tag{6.121}$$

To find the undetermined constants C_1, C_2 and the eigenvalues, one should substitute Eq. (6.119) for E_x and E_y into boundary conditions (6.35) to yield

$$\frac{dq_n}{dz} = \pm i \omega \mu_0 \Sigma_P^{\pm} q_n. \tag{6.122}$$

Here the sign plus and Σ_P^+ correspond to the northern ionosphere, i.e., $z = l_1$, while the sign minus and Σ_P^- correspond to the southern ionosphere, i.e., $z = 0$. Substituting Eq. (6.121) for q_n into Eq. (6.122) and rearranging, we come to the set of algebraic equations for the constants C_1 and C_2. These equations have nontrivial solutions under the requirement that

$$\exp(-2i k_n l_1) = \frac{(1 - X_+)(1 - X_-)}{(1 + X_+)(1 + X_-)}, \tag{6.123}$$

where

$$X_{\pm} = \frac{\mu_0 \omega \Sigma_P^{\pm}}{k_n}. \tag{6.124}$$

Consider first two opposite extreme cases of zeroth and infinite Pedersen conductivities. If $\Sigma_P = 0$ at both the conjugate ionospheres, the right-hand side of Eq. (6.123) is equal to unity. The same is true in the inverse case of the perfectly conducting ionosphere when both $\Sigma_P^{\pm}$ and $X_{\pm}$ tend to infinity. In these extreme cases Eq. (6.123) has only real roots

$$k_n = \pi n / l_1, \tag{6.125}$$

where $n = 1, 2, 3\ldots$ Moreover, these eigenvalues are independent of both the frequency ω and coordinate x.

In a general case of finite $\Sigma_P^{\pm}$ Eq. (6.123) has a discrete spectrum of complex eigenvalues k_n. The normalized eigenfunctions of the problem that obey Eqs. (6.120) and (6.122) can be written as

$$q_n = \frac{\sin k_n z}{k_n} + \frac{i \cos k_n z}{\mu_0 \omega \Sigma_P^-}. \tag{6.126}$$

First of all we note that these functions satisfy both Eq. (6.120) and the boundary condition (6.122) at $z = 0$. Moreover, substituting Eq. (6.126) for q_n into the boundary condition (6.122) at $z = l_1$ we come to the identity taking into account Eq. (6.123).

It can be shown that these eigenfunctions form a set of orthonormal functions in the sense that

$$\int_0^{l_1} q_n(z) q_m(z) \, dz = \delta_{nm}, \tag{6.127}$$

where δ_{nm} denotes the Kronecker symbol

$$\delta_{nm} = \begin{cases} 1, & n = m; \\ 0, & n \neq m. \end{cases} \tag{6.128}$$

To find the functions $a_n(x)$ and $b_n(x)$ we substitute Eq. (6.119) for E_x and E_y into Eqs. (6.117) and (6.118)

$$\sum_m \left(k_A^2 - k_m^2 \right) a_m q_m = i k_y \sum_m \left(b_m' - i k_y a_m \right) q_m - i \mu_0 \omega J_x, \tag{6.129}$$

$$\sum_m \left\{ b_m'' + \left(k_A^2 - k_m^2 \right) b_m \right\} q_m = i k_y \sum_m a_m' q_m - i \mu_0 \omega J_y, \tag{6.130}$$

where the prime denotes derivative with respect to x.

Now we consider free oscillations of the electromagnetic field in the MHD box. In other words, the sources of the driving/external currents and perturbations are assumed to be "turn off" so that $J_x = J_y = 0$. Multiplying both sides of Eqs. (6.129) and (6.130) by $q_n(z)$, integrating these equations over z from 0 to l_1 and using the orthogonality condition (6.127) we are thus left with the set

$$i k_y b_n' = \left(k_A^2 - k_n^2 - k_y^2 \right) a_n, \tag{6.131}$$

$$i k_y a_n' = b_n'' + \left(k_A^2 - k_n^2 \right) b_n. \tag{6.132}$$

Eliminating a_n from Eqs. (6.131) and (6.132) and rearranging, we obtain

$$b_n'' - \frac{k_y^2 \left(k_A^2 \right)' b_n'}{\left(k_A^2 - k_n^2 \right) \left(k_A^2 - k_n^2 - k_y^2 \right)} + \left(k_A^2 - k_n^2 - k_y^2 \right) b_n = 0. \tag{6.133}$$

The solution of Eq. (6.133) depends on the boundary conditions at $x = 0$ and $x = l_2$. Without specifying these boundary conditions, we now will study the differential equation (6.133) to find some features which are common to all solutions. This equation exhibits strong singularities found in the denominator of its second term. Assuming for the moment that $k_A(x) = \omega/V_A(x)$ is a monotonic function of x, then Eq. (6.133) may have two regular singular points, say $x = \xi$, where

$$k_A^2(\xi) = k_n^2, \qquad (6.134)$$

and $x = \eta$ where

$$k_A^2(\eta) = k_n^2 + k_y^2. \qquad (6.135)$$

Following Southwood (1974), we suppose that the value of η in Eq. (6.135) is real and consider a small neighborhood of the singular point $x = \eta$ where the function k_A^2 can be expanded in a power series of $x - \eta$, that is, $k_A^2 \approx k_n^2 + k_y^2 + (k_A^2)'(x - \eta) + o(x - \eta)$. Here the derivative $(k_A^2)'$ is taken at $x = \eta$. In the first approximation Eq. (6.133) is thus reduced to

$$b_n'' - \frac{b_n'}{x - \eta} + (k_A^2)'(x - \eta)\, b_n = 0. \qquad (6.136)$$

Southwood (1974) has shown that two independent solutions of Eq. (6.136) are finite at the singular point $x = \eta$. The implication of this singularity is that $x = \eta$ corresponds to a turning point, where the solutions change in character. We cannot come close to exploring this problem in any detail, but we need to note that this point divides a space of MHD box into two regions. In the first region the solutions are quasioscillatory in nature, whereas in the next one the solutions are monotonically increasing or decreasing functions of distance. In other words, the point $x = \eta$ corresponds to a turning point where solutions change from being oscillatory in nature to characteristically growing or decaying with coordinate x.

More importantly, the solution can be infinite at the next singular point, $x = \xi$, which corresponds to the FLR conditions. Indeed, substituting $k_n = k_A$ into Eqs. (6.123) and (6.124) we come to Eq. (6.37), which describes the resonance frequencies of the Alfvén oscillations. The implication here is that if $k_A^2(\xi) = k_n^2$ then the field line at $x = \xi$ will resonate with the shear Alfvén wave since the wave frequency ω equals to one of the Alfvén resonance frequencies.

If the energy dissipation in the conjugate ionospheres is neglected, the eigenvalues are given by $k_n = \pi n/l_1$, that is, they are real whence it follows that the resonant point ξ is real as well. In the dissipative case the parameter k_n^2 is a complex value so that the roots of Eq. (6.134) are in the complex plane x. Decomposing the roots x_n into its real and imaginary parts, we obtain $x_n = \xi_n + i\delta_n$. In what follows we do not specify the value and sign of δ_n.

Expanding k_A^2 in a power series of $x - (\xi_n + i\delta_n)$ we can reduce Eq. (6.133) to the form

$$b_n'' + \frac{b_n'}{x - \xi_n - i\delta_n} - k_y^2 b_n = 0. \tag{6.137}$$

The general solution of Eq. (6.137) is given by

$$b_n = A I_0(\tilde{x}) + B K_0(\tilde{x}), \tag{6.138}$$

where $\tilde{x} = k_y(x - \xi_n - i\delta_n)$ is a dimensionless variable, I_0 and K_0 are modified Bessel functions of order zero, and A and B denote arbitrary constants. We recall that this solution is valid only near the singular points $x_n = \xi_n + i\delta_n$.

It should be noted that the function K_0 has a logarithmic singularity in the neighborhood of zero point, i.e., $K_0(\tilde{x}) \propto -\ln\tilde{x}$ as $\tilde{x} \to 0$. This means that if the energy dissipation is negligible, that is to say $\delta_n = 0$, then the function K_0 is logarithmically infinite at $x = \xi_n$. Not surprisingly, the amplitude of the resonance may tend to infinity as the energy loss is ignored. Actually, the dissipation factor $\delta_n \neq 0$ so that for real x the solution (6.138) exhibits finite behavior near the point $x = \xi_n$. This notation has concerns with the electric field given by Eq. (6.119). In the vicinity of the resonant shell at $x = \xi_n$ we obtain the following asymptotic formula:

$$E_y \propto (\ln\tilde{x}) q_n(z), \tag{6.139}$$

Substituting $k_A^2 = k_n^2$ into Eq. (6.131) gives $a_n = -i b_n'/k_y$ whence it follows that

$$E_x \propto q_n(z)/\tilde{x}. \tag{6.140}$$

Combining Eqs. (6.114) and (6.115) with Eqs. (6.139) and (6.140), we obtain

$$\delta B_x \propto \ln\tilde{x}\,\frac{dq_n(z)}{dz}, \tag{6.141}$$

$$\delta B_y \propto \frac{1}{\tilde{x}}\frac{dq_n(z)}{dz}. \tag{6.142}$$

In order to find δB_z one should use the series development of $K_0(\tilde{x})$ up to the terms $\sim \tilde{x}^2 \ln\tilde{x}$. As a result we get

$$\delta B_z \propto \left(1 + \frac{\tilde{x}^2}{2}\ln\tilde{x}\right) q_n(z). \tag{6.143}$$

As is seen from the above equations, the field components E_x and δB_y, which contain the factor $\tilde{x}^{-1}$, reach a peak value near the point $x = \xi_n$ whereas the components E_y and δB_x have logarithmic, that is, weak singularities, and δB_z is a slowly varying function of $\tilde{x}$ in this region.

To study the shape of resonance components, E_x and δB_y, in a little more detail we use the representation of $\tilde{x}^{-1}$ in the form

$$\frac{1}{\tilde{x}} = \frac{\exp\left(i\varphi\right)}{\left\{(x - \xi_n)^2 + \delta_n^2\right\}^{1/2}}, \tag{6.144}$$

where the argument φ of complex number is determined via

$$\tan\varphi = \frac{\delta_n}{x - \xi_n}. \tag{6.145}$$

It is clear from Eq. (6.144) that near the resonant point the dependence $|E_x|^2$ and $\left|\delta B_y\right|^2$ on x has a form of Lorentz's curve outlined in Fig. 6.3. The normalized component $\left|\delta B_y\right|^2$ versus normalized variable x/ξ_n is sketched in this figure with solid line 1. The Lorentz's curve has a maximum at $x = \xi_n$ and the parameter $|\delta_n|$ is the characteristic half-width of this maximum. The argument/phase φ of E_x and δB_y changes by π when crossing the maximum. A schematic plot of the component δB_x is shown in Fig. 6.3 with dashed line 2.

In conclusion it should be noted that if $V_A(x)$ is not a monotonic function there may be more turning and resonance points.

References

Alperovich LS, Fedorov EN (2007) Hydromagnetic waves in the magnetosphere and the ionosphere. Springer, Berlin, 426 pp

Anderson RV (1982) The dependence of space charge spectra on Aitken nucleus concentrations. J Geophys Res 87:1216–1218

Anisimov SV, Mareev EA, Bakastov SS (1999) On the generation and evolution of aeroelectric structures in the surface layer. J Geophys Res D 104:14359–14367

Anisimov SV, Mareev EA, Shikhova NM, Dmitriev EM (2002) Universal spectra of electric field pulsations in the atmosphere. Geophys Res Lett 29(24):2217. doi:10.1029/2002GL015765

Annexstad JO, Wilson ChR (1968) Characteristics of Pg micropulsations at conjugate points. J Geophys Res 73:1805–1818

Atkinson G, Watanabe T (1966) Surface waves on the magnetospheric boundary as a possible origin of long period micropulsations. Earth Planet Sci Lett 1:89–91

Baransky LN, Borovkov YuE, Gokhberg MB, Krylov SM, Troitskaya VA (1985) High resolution method of direct measurement of the magnetic field lines eigenfrequencies. Planet Space Sci 33:1369–1374

Baransky LN, Belokris SP, Borovkov YuE, Green CA (1990) Two simple methods for the determination of the resonance frequencies of magnetic field lines. Planet Space Sci 38:1573–1576

Baumjohann W, Glassmeier KN (1984) The transient response mechanism and Pi2 pulsations at substorm onset: review and outlook. Planet Space Sci 32:1361–1370

Burov VA, Lapshin VB, Syroezhkin AV (2013) Space weather and air traffic. World of measurements (Mir Izmereniy), No. 2. pp 11–16 (in Russian)

Chalmers JA (1967) Atmospheric electricity, 2nd edn. Pergamon, New York

Chen L, Hasegawa A (1974) Plasma heating by spatial resonance of Alfvén wave. Phys. Fluids 17:1399

Cornwall JM (1965) Cyclotron instabilities and electromagnetic emission in the ultra low frequency ranges. J Geophys Res 70:61–69

Crowley G, Hughes WJ, Jones TB (1987) Observational evidence of cavity modes in the Earth's magnetosphere. J Geophys Res 92:12233–12240

Cummings WD, O'Sullivan RJ, Coleman PJ Jr (1969) Standing Alfvén waves in the magnetosphere. J Geophys Res 74:778–793

Davydenko SS, Mareev EA, Marshall TC, Stolzenburg M (2004) On the calculation of electric fields and currents of mesoscale convective systems. J Geophys Res 109:D11103. doi:10.1029/2003JD003832

Dowden RL (1966) Micropulsation "nose whistlers". A helium explanation. Planet Space Sci 14:1273–1274

Dungey JW (1954) Electrodynamics of the outer atmospheres. Ionosphere Scientific Report, vol 69. Ionosphere Research Labaratory, Cambridge

Dungey JW (1963) The structure of the exosphere, or, adventures in velocity in geophysics. In: DeWitt C, Hieblot J, Lebeau A (eds) The earth's environment proceedings of the 1962 Les Houches Summer School. Gordon and Breach, New York, pp 503–550

Engebretson MJ, Zanetti LJ, Potemra TA (1986) Harmonically structured ULF pulsations observed by the AMPTE/CCE magnetic field experiment. Geophys Res Lett 13:905–908

Engebretson MJ, Zanetti LJ, Potemra TA, Baumjohann W, Lühr H, Acua MN (1987) Simultaneous observation of Pc3–4 pulsations in the solar wind and the Earth's magnetosphere. J Geophys Res 92:10053–10062

Fraser-Smith AC (1995) Low-frequency radio noise. In: Volland H (ed) Handbook of atmospheric electrodynamics, vol 1. CRC Press, Boca Raton, pp 297–310

Ginzburg VL (1970) The propagation of electromagnetic waves in plasmas. Pergamon Press, Oxford

Glassmeier K-H (1980) Magnetometer array observations of a giant pulsation event. J Geophys 48:127–138

Glassmeier K-H (1995) ULF pulsations. In: Volland H (ed) Handbook of atmospheric electrodynamics, vol 2. CRC Press, Boca Raton, pp 463–502

Guglielmi AV (1974) Diagnostics of the magnetosphere and interplanetary medium by means of pulsations. Space Sci Rev 16:331

Guglielmi AV (1989) Diagnostics of the plasma in the magnetosphere by means of measurement of the spectrum of Alfvén oscillations. Planet Space Sci 37:1011–1012

Guglielmi AV, Troitskaya VA (1973) Geomagnetic pulsations and diagnostics of the magnetosphere. Nauka, Moscow (in Russian)

Higbie PR, Baker DN, Zwicki RD, Belian RD, Asbridge JR, Fennell JF, Wilken B, Arthur CW (1982) The global Pc5 events of November 14–15. J Geophys Res 87:2337–2345

Hoppel WA, Anderson RV, Willet JC (1986) In earth's electrical environment. National Academic Press, Washington, DC, pp 149–165

Jacobs JA (1970) Geomagnetic micropulsations. Springer, Berlin

Junginger H, Geiger G, Haerendel G, Melzner F, Amata E, Higel B (1984) A statistical study of dayside magnetospheric field fluctuations with periods between 150 and 600 s. J Geophys Res 89:10757–10762

Karpman VI, Meerson BI, Mikhailovsky AB, Pokhotelov OA (1977) The effects of bounce resonance on wave rates in the magnetosphere. Planet Space Sci 25:573–585

Kato Y (1962) Geomagnetic pulsations and hydromagnetic oscillations of exosphere. J Phys Soc Jpn 17:71

Kivelson MG, Southwood DJ (1985) Resonant ULF waves: a new interpretation. Geophys Res Lett 12:49–52

Kivelson MG, Etcheto J, Trotignon JG (1984) Global compressional oscillations of the terrestrial magnetosphere: the evidence and a model. J Geophys Res 89:9851–9856

Krylov AL, Fedorov EN (1976) Concerning eigen oscillations of bounded volume of a cold magnetized plasma. Rep USSR Acad Sci (Dokl Akad Nauk SSSR) 231:68–70

Krylov AL, Lifshitz AF (1984) Quasi-Alfvén oscillations of magnetic surfaces. Planet Space Sci 32:481–489

Kwok YC, Lee LC (1984) Transmission of magnetohydrodynamic waves through the rotational discontinuity of Earth's magnetopause. J Geophys Res 89:10697–10708

Landau LD, Lifshitz EM (1982) Electrodynamics of continuous media, 2nd edn. Nauka, Moscow (in Russian)

Lanzerotti LJ (1978) Studies of geomagnetic pulsations. In: Lanzerotti LJ, Park CG (eds) Upper atmosphere research in Antarctica. American Geophysical Union as part of the Antarctic Research Series, vol 29, pp 130–156. doi:10.1029/AR029

Lanzerotti LJ, Fukunishi H (1974) Modes of MHD waves in the magnetosphere. Rev Geophys Space Phys 12:724–729

Lanzerotti LJ, Maclennan CG, Fraser-Smith AC (1990) Background magnetic spectra: $\sim 10^{-5}$ to 10^5 Hz. Geophys Res Lett 17:1593–1596

Lee D-L, Lysak RL (1989) Magnetospheric ULF waves coupling in the dipole model: the impulsive excitation. J Geophys Res 94:17097–17103

Lee D-L, Lysak RL (1990) Effects of azimuthal asymmetry on ULF waves in the dipole magnetosphere. Geophys Res Lett 17:53–56

Leonovich AS (2000) Magnetospheric MHD response to a localized disturbance in the ionosphere. J Geophys Res 105:2507–2519

Leonovich AS, Mazur VA (1993) A theory of transverse small-scale standing Alfvén waves in an axially symmetric magnetosphere. Planet Space Sci 41:697–717

McHenry MA, Clauer CR (1987) Modelled ground magnetic signatures of flux transfer events. J Geophys Res 92:11231–11240

Nishida A (1978) Geomagnetic diagnostics of the magnetosphere. Springer, New York

Obayashi T (1958) Geomagnetic pulsations and the Earth's outer atmosphere. Ann Geophys 14:464–474

Ogden TL, Hutchinson WCA (1970) Electric space-charge pulses near the ground in sunny weather. J Atmos Terr Phys 32:1131–1138

Ohl AI (1962) Pulsations during sudden commencements of magnetic storms and long period pulsations in high latitudes. J Phys Soc Jpn 17(supp. A-2, paper II-1B-3):24–26

Ohl AI (1963) Long period pulsations of the geomagnetic field. Geomagn Aeron 3:113–120

Pilipenko V, Vellante M, Fedorov E (2000) Distortion of the ULF wave spatial structure upon transmission through the ionosphere. J Geophys Res A105:21225–21236

Pilipenko V, Yagova N, Romanova N, Allen J (2006) Statistical relationship between satellite anomalies at geostationary orbit and high-energy particles. Adv Space Res 37:1192–1205

Radoski HR (1966) Magnetic toroidal resonances and vibrating field lines. J Geophys Res 71:1891–1893

Radoski HR (1967a) Highly asymmetric MHD resonances: the guided poloidal mode. J Geophys Res 72:4026–4033

Radoski HR (1967b) A note on oscillating field lines. J Geophys Res 72:418–419

Radoski HR, Carovillano RR (1969) Axisymmetric plasmaspheric resonances: toroidal mode. Phys Fluids 9:285–291

Rytov SM, Kravtsov YaA, Tatarsky VI (1978) Introduction to statistical radiophysics, P. 2, random fields. Nauka, Moscow (in Russian)

Samson JC, Jacobs JA, Rostoker G (1971) Latitude-dependent characteristics of long-period geomagnetic micropulsations. J Geophys Res 76:3675–3683

Sato T (1978) A theory of quiet auroral arcs. J Geophys Res 83:1042–1048

Singer HJ, Kivelson MG (1979) The latitude structure of Pc5 waves in space: magnetic and electric field observations. J Geophys Res 84:7213–7222

Singer HJ, Russell CT, Kivelson MG, Fritz TA, Lennartsson W (1979) Satellite observations of the spatial extent and structure of Pc3, 4, 5 pulsation near the magnetospheric equator. Geophys Res Lett 6:889–892

Southwood DJ (1974) Some features of field line resonances in the magnetosphere. Planet Space Sci 22(3):483–491

Southwood DJ (1980) Low frequency pulsation generation by energetic particles. J Geomagn Geoelectr 32(Suppl. II):75–88

Southwood DJ, Kivelson MG (1982) Charged particle behaviour in low-frequency geomagnetic pulsations, 2. Graphical approach. J Geophys Res Space Phys 87(A3):1707–1710

Surkov VV, Hayakawa M (2007) ULF electromagnetic noise due to random variations of background atmospheric current and conductivity. J Geophys Res 112:D11116. doi:10.1029/2006JD007788

Surkov VV, Hayakawa M (2008) Natural electromagnetic ULF noise due to fluctuations of ionospheric currents. J Geophys Res Space Phys 113:A11310. doi:10.1029/2008JA013196

Takada M, Nunomiya T, Ishikara T, Nakamura T, Levis BJ, Bennett LJ, Getley IL, Bennett BH (2012) Measuring cosmic-ray exposure in aircraft using real-time personal dosimeters. Radiat Prot Dosim 149(2):169–176

Takahashi K (1988) Multisatellite studies of ULF waves. Adv Space Res 8(9–10), 427–436

Takahashi KT, McPherron RL, Terasawa T (1984) Dependence of the spectrum of Pc3–4 pulsations on the interplanetary magnetic field. J Geophys Res 89:2770–2780

Trakhtengerts VY, Rycroft MJ (2008) Whistler and Alfvén mode cyclotron masers in space. Cambridge University Press, Cambridge, 354 p

Voelker H (1962) Zur Breitenabhä ngigkeit der Perioden erdmagnetischer Pulsationen. Naturwissenschaften 49:8–9

Weissman MB (1988) $1/f$ noise and other slow, nonexponential kinetics in condensed matter. Rev Mod Phys. 60:537–571

Yasuda H, Lee J, Yajima K, Hwang JA, Sakai K (2011) Measurement of cosmic-ray neutron dose onboard a polar route flight from New York to Seoul. Radiat Prot Dosimetry 146(1–3):213–216

Yerg DG, Johnson KR (1974) Short period fluctuations in the fair-weather electric field. J Geophys Res 79:2177–2184

Yumoto K (1984) Long-period magnetic pulsations generated in the magnetospheric boundary layer. Planet Space Sci 32:1205–1218

Zibyn KYu (1965) Properties and nature of geomagnetic pulsations with periods from 10 s to several minutes. Geomagn Aeron 5:494

Ziesolleck CWS, Menk FW, Fraser BJ, Webb PW (1993) Spatial characteristics of low Pc3-4 geomagnetic pulsations. J Geophys Res 98:197–207

Part III

Electromagnetic Fields Due to Rock Deformation and Fracture

Chapter 7
Geomagnetic Perturbations (GMPs)

Abstract In previous chapters we discussed the magnetospheric, ionospheric, and atmospheric sources of the ULF electromagnetic fields. The main emphasis has been put on studies of the global lightning activity, which results in an energy storage inside the resonators followed by the excitation of global electromagnetic resonances such as Schumann and IAR resonances. Furthermore, there are a variety of other terrestrial and atmospheric causes for the generation of ULF electromagnetic fields: ocean waves, tsunami, volcanic eruptions, earthquakes (EQs), meteoritic falls to the atmosphere, as well as the man-made sources such as the stray current, atmospheric and underground explosions, and so on. It is usually the case that the large-scale ULF perturbations resulted from the magnetospheric sources and lightning activity can be easily distinguished from the terrestrial fields and different kinds of local noises but sometimes there are a few problems because the terrestrial fields may be very weak.

In this chapter we focus our attention on the study of low-frequency electromagnetic fields associated with large-scale tectonic processes. In the course of this text, some of the phenomena related to natural disasters will be treated in detail and others in a more sketchy fashion. We also consider in-situ measurements and laboratory tests of the low-frequency electromagnetic fields generated during the deformation and fracture of solid.

We start with the problem of generation and propagation of the geomagnetic field perturbations caused by the seismic waves travelling in the ground. These perturbations are of special interest in applied geophysical studies because they provide us with additional information about the depth and energy of seismic source as well as about the conductivity and other rock parameters taken along the seismic rays. The signals due to perturbations of the Earth's magnetic field may be useful for the analysis and interpretation of co-seismic phenomena associated with the EQs. A theory of these phenomena is necessary in order to study both the signal-to-noise ratio and the electromagnetic interferences due to device vibrations caused by seismic wave propagation.

V. Surkov and M. Hayakawa, *Ultra and Extremely Low Frequency Electromagnetic Fields*, Springer Geophysics, DOI 10.1007/978-4-431-54367-1_7,
© Springer Japan 2014

Keywords Co-seismic signals • Diffusion and seismic zones • Electromagnetic forerunner • Geomagnetic perturbations (GMPs) • Rayleigh surface wave

7.1 Two Source Mechanisms for ULF Electromagnetic Field Generation

In what follows we frequently meet different models of low-frequency underground sources which are assumed to be responsible for a variety of electromagnetic phenomena associated with the rock deformation. In deriving simple estimates for the amplitudes of these phenomena, all the sources can be split into two classes depending on whether the extrinsic/source current is closed or not (e.g., see Landau and Lifshits 1982; Jackson 2001).

Since the total current is always closed, the open-ended extrinsic current I_e must be closed through displacement currents in the insulator or through conduction currents I_c in the conducting medium. Far away from the source the distribution of the extrinsic current density $\mathbf{j}_e$ becomes unimportant so that the actual source can be replaced by an equivalent point current element/electric dipole whose basic characteristic is the current dipole moment

$$\mathbf{d} = \int_{\Delta V} \mathbf{j}_e \, dV. \tag{7.1}$$

In many cases Eq. (7.1) can be simplified to $d = I_e \Delta l = j_e \Delta V$, where Δl denotes the typical length/size of the current element and ΔV stands for its volume. The current element and the corresponding configuration of the currents connecting this element are sketched in Fig. 7.1. This kind of source results in the generation of axially symmetric toroidal magnetic field $\mathbf{B}$ shown with dotted blue lines.

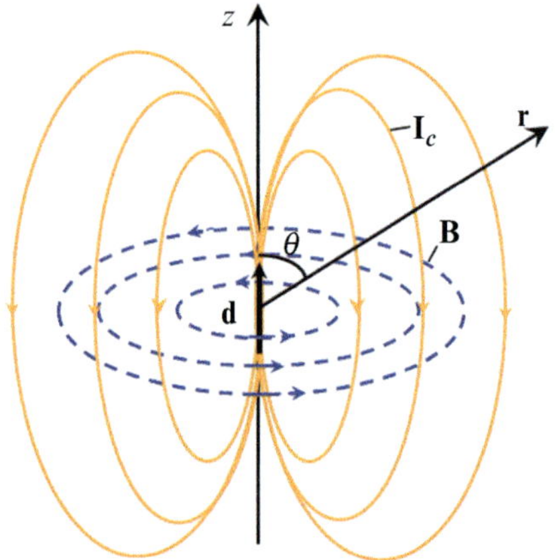

Fig. 7.1 The current configuration caused by an electric dipole **d**. The open-ended extrinsic current shown with vertical *black arrow* is closed through the conduction current I_c or through the displacement current that results in the generation of toroidal magnetic field **B** shown with *dotted blue lines*

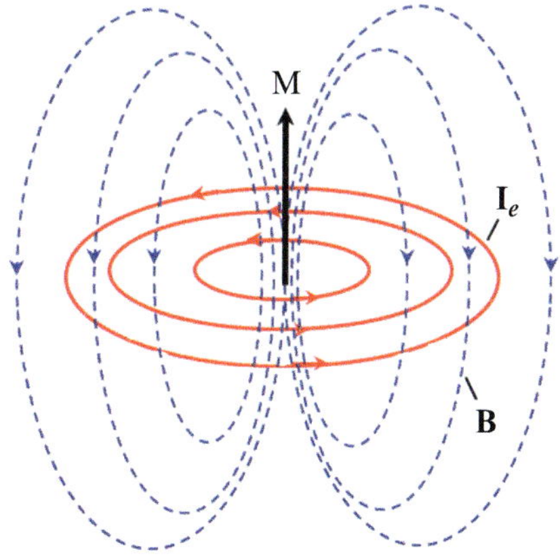

Fig. 7.2 The current configuration caused by a magnetic dipole **M**. The closed extrinsic current I_e generates the poloidal magnetic field **B** shown with *dotted blue lines*

The quasi-steady electric field of the current element in a homogeneous conducting medium is given by

$$\mathbf{E} = -\frac{1}{4\pi\sigma}\nabla\frac{\mathbf{d}\cdot\mathbf{r}}{r^3},\tag{7.2}$$

where σ is the medium conductivity and r is the distance from the dipole. The quasi-steady magnetic field of the current element has only an azimuthal component given by

$$B_\varphi = \frac{\mu_0 d\,\sin\theta}{4\pi r^2},\tag{7.3}$$

where θ is the polar angle shown in Fig. 7.1.

The electromagnetic field of the closed extrinsic current has another pattern shown in Fig. 7.2. Far away from the source the field is described by the vector of magnetic moment given by (e.g., see Landau and Lifshits 1982; Jackson 2001)

$$\mathbf{M} = \frac{1}{2}\int_{\Delta V}(\mathbf{r}\times\mathbf{j}_e)\,dV',\tag{7.4}$$

where ΔV denotes the space occupied by the currents. The closed extrinsic current $\mathbf{j}_e$ generates the poloidal magnetic field **B** shown with dotted blue lines. Far away from the source the quasi-steady magnetic field of the magnetic dipole **M** in a homogeneous conducting medium is given by

$$\mathbf{B} = -\frac{\mu_0}{4\pi}\nabla\frac{\mathbf{M}\cdot\mathbf{r}}{r^3}.\tag{7.5}$$

Certainly the presence of the boundary between the conducting earth and the atmosphere can distort the simple symmetrical pictures given by above equations. Sometimes a simple solution of the boundary problem can be found by using the mirror images method. To illustrate this, in Fig. 7.3 we plot the current element **d**

Fig. 7.3 Schematic plot
of a point current element, **d**,
immersed in a conducting
half-space

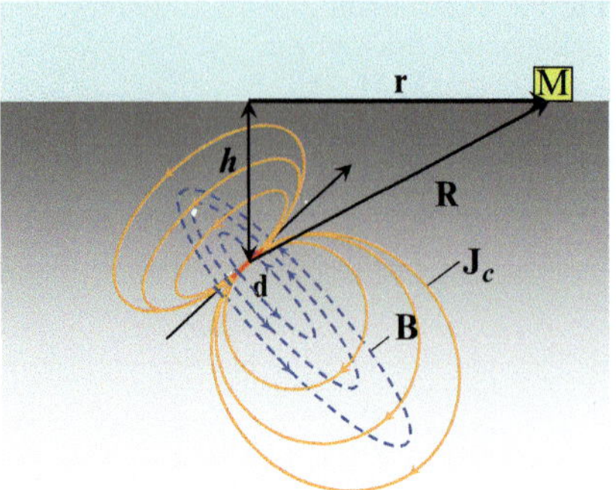

located in the conducting ground ($z < 0$). The total field in the ground is the sum of
the fields generated by the current element and its mirror image. Since more usually
we measure electromagnetic perturbations on the ground surface we shall exemplify
the field of current moment at $z = 0$. Let the point current element **d** be in the xz
plane at the depth h in the homogeneous half-space with constant conductivity σ as
shown in Fig. 7.3. Using the polar coordinate ρ and ψ, where the angle is measured
from x axis, the field components are written as follows (e.g., see Surkov et al. 2002)

$$E_\rho = \frac{3d_z\rho h + d_x\left(2\rho^2 - h^2\right)\cos\psi}{2\pi\sigma R^5}, \quad E_\psi = \frac{d_x\sin\psi}{2\pi\sigma R^3},$$

$$E_z(0+) = \frac{d_z\left(2h^2 - \rho^2\right) + 3d_x\rho h\cos\psi}{2\pi\sigma R^5}, \quad E_z(0-) = 0,$$

$$B_\rho = \frac{\mu_0 d_x\sin\psi}{4\pi R^2}\left(\frac{R}{R+h} - \frac{h}{R}\right), \quad B_\psi = -\frac{\mu_0 d_x\cos\psi}{4\pi R(R+h)},$$

$$B_z = \frac{\mu_0 d_x\rho\sin\psi}{4\pi R^3}, \quad R = \left(\rho^2 + h^2\right)^{1/2}, \tag{7.6}$$

where d_x and d_z are projections of the vector **d** on the axes x and z, respectively.
Here $E_z(0+)$ and $E_z(0-)$ denote the vertical electric fields just above and below
the surface $z = 0$.

However, the simple formulas (7.1)–(7.6) can be used for order-of-magnitude
estimates of quasi-steady electromagnetic field generated by underground sources.

In Chap. 7 we study perturbations of the Earth's magnetic field caused by the
currents generated in the conducting rock or in the ionosphere. As we shall see, these
perturbations are related to the class of magneto-dipole type fields. By contrast, the
electrokinetic effects studied in Chap. 8 is usually related to the fields of the electric
dipole type.

7.2 Local GMPs Due to Seismic Waves in Conductive Ground

7.2.1 In-Situ Measurements

The seismic waves in the Earth's crust caused by EQs and explosions, the heavy sea, tsunami can generate the local perturbations of the Earth's magnetic field. In the case under consideration the geomagnetic field plays the role of an external magnetic field and the ground serves as a conducting medium. The electric currents in the ground are developed due to the conductor motion in the seismic wave. The electric currents give rise to the geomagnetic field perturbations. In other words, this MHD effect arises from the movement of a conducting layer of the Earth's crust or seawater in the presence of the Earth's magnetic field. In geophysical studies this effect is frequently termed the inductive seismomagnetic effect. One more phenomenon termed the seismoelectric effect in moist soil will be examined in more detail in Chap. 8. Both of these phenomena associated with seismic waves propagation, i.e., seismoelectric and inductive seismomagnetic effects, are referred to as the class of co-seismic phenomena.

In this section we deal with the MHD effects, i.e., the perturbations of the Earth's magnetic field, which are capable of explaining the electromagnetic perturbations in the atmosphere as observed by ground-based magnetometers and antennas during and just after the seismic wave arrival at a measurement point. The great majority of signals routinely detected are observed far away from large-scale explosions. As one example, let us consider the ground-based magnetic and electric measurements made at distances 1.5–5.5 km from the explosion point during a series of industrial explosions in Khorezmskiy region of the USSR (Anisimov et al. 1985). It was found that the seismic, electric, and magnetic perturbations take place practically simultaneously. The total duration of both seismic and electromagnetic signals was about 7–11 s. Additionally, the frequencies of the electromagnetic and seismic vibrations are about 1 Hz and correlate with each other. Typically, the magnitudes of the electric field vary within 1–10 μV/m while the magnetic field variations are about several nT.

The similar ULF phenomena have been observed during underground nuclear explosions (e.g., see the paper by Sweeney (1989)). The measurements were made at the ground-recording stations located at the distances 5 and 10 km from the epicenter of explosion. At the first station the electromagnetic signal has been detected 1.5 s after the explosion and it has been observed at the second station after 3 s. The amplitudes of the magnetic and electric components exceed hundreds pT and tens μV/m, respectively. These signals are usually observed at the background of the so-called electromagnetic pulse (EMP) appearing practically simultaneously with the explosion moment. In Chap. 11 we show that the EMP is caused by products of nuclear detonation gamma quantum emission and so on. In this picture, the time lag of the recorded signal for several seconds can be explained by the action of another mechanism such as the local GMPs or seismoelectric effect because the beginning of both the effects can be associated with the moment of seismic wave arrival at the ground-recording stations.

Nevertheless such interpretation of the phenomena should be made with some care because there is no information on how the ground shaking due to seismic vibration affects the sensor operation. In the above experiments (Sweeney 1989) the magnetic and electric sensors were embedded in the rock, so that the factor of the rock vibrations should be taken into account for correct interpretation of the experimental data. Here we mean the so-called microphone effect, vibration of transistors, ferrite cores, changes of the effective capacity and inductance of the electric circuits of the instruments and etc. This problem has been discussed in more detail by Eleman (1965).

In the last decades some empirical evidence of the electromagnetic signals possibly associated with EQs has been reported by Eleman (1965) and Belov et al. (1974). These signals lie in the ULF band, i.e., 0.1–1 Hz or lower, which is close to typical frequencies of the seismic waves radiated by EQs. The co-seismic signals have been observed practically simultaneously with the moment of the seismic wave arrival (Nagao et al. 2000; Skordas et al. 2000; Huang 2002) or within several minutes around this moment (Takeuchi et al. 1997; Molchanov and Hayakawa 2008). In the early study it was supposed that such low-frequency variations can be carried out by seismic waves at the distances of thousands kilometers from the EQ hypocenters (Gogatishvili 1983, 1984; Sorokin and Fedorovich 1982). Much more treatise must be done to either validate or disprove this hypothesis because of a small value of signal-to-noise ratio (Surkov 2000a). It seems more plausible that tsunamis are able to give rise a detectable value of the ULF fields due to the fact that the magnitude of tsunami and the conductivity of the sea-water are much greater than the magnitude of the seismic wave and the ground conductivity, respectively (Pavlov and Sukhorukov 1987; Gershenzon and Gokhberg 1992; Chave and Luther 1990).

7.2.2 Basic Equations

The upper layer Earth's crust is a good conductor despite low conductivity of dry rocks. Basically, this is because of the presence of groundwater in pores and channels that occur up to several km depth. The ground conductivity varies with depth and regions. In the range of seismic frequencies the mean conductivity of continental crust (the layer of sedimentary rocks) is about $\sigma \sim 10^{-2}\text{--}10^{-3}$ S/m and the dielectric permittivity of the ground is about $\varepsilon \sim 10$.

To model the conducting layers of the Earth's crust, consider first a conductive elastic space with the specific conductivity σ immersed in an external constant geomagnetic field with induction $\mathbf{B}_0$. Actually the value and inclination angle of the Earth's magnetic field depend on magnetic latitude. In what follows we consider an acoustic/seismic wave propagating in this conducting medium. The field of the mass velocities $\mathbf{V} = \mathbf{V}(\mathbf{r}, t)$ is supposed to be a given function of the position vector $\mathbf{r}$

and time t. This indicates that variations of acoustic pressure are much greater than those of magnetic pressure due to perturbations of the external magnetic field $\mathbf{B}_0$ so that the electromagnetic perturbations do not influence acoustic ones. In practice this is a good approximation, which can be applied to real seismic process (Kaliski 1960; Kaliski and Rogula 1960; Viktorov 1975, 1981).

There is an exotic case which deserves mention, that is, as the magnitude of the magnetic perturbation amounts to so large a value that the magnetic pressure become comparable with the acoustic one. In this extreme case an MHD wave, which depends on acoustic and electromagnetic parameters, can propagates as it follows from hydrodynamical and Maxwell equations. A number of researchers have assumed that such MHD modes can appear in the vicinity of the Earth kernel boundary (Knopoff 1955; Keilis-Borok and Monin 1959), where the elastic and magnetic pressures may be much stronger than that at the ground surface.

To study the generation and propagation of electromagnetic field in the Earth's crust due to motion of the conductive layer of the ground, we need Maxwell's electrodynamics equations, which are given by Eqs. (1.1)–(1.4). First of all let us compare the displacement current with the conduction one in the conducting ground. The displacement current can be estimated as $|\partial_t \mathbf{D}| \sim \varepsilon \varepsilon_0 \omega E$, where ε is the dielectric permittivity of the ground. This current is small compared to the conduction one if $\sigma E \gg \varepsilon \varepsilon_0 \omega E$. The last inequality is valid if $\omega \ll \sigma / (\varepsilon \varepsilon_0) \sim 10^7$–$10^8$ Hz. This means that in the seismic frequency range (0.1–10 Hz) the displacement current is much smaller than the conduction one.

Considering seismic waves travelling through ground-based stations, we note that in practice the sensors of acoustic and electromagnetic fields are situated in the reference frame, K', moving with the mass velocity $\mathbf{V}$ of the ground. As a first step we start with consideration of the electromagnetic fields in an earth-fixed or motionless coordinate system, K. In such a case Maxwell equation (1.1) is reduced to the form given by Eq. (1.9).

Let $\delta \mathbf{B}$ be the perturbation of the Earth's/external magnetic field $\mathbf{B}_0$, so that $\mathbf{B} = \mathbf{B}_0 + \delta \mathbf{B}$. Usually the GMPs caused by the seismic waves are as much as 1–100 nT, that is much smaller than the typical value of the undisturbed geomagnetic field $\mathbf{B}_0 \approx 50\,\mu\text{T}$. We consider therefore only weak perturbations, i.e., $|\delta \mathbf{B}| \ll |\mathbf{B}_0|$. Hence the last term on the right-hand sides of Eq. (1.9) can be reduced to the form $\mathbf{V} \times \mathbf{B} = \mathbf{V} \times (\mathbf{B}_0 + \delta \mathbf{B}) \approx \mathbf{V} \times \mathbf{B}_0$. So, the basic equations of the problem are given by

$$\nabla \times \delta \mathbf{B} = \mu_0 \sigma \left(\mathbf{E} + \mathbf{V} \times \mathbf{B}_0 \right), \tag{7.7}$$

$$\nabla \times \mathbf{E} = -\partial_t \delta \mathbf{B}. \tag{7.8}$$

To find the variations of magnetic induction $\delta \mathbf{B}$ we first divide Eq. (7.7) by σ and then take the cross product of the operator ∇ with both sides of Eq. (7.7). Using

$$\nabla \times ((1/\sigma)\, \nabla \times \delta \mathbf{B}) = \nabla\,(1/\sigma) \times (\nabla \times \delta \mathbf{B}) + (1/\sigma)\, \nabla \times (\nabla \times \mathbf{B})$$

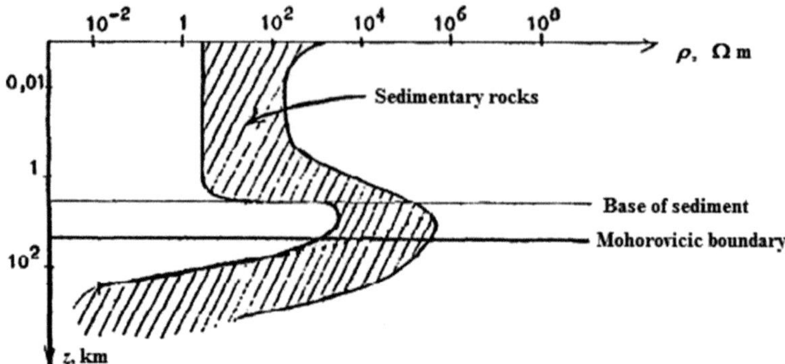

Fig. 7.4 Variations of the ground-specific resistance with depth as observed in the continental landmass. Adapted from Schwarz (1990)

and

$$\nabla \times (\nabla \times \delta\mathbf{B}) = \nabla (\nabla \cdot \delta\mathbf{B}) - \nabla^2 \delta\mathbf{B}$$

one obtains

$$\left(\nabla \frac{1}{\sigma}\right) \times (\nabla \times \delta\mathbf{B}) + \frac{1}{\sigma} \left(\nabla (\nabla \cdot \delta\mathbf{B}) - \nabla^2 \delta\mathbf{B}\right) = \mu_0 \left[\nabla \times \mathbf{E} + \nabla \times (\mathbf{V} \times \mathbf{B}_0)\right].$$

$$(7.9)$$

Combining Eqs. (7.7), (7.8), and (7.9) we finally come to

$$\partial_t \delta\mathbf{B} = \frac{1}{\mu_0\sigma} \nabla^2 \delta\mathbf{B} + \frac{1}{\mu_0} (\nabla \times \delta\mathbf{B}) \times \left(\nabla \frac{1}{\sigma}\right) + \nabla \times (\mathbf{V} \times \mathbf{B}_0). \qquad (7.10)$$

This is the so-called quasi-stationary equation for magnetic field that can be applied to the moving conductive medium. For example, this equation can describe the co-seismic phenomena in the earth-fixed reference frame.

As a solution of Eq. (7.10), i.e., $\delta\mathbf{B}$, is known, the electric field, $\mathbf{E}$, in the motionless reference frame can be found from Eq. (7.7)

$$\mathbf{E} = \nu_m (\nabla \times \delta\mathbf{B}) - \mathbf{V} \times \mathbf{B}_0. \qquad (7.11)$$

The Earth's conductivity at the continents varies with depth as schematically shown in Fig. 7.4. The conductivity of the upper sedimentary rock is higher than that of lower layers of rock basically due to the presence of groundwater. The enhancement of the rock conductivity at the higher depth is believed to be due to the increase of the ionic conductivity of the rock, which in turn results from the increase of the pressure and temperature with depth.

Now we consider the seismic waves propagating in the upper layers of the Earth's crust. To simplify the problem, we shall ignore the conductivity variations with depth as well as the local inhomogeneities by considering the case of a homogeneous medium with constant conductivity σ. This means that the second term on the right-hand side of Eq. (7.10) is equal to zero, so that we arrive at the following equation, which has the form similar to Eq. (1.14),

$$\partial_t \delta\mathbf{B} = \nu_m \nabla^2 \delta\mathbf{B} + \nabla \times (\mathbf{V} \times \mathbf{B}_0), \tag{7.12}$$

where $\nu_m = (\mu_0 \sigma)^{-1}$ stands for the coefficient of magnetic diffusion or magnetic viscosity in the conducting ground. This coefficient is measured in m^2/s.

Equation (7.12) is referred to as the diffusion-type equations. Here the second term on the right-hand side of equation plays a role of the source of the GMPs. In the case of ideal magnetohydrodynamics, that is, the case of perfectly conductive medium when $\sigma \to \infty$, Eq. (7.12) reduces to the form similar to Eq. (1.18)

$$\partial_t \delta\mathbf{B} = \nabla \times (\mathbf{V} \times \mathbf{B}_0). \tag{7.13}$$

This means that the magnetic field lines are "frozen in a conducting medium" and can be considered to move with the medium. This point is studied in any detail in Sect. 1.1.3. If the conductivity is finite, the magnetic field perturbations can diffuse through the conducting medium.

More usually we measure the electric and magnetic field variations in the reference frame moving together with ground and instruments, that is at the background of vibrations caused by the seismic wave propagation. One can use Eqs. (1.6)–(1.8) for transformation to the moving reference frame. Substitution Eq. (7.11) for $\mathbf{E}$ into Eq. (1.6) gives the electric field $\mathbf{E}'$ in the moving reference frame

$$\mathbf{E}' = \nu_m \left(\nabla \times \delta\mathbf{B}\right). \tag{7.14}$$

These equations should be supplemented by the proper boundary conditions at the ground–atmosphere interfaces.

7.2.3 Diffusion and Seismic Zones

The EQs, volcano eruptions, and underground explosions are the sources of the most intense seismic waves propagating in the Earth's crust. The seismic waves radiated by the underground acoustic sources can interact with rock inhomogeneities and ground surface forming a variety of scattering waves. The net field of the elastic displacements is very complex because it includes the primary longitudinal and transverse waves coming from the source and a number of reflected waves. Among them are the surface wave modes such as Rayleigh and Love modes (Love 1911), which are formed at the large distance (Aki and Richards 2002; Scholz 1990). In spite of complexity of the seismic wave field, modern seismometers are capable of separating and distinguishing these waves at teleseismic distances due to the different properties of the longitudinal, transverse, and surface waves such as polarizations and different speed of propagation.

Here we start with qualitative estimations and analysis of Eq. (7.12). For simplicity consider only a longitudinal seismic wave propagating in the homogeneous elastic wave with constant velocity C_l. Far away from the seismic source the shape of the longitudinal wave front is close to spherical one. This approximation is valid for a distance, which is much greater than a typical size of the explosion chamber or earthquake focal zone. In this notation we can consider a point seismic source which radiates the longitudinal seismic wave followed by the GMPs. If the seismic event happens at the origin of coordinate system at the moment $t = 0$, then the radius, r_l, of the spherical wave front increases with time t as

$$r_l \sim C_l t. \tag{7.15}$$

As we have noted above, the GMPs and terrestrial electric currents can propagate in the ground due to diffusion in the conductive medium. To estimate the characteristic velocity and the size of the region covered by diffusion process, we compare the left-hand side of Eq. (7.12) and the first term on the right-hand side of this equation by the order of amplitude

$$\frac{\delta B}{t} \sim v_m \frac{\delta B}{r_d^2}. \tag{7.16}$$

Here r_d is the characteristic distance at which the diffusion front spreads for the characteristic time t. It follows from Eq. (7.16) that $r_d \sim (v_m t)^{1/2}$. A more accurate analysis produces the similar relationship which differs from above estimate by a factor of 2 (Surkov 1989a,b, 2000a), i.e.,

$$r_d \sim 2 (v_m t)^{1/2}. \tag{7.17}$$

The diffusion front serves as the effective boundary surface between the inner region covered by the diffusion process and the rest medium surrounding the perturbed region. In the diffusion region the GMP reaches a peak value at least at the initial stage of the process. It should be noted that there is not a distinct surface of separation between these two regions. However the boundary is abrupt enough so that outside the diffusion zone the GMP decreases exponentially with distance, say, as $\exp\left(-r^2/r_d^2\right)$ or anything else. The effective velocity of the diffusion front propagation is

$$\frac{dr_d}{dt} \sim \left(\frac{v_m}{t}\right)^{1/2}. \tag{7.18}$$

At the initial moments, while as $t \to 0$, the velocity of the GMP and electric current diffusion is much greater than the seismic/acoustic velocity C_l. This means that

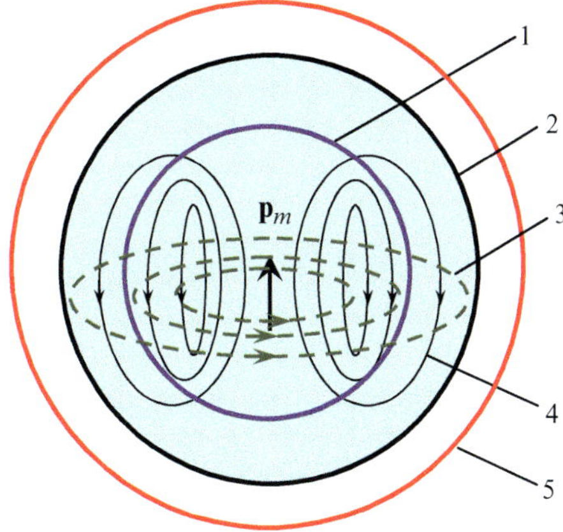

Fig. 7.5 A physical mechanism of geomagnetic field perturbations due to the seismic wave propagation in a conducting rock. *1*—seismic wave front $r_l = C_l t$, *2*—diffusion front $r = r_d$ confining the region where electromagnetic perturbations occur, *3*—electric current lines, *4*—magnetic field variation lines, *5*—radius r_* at which the seismic wave comes up the diffusion front

the GMP and currents propagate ahead of the acoustic wave front. The diffusion velocity by Eq. (7.18) falls off with time and distance whereas the acoustic velocity keeps constant, and so at certain moment the acoustic wave comes up the diffusion front. Equating the radii r_l and r_d in Eq. (7.15) and (7.17), we can estimate this moment

$$t_* \sim \frac{4v_m}{C_l^2} = \frac{4}{\mu_0 \sigma C_l^2}. \tag{7.19}$$

Taking into account that the radii of the acoustic and diffusion fronts are equal at the distance $r_* \sim C_l t_*$ yields

$$r_* \sim \frac{4}{\mu_0 \sigma C_l}. \tag{7.20}$$

In the region $r < r_*$, referred to the diffusion zone (Surkov 1989a,b, 2000a), the diffusion-type propagation of the GMP and terrestrial electric currents is thus predominant. While the region $r > r_*$ is called the seismic zone.

The diffusion and seismic zones are sketched in Fig. 7.5. Below we will show that the effective magnetic moment **M** of electric currents is directed oppositely to the vector **B**$_0$ of the Earth's magnetic field.

The ground conductivity usually varies from 10^{-2} S/m for the upper layer of sedimentary rocks (1–2 km of the depth) to 10^{-3} S/m for the basement rocks (lower boundary is about 10 km depth). The longitudinal seismic wave velocity changes within 4.5–6.2 km/s (granite, marble, basalt, limestone) (Babichev et al. 1991). Using these parameters we obtain an estimate for the diffusion zone size: $r_* \sim 50$–700 km and the time scale $t_* \sim 8$–160 s.

At the time $t > t_*$ and the distance $r > r_*$ other tendency begins to prevail. At this stage the GMPs are generated in the vicinity of the seismic wave front and propagate together with the wave front at the velocity C_l. In such a case the wave front, where the mass velocity reaches a peak, plays a role of the effective moving source, which "radiates" the GMP in all directions. The region $r > r_*$ is referred to as seismic/acoustic zone. We shall consider these zones separately because the attenuation factor and other properties of the GMP in these two zones are different.

7.2.4 Electromagnetic Forerunner of Seismic Wave

Considering Eq. (7.12), we note that the term $\nabla \times (\mathbf{V} \times \mathbf{B}_0)$, playing a role of the GMP source, is nonzero only behind the seismic wave front, i.e., in the region $r < r_l$ covered by seismic wave. Nevertheless, due to the diffusion the GMP and electric currents penetrate into the region ahead of seismic wave front where the medium is at rest. Certainly, the radial field diffusion in the direct and reverse directions is developed nonsymmetrically first because of the geometrical factor and second due to the motion of diffusion source, i.e., the acoustic wave. This fact is analogous, in part, to the known Doppler effect arisen from the motion of the radiation source. We remind that the characteristic wavelength in the direction of the radiation source motion is shortened and vice versa it becomes longer in the reverse directions. A similar effect can take place in our case in spite of the fact that the diffusion and radiation are quite different processes.

The accumulation of the GMP ahead of acoustic wave results in the formation of the so-called electromagnetic forerunner of the acoustic/seismic wave that propagates in front of seismic wave (Surkov 1989a). Here we can only estimate the characteristic size of the electromagnetic forerunner. If the decrease in the amplitude due to the divergence of acoustic energy in the spherical wave is ignored, the quasistationary stage occurs so that the characteristic length and duration of the forerunner keep approximately constant. In such a case the forerunner speed C_l has to be on the same order of magnitude as that of the diffusion front. Equating the diffusion velocity (7.18) to C_l gives the characteristic duration, $t_m \sim v_m/C_l^2$, and size, $\lambda_m \sim v_m/C_l \sim 13$–$180$ km, of the electromagnetic forerunner. It is not a surprise that λ_m is on the same order of magnitude as the diffusion zone size r_* and $t_m \sim t_*$.

It follows from above estimate that the electromagnetic forerunner can propagate not far as several seconds ahead of seismic waves. In the case of a solitary compression impulse, the forerunner signal increases in time up to the moment of the seismic wave arrival, and then the signal changes the sign (Surkov 1997a,b). It appears that the electromagnetic forerunner can be detected only at short distances from the earthquake epicenter or explosion point. For example, the theory predicts that at the distances about tens kilometers from the earthquake epicenter the forerunner amplitude decreases to the value from several pT to 1 nT for magnetic component and from several nV/m to 1 μV/m for electrical component, correspondingly.

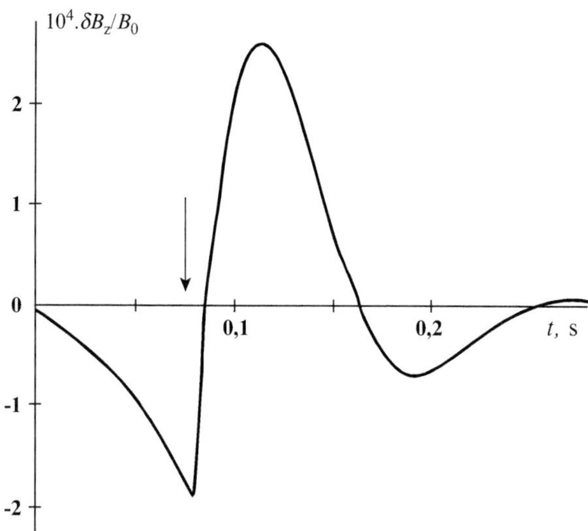

Fig. 7.6 Normalized vertical component of the GMPs resulted from the longitudinal seismic wave propagation. The numerical calculations are made for a distance 500 km from the seismic source (Surkov 1989b). The ground conductivity is taken as $\sigma = 1$ S/m. The *arrow* shows the moment of seismic wave arrival at the ground-recording station

The typical time-dependence of the co-seismic signal calculated by Surkov (1989b) for the seismic zone is shown in Fig. 7.6. Detonation on the ground surface was served as a source of seismic waves. The calculations were performed for a distance 500 m from a point of the detonation. The arrow corresponds to the moment of seismic wave arrival at the ground-recording station. The magnetic forerunner is seen in the initial part of the signal before the arrow.

Notice that, in practice, the signal caused by the detonation of high explosive charges on the ground surface can be much more complicated due to the effect of electric charges and currents appearing inside dust clouds and explosive products (e.g., see Soloviev and Surkov 2000; Soloviev et al. 2002). Aerial shock wave propagation is accompanied by the perturbation of heavy ions and charged aerosol densities in the atmospheric surface layer which in turn results in local perturbations of the Earth' electric field (Soloviev and Surkov 1994). One more effect is due to an impact of the aerial shock wave (SW) on instruments.

7.2.5   Estimates of Typical Amplitude and Frequencies of Co-seismic Signals

As is seen from Fig. 7.6, the signal is not a monotonic one behind the front of seismic wave. This part of the electromagnetic signal is similar, in character, to a pattern of vibrations caused by seismic wave. It is usually the case that the frequency and the characteristic size/wavelength of the electromagnetic vibrations are close to those observed in seismic waves, although the electromagnetic signals are shifted in phase with respect to seismic signals. A peak of the electromagnetic signal should be expected in the vicinity of the seismic front where the mass velocity of the medium amounts to the maximum value.

Consider first the limit when $|\partial_t \delta \mathbf{B}| \ll \nu_m |\nabla^2 \delta \mathbf{B}|$ that means that the diffusion of the electromagnetic perturbations dominates over the effect of "frozen-in" magnetic field lines. Let λ_a and $k_a = 2\pi/\lambda_a$ be the characteristic size/wavelength and acoustic wave number of the seismic wave, respectively. Using these parameters, the above inequality reads as

$$\omega |\delta \mathbf{B}| \ll \nu_m |\delta \mathbf{B}| k_a^2. \tag{7.21}$$

Substituting the frequency of the acoustic wave $\omega = k_a C_l$ into Eq. (7.21), we obtain

$$f \gg C_l^2 / (2\pi \nu_m) = \mu_0 \sigma C_l^2 / (2\pi), \tag{7.22}$$

where the frequency f is related to the cyclic frequency ω via $f = \omega/(2\pi)$. Taking numerical values $C_l = 5$ km/s and $\sigma = 10^{-3}$–10^{-2} S/m we find that inequality (7.22) keeps as $f \gg (5$–$50)$ mHz.

In the same manner we can estimate the source function; that is, the last term on the right-hand side of Eq. (7.12) by the order of magnitude

$$|\nabla \times (\mathbf{V} \times \mathbf{B}_0)| \sim V_0 B_0 k_a, \tag{7.23}$$

where V_0 denotes the magnitude of the mass velocity. So, neglecting the term on the left-hand side of Eq. (7.12) and equating the terms on the right-hand side of this equation gives

$$\nu_m |\delta \mathbf{B}| k_a^2 \sim V_0 B_0 k_a. \tag{7.24}$$

The value $|\delta \mathbf{B}|$ in Eq. (7.24) can be considered as the order-of-magnitude estimation of the perturbation of the Earth's magnetic field caused by seismic wave propagation. So we finally come to

$$\delta B_{\max} \sim \frac{V_0 \lambda_a B_0}{2\pi \nu_m} \sim \frac{V_0 C_l B_0}{2\pi f \nu_m}. \tag{7.25}$$

Thus, such parameters as the mass velocity, seismic frequency, and ground conductivity may greatly affect the magnitude of magnetic perturbations. For example, taking a typical seismic frequency $f = 1$ Hz, $V_0 = 1$–10 cm/s, $B_0 = 5 \times 10^{-5}$ T and the abovementioned values of the parameters we obtain that $\delta B_{\max} \sim 3$–30 pT.

The above estimate for magnetic field variations is valid not only for the earth-fixed, that is, for the motionless reference frame but also for the reference frame moving together with instruments. The electric field in the latter reference frame can be found from Eq. (7.14)

$$E'_{\max} \sim k_a v_m \delta B_{\max} \sim V_0 B_0, \tag{7.26}$$

that is, given above parameters, we obtain that $E'_{\max} \sim 0.5\text{--}5\ \mu\text{V/m}$.

The electric field magnitude, $E_{\max}$, in the earth-fixed reference frame cannot be estimated from Eq. (7.11) because both terms on the right-hand side of Eq. (7.11) are of the same order of magnitude, which means that substitution of Eq. (7.25) in this equation results in the cancellation of these terms. In this notation Eq. (7.8) is best suited for our purpose. It follows from this equation that $k_a E_{\max} \sim \omega \delta B_{\max}$. Hence

$$E_{\max} \sim C_l \delta B_{\max} \sim \frac{V_0 C_l^2 B_0}{2\pi f v_m}. \tag{7.27}$$

Comparing the electric fields in both reference frames, we obtain

$$\frac{E'_{\max}}{E_{\max}} \sim \frac{2\pi f v_m}{C_l^2} \gg 1. \tag{7.28}$$

The electric field in the moving frame is thus greater than that in the earth-fixed frame.

Now we shall consider an opposite case and derive alternative estimates of the electromagnetic perturbations. In the low-frequency limit, i.e., $f \ll \mu_0 \sigma C_l^2 / (2\pi)$, the inequality (7.21) must be replaced by reverse one.

In such a case the first term on the right-hand side of Eq. (7.12) can be omitted and this equation, in fact, would take the form of Eq. (7.13). As has already been stated, this case corresponds to the "frozen in" magnetic field. For this frequency band, in a similar manner, one can find the following estimation for the GMP

$$\delta B_{\max} \sim \frac{V_0}{C_l} B_0, \tag{7.29}$$

$$E_{\max} \sim V_0 B_0, \tag{7.30}$$

$$E'_{\max} \sim \frac{2\pi v_m f}{C_l^2} V_0 B_0 \ll E_{\max}. \tag{7.31}$$

Equation (7.31) shows that the electric field $E_{\max}$ in a reference frame fixed to the earth is greater than that in the moving reference frame, in contrast to the inequality (7.28). Moreover, as is seen from Eqs. (7.29)–(7.31), the amplitudes of magnetic and electric perturbations are independent of the frequency. We are reminded that all the estimations are valid only for the ULF range, i.e., for $f \ll (5\text{--}50)$ mHz or $\lambda_a \gg 10^2\text{--}10^3$ km. In practice, we encounter such frequencies and wavelengths extremely seldom.

To summarize, we note that all the electromagnetic amplitudes given by Eqs. (7.25)–(7.27) and (7.29)–(7.31) are proportional to the amplitude of the medium mass velocity V_0. Hence we conclude that the attenuation of the GMP in the seismic zone is the same as that of the mass velocity. For example, once the amplitude of the longitudinal elastic wave radiated by the spherical seismic source decreases inversely proportional to the distance r, the amplitudes of the magnetic and electric perturbations originated from the seismic wave propagation decrease in a similar manner, i.e., as r^{-1}. This conclusion is not valid for non-perfect elastic or dissipative media, which require a special consideration (Dunin and Surkov 1992).

7.2.6 Spherically Symmetric Seismic Wave

Far away from the source a seismic longitudinal wave can be considered as approximately spherically symmetric one at least as a first approximation. This approach is applied to longitudinal waves radiated from expansion of the spherical pore in the ground and from an underground explosion at least up to the moment of the wave reflection from the boundary with atmosphere. The EQ focal zone, as is often the case, is equivalent to a dislocation/shear crack, whose seismic radiation is nonsymmetric (Aki and Richards 2002). In the analysis that follows, we consider, however, the spherically symmetric wave propagating in a homogeneous medium in order to analyze a rigorous solution of the problem, which are valid for both the diffusion and acoustic zones.

So, let us consider an infinite homogeneous conductive medium immersed in the constant magnetic field with induction $\mathbf{B}_0$. A spherically symmetric source begins to radiate a longitudinal acoustic/seismic wave at the moment $t = 0$. The perturbations, $\delta\mathbf{B}$, of the external magnetic field caused by the acoustic wave propagation are supposed to be small, so that $|\delta\mathbf{B}| \ll |\mathbf{B}_0|$ as before. In such a case the magnetic perturbation $\delta\mathbf{B}$ and the electric field $\mathbf{E}$ in an earth-fixed reference frame are defined by Eqs. (7.11) and (7.12). The origin of the coordinate system is chosen to be in the center of symmetry of the acoustic source. We use the coordinate system whose z axis is directed along the external field $\mathbf{B}_0$. We apply spherical coordinates r, θ, and ϕ, where the polar angle θ is measured from the z axis and the azimuthal angle ϕ is measured from the positive direction of x axis. The mass velocity of the medium is radially directed, i.e., $\mathbf{V} = V(r, t)\,\hat{\mathbf{r}}$, where $\hat{\mathbf{r}}$ is the radially directed unit vector. In such a case all functions are independent of the azimuthal angle ϕ. Taking into account the symmetry of the problem, only three components of the electromagnetic perturbations are nonzero, namely δB_r, δB_θ, and E_ϕ. Using the spherical coordinates, Eqs. (7.11) and (7.12) can be thus rewritten as (Landau and Lifshits 1982)

$$\partial_t \delta B_r = \frac{\nu_m}{r^2} \left\{ \partial_r \left(r^2 \partial_r \delta B_r \right) - 2\delta B_r \right.$$

$$\left. + \frac{1}{\sin\theta} \partial_\theta \left[\sin\theta \left(\partial_\theta \delta B_r - 2\delta B_\theta \right) \right] \right\} - \frac{2VB_0 \cos\theta}{r}, \qquad (7.32)$$

$$\partial_t \delta B_\theta = \frac{\nu_m}{r^2} \left\{ \partial_r \left(r^2 \partial_r \delta B_\theta \right) + \frac{1}{\sin \theta} \partial_\theta \left(\sin \theta \partial_\theta \delta B_\theta \right) \right.$$

$$\left. - \frac{\delta B_\theta}{\sin^2 \theta} + 2 \partial_\theta \delta B_r \right\} + \frac{B_0 \sin \theta}{r} \partial_r \left(rV \right), \tag{7.33}$$

$$E'_\phi = \frac{\nu_m}{r} \left\{ \partial_r \left(r \delta B_\theta \right) - \partial_\theta \delta B_r \right\}. \tag{7.34}$$

where ν_m is the magnetic diffusion coefficient, and ∂_t and ∂_r denote partial derivatives with respect to time and radius, respectively. The mass velocity, $V = V(r,t)$, is supposed to be a given function.

The set of Eqs. (7.32)–(7.34) should be supplemented by the proper initial and boundary conditions at the infinity and at the origin of the coordinate system. At the initial moment $t = 0$ there is a uniform constant magnetic field $\mathbf{B}_0$ and therefore the initial conditions are $\delta B_r(r,0) = \delta B_\theta(r,0) = 0$ and $E'_\phi(r,0) = 0$. All the functions have to tend to zero as $r \to \infty$, besides they have to be finite as $r \to 0$.

A rigorous solution of this problem is found in Appendix G. The components of the electromagnetic perturbations are given by

$$\delta B_r = \frac{B_0 \cos \theta}{r^3} \int_0^r r'^2 g_1\left(r',t\right) dr', \tag{7.35}$$

$$\delta B_\theta = \frac{B_0 \sin \theta}{2} \left(\frac{1}{r^3} \int_0^r r'^2 g_1\left(r',t\right) dr' - g_1\left(r,t\right) \right), \tag{7.36}$$

$$E'_\phi = -\frac{\nu_m B_0 \sin \theta}{2} \partial_r g_1\left(r,t\right), \tag{7.37}$$

where the function g_1 can be written as

$$g_1\left(r,t\right) = \frac{1}{\left(\pi \nu_m\right)^{1/2} r} \int_0^t \frac{dt'}{\left(t-t'\right)^{1/2}} \int_0^\infty \left\{ \exp\left(-\frac{\left(r+r'\right)^2}{4\nu_m\left(t-t'\right)} \right) - \right.$$

$$\left. \exp\left(-\frac{\left(r-r'\right)^2}{4\nu_m\left(t-t'\right)} \right) \right\} \left(r'\partial_{r'} V + 2V \right) dr'. \tag{7.38}$$

This solution can be applied for an arbitrary function $V = V(r,t)$.

Large-scale seismic sources such as crust and deep-focus EQs and underground explosions differ essentially from one another by the duration, spectrum, and etc. During powerful underground explosions the most part of the seismic wave energy is radiated by a crushing wave moving at supersonic velocity in the ground (e.g., see Chadwick et al. 1964; Rodionov et al. 1971). Due to the large mechanical stress behind the front of the crushing wave, the rock fails in shock compression and is broken to pieces and fragments. The speed of such a wave decreases gradually owing to dissipation of the mechanical energy, and the crushing wave stops abruptly as the stress at the front falls off below the crushing strength of the rock. For nuclear underground explosions the final radius of the crushing wave or the so-called radius of the crushing zone is of the order of several tens or hundreds meters that may characterize a size of seismic source associated with the underground explosion.

A peculiarity of EQs is that the shear waves make larger contribution to the total seismic energy as compared to the underground explosions. In the case of earthquakes the effective seismic radiator is situated at the depth of several tens or hundred kilometers and its radius is determined by the typical scale of a focal zone. For the EQs with magnitudes $M > 6$ the typical size of the seismic source exceeds 10 km (Scholz 1990).

We next derive an order-of-magnitude estimate of the mass velocity for the case of a large-scale seismic source. The acoustic wave equation for a uniform elastic medium can be written as (Landau and Lifshits 1987)

$$\partial_t^2 \mathbf{u} = C_l^2 \nabla (\nabla \cdot \mathbf{u}) - C_t^2 \nabla \times (\nabla \times \mathbf{u}), \tag{7.39}$$

where $\mathbf{u}$ is the vector of medium displacement, and C_l and C_t are the longitudinal and transverse sound velocities. Let us consider a spherically symmetric acoustic source with a radius R_0. The source radiates the longitudinal acoustic wave in which the displacement vector $\mathbf{u}$ has only radial component $u_r = u_r(r, t)$.

A general solution of this problem in the form of outgoing wave is found in Appendix G. This solution can be expressed through the so-called normalized potential of the elastic displacement $f = f(t_1)$ which depends on the variable $t_1 = t - (r - R_0)/C_l$. The mass velocity, $\mathbf{V}(r, t) = \partial_t u_r(r, t)\,\hat{\mathbf{r}}$, is represented by the potential as follows:

$$\mathbf{V} = \frac{R_0^3 \hat{\mathbf{r}}}{r C_l} \left(\partial_t^2 f(t_1) + \frac{C_l}{r} \partial_t f(t_1) \right). \tag{7.40}$$

To find the explicit expression of the normalized potential we require the boundary and initial conditions. At first consider the conventional formulation of the problem. The radial displacement of the medium at the radius $r = R_0$ is supposed to be a given function, i.e., $u_r(R_0, t) = u(t)$. At the initial moment $u_r(r, 0) = 0$ everywhere and $u(0) = 0$ and $\partial_t u(0) = 0$. In such a case the normalized potential is given by

$$f(t) = \frac{C_l}{R_0^2} \int_0^t u(t') \exp\left[\frac{C_l}{R_0}(t' - t) \right] dt'. \tag{7.41}$$

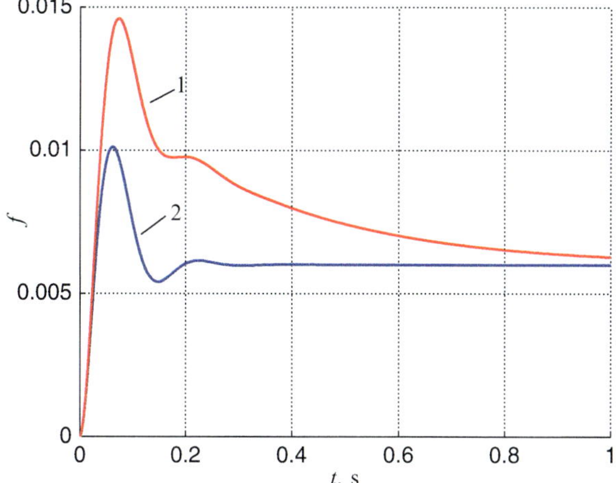

Fig. 7.7 A model calculation of temporal variations of the normalized potential of elastic displacements (Eq. (7.106)) resulted from an underground explosion. The lines 1 and 2 correspond to the values of stress relaxation time $t_r = 0.3$ s and 0.03 s, respectively

In the theory of underground explosions another formulation of the boundary problem is accepted. In general case the estimation of the mechanical effect of the underground explosion is based on the set of equations describing the fracture process and the motion of broken rocks in the vicinity of the explosion and together with equations for dynamics of the surrounding elastic media. The method of solving such a problem is rather complicated and we do not go into details. The interested reader is referred to the texts by Chadwick et al. (1964) and Rodionov et al. (1971) or other books for a more complete treatise on mechanical processes associated with underground explosions.

In the simple model the radial component of the stress tensor at the boundary of the crushing zone is considered as a given function which depends on mechanical strength of the rock, depth of explosion, and other parameters. For example, the temporal variations of the radial stress at the boundary $r = R_0$ can be taken in the following form:

$$s_{rr}\left(R_0, t\right) = -\left[P_0 + \left(P_* - P_0\right)\exp\left(-t/t_r\right)\right], \tag{7.42}$$

where P_* stands for the constant of the order of tensile or compression strength, P_0 is approximately equal to the lithostatic pressure at the depth of the explosion and t_r is the stress relaxation time. Notice that the minus sign in Eq. (7.42) corresponds to the compression stress.

The components of the stress tensor can be expressed through the radial displacement at the boundary $r = R_0$. The solution of this problem is found in Appendix G. In Fig. 7.7, we plot the time dependence of the normalized potential given by Eq. (7.106). This function has a sharp front followed by damped vibrations

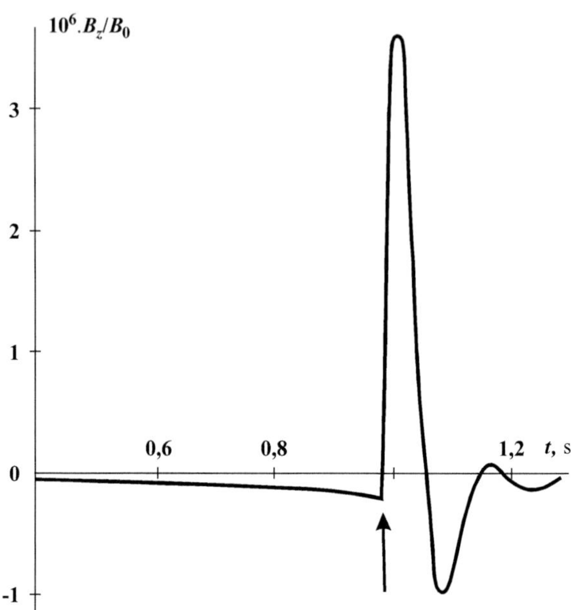

Fig. 7.8 Normalized vertical component of the GMPs resulted from the longitudinal seismic wave propagation (Surkov 1989b). The *arrow* shows the moment of seismic wave arrival at the ground-recording station. The numerical calculations are made for a distance 5 km from the seismic source

with period of several tenth of seconds. The normalized potential of the elastic displacements shown in Fig. 7.7 was used for calculation of the mass velocity (7.40) and the electromagnetic perturbations (7.35)–(7.37) caused by the underground explosion (Surkov 1989b). The numerical calculations are based on the following parameters: $R_0 = 100$ m is radius of the crushing zone, $P_* = 5 \cdot 10^8$ Pa is the rock compression strength, $P_0 = 1.5 \cdot 10^8$ Pa is the lithostatic pressure at the depth of the explosion, $C_l = 5$ km/s is the velocity of longitudinal seismic wave, $\rho_m C_l^2 = 5 \cdot 10^{10}$ Pa, $\gamma = C_t^2 / C_l^2 = 0.2$ where ρ_m and C_t are the rock density and velocity of transverse wave, respectively. A plot of the vertical component of the magnetic perturbations on the ground surface in the acoustic zone is shown in Fig. 7.8. The rock and explosion parameters are the same as in Fig. 7.7. The numerical calculations are made for a distance 5 km from the seismic source. The ground conductivity is taken as $\sigma = 0.1$ S/m, that is a typical value for the upper layer of moistened sedimentary rocks. Notice that the damped vibrations, arising after the seismic wave arrival (shown with arrow), correlate in frequency with the seismic vibrations.

7.2.7 *Magneto-Dipole Approximation for the Diffusion Zone*

To achieve better understanding of the phenomenon we shall now simplify the general solution (7.35)–(7.37) for a spherically symmetric acoustic source and then extend them to the source of arbitrary shape. First of all, notice that the factor

$r\partial_r V + 2V$ in Eq. (7.38) depends on the distribution of the mass velocity, which becomes zero ahead of the acoustic wave front. So, the region of integration is restricted, in fact, by the radii within interval $R_0 < r < R_0 + C_l t$. Moreover, far away from the seismic source, when $C_l t \gg R_0$, the mass velocity V reaches a peak value within a short distance/interval near the wave front. The characteristic longitudinal scale/wavelength of this interval is supposed to be much smaller than the distance r and the diffusion radius r_d. In fact, this means the validity of the short wavelength approximation in evaluating the integral in Eq. (7.38). It follows from Appendix G that the approximate expressions for the GMP are given by (Surkov 2000a,b)

$$\delta B_r = \frac{u_0 S B_0 \cos\theta}{4\pi r} \partial_r \left(\frac{G_1}{r}\right), \tag{7.43}$$

$$\delta B_\theta = \frac{u_0 S B_0 \sin\theta}{8\pi r} \partial_r \left(\frac{G_1}{r} - \partial_r G_1\right), \tag{7.44}$$

$$E'_\phi = -\frac{u_0 S B_0 v_m \sin\theta}{8\pi} \partial_r \left(\frac{1}{r}\partial_r^2 G_1\right) \tag{7.45}$$

where

$$G_1(r,t) = \exp\frac{r_l + r}{\lambda_m}\mathrm{erfc}\left(\frac{r + 2r_l}{r_d}\right) + \exp\left(\frac{r_l - r}{\lambda_m}\right)\mathrm{erfc}\left(\frac{r - 2r_l}{r_d}\right) -$$
$$2\mathrm{erfc}\left(\frac{r}{r_d}\right) + 2\left[1 - \exp\left(\frac{r_l - r}{\lambda_m}\right)\right]\eta\,(r_l - r). \tag{7.46}$$

Here u_0 is the static displacement, S is the area of source surface, $r_d = 2\,(v_m t)^{1/2}$ is the radius of the diffusion front, $r_l = C_l t$ is the radius of the seismic wave front, and $\lambda_m = v_m / C_l$ is the length of the magnetic forerunner, $\eta\,(x)$ is the step-function, i.e., $\eta\,(x) = 1$ if $x \geq 0$, otherwise $\eta\,(x) = 0$, and $\mathrm{erfc}\,(x) = 1 - \mathrm{erf}\,(x)$ where $\mathrm{erf}\,(x)$ denotes the error function, i.e.,

$$\mathrm{erf}\,(x) = \frac{2}{\pi^{1/2}} \int_0^x \exp\left(-y^2\right) dy. \tag{7.47}$$

As is seen from Eqs. (7.43)–(7.45), all the components of electromagnetic perturbations are proportional to the value of $u_0 S$, that equals the increment of the source volume. The last value is in turn proportional to the seismic energy emitted by the source in the form of acoustic waves. So, the magnitude of electromagnetic perturbations caused by the spherically symmetric longitudinal wave is directly proportional to the seismic energy. In what follows we will show that this conclusion is valid for the sources with other, i.e., nonspherical, shapes.

The approximate analytical solution given by Eqs. (7.43)–(7.45) is more convenient for analysis than the general solution in the form of Eqs. (7.35)–(7.37). Now we use the short wave approximation for the diffusion zone ($r < r_*$), in which the diffusion front $r_d = 2\,(\nu_m t)^{1/2}$ propagates ahead of the seismic wave. The observation point is assumed to be located within interval

$$C_l t \ll r \ll r_d. \tag{7.48}$$

This implies that the electromagnetic perturbations have already reached the observation point whereas the seismic wave is still far from this point. This situation occurs for the time interval

$$r^2 / (4\nu_m) \ll t \ll r/C_l. \tag{7.49}$$

It follows from these inequalities that the arguments of the exponential and error functions included in Eq. (7.46) for the function G_1 are small in comparison with unity. Expanding these functions in terms of power series of the small parameters r_l/λ_m, r_l/r_d, r/λ_m and r/r_d and taking into account that step a function in Eq. (7.46) is equal to zero, we obtain

$$G_1 = \frac{2r_l}{\lambda_m} + \frac{\left(r^2 + r_l^2\right)}{\lambda_m^2} - \frac{8rr_l}{\pi^{1/2}\lambda_m r_d} + \cdots \tag{7.50}$$

The first term in Eq. (7.50) is the largest one while the second and the third terms are necessary for correct calculations of derivatives of the function G_1. Substituting Eq. (7.50) for G_1 into Eqs. (7.43)–(7.45) gives in the first approximation

$$\delta B_r = -\frac{u_0 S B_0 C_l^2 t \cos\theta}{2\pi \nu_m r^3}, \tag{7.51}$$

$$\delta B_\theta = \frac{\tan\theta}{2} B_r, \tag{7.52}$$

$$E'_\phi = \frac{u_0 S B_0 C_l^2 \sin\theta}{4\pi \nu_m r^2}. \tag{7.53}$$

These equations should be considered as an approximation which is valid only within distance and time intervals (7.48) and (7.49).

A simple interpretation of the above obtained results can be achieved, as we calculate the effective magnetic moment of the extrinsic currents caused by the motion of conductive medium in the region covered by seismic wave. It follows from Eq. (7.7) that the density, $\mathbf{j}_e$, of the extrinsic currents is

$$\mathbf{j}_e = \sigma\,(\mathbf{V} \times \mathbf{B}_0). \tag{7.54}$$

These currents play a role of the source function which generates the conduction current $\mathbf{j} = \sigma \mathbf{E}$ in the medium. The effective magnetic moments, $\mathbf{M}$, of the extrinsic current system is given by (7.4). Substituting Eq. (7.54) for $\mathbf{j}_e$ into Eq. (7.4) for $\mathbf{M}$ and taking the triple cross product in the integral (7.4), we obtain

$$\mathbf{M} = \frac{\sigma}{2} \int_{V'} \left(\mathbf{r} \times (\mathbf{V} \times \mathbf{B}_0) \right) dV' = \frac{\sigma}{2} \int_{V'} \left\{ \mathbf{V} (\mathbf{r} \cdot \mathbf{B}_0) - \mathbf{B}_0 (\mathbf{r} \cdot \mathbf{V}) \right\} dV'. \tag{7.55}$$

As before we use the coordinate system whose z axis is directed along the external field $\mathbf{B}_0$. In calculating the integral in Eq. (7.55), we apply spherical coordinates r, θ and ϕ, where the polar angle θ is measured from the z axis and the azimuthal angle ϕ is measured from the positive direction of x axis. A small element of volume has the form $dV' = r^2 \sin \theta dr d\theta d\phi$. Taking into account the symmetry of the problem, only z-component of the vector $\mathbf{M}$ has to be nonzero. Substituting Eq. (7.40) for $\mathbf{V} = V(r, t) \hat{\mathbf{r}}$ into Eq. (7.55), integrating over ϕ and over θ, and taking into account that $\partial_t f = -C_l \partial_r f$, yields

$$\mathbf{M} = -\mathbf{B}_0 \frac{\sigma S C_l R_0}{3} \int_0^{r_l} r \left(r \partial_r^2 f - \partial_r f \right) dr. \tag{7.56}$$

Integrating Eq. (7.56) in parts we come to

$$\mathbf{M} = -\mathbf{B}_0 \sigma S C_l R_0 \int_0^{r_l} f dr \approx -\mathbf{B}_0 u_0 S \sigma C_l^2 t. \tag{7.57}$$

As is seen from Eq. (7.57) the magnetic moment increases with time directly proportional to t. To gain better understanding of this result let us recall that the magnetic moment is proportional, first, to the magnitude of the mass velocity and, second, to the volume occupied by the extrinsic current. The magnetic field and current distribution around the acoustic source is sketched in Fig. 7.5. Actually the azimuthal currents shown in this figure with dashed lines 3 are predominantly concentrated in the vicinity of the acoustic wave front shown with line 1. Since the characteristic size or wavelength, λ_a, is supposed to be short, the volume occupied by the currents is as much as $4\pi r_l^2 \lambda_a$. Due to the fact that $r_l = C_l t$, the volume increases with time as t^2. On the other hand, far away from the seismic source the amplitude of mass velocity falls off inversely proportional to distance. Hence, at the front of seismic wave the velocity magnitude changes as r_l^{-1}, i.e., inversely proportional to t. Taking together these two factors results in an enhancement of the magnetic moment in time according to Eq. (7.57). Notice that this conclusion holds true for the time interval (7.49).

As is evident from Eq. (7.57), the effective magnetic moment $\mathbf{M}$ is directed oppositely to the vector of external magnetic field. This so-called diamagnetic effect occurs whenever the conductor moves in the external magnetic field (e.g., see Fedorovich 1969 and Landau and Lifshits 1982).

A quasistatic magnetic field originated from the magnetic moment is given by Eq. (7.5). Substituting Eq. (7.57) for $\mathbf{M}$ into Eq. (7.5) gives the expressions for components of the magnetic perturbations, δB_r and δB_θ, that just coincide with the expressions given by Eqs. (7.51) and (7.52).

So, we have shown that the magneto-dipole approximation can be applied for the diffusion zone in the distance range given by Eq. (7.48). We will make use of this approximation in the next section more than once.

In conclusion, let us estimate the magnitude of the electromagnetic perturbations in the diffusion zone. Since the magnetic perturbations in Eq. (7.51) and (7.52) increase in time, the maximum value $t \sim r/C_l$ should be taken from the time interval (7.49) in order to obtain this order-of-magnitude estimate

$$\delta B_{\max} \sim \frac{u_0 S B_0 C_l}{2\pi v_m r^2}, \quad E'_{\max} \sim C_l \delta B_{\max}. \tag{7.58}$$

In contrast to the seismic zone, the magnitude of the electromagnetic signals falls off with distance more rapidly, i.e., as r^{-2}, while in the seismic zone $\delta B_{\max} \propto E'_{\max} \propto V_0$, that is, the magnitude decreases as r^{-1}. This conclusion has been confirmed by numerical calculations (Surkov 2000b; Molchanov et al. 2002).

7.2.8 Rayleigh Surface Wave in a Conductive Half-Space

Among all types of seismic waves detected at teleseismic distances from the source the surface seismic waves have the most intense amplitude. In a theory the amplitude of primary/longitudinal and secondary/transverse seismic waves in perfectly elastic media decreases inversely proportional to distance r from the seismic source whereas the amplitude of Rayleigh surface wave varies with distance as $r^{-1/2}$ (e.g., see Aki and Richards 2002).

Consider a quasi-harmonic Rayleigh wave propagating along horizontal x-axis in a perfectly elastic half-space $z < 0$. The origin of coordinate system is placed on the boundary of the half-space. The z-axis points vertically upward while x-axis is positive parallel to the velocity C_R of the Rayleigh wave. In such a case the elements of medium move on elliptic trajectories in the vertical zx plane. The components of mass velocity is given by (e.g., Viktorov 1975)

$$V_x = \frac{V_0}{q} \left\{ \exp\left(q k_R z\right) - \frac{2qs}{1+s^2} \exp\left(s k_R z\right) \right\} \exp\left(i\psi\right), \tag{7.59}$$

$$V_z = i V_0 \left\{ \exp\left(q k_R z\right) - \frac{2}{1+s^2} \exp\left(s k_R z\right) \right\} \exp\left(i\psi\right), \tag{7.60}$$

where V_0 stands for the amplitude of the component V_z, i is imaginary unity, $k_R = \omega/C_R$ is acoustic wave number, ω is the wave frequency and $\psi = k_R x - \omega t$. Here we made use of the following abbreviations:

$$q = \left(1 - \frac{C_R^2}{C_l^2}\right)^{1/2}, \quad s = \left(1 - \frac{C_R^2}{C_t^2}\right)^{1/2}. \tag{7.61}$$

As is seen from Eqs. (7.59) and (7.60) the amplitude of the mass velocity exponentially decreases with depth. Thus, the elastic energy of Rayleigh wave is mainly concentrated in the near surface layer with thickness of the order of $(qk_R)^{-1}$ or $(sk_R)^{-1}$.

The half-space $(z < 0)$ is assumed to be a conductive medium with constant conductivity σ. In such a case the electromagnetic perturbations of the external/Earth's magnetic field $\mathbf{B}_0$ are described by Eqs. (7.11) and (7.12). In the atmosphere $(z > 0)$ which is considered as an insulator, Eq. (7.12) reduces to the form $\nabla^2 \delta \mathbf{B} = 0$, while the electric field $\mathbf{E}$ can be found from Eq. (7.8).

We seek for the solution of these equations in the form

$$\delta \mathbf{B} = \mathbf{b}(z) \exp(i\psi), \quad \mathbf{E} = \mathbf{e}(z) \exp(i\psi). \tag{7.62}$$

Taking into account that all the perturbations are constant as a function of y, substituting Eqs. (7.59)–(7.62) for $\mathbf{V}$, $\delta \mathbf{B}$, and $\mathbf{E}$ into Eq. (7.12), and rearranging, we come to the set of equations

$$
b_x'' - p^2 b_x = \frac{k_R V_0}{v_m} \left\{ \left(B_{0z} - iqB_{0x}\right) \exp(qk_R z) \right.
$$
$$
\left. + \left(iB_{0x} - sB_{0z}\right) \frac{2s}{1 + s^2} \exp(sk_R z) \right\}, \tag{7.63}
$$

$$
b_z'' - p^2 b_z = \frac{k_R V_0}{qv_m} \left\{ \left(qB_{0x} + iB_{0z}\right) \exp(qk_R z) \right.
$$
$$
\left. - \left(is_m B_{0z} + B_{0x}\right) \frac{2q}{1 + s^2} \exp(sk_R z) \right\}, \tag{7.64}
$$

$$
b_y'' - p^2 b_y = \frac{ik_R V_0 \left(1 - q^2\right)}{qv_m} B_{0y} \exp(qk_R z), \tag{7.65}
$$

where B_{0x}, B_{0y}, and B_{0z} are projections of $\mathbf{B}_0$ on the coordinate axes, and $p^2 = k_R^2 - i\omega/v_m$. Here the primes denote derivatives with respect to z.

The similar equations can be derived for the atmosphere $(z > 0)$

$$b_j'' - k_R^2 b_j = 0, \quad j = x, y, z. \tag{7.66}$$

The solution of these equations should match with that of Eqs. (7.63)–(7.65) through standard boundary conditions at $z = 0$, as shown in Appendix G. As it follows from equations, the solution of the problem for the atmosphere can be written as follows:

$$\delta B_x = \frac{k_R V_0}{v_m (k_R + p)} \left\{ \frac{\gamma_1}{q} (q k_R - p) + \frac{\gamma_2}{s} (s k_R - p) \right\} \exp(-k_R z + i\psi),$$

$$\delta B_z = -i\delta B_x, \quad \delta B_y = 0, \tag{7.67}$$

where $\mathrm{Re}\, p > 0$ and

$$\gamma_1 = (iqB_{0x} - B_{0z}) \left(\frac{\omega^2}{C_l^2} - \frac{i\omega}{v_m} \right)^{-1},$$

$$\gamma_2 = \frac{2s (sB_{0z} - iB_{0x})}{(1 + s^2)} \left(\frac{\omega^2}{C_t^2} - \frac{i\omega}{v_m} \right)^{-1}. \tag{7.68}$$

The electric field components are given by

$$E_x = iB_{0y} V_0 \frac{C_l}{C_R} \left\{ \frac{\omega p v_m - iq k_R C_l^2}{q k_R (\omega v_m - iC_l^2)} - \frac{2}{1 + s^2} \right\} \exp(-k_R z + i\psi),$$

$$E_z = -\frac{iC_R}{C_l} E_x, \quad E_y = -iqC_R\delta B_x. \tag{7.69}$$

As is seen from Eqs. (7.67)–(7.69), the amplitudes of the electromagnetic perturbations decrease with altitude. A typical vertical scale of the field attenuation in the atmosphere is of the order of k_R^{-1}, that is about Rayleigh wavelength.

If the vector of the Earth's magnetic field is parallel to the vertical plane in which the medium particles move; that is, if $B_{0y} = 0$, then only three components δB_x, δB_z and E_y are nonzero. Moreover, the density of electric charges is equal to zero everywhere. This implies that the field is a vortex one in nature. By contrast, if $B_{0y} \neq 0$, then there is the vertical component of electric current $j_z = \sigma E_z$, which leads to the generation of uncompensated electric charges in the conducting half-space and on its surface.

The magnitude of the electromagnetic field variations in the region $z < 0$ decreases with depth due to energy absorption depending on the relation between seismic wavelength, $\lambda_R = 2\pi/k_R$, and skin-layer depth in the conducting media, $\delta = (\mu_0\sigma\omega/2)^{-1/2}$. It follows from Eqs. (7.115)–(7.117) that if $\delta^2 k_R^2 \gg 1$ which is valid under the requirement $\omega \gg C_R^2/v_m$, then the typical scale of the electromagnetic field damping is about seismic wavelength λ_R. In such a case the amplitudes of the magnetic field perturbations on the ground surface can be estimated as follows: $\delta B_x \sim \delta B_z \sim V_0 B_0/ (k_R v_m) \gg \delta B_y \sim V_0 B_0/C_R$ while the electric field estimates are $E_y \sim C_R\delta B_x \gg E_x \sim E_z \sim V_0 B_0$. Not surprisingly, the estimates for amplitudes δB_x, δB_z and E_y are similar to those given by Eqs. (7.25)–(7.27) because in both cases the diffusion of the electromagnetic perturbations dominates over the effect of "frozen-in" magnetic field lines.

Under the inverse requirement $\delta^2 k_R^2 \ll 1$ that means that the magnetic field is practically frozen in the conducting media, all the magnetic components are of the order of $V_0 B_0 / C_R$ while the electric field variations are as much as $V_0 B_0$.

Gorbachev and Surkov (1987) have studied Earth's magnetic field perturbations due to Rayleigh surface waves generated by linear and point acoustic sources. In these cases the surface wave fields radiated by seismic sources can be represented by a superposition of quasi-harmonic Rayleigh waves given by Eqs. (7.59) and (7.60). Far away from the point source the Rayleigh wavepacket decreases in inverse proportion to the square root of distance r. In the model of perfectly conducting half-space the same tendency keeps for the magnitude of the GMPs.

To summarize, we recall that during the diffusion regime the amplitude of GMPs falls off more rapidly with distance due to the electromagnetic energy absorption and dissipation in conducting media. At a later moment the GMPs are localized in the vicinity of seismic wave front that results in a slow decrease of amplitude with distance. In this case the seismic and electromagnetic perturbations depend on distance in the same manner; that is, for the primary/longitudinal wave they decrease as r^{-1} whereas for the Rayleigh surface wave they vary as $r^{-1/2}$. A strong EQ and an underground explosion are accompanied by a variety of seismic waves including the primary, secondary/transverse, Rayleigh, and Love surface waves, which can be detected at the distances about several thousands of kilometers. This implies that the co-seismic GMPs caused by these large-scale tectonic phenomena may be detectable at such distances.

It is interesting to note also that the dispersion-dissipative properties of actual geological media resulted from viscosity, nonuniform inclusions, and so on may have different effects on seismic and electromagnetic perturbations (Dunin and Surkov 1992). In the theory, the mass velocity amplitude V_0 of seismic surface wave propagating in dissipative media decreases with distance as r^{-n}, where index n can vary from 0.5 to 1.75–2.25 whereas the seismic wavelength λ_R increases as $r^{1/2}$. In the frequency range $f \gg (5$–$50)$ mHz we can use estimates (7.25) and (7.26) for GMPs according to which $\delta B_{\max} \sim V_0 \lambda_R B_0 / v_m$ and $E'_{\max} \sim V_0 B_0$. Whence it follows that $\delta B_{\max} \sim r^{-n+1/2}$ and $E'_{\max} \sim r^{-n}$. Thus, in the dissipative media the magnetic perturbations $\delta B_{\max}$ can fall off slowly as compared to the seismic ones. The interpretation we make is that the mass velocity of conducting media determines only the local current density whereas the magnetic perturbations are an integral effect resulted from the net action of all the currents induced in the region with typical scale $\sim r^{1/2}$.

7.3   ULF Electromagnetic Noise Due to Crack Formation in a Conductor

7.3.1   GMPs Due to Expansion of Tension Cracks

This section deals with GMPs caused by rock cracking and fracture. Acoustic waves emitted by cracks are accompanied by generation of electric currents in conductive layers of the ground. These currents develop because the conductive medium moves in the geomagnetic field. Since a tectonic activity triggers the crack formation in the rock, it produces the GMPs, which can contribute to the electromagnetic noise occasionally observed prior to the occurrence of strong EQs (e.g., see Surkov et al. 2003). The detailed review of these phenomena is found in Chap. 10.

It is usually the case that magnetometers are situated far away from the source electromagnetic noise. So, we shall consider the diffusion zone $r < r_*$ and the intermediate range of distances, that is, $C_l t \ll r \ll r_d$. As we have shown above, in such a case the system of extrinsic/induction can be replaced by an effective magnetic dipole.

At first we examine the electromagnetic effect caused by individual tension cracks. The flat crack is located in the plane x', y' as shown in Fig. 7.9. The vector $\mathbf{B}_0$ of the Earth's magnetic field makes an angle θ_0 with axis z' which is perpendicular to the crack plane. Let $[u_z(t)] = u_z(t, z' = 0+) - u_z(t, z' = 0-)$ be a given function, which defines the displacement discontinuity/jump, and which is normal to the crack surface and parallel to the z'-axis. Far away from the crack, the displacement field due to the crack expansion can be expressed through the

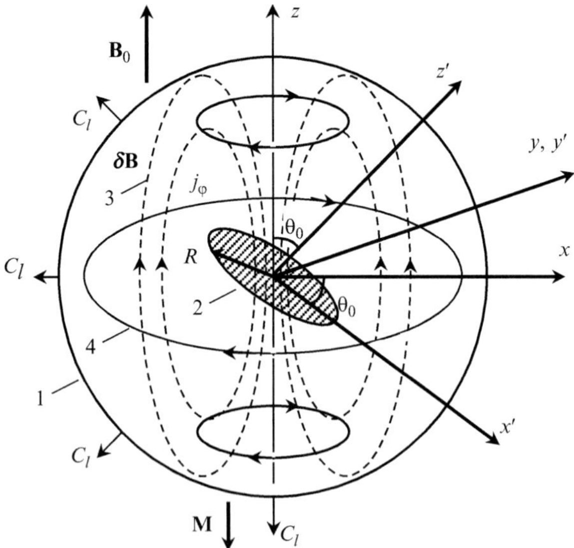

Fig. 7.9 A system of electric currents and GMPs generated by tension cracks. _1_—longitudinal acoustic wave radiated by crack 2 with radius R; _3_—field lines of geomagnetic field perturbations $\delta\mathbf{B}$; _4_—field lines of extrinsic/induction current j_φ; θ_0 is the angle between the vector $\mathbf{B}_0$ of the Earth's magnetic field and axis z' which is perpendicular to the crack plane x', y'. The effective magnetic moment $\mathbf{M}$ of extrinsic currents is antiparallel to the vector $\mathbf{B}_0$

so-called seismic moment tensor, $\hat{\mathbf{M}}$, which is considered in Appendix H. In the case of tension cracks only three diagonal components of the seismic moment tensor are nonzero (Aki and Richards 2002). The components of medium displacement around the crack are given by Eq. (10.59). For convenience, we also introduce the spherical coordinate system r, θ, φ, where the polar angle θ and azimuthal angle φ are measured from axes z' and x', respectively. Taking time-derivative of Eq. (10.59) and rearranging this equations, we obtain the radial and transverse components of mass velocity:

$$V_r = \frac{Aw}{r} \left\{ g_1(r,t) + 2w^2 g_2(r,t) \cos^2 \theta \right\},$$

$$V_\theta = -\frac{A}{r} g_\theta(r,t) \sin 2\theta, \quad V_\varphi = 0. \tag{7.70}$$

Here we made use of the following abbreviations:

$$g_1 = [\ddot{u}_z^l](1 - 2w^2) + [\dot{u}_z^l](1 - 4w^2)\frac{C_l}{r} + [\dot{u}_z^t]\frac{2C_t}{rw}, \quad w = \frac{C_t}{C_l},$$

$$g_2 = [\ddot{u}_z^l] + [\dot{u}_z^l]\frac{4C_l}{r} - [\dot{u}_z^t]\frac{3C_t}{rw^3},$$

$$g_\theta = [\ddot{u}_z^t] + [\dot{u}_z^t]\frac{3C_t}{r} - [\dot{u}_z^l]\frac{2w^3 C_l}{r},$$

$$[u_z^l] = \left[u_z\left(t - \frac{r}{C_l}\right)\right], \quad [u_z^t] = \left[u_z\left(t - \frac{r}{C_t}\right)\right], \quad A = \frac{S}{4\pi C_t}, \tag{7.71}$$

where S is the crack area and the dots above the symbols denote the time-derivatives. These equations are valid in wave and intermediate acoustic zones. In order to take into account the attenuation of acoustic waves we multiply the velocity components by the acoustic damping factor $T_a(r, R) = \exp(-r/L(R))$, which depends on the distance r and the crack radius R. The characteristic length of the acoustic waves attenuation, L, is given by Eq. (10.16).

Substituting Eqs. (7.70) and (7.71) into Eq. (7.55), we obtain the effective magnetic moment $\mathbf{M}$ of the induction currents generated in the conducting medium. Performing integration over the region with radius $r_l = C_l t$ and ignoring the near-field contribution to the integral yields

$$\mathbf{M} = -\mathbf{B}_0 \sigma A \frac{4\pi}{3} \int_0^{r_l} \left\{ w g_1 + \frac{2w^3 g_2}{5}\left(1 + \sin^2 \theta_0\right) \right.$$

$$\left. - \frac{g_\theta}{5}\left(2\cos^2 \theta_0 - \sin^2 \theta_0\right) \right\} T_a r^2 dr. \tag{7.72}$$

It should be noted that the effective magnetic moment is antiparallel to Earth's magnetic field irrespective of crack plane orientation. Notice the same property follows from Eq. (7.56).

Considering a random crack population, one should average the magnetic moment over the random orientations of the vector normal to the crack plane. As the normal may take any directions in space with equal probability then we obtain the following average values: $\langle \cos^2 \theta_0 \rangle = 1/3$ and $\langle \sin^2 \theta_0 \rangle = 2/3$. Whence it follows that the average magnetic moment reads

$$\langle \mathbf{M} \rangle = -\mathbf{B}_0 \sigma A w \frac{4\pi}{3} \int_0^{r_l} \left(g_1 + \frac{2w^2 g_2}{3} \right) T_a r^2 dr. \tag{7.73}$$

As is seen from this equation, the function g_θ does enter this equation. This means that the contribution of transverse velocity component V_θ to the average magnetic moment is equal to zero.

Substituting Eq. (7.71) for g_1, g_2 and A into Eq. (7.73) and performing integration by parts, gives

$$\langle \mathbf{M} \rangle = -\frac{\mathbf{B}_0 S \sigma}{3} \left(1 - \frac{4w^2}{3} \right) \int_0^{r_l} [\dot{u}_z^l] \left(r \frac{dT_a}{dr} + 3T_a \right) r dr. \tag{7.74}$$

The average magnetic moment thus depends only on the radial motion due to longitudinal waves radiated by the cracks.

Now we assume that the discontinuity of the vertical displacement at the crack surface, $[u_z(t)]$, is the increasing or slightly oscillating function that tends to a constant value u_0 as $t \to \infty$. The rise time of this function and typical time of its oscillations are supposed to be much smaller than the arrival time $t = r/C_l$ of the longitudinal wave. This implies that the function $[\dot{u}_z^l] = [\dot{u}_z(t - r/C_l)]$ under the integral sign in Eq. (7.74) is close to zero everywhere except for a short interval in the vicinity of point $r = r_l$. Consequently, one may take the factor $(r dT_a/dr + 3T_a) r$ at the point $r = r_l$ and then move it through the integral sign. As a result, we come to the following estimate for the average value of the effective magnetic moment

$$\langle \mathbf{M} \rangle \approx -\frac{\mathbf{B}_0 S C_l \sigma u_0}{3} \left(1 - \frac{4w^2}{3} \right) \left(r_l \frac{dT_a(r_l)}{dr} + 3T_a \right) r_l. \tag{7.75}$$

Since the average magnetic moment is proportional to the volume $S u_0$ arising due to the crack opening, the net electromagnetic effect produced by the tension crack population can vary in direct proportion to the total volume of all the cracks and pores generated due to the rock fracture and dilatancy effect (Surkov 1999). The GMPs caused by tension cracks can be roughly estimated by substituting $\langle \mathbf{M} \rangle$ into Eq. (7.5). Additionally one should take into account the crack distribution over sizes. We study this problem in Chap. 10 in more detail.

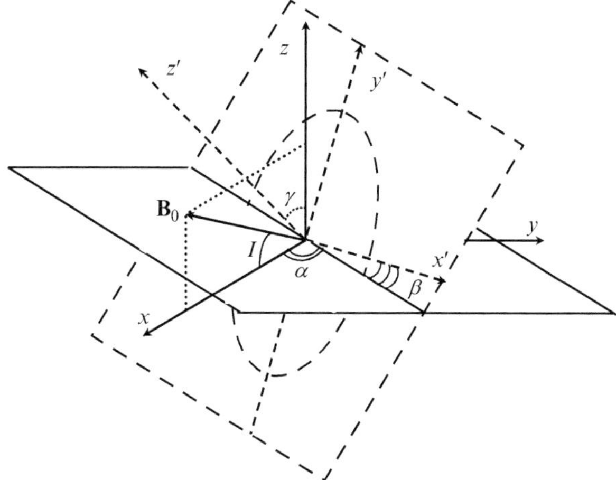

Fig. 7.10 A schematic plot of general coordinate systems (x, y, z) and local coordinate systems (x', y', z') in which the shear crack is in the plane (x', y')

7.3.2   GMPs Due to Shear Cracks

The mechanics of shear cracks is a crucial factor in the theory of faulting and EQ mechanics, since the growth and interaction of shear cracks are assumed to play a key role in both the rock strength and development of fault zones (Scholz 1990). In a sense, the fault zone as a whole can be considered as a gigantic shear crack which is described by its seismic moment tensor (Aki and Richards 2002). In what follows we show that not only tension but also the shear cracks can generate GMPs. Since the displacement field around a shear crack is more complicated than that originated from a tension crack, this results in a more complicated theory of electromagnetic phenomena associated with shear crack growth (Surkov 2001).

As before, we consider the ground as a uniform elastic half-space with constant conductivity σ. Let x-axis be horizontally directed along the magnetic meridian while z-axis is vertically upward. The vector $\mathbf{B}_0$ of geomagnetic field makes an angle I with x-axis ($I < 0$ in the northern hemisphere) in the vertical plane containing the magnetic meridian. Additionally, we use the local coordinate system (x', y', z') in which the shear crack is in the plane (x', y') as shown in Fig. 7.10. The x'-axis is perpendicular to the shear crack plane and makes an angle γ with z-axis. In general, the orientation of the axes x', y', and z' is defined by Euler angles α, β and γ.

Assuming for the moment that the rock shift points parallel to x'-axis, then the seismic moment tensor of the shear crack has only two components $m_{13} = m_{31} = \mu S [u_x]$, where μ is the shear modulus of the matter, and S is the crack area. The displacement discontinuity at the crack surface, $[u_x (t)] = u_x (t, z' = 0+) - u_x (t, z' = 0-)$, is considered as a given function of time. In the wave and intermediate zones, that is far away from the shear crack, the components of mass velocity are given by (Aki and Richards 2002)

$$V_r = \frac{Aw^3}{r} g_2\,(r,t)\sin 2\theta'\cos\varphi', \quad V_\theta = \frac{A}{r} g_\theta\,(r,t)\cos 2\theta'\cos\varphi',$$

$$V_\varphi = -\frac{A}{r} g_\theta\,(r,t)\cos\theta'\sin\varphi', \tag{7.76}$$

where r, θ', and φ' are spherical coordinates. The angles θ' and φ' are measured from axis z' and x', respectively. The functions g_2, g_θ and constant A are given by Eq. (7.71) where $[u_z]$ should be replaced by $[u_x]$. The expressions for the velocity components become more complicated in the general coordinate system (x, y, z). The velocity transform from the local coordinate system to the general one can be written as follows:

$$\left(V_x, V_y, V_z\right) = \hat{\mathbf{F}} \cdot \left(V_r, V_{\theta'}, V_{\varphi'}\right). \tag{7.77}$$

Here the components of the transfer matrix $\hat{\mathbf{F}}$ are given by

$$F_{11} = F_1\sin\theta' - \cos\theta'\sin\alpha\sin\gamma,$$
$$F_{12} = F_2\sin\theta' + \cos\theta'\cos\alpha\sin\gamma,$$
$$F_{13} = \cos\theta'\cos\gamma + \sin\theta'\sin\gamma\sin\left(\beta - \varphi'\right),$$
$$F_{21} = F_1\cos\theta' + \sin\theta'\sin\alpha\sin\gamma,$$
$$F_{22} = F_2\cos\theta' - \sin\theta'\cos\alpha\sin\gamma,$$
$$F_{23} = \cos\theta'\sin\gamma\sin\left(\beta - \varphi'\right) - \sin\theta'\cos\gamma,$$
$$F_{31} = \cos\alpha\sin\left(\beta - \varphi'\right) - \sin\alpha\cos\gamma\cos\left(\beta - \varphi'\right),$$
$$F_{32} = \cos\alpha\cos\gamma\cos\left(\beta - \varphi'\right) + \sin\alpha\sin\left(\beta - \varphi'\right),$$
$$F_{33} = -\cos\gamma\cos\left(\beta - \varphi'\right), \tag{7.78}$$

where

$$F_1 = \cos\alpha\cos\left(\beta - \varphi'\right) + \sin\alpha\cos\gamma\sin\left(\beta - \varphi'\right),$$
$$F_2 = \sin\alpha\cos\left(\beta - \varphi'\right) - \cos\alpha\cos\gamma\sin\left(\beta - \varphi'\right). \tag{7.79}$$

In order to find the effective magnetic moment of the induction currents generated by the shear crack one should substitute Eqs. (7.76)–(7.78) for the mass velocity into Eq. (7.55). Performing integration with respect to angles θ' and φ' and neglecting the near-field contribution to the integral, we obtain

$$M_x = \{\eta\sin I - 2\cos I\,(\cos\alpha\cos\beta + \sin\alpha\sin\beta\cos\gamma)\}\,\Psi,$$
$$M_y = \{\sin I\,(\sin\alpha\cos\beta\cos\gamma - \cos\alpha\sin\beta\cos 2\gamma)$$
$$+ \cos I\sin\gamma\,(\sin 2\alpha\sin\beta\cos\gamma + \cos 2\alpha\cos\beta)\}\,\Psi,$$

$$M_z = \left(\sin I \, \sin \beta \, \sin 2\gamma + \eta \cos I\right) \Psi,$$

$$\eta = \cos \alpha \, \cos \beta \, \cos \gamma + \sin \alpha \, \sin \beta \, \cos 2\gamma, \tag{7.80}$$

where

$$\Psi(t) = \frac{\sigma B_0 S}{10 C_t} \int_0^{r_l} \left\{ \left[\ddot{u}_x^t\right] + \left[\dot{u}_x^t\right] \frac{C_t}{r} + \frac{2w^3}{3} \left(\left[\ddot{u}_x^l\right] + \left[\dot{u}_x^l\right] \frac{C_l}{r} \right) \right\} r^2 dr. \tag{7.81}$$

Here the functions $\left[\ddot{u}_x^t\right]$ and $\left[\dot{u}_x^t\right]$ become zero if $r > C_t t$. In a similar fashion we assume that the discontinuity of the shear displacement at the crack surface, $[u_x(t)]$, is an increasing or slightly oscillating function that tends to constant value u_0 as $t \to \infty$. Then, performing integration in Eq. (7.81) by parts several times, we come to

$$\Psi(t) \approx \frac{B_0}{2} \sigma S u_0 C_t^2 t. \tag{7.82}$$

Absolute value of the effective magnetic moment can be written as $M = \Psi(t) f(\alpha, \beta, \gamma, I)$ where f is a complicated function of the angles. For example, at the polar latitudes, when $I \approx \pi/2$, this function is reduced to $f = \left(\sin^2 \beta + \cos^2 \beta \, \cos^2 \gamma\right)^{1/2}$ whence it follows that the peak value of magnetic moment $M = \Psi(t)$ is achieved as the vector $\mathbf{B}_0$ is perpendicular to the crack plane ($\gamma = 0$).

A population of the shear cracks produce the random GMPs which can contribute to the ULF electromagnetic noise associated with the rock fracturing. A random population of the shear cracks can be characterized by the mean magnetic moment. If all orientations of the crack planes are equiprobable, then the mean value of magnetic moment of (7.80) is equal to zero. However, we cannot ignore the shear crack effect because there may be certain correlation between the crack locations in the fault zone. The planes of the shear cracks are predominantly distributed parallel to that plane where the shear stress in the rock is at its maximum. One may expect that the plane of maximal shear stress is approximately parallel to the fault surface. This implies in turn that the crack distribution function over angles has a peak around the angles α and γ, which determine the orientation of the fault surface. The distribution over angle β, which determines the slip direction in the crack plane, may be anisotropic as well. For example, the upward crack slipping may prevail over downward one due to the difference in lithostatic pressure in the upper and lower crack tips.

As is seen from Eq. (7.80), the direction of the vector $\langle \mathbf{M} \rangle$ can be a complicated function of angles between $\mathbf{B}_0$ and the slip plane of the cracks. Contrary to the case of tension cracks, the average magnetic moment of shear crack ensemble is not directed parallel to the Earth' magnetic field. Taking into account the crack distribution over their size and combining the vector $\langle \mathbf{M} \rangle$ with Maxwell equations, one may estimate the GMPs provided by the random population of the shear cracks.

Appendix G: Earth's Magnetic Field Perturbation by Acoustic Waves Propagating in Conductive Ground

General Solution for the Spherically Symmetric Acoustic Wave

This section deals with the spherically symmetric longitudinal acoustic wave propagating in an infinite homogeneous conductive medium immersed in the constant magnetic field with induction $\mathbf{B}_0$. The electromagnetic perturbations caused by the acoustic wave propagation are described by Maxwell equations which are given by Eqs. (7.32)–(7.34). For the convenience, these equations can be supplemented by the equation $\nabla \cdot \mathbf{B} = 0$, which takes the form

$$\frac{1}{r^2}\partial_r \left(r^2 \delta B_r\right) + \frac{1}{r \sin \theta}\partial_\theta \left(\sin \theta \delta B_\theta\right) = 0. \tag{7.83}$$

The solution of the set of Eqs. (7.32)–(7.34) and (7.83) proves to be found in the form (Surkov 1989a)

$$\delta B_r = B_1\left(r,t\right)\cos\theta, \quad \delta B_\theta = -B_2\left(r,t\right)\sin\theta \quad \text{and} \quad E'_\phi = E_1\left(r,t\right)\sin\theta. \tag{7.84}$$

where $B_1\left(r,t\right)$, $B_2\left(r,t\right)$, and $E_1\left(r,t\right)$ are unknown functions. Substituting Eq. (7.84) into Eqs. (7.32)–(7.34) and (7.83) and rearranging, we obtain

$$\partial_t B_1 = v_m \left(\partial_r^2 B_1 + \frac{2\partial_r B_1}{r} - \frac{4\left(B_1 - B_2\right)}{r^2}\right) - \frac{2B_0 V}{r}, \tag{7.85}$$

$$\partial_t B_2 = v_m \left(\partial_r^2 B_2 + \frac{2\partial_r B_2}{r} + \frac{2\left(B_1 - B_2\right)}{r^2}\right) - \frac{B_0}{r}\partial_r\left(rV\right), \tag{7.86}$$

$$\partial_r B_1 + \frac{2\left(B_1 - B_2\right)}{r^2} = 0, \tag{7.87}$$

$$E_1 = \frac{v_m}{r}\left\{\partial_r\left(rB_2\right) + B_1\right\}. \tag{7.88}$$

Below we will argue that only two equations are independent among Eqs. (7.85)–(7.88).

The set of these equations should be supplemented by the initial and boundary conditions given in Sect 7.2.6. At the initial moment $t = 0$ there is a uniform constant magnetic field $\mathbf{B}_0$ and therefore the initial conditions are $B_1\left(r,0\right) = B_2\left(r,0\right) = 0$ and $E_1\left(r,0\right) = 0$. All the functions have to tend to zero as $r \to \infty$, besides they have to be finite as $r \to 0$.

It is convenient to introduce new dimensionless functions $g_1 = (B_1 + 2B_2)/B_0$ and $g_2 = (B_1 - B_2)/B_0$. Summing and subtracting Eq. (7.85) and (7.86), and rearranging Eq. (7.87), leads to

$$\partial_t g_1 = \nu_m \left(\partial_r^2 g_1 + \frac{2\partial_r g_1}{r} \right) - 2 \left(\partial_r V + 2\frac{V}{r} \right), \tag{7.89}$$

$$\partial_t g_2 = \nu_m \left(\partial_r^2 g_2 + \frac{2\partial_r g_2}{r} - \frac{6g_2}{r^2} \right) + \partial_r V - \frac{V}{r}, \tag{7.90}$$

$$\partial_r (g_1 + 2g_2) + \frac{6g_2}{r} = 0. \tag{7.91}$$

Such a form of the equations are preferable because either of Eqs. (7.89) or (7.90) includes only a single unknown function. Note that Eq. (7.89) can be transformed to a classical 1D diffusion equation via changing the unknown function $g_1 (r, t)$ by the function $g_1' (r, t) = g_1 (r, t) / r$. The solution of this equation at zero initial and boundary conditions are known (Surkov 1989a)

$$g_1 (r, t) = -\frac{2}{\pi^{1/2} r} \int\limits_0^t \frac{dt'}{r_0} \int\limits_0^\infty g_3 (r, r', r_0) (r' \partial_{r'} V + 2V) \, dr', \tag{7.92}$$

where $V = V (r', t')$ is the mass velocity of the medium. Here we made use of the following abbreviations:

$$g_3 (r, r', r_0) = \exp \left(-\frac{(r - r')^2}{r_0^2} \right) - \exp \left(-\frac{(r + r')^2}{r_0^2} \right), \tag{7.93}$$

and

$$r_0 = 2 \left[\nu_m (t - t') \right]^{1/2}. \tag{7.94}$$

Substituting Eq. (7.92) for g_1 into Eq. (7.91) we come to a differential equation of the first order with respect to the function g_2. The solution of this equation takes the form

$$g_2 (r, t) = -\frac{1}{2r^3} \int\limits_0^r r'^3 \partial_{r'} g_1 (r', t) \, dr'. \tag{7.95}$$

So the functions g_1 and g_2 given by Eq. (7.92) and Eq. (7.95) are the solutions of Eq. (7.89) and (7.91).

Now we will establish that the function Eq. (7.95) is a solution of not only Eq. (7.91) but also Eq. (7.90). Taking the time-derivative of the both sides of Eq. (7.95) and substituting Eq. (7.89) for $\partial_t g_1$ into the integral (7.95), we obtain

$$\partial_t g_2 = -\frac{1}{2r^3} \int_0^r r'^3 \partial_{r'} \left[v_m \left(\partial_{r'}^2 g_1 + \frac{2\partial_{r'} g_1}{r'} \right) - 2 \left(\partial_{r'} V + 2\frac{V}{r'} \right) \right] dr'. \quad (7.96)$$

Integrating Eq. (7.96) by parts and taking into account that V and $\partial_r V$ are equal to zero at $r = 0$, we come to Eq. (7.90). This completes the proof of the above statement.

The sought functions δB_r and δB_θ can be expressed through the functions g_1 and g_2

$$\delta B_r = \frac{B_0}{3} (g_1 + 2g_2) \cos \theta \text{ and } \delta B_\theta = \frac{B_0}{3} (g_2 - g_1) \sin \theta. \quad (7.97)$$

Substituting Eq. (7.92) for g_1 and Eq. (7.95) for g_2 into Eq. (7.97), taking integral (7.95) by parts and rearranging we come to Eqs. (7.35) and (7.36).

The electric field can be found from Eq. (7.84) and (7.88) that leads to Eq. (7.37).

The general solution of the problem given by these equations and Eq. (7.92) can be applied to an arbitrary function $V = V(r, t)$.

Normalized Potential of Elastic Displacement

In this section we deal with a spherically symmetric acoustic wave which results in the radial displacement of elastic medium. Since all the functions are dependent only on radius r, the acoustic wave equation (7.39) is reduced to the form:

$$\partial_t^2 u_r = C_l^2 \partial_r \left(\frac{1}{r^2} \partial_r \left(r^2 u_r \right) \right), \quad (7.98)$$

where $u_r = u_r(r, t)$ denotes the radial displacement of the medium and C_l is longitudinal wave velocity.

Equation (7.98) should be supplemented by the proper boundary and initial conditions. Let R_0 be the radius of the effective acoustic source and a given function $u_r(R_0, t) = u(t)$ be the radial displacement of the medium at the radius $r = R_0$. This function must satisfy the following conditions: $u(0) = 0$ and $\partial_t u(0) = 0$. At the initial moment the medium is at rest, that is, $u_r(r, 0) = 0$. Since the radial displacement has to be continuous at the front of acoustic wave, it is necessary that $u_r(r_l, t) = 0$, where $r_l = R_0 + C_l t$ is the radius of acoustic wave front.

It is convenient to introduce a new auxiliary unknown function $f(r, t)$ instead of the function $u_r(r, t)$ as follows: $u_r = -R_0^3 \partial_r (f/r)$. Substituting the last expression into Eq. (7.98) and rearranging, we arrive at the conventional wave equation for the function f

$$\partial_t^2 f = C_l^2 \partial_r^2 f. \tag{7.99}$$

The solution of Eq. (7.99) in the form of outgoing wave is an arbitrary twice differentiable function depending on the variable $t_1 = t - (r - R_0)/C_l$, i.e., $f = f(t_1)$. So, the solution of the problem has the form

$$u_r = -R_0^3 \partial_r \left(\frac{f(t_1)}{r} \right) = \frac{R_0^3}{rC_l} \left(\partial_t f(t_1) + \frac{C_l}{r} f(t_1) \right). \tag{7.100}$$

In rearranging this equation we have used that $\partial_t f = -C_l \partial_r f$. The mass velocity, $\mathbf{V}(r,t) = \partial_t u_r(r,t)\,\hat{\mathbf{r}}$, can be expressed through Eq. (7.100) in the form given by Eq. (7.40).

The dimensionless function $f(t_1)$ is usually referred to as the so-called normalized potential of the elastic displacement. The normalized potential can be extracted from seismic recordings. In situ measurements the instruments are arranged far away from the seismic source so that the distances are much greater than the characteristic seismic wavelength, i.e., $r \gg \lambda_a$. In such a region, referred to as a wave zone, the last term on the right-hand side of Eq. (7.100) can be neglected in comparison with the first term. This implies that the recording of the displacement is proportional to the time-derivative of the normalized potential, so that the function $f(t)$ can be found through the data processing of the seismic recordings.

In a theory the analytical form of the normalized potential can be found from Eq. (7.100) and the initial and boundary conditions. For example, as the displacement at the surface of the source is a given function $u(t)$, taking Eq. (7.100) at the boundary $r = R_0$ we come to the following differential equation for the function $f(t)$

$$u(t) = R_0 \left(\frac{R_0}{C_l} d_t f(t) + f(t) \right), \tag{7.101}$$

where the symbol d_t stands for time-derivative. The solution of Eq. (7.101) is given by Eq. (7.40).

In the theory of underground explosion the crushing zone can serve as an effective source of seismic waves (Rodionov et al. 1971). In such a case the radial component of the stress tensor s_{rr} at the crushing zone boundary is considered as a given function. For example, the radial stress can be approximated by Eq. (7.42).

In accordance with Hooke law the radial component of the stress tensor can be expressed through the radial displacement (Landau and Lifshits 1987)

$$s_{rr} = \rho_m C_l^2 \left[(1 - v)\, \partial_r u_r + 2v u_r/r \right], \tag{7.102}$$

where ρ_m is medium density and v is *Poisson*'s coefficient which defines the ratio of the transverse and longitudinal components of the strain tensor. The *Poisson*'s coefficient can be expressed by the ratio C_t/C_l of the transverse and longitudinal wave velocities:

$$\nu = \frac{1 - 2C_t^2/C_l^2}{2\left(1 - C_t^2/C_l^2\right)}. \tag{7.103}$$

Substituting Eq. (7.100) for u_r into Eq. (7.102) gives

$$s_{rr} = -\rho_m R_0^2 \left[(1 - \nu)\, \partial_t^2 f + \frac{2C_l}{R_0} (1 - 2\nu) \left(\partial_t f + \frac{C_l}{R_0} f \right) \right], \tag{7.104}$$

where $f = f(t_1)$. Equating Eq. (7.42) for s_{rr} and Eq. (7.104) taken at $r = R_0$ yields

$$d_t^2 f + \frac{2C_l (1 - 2\nu)}{R_0 (1 - \nu)} \left(d_t f + \frac{C_l}{R_0} f \right) = \frac{P_0 + (P_* - P_0) \exp\left(-t/t_r\right)}{\rho_m R_0^2 (1 - \nu)}, \tag{7.105}$$

where $f = f(t)$. The initial condition for Eq. (7.105) is as follows: $f(0) = 0$ and $d_t f(0) = 0$. The solution of this problem can be written as

$$f(t) = a_1 + a_2 \exp\left(-\frac{t}{t_r}\right) - \left[(a_1 + a_2) \cos\left(\frac{2\gamma_2 C_t t}{R_0}\right) \right.$$

$$\left. + \frac{(2\gamma_1 (a_1 + a_2) - a_2 a_3)}{2\gamma_1^{1/2} \gamma_2} \sin\left(\frac{2\gamma_2 C_t t}{R_0}\right) \right] \exp\left(-\frac{2\gamma_1 C_l t}{R_0}\right). \tag{7.106}$$

Here the following designations are introduced

$$a_1 = \frac{P_0 (1 - \gamma_1)}{2\gamma_1 \rho_m C_l^2}, \quad a_2 = \frac{2 (P_* - P_0) (1 - \gamma_1)}{\rho_m C_l^2 \left(a_3^2 - 4\gamma_1 a_3 + 4\gamma_1\right)},$$

$$a_3 = \frac{R_0}{C_l t_r}, \quad \gamma_1 = \left(\frac{C_t}{C_l}\right)^2, \quad \gamma_2 = (1 - \gamma_1)^{1/2}. \tag{7.107}$$

It should be noted that we have corrected some errors in the coefficients a_1 and a_2 which were made in the work by Surkov (1989b).

The normalized potential (7.106) versus time is illustrated in Fig. 7.7 with lines 1 and 2, which correspond to $t_r = 0.3$ and 0.03 s, respectively. In making these plots the following parameters have been used: $R_0 = 100$ m, $P_* = 5 \cdot 10^8$ Pa, $P_0 = 1.5 \cdot 10^8$ Pa, $C_l = 5$ km/s , $\rho_m C_l^2 = 5 \cdot 10^{10}$ Pa, and $\gamma_1 = 0.2$.

Approximation for Short Acoustic Wavelength

For simplicity, the radius of seismic wave front r_l is assumed to be much greater than the seismic source radius R_0. This implies that the region of integration in Eq. (7.92) is restricted, in fact, by the radii within interval $0 < r < C_l t$. The mass velocity V of the medium reaches a peak value in the acoustic wave zone, which is located at the distance r much greater than the acoustic wavelength. In this zone the last term on the right-hand side of Eq. (7.40) can be neglected. In this approach, substituting Eq. (7.40) for the mass velocity in Eq. (7.92), and taking into account that

$$r\partial_r V + 2V = -\frac{R_0^3}{C_l^2}\partial_t^3 f\,(t_1) = C_l R_0^3 \partial_r^3 f\,(t_1), \tag{7.108}$$

we obtain

$$g_1\,(r,t) = -\frac{2C_l R_0^3}{\pi^{1/2} r} \int\limits_0^t \frac{dt'}{r_0} \int\limits_0^{C_l t'} g_3\,(r,r',r_0)\,\partial_{r'}^3 f\,(t_1')\,dr', \tag{7.109}$$

where $t_1' = t' - r'/C_l$ and the functions g_3 and r_0 are defined by Eq. (7.93) and (7.94). Suppose that the profile of mass velocity is described by a continuous smoothing function so that the first and second derivatives of the function $f\,(t_1')$ are zero at the wave front, i.e., at $r' = C_l t'$ or at $t_1' = 0$. Then we can transform Eq. (7.109) integrating with respect to r' by parts twice. Taking into account that $g_3 = 0$ at $r' = 0$, we get

$$g_1\,(r,t) = -\frac{2C_l R_0^3}{\pi^{1/2} r} \int\limits_0^t \frac{dt'}{r_0} \int\limits_0^{C_l t'} \partial_{r'}^2 g_3\,(r,r',r_0)\,\partial_{r'} f\,(t_1')\,dr'. \tag{7.110}$$

Recall that the normalized potential $f\,(t_1)$ can be expressed through the radial displacement $u\,(t)$ at the source surface. Suppose that the radial displacement at the source surface increases gradually and tends to a constant value u_0 as $t \to \infty$. In other words, $u\,(t)$ is an increasing or weakly oscillating function, which tends to the value of static displacement u_0 as $t \to \infty$. As it follows from Eq. (7.106), the characteristic time of the displacement and potential variations depend on the parameters t_r, R_0/C_l and R_0/C_t which define the stress relaxation time in the source and the typical periods of the seismic vibrations and relaxation. At far distances all of these parameters are much smaller than the time $t = r/C_l$ of seismic wave arrival to the observation point. In such a case we may simplify Eq. (7.41) for the function $f\,(t)$ by replacing $u\,(t')$ under the integral sign by the constant value u_0. Calculating this integral we find that $f\,(t) \to u_0/R_0$ as $t \to \infty$ in accordance with the plot shown in Fig. 7.6. It follows from this fact that the derivative of the normalized potential under the integral sign in Eq. (7.110) changes considerably within a short interval near the point $r' = C_l t'$ and it tends to zero outside this interval. Note that the function $\partial^2_{r'} g_3$ changes more slowly and thus can be approximately considered as a constant within this interval. In this approximation the function $\partial^2_{r'} g_3$, taken at the point $r' = C_l t'$, can be moved through the integral sign in Eq. (7.110). Integrating the remaining function $\partial_{r'} f\left(t'_1\right)$ with respect to r', we obtain

$$g_1\left(r, t\right) \approx \frac{u_0 S C_l}{2\pi^{3/2} r} \int_0^t \partial^2_{r'} g_3\left(r, C_l t', r_0\right) \frac{dt'}{r_0}, \tag{7.111}$$

where the function $\partial^2_{r'} g_3\left(r, r', r_0\right)$ is taken at $r' = C_l t'$. Here $S = 4\pi R_0^2$ stands for the area of source surface. Since $\partial^2_{r'} g_3 = \partial^2_r g_3$, as it follows from Eq. (7.93), the expression (7.111) can be transformed to the form

$$g_1\left(r, t\right) = \frac{S u_0}{4\pi r} \partial^2_r G_1\left(r, t\right), \tag{7.112}$$

where

$$G_1\left(r, t\right) = \frac{2 C_l}{\pi^{1/2}} \int_0^t g_3\left(r, C_l t', r_0\right) \frac{dt'}{r_0}. \tag{7.113}$$

The integrals in Eqs. (7.35) and (7.36) for the GMP can be expressed via the function G_1 as follows:

$$\frac{1}{r^3} \int_0^r r'^2 g_1\left(r', t\right) dr' = \frac{S u_0}{4\pi r^3}\left(r \partial_r G_1 - G_1\right) = \frac{S u_0}{4\pi r} \partial_r\left(\frac{G_1}{r}\right). \tag{7.114}$$

Substituting Eqs. (7.112) and (7.114) into Eqs. (7.35)–(7.37) we arrive at Eqs. (7.43)—(7.45).

Substituting Eq. (7.93) for g_3 and Eq. (7.94) for r_0 into Eq. (7.113), and performing integration, one can reduce the function G_1 to the form given by Eq. (7.46).

Rayleigh Surface Wave in a Conductive Half-Space

The perturbations of the Earth's magnetic field resulted from Rayleigh surface wave propagation in a conducting ground ($z < 0$) are described by the set of Eqs. (7.63)–(7.65) for the amplitudes b_x, b_y, and b_z of magnetic perturbations. These amplitudes as functions of vertical coordinates z have to tend to zero when $z \longrightarrow -\infty$. The solution of the problem can be written as

$$b_x = \frac{k_R V_0}{v_m}\left\{a_1 \exp\left(pz\right) + \gamma_1 \exp\left(q k_R z\right) + \gamma_2 \exp\left(s k_R z\right)\right\}, \tag{7.115}$$

$$b_y = \frac{i k_R V_0 \left(1 - q^2\right)}{q v_m} \left\{a_2 \exp\left(pz\right) - \gamma_3 \exp\left(q k_R z\right)\right\}, \tag{7.116}$$

$$b_z = \frac{k_R V_0}{v_m} \left\{a_3 \exp\left(pz\right) + \frac{i \gamma_1}{q} \exp\left(q k_R z\right) + \frac{i \gamma_2}{s} \exp\left(s k_R z\right)\right\}, \tag{7.117}$$

where a_1, a_2, and a_3 are undermined coefficients while the coefficient γ_1 and γ_2 are determined by Eq. (7.68) and the coefficient γ_3 is given by

$$\gamma_3 = B_{0y} \left(\frac{\omega^2}{C_l^2} - \frac{i\omega}{v_m}\right)^{-1}. \tag{7.118}$$

Besides the coefficients q and s are given by Eq. (7.61) and $p = \left(k_R^2 - i\omega/v_m\right)^{1/2}$ is defined in such a way that $\mathrm{Re}\, p > 0$.

In a similar manner one can find the solutions of Eq. (7.66) for the atmosphere $(z > 0)$

$$b_j = d_j \exp\left(-k_R z\right), \quad j = x, y, z, \tag{7.119}$$

where d_j denote the undefined coefficients.

It is easy to show that the solutions (7.115)–(7.117) and (7.119) satisfy Maxwell equation $\nabla \cdot \delta \mathbf{B} = 0$ under the requirements that $i k_R a_1 = p a_3$ and $d_3 = -i d_1$.

All the components of electromagnetic perturbations must be continuous at the boundary $z = 0$. In addition, the normal component of the conduction current is equal to zero at $z = 0$ because the atmosphere is supposed to be an insulator. It follows from these boundary conditions that

$$a_1 = -\frac{p}{k_R + p} \left\{(1 + q) \frac{\gamma_1}{q} + (1 + s) \frac{\gamma_2}{s}\right\}, \tag{7.120}$$

$$d_1 = \frac{k_R V_0}{v_m} \left\{(q k_R - p) \frac{\gamma_1}{q} + (s k_R - p) \frac{\gamma_2}{s}\right\}, \tag{7.121}$$

$$a_2 = \gamma_3, \quad d_2 = 0. \tag{7.122}$$

We thus have just found all the coefficients in Eqs. (7.115)–(7.117) for the amplitudes b_x, b_y, and b_z of magnetic perturbations. Substituting these amplitudes into Eq. (7.62) gives the vector $\delta \mathbf{B}\left(x, z, t\right)$, that is, the solution of problem. The electric field $\mathbf{E}$ is related to $\delta \mathbf{B}$ through Eq. (7.11).

References

Aki K, Richards P (2002) Quantitative seismology, 2nd edn. University Science Books, Sausalito, 932 p

Anisimov SV, Gokhberg MB, Ivanov EA, Pedanov MV, Rusakov NN, Troitskaya VA, Goncharov VI (1985) Short period vibrations of the Earth's electromagnetic field due to industrial explosion. Rep USSR Acad Sci (Dokl Akad Nauk SSSR) Engl Transl 281(3):556–559

Babichev AP, Babushkina NA, Bratkovsky AM, et al (1991) Handbook of physical values. In: Grigorieva IS, Meilikhova EZ (eds) Energoatomizdat, Moscow (in Russian)

Belov SV, Migunov NI, Sobolev GA (1974) Magnetic effects accompany strong earthquakes in Kamchatka. Geomagn Aeron Engl Transl 14(2):380–382

Chadwick P, Cox AD, Hopkins HG (1964) Mechanics of deep underground explosion. Phil Trans Roy Soc Lond A256:235–300

Chave AD, Luther DS (1990) Low-frequency, motionally induced electromagnetic fields in the ocean. 1. Theory J Geophys Res 95(C5):7185–7200

Dunin SZ, Surkov VV (1992) Perturbation of an external magnetic field by a Rayleigh wave propagating through a conducting elastic dissipative half space. Magnetohydrodynamics (Magnitnaya gidrodinamica) Engl Transl 28(2):105–111

Eleman F (1965) The response of magnetic instruments to earthquake waves. J Geomagn Geoelectr 18(1):43–72

Fedorovich GV (1969) Diamagnetism of conductors moving in a magnetic field. Appl Mech Tech Phys (Prikladnaya Mekhanika i Tekhnicheskaya Fizika) Engl Transl 10(2):55–61

Gershenzon NI, Gokhberg MB (1992) Electromagnetic forecasting of tsunami. Phys Solid Earth (Izvestiya Akad Nauk SSSR Fizika Zemli) Engl Transl 28(2):38–42

Gogatishvili YaM (1983) Seismomagnetic effect in spectrum of geomagnetic pulsations. Rep Georgian SSR Acad Sci (Soobscheniya Akad. Nauk Gruzinskoi SSR) 1:73–76 (in Russian)

Gogatishvili YaM (1984) Geomagnetic precursors of strong earthquakes in spectrum of geomagnetic pulsations. Geomagn Aeron 24(4):697–700 (in Russian)

Gorbachev LP, Surkov VV (1987) Perturbation of an external magnetic field by a Rayleigh surface wave. Magnetohydrodynamics 23:117–125

Huang Q (2002) One possible generation mechanism of co-seismic electric signals. Proc Jpn Acad 78:173–178

Jackson JD (2001) Classical electrodynamics, 3rd edn. Wiley, New York, 808 pp

Kaliski S (1960) Solution of the equation of motion in a magnetic field for an isotropic body in an infinite space assuming perfect electric conductivity. Proc Vibr Probl 1(3):53–67

Kaliski S, Rogula D (1960) Rayleigh waves in a magnetic field in the case of a perfect conductor. Proc Vibr Probl 1(5):63–80

Keilis-Borok VI, Monin AS (1959) Magnetoelastic waves and boundary of the Earth's core. Izvestiya Akad Nauk SSSR Seriya Geofizika (Proc USSR Acad Sci Ser Geophys) 11:1529–1541 (in Russian)

Knopoff EL (1955) The interaction between elastic waves motion and a magnetic field in electrical conductors. J Geophys Res 60:617–629

Landau LD, Lifshits EM (1982) Electrodynamics of continuous media, Nauka, Moscow (in Russian)

Landau LD, Lifshits EM (1987) Theory of elasticity. Nauka, Moscow (in Russian)

Love AEH (1911) Some problems of geodynamics. Cambridge University Press, Cambridge

Molchanov O, Kulchitsky A, Hayakawa M (2002) ULF emission due to inductive seismo-electromagnetic effect. In: Hayakawa M, Molchanov OA (eds) Seismo electromagnetics: lithosphere-atmosphere-ionosphere coupling. TERRAPUB, Tokyo, pp 153–162

Molchanov OA, Hayakawa M (2008) Seismo-electromagnetics and related phenomena: history and latest results. TERRAPUB, Tokyo, 189 pp

Nagao T, Orihara Y, Yamaguchi T, Takahashi I, Hattori K, Noda Y, Sayanagi K, Uyeda S (2000) Co-seismic geoelectric potential changes observed in Japan. Geophys Res Lett 27:1535–1538

Pavlov VI, Sukhorukov AI (1987) About moving ionospheric perturbations caused by propagation of tsunami wave in ocean. Lett J Tech Phys (Pisma v jurnal Tekhicheskoy Fiziki) 13(6):351–354 (in Russian)

Rodionov VN, Adushkin VV, Kostyuchenko VN, Nikolaevskiy VN, Romashev AN, Tsvetkov VM (1971) Mechanical effect of underground explosion. In: Sadovsky MA (ed) Nedra, Moscow (in Russian)

Scholz CH (1990) The mechanics of earthquakes and faulting. Cambridge University Press, Cambridge, 439 pp

Schwarz G (1990) Electrical conductivity of the Earth's crust and upper mantle. Surv Geophys 11:133–161

Skordas E, Kapiris P, Bogris N, Varotsos P (2000) Field experimentation on the detectability of co-seismic electric signals. Proc Jpn Acad 76:51–56

Soloviev SP, Surkov VV (1994) Electric perturbations in the atmospheric surface layer caused by an aerial shock wave. Combustion Explosion Shock Waves 30(Iss. 1):117–121

Soloviev SP, Surkov VV (2000) Electrostatic field and lightning generated in the gaseous dust cloud of explosive products. Geomagn Aeron 40(1):61–69

Soloviev SP, Surkov VV, Sweeney JJ (2002) Quadrupolar electromagnetic field from detonation of high explosive charges on the ground surface. J Geophys Res 107(B6):2119–2130

Sorokin VM, Fedorovich GM (1982) Propagation of short-period waves in the ionosphere. Izvestiya VYZov Radiofizika (Proc Higher Institutes Educ Radiophys) 25(5):495–507 (in Russian)

Surkov VV (1989a) Perturbation of an external magnetic field by a longitudinal acoustic wave. Magnetohydrodynamics 25(2):145–148

Surkov VV (1989b) Geomagnetic perturbations in a stratified medium, caused by propagation of a longitudinal spherical wave. J Appl Mech Tech Phys 30(5):687–696

Surkov VV (1997a) The nature of electromagnetic forerunners of earthquakes. Trans (*Doklady*) Russian Acad Sci Earth Sci Sect 355:945–947

Surkov VV (1997b) The electromagnetic precursor of a seismic wave. Geomagn Aeron 37: 792–796

Surkov VV (1999) ULF electromagnetic perturbations resulting from the fracture and dilatancy in the earthquake preparation zone. In: Hayakawa M (ed) Atmospheric and ionospheric phenomena associated with earthquakes. TERRAPUB, Tokyo, pp 371–382

Surkov VV (2000a) Electromagnetic effects caused by earthquakes and explosions. MEPhI, Moscow, 448 pp (in Russian)

Surkov VV (2000b) On the nature of ULF electromagnetic noise anticipating some earthquakes. Phys Solid Earth 12:61–66

Surkov VV (2001) The role of shear cracks in the formation of electromagnetic noise preceding some earthquakes. Doklady Earth Sci (Translated from Doklady Akad Nauk Russian Acad Sci) 377A:349–355

Surkov VV, Uyeda S, Tanaka H, Hayakawa M (2002) Fractal properties of medium and seismoelectric phenomena. J Geodyn 33:477–487

Surkov VV, Molchanov OA, Hayakawa M (2003) Pre-earthquake ULF electromagnetic perturbations as a result of inductive seismomagnetic phenomena during microfracturing. J Atmos Solar Terr Phys 65:31–46

Sweeney JJ (1989) An investigation of the usefulness of extremely low-frequency electromagnetic measurements for treaty verification. Lawrence Livermore National Laboratory Report UCRL-53899, 60 pp, January 1989

Takeuchi N, Chubachi N, Narita K (1997) Observation of earthquake waves by the vertical earth potential difference method. Phys Earth Planet Inter 101:157–161

Viktorov IA (1975) Elastic waves in a solid semi-space immersed in magnetic field. Doklady Akad Nauk SSSR (Rep USSR Acad Sci) 221(5):1069–1072 (in Russian)

Viktorov IA (1981) Sound surface waves in a solid. Nauka, Moscow, 287 pp (in Russian)

Chapter 8
Electrokinetic Effect in Water-Saturated Rock

Abstract Basics of electrokinetic and seismoelectric phenomena in water-saturated rocks are discussed in this chapter. We study electrochemical and hydrodynamical processes in the fluid flowing in pores space in order to derive the relationship between the electrokinetic current and pore pressure gradient. Then we focus on the electrokinetic effect in anisotropical and fractal media. The remainder of this chapter covers seismoelectric effect caused by seismic waves.

Keywords Electrokinetic effect • Pore pressure gradient • Porosity • Seismo-electric effect • Streaming potential coefficient

8.1 Theory of Electrokinetic Effect

8.1.1 Basic Equations: Laboratory Study

A contact potential difference and electric charges are known to usually arise on the interface area in multiphase heterogenous media. For example, double electrical layers (DELs) are formed at the surfaces of pores, capillaries, and cracks in the porous water-saturated rocks. The basic reason of this phenomenon is that the groundwater is similar in content to the electrolytic solution which contains different kinds of ions and dissociated molecules. The surfaces of pores, capillaries, and cracks absorb the ions of specified sign from the water solution that results in charge buildup at the interphase boundaries. The excess ions of opposite sign keep in the groundwater so that the crack surfaces and the groundwater are charged oppositely thereby forming DELs at the crack surfaces.

It is common knowledge that there are a few mechanisms of ion absorption (Parks 1965; Fuerstenau et al. 1970; Wiese et al. 1971; James and Healy 1972). One possibility is that the surface of solid phase gains the negative charges due to acidic dissociation of surface hydroxyl groups (Parks 1965):

V. Surkov and M. Hayakawa, *Ultra and Extremely Low Frequency Electromagnetic Fields*, Springer Geophysics, DOI 10.1007/978-4-431-54367-1_8,
© Springer Japan 2014

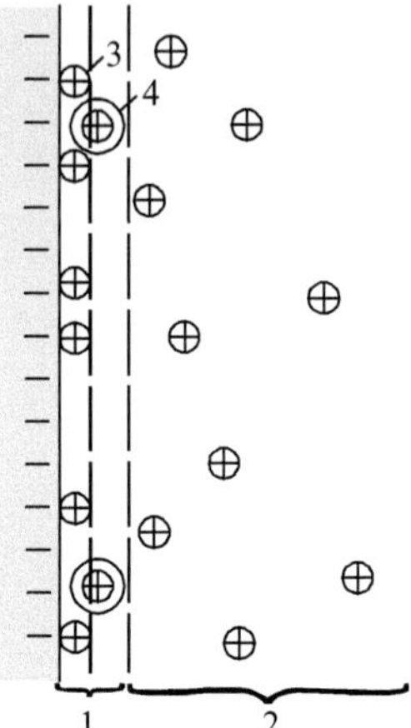

Fig. 8.1 A model of surface diffuse layer near the wall of a capillary filled with groundwater. *1*—Stern layer which consists of two sublayers; *2*—A region of mobile ions; *3*—An adsorbed ion; *4*—hydrated ion

$$M\,(OH)_n \leftrightarrow [M\,(OH)_{n-1}\,O]^- + H^+. \tag{8.1}$$

In this model the positive charge of the solution consists of the bound H^+ ions fixed at the surface and of the diffuse layer of mobile H^+ ions. A model of this layer is displayed in Fig. 8.1.

The layer of bound charges referred to as Stern layer consists of two sub-layers (Overbeek 1952). The first sub-layer contains the positive ions connected with the negatively charged wall by an absorption force. The sub-layer of hydrated/aquated ions, which weakly interact with the solid phase, joints the flank of the first sub-layer. In the region of mobile ions there are the diffusion flow of excess ions and electric current flowing in opposite direction. At the state of dynamical equilibrium there is a balance between these two flows.

The diffusion layer has a thickness of the order of Debye screening radius of ions. Figure 8.2 shows the approximate dependence of electric potential on distance near the capillary wall. Usually the Stern layer has a negative potential with respect to that of bulk electrolyte as shown with curve 1 in Fig. 8.2. However, if the number of positive ions in the Stern layer is greater than that of the negative charges induced on the surface of capillary, then the potential of Stern layer becomes positive as shown in Fig. 8.2 with the curve 2.

In the presence of pressure gradient the capillary fluid becomes to move. However, the fluid flow is not excited in the boundary layer which consists of ions and molecules absorbed at the capillary walls. This layer is separated from the moving fluid by the so-called slip plane which approximately coincides with the external boundary of the Stern layer shown in Fig. 8.1. It is customary to introduce the electrokinetic potential or ς-potential as the potential difference between the slip

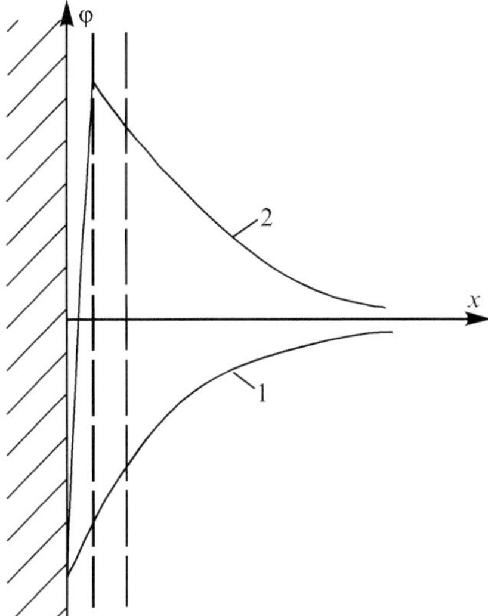

Fig. 8.2 An approximate dependence of electric potential on distance near the capillary wall. *1*—Usual character of the dependence; *2*—A case when the total charge of ions adsorbed by the capillary wall and of hydrated ions is positive

layer and the DEL boundary (or the capillary axis). In Fig. 8.2, in one case the ς-potential is negative (curve 1) and in the next one it is positive (curve 2).

The moving fluid drags the ion excess contained in the bulk fluid thereby exciting the convective current of ions. This gives rise to the formation of both the potential difference between the ends of capillary and the electric field which in turn result in the generation of the reverse electric current that compensates the convective current of ions.

We now consider a laminar flow of viscous fluid in a cylindrical channel with constant radius r_0. The projection of fluid velocity V_x on the channel axis x is described by Poiseuille formula

$$V_x = -\frac{r_0^2 - r^2}{4\eta}\frac{dP}{dx},\tag{8.2}$$

where η is the viscosity coefficient of the fluid, P is the pressure in the fluid, and r is the radius counted from the channel axis. Since Debye radius r_D is small, that is, $r_D \ll r_0$, the DEL can be considered as a narrow layer. The number density of mobile ions is maximal in the vicinity of the external DEL boundary; that is, near the surface of channel. In this region we can use the following approximation: $r_0^2 - r^2 \approx 2r_0 y$ where $y = r_0 - r$. Here we dwell on the simple model which assumes the non-conducting channel walls and conducting fluid. The electric potential Φ in the fluid obeys Poisson equation

$$\partial_x^2 \Phi + \partial_y^2 \Phi = -\frac{e\Delta n}{\varepsilon\varepsilon_0}, \tag{8.3}$$

where Δn denotes the ions excess in the solution, and e is their charge. The dielectric permittivity ε of the fluid is assumed to be constant. Near the channel surface the potential changes rapidly along the normal to the surface that is along the y-axis so that we can neglect the derivative with respect to x in Eq. (8.3).

In order to calculate the extrinsic/fluid-driven current I_e caused by convective transfer of the ions one should integrate the extrinsic current density $j_e = e\Delta n V_x$ over the channel cross section. Taking into account of Eq. (8.2) for V_x and Eq. (8.3) for Δn, we get

$$I_e \approx 2\pi r_0 e \int_0^{r_0} \Delta n V_x dy = \frac{\pi r_0^2 \varepsilon\varepsilon_0}{\eta}\frac{dP}{dx}\int_0^{r_0}\frac{d^2\Phi}{dy^2}y\,dy. \tag{8.4}$$

Performing integration by parts, we arrive at

$$I_e \approx -\frac{\pi r_0^2 \varepsilon\varepsilon_0\varsigma}{\eta}\frac{dP}{dx}, \tag{8.5}$$

where $\varsigma = \Phi(0) - \Phi(r_0)$. Here we have taken into account that the potential Φ is a rapidly decreasing function whose typical y-scale is much smaller than r_0. The parameter ς in Eq. (8.5) is, in fact, the potential jump across the DEL that is the ς-potential/zeta potential which we have introduced above.

The compensated conduction current is $I = \pi r_0^2 \sigma_f E_x$ where σ_f denotes the fluid conductivity and E_x is the projection of electric field on x-axis. At the state of equilibrium this current is equal to the electrokinetic current given by Eq. (8.5). Equating these currents we can find the effective field of extrinsic force $E_{\mathrm{eff}} = -E_x$:

$$E_{\mathrm{eff}} = -C\frac{dP}{dx}, \quad C = \frac{\varepsilon\varepsilon_0\varsigma}{\eta\sigma_f}. \tag{8.6}$$

Here C is the so-called streaming potential coefficient. This formula was first derived by Helmholtz and then by Smoluchowski on the basis of more rigorous treatments. In practice, Eq. (8.6) serves as a basic equation of the theory of electrokinetic effect.

The value of parameter σ_f is controlled by both intrinsic and extrinsic/impurity conductivity of the water depending on the contents of mineral salts in the solution. For the case of high mineralized groundwater the conductivity σ_f can reach a value of several unities or tens S/m, whereas for the sweet water the groundwater conductivity does not exceed several thousandth S/m (e.g., Semenov 1974).

Assuming for the moment that the uniaxial stress is applied to the porous rock sample with the length l, cross section S, and mean rock conductivity σ, then the electromotive force $E_{\mathrm{eff}}l$ arises in the sample. According to Ohm's law, it produces the electrokinetic current $I_{ek} = E_{eff}l/R$ where $R = l/(\sigma S)$ is the sample resistance. Combining these formulas and taking the notice of $j_{ek} = I_{ek}/S$, we obtain the electric current density j_{ek} due to the electrokinetic effect. The result can be written in the vector form as follows (Frenkel 1944; Mizutani and Ishido 1976; Mizutani et al. 1976; Pride 1994)

$$\mathbf{j}_{ek} = -\sigma C \nabla P. \tag{8.7}$$

The total current density which includes both the conduction and electrokinetic currents can thus be written as

$$\mathbf{j} = \sigma \mathbf{E} - \sigma C \nabla P. \tag{8.8}$$

In the case of narrow cracks and capillaries ($r_0 \sim r_D$) the DEL can contribute to the total conductivity due to the presence of ion excess in the DEL. The electric current in the surface layer with thickness r_D can be estimated as $I_s = 2\pi r_0 r_D \sigma_f E_x$ while the surface current density is $j_s = I_s/(\pi r_0^2)$. In order to take into account the surface current in the capillaries one should replace σ_f in Eq. (8.6) by $\sigma_{eff} = \sigma_f + 2\sigma_f r_D/r_0$ (Watillon and de Backer 1970; Dukhin 1975). If we take the values $r_0 = 10^{-7} - 5 \times 10^{-6}$ m which is typical for granites Westerly and Sherman (Brace et al. 1968; Brace 1977) and take into account the parameters $\sigma_f = 0.025$ S/m and $2\sigma_f r_D = 10^{-8}$ S reported by Watillon and de Backer (1970), then we come to the following numerical estimate for surface conductivity: $2\sigma_f r_D/r_0 = 0.002 - 0.1$ S/m. The typical values of the streaming potential coefficient C are as follows: 0.8 μV/Pa for granite Westerly, -4.2 μV/Pa for sandstones, and -4.7 μV/Pa for porous rocks ($r_0 > 10^{-5}$ m).

Since the solid matrix/dry rock conductivity is much smaller than that for the fluid, the average rock conductivity in Eq. (8.8) is mainly determined by the fluid content. In the simple model the rock conductivity and permeability are proportional to the rock porosity n, which is equal to the volume fraction of fluid-filled pores and cracks (Frenkel 1944; Mizutani and Ishido 1976). As the non-conductive matrix approximation is assumed, the single pores and cracks cannot conduct the electric current and thus only those cracks and channels, which create a connected system or cluster are able to contribute to the conductivity σ. In the percolation theory the effective conductivity of the medium near the percolation threshold depends on the porosity by a power law (Snante and Kirkpatrick 1971; Staüffer 1979; Feder 1988):

$$\sigma = \sigma_0 (n - n_c)^t, \tag{8.9}$$

where n_c is the percolation threshold, σ_0 is a constant with dimension of conductivity, and t is the transport critical exponent. The numerical modeling based on 3D grid has shown that $t = 1.6$ (Snante and Kirkpatrick 1971; Staüffer 1979).

Actually the dry rock conductivity is never equal to zero because there is ion conductivity of the solid matrix, i.e., the percolation threshold of the rock conductivity is absent. However, the nonlinear character of the $\sigma(n)$ dependence,

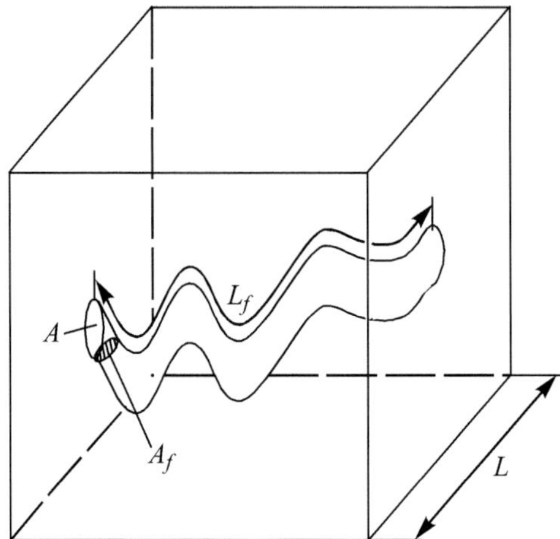

Fig. 8.3 A schematic plot of tortuous channel entering the sample on one side and coming out the sample on the other side. This figure illustrates the capillary model of a porous medium by De Groot and Mazur (1962) and Pfannkuch (1972)

such as $\sigma \propto n^2$, has been demonstrated by Brace et al. (1965) and Ishido and Mizutani (1981) for actual rocks. Taking the notice of $n < 1$, we conclude that the earlier assumption $\sigma \propto n$ can lead to an overestimated value of the electrokinetic current.

In the capillary model of porous media proposed by De Groot and Mazur (1962) and Pfannkuch (1972) the porosity, tortuosity, and specific surface of pores are taken into account. For illustrative purposes, Fig. 8.3 shows one tortuous channel entering the sample on one side and coming out the sample on the other side. The area of intersection between the channel and the sample surface is indicated by A while the channel cross section is symbolized by A_f. The channel length and the sample thickness are indicated by L_f and L, respectively. The basic characteristics of the capillary model are introduced as follows: the porosity $n = A_f L_f / (AL)$, tortuosity $b = L_f / L$, and specific pore surface $S = S_f / (AL)$ where S_f denotes the area of internal capillary surface. According to this model the current density is given by Eq. (8.8) where $\sigma_r = nb^{-2}\sigma_f + Sb^{-2}\sigma_s$, and $\sigma C = nb^{-2}\varepsilon\varepsilon_0\varsigma/\eta$. It should be noted that the conductivity σ given by above equation is consistent with observations if only $b \sim n^{-1/2}$ that can hardly be conceived (Brace et al. 1965).

It should be noted that Eq. (8.8) and other similar equations can be derived from the general Onsager relations which connect the electrokinetic and electroosmosis properties of porous media through the following set of coupled equations:

$$\mathbf{J} = -L_{11}\nabla P + L_{12}\mathbf{E},$$

$$\mathbf{j} = -L_{21}\nabla P + L_{22}\mathbf{E}. \tag{8.10}$$

Here $\mathbf{J}$ stands for the mean fluid flux density in the porous channels. In our case the Onsager coefficients are $L_{12} = L_{21} = \sigma C$, $L_{22} = \sigma$, and $L_{11} = k/\eta$ where k is the rock permeability and η is the viscosity of underground fluid.

The laboratory tests on zeta potential measurement are usually based on measurements of streaming potential coefficient and using Eq. (8.8) or other analogous dependences in order to extract the value of zeta potential from these measurements (e.g., see Wiese et al. 1971; Ishido and Mizutani 1981; Hunter 1981; Jouniaux and Pozzi 1999; Jouniaux and Bordes 2012; Luong and Sprik 2013). The experiments with granite, quartz, and other water-saturated rocks have shown that the zeta potential is usually negative in sign. Typically the zeta potential varies from several unities to several tens mV and these values can increase with the enhancement of pH or temperature of electrolyte. The observed values $\varsigma = 100\text{--}120$ mV at the temperature $T = 300$ K (Ishido and Mizutani 1981) seem to have been overestimated because the maximal value of the zeta potential corresponds to the condition $e\varsigma_{\max} \approx k_B T$ whence it follows that $\varsigma_{\max} \approx 50$ mV.

To estimate the streaming potential coefficient we choose such groundwater parameters as $\varepsilon = 80$, $\eta = 10^{-4}$ Pa·s, $\sigma_f = 1 - 15$ S/m and $\varsigma = 10^{-2}\text{--}10^{-1}$ V. Substituting these parameters into Eq. (8.6) we obtain the numerical estimate $C \approx 0.01\text{--}1\,\mu$V/Pa which is close to the abovementioned values measured for granites, sandstones, and other porous rocks.

8.1.2 Electrokinetic Effect in Homogeneous Media

The low-frequency electric field is derivable from a potential function Φ through $\mathbf{E} = -\nabla\Phi$. Substituting this relation into Eq. (8.8) and taking the notice of continuity equation for the electric current; that is $\nabla\cdot\mathbf{j} = 0$, we come to the following equation:

$$\nabla \cdot \sigma \left(\nabla\Phi + C\nabla P\right) = 0. \tag{8.11}$$

Here P stands for the excess fluid pressure in pores with respect to hydrostatic fluid pressure. Assuming that the ground is a homogeneous conductor with constant σ and C, one can reduce Eq. (8.11) to the following:

$$\nabla^2\Psi = 0, \tag{8.12}$$

where $\Psi = \Phi + CP$. Let z be the upward-directed axis perpendicular to the ground surface $z = 0$. As the atmosphere is considered as an insulator, the vertical component of the total current has to be equal to zero at $z = 0$. Equation (8.12) under the boundary conditions $j_z = -\sigma\partial_z\Psi = 0$ and $\Psi = 0$ has the trivial solution $\Psi = 0$ everywhere. Whence we get

$$\Phi = -CP. \tag{8.13}$$

Since the pore fluid pressure P at the ground surface is close to zero, the potential $\Phi \approx 0$ at $z = 0$ whereas the electric field has only vertical component. As is evident from Eq. (8.13), the total current $\mathbf{j} = -\sigma \nabla \Psi$ in the uniform medium is equal to zero everywhere. This means that the magnetic field due to electrokinetic effect is absent as well. In Chap. 10 we show that the significant variations of electric potential can only be expected around the inhomogeneities of rock conductivity σ and of streaming potential coefficient C.

8.1.3   Electrokinetic Effect in Anisotropical Media

However the magnetic effects can appear in a homogeneous but a macro-anisotropical medium (Surkov 2000). As one example, let us assume that the rock has a plane-stratified structure in which the water-saturated layers are sandwiched between the low-permeable layers. The underground fluid can filtrate along the networks of clustered channels located inside the water-saturated layers but it cannot flow across these layers. For simplicity, we suppose that all the layers are parallel and make the same angle α with the vertical axis z.

The medium conductivity along the high-conducting water-saturated layers is greater than that of the dry rock. So, the average rock conductivity can exhibit anisotropic properties as well because of the higher conductivity along the layers as compared to that across these layers. However, at first the average rock conductivity σ is considered as a scalar constant value.

The origin of coordinate system x, y, z is situated at the ground surface while the axes x', y', z' of the auxiliary coordinate system fixed to the high-permeable layers are shown in Fig. 8.4. The components of electrokinetic current density in the coordinate system x', y', z' are given by

$$j_{x'}^{ek} = -\sigma C \partial_{x'} P, \quad j_{y'}^{ek} = -\sigma C \partial_{y'} P, \quad j_{z'}^{ek} = 0. \tag{8.14}$$

The same vector in the basic coordinate system x, y, z can be expressed through pore pressure gradient in the following manner

$$\mathbf{j}_{ek} = -\sigma \hat{C} \nabla P, \tag{8.15}$$

where $\hat{C}$ denotes the tensor with the following components

$$\hat{C} = \begin{pmatrix} 1 & 0 & 0 \\ 0 & \sin^2 \alpha & \sin 2\alpha/2 \\ 0 & \sin 2\alpha/2 & \cos^2 \alpha \end{pmatrix}. \tag{8.16}$$

A cylindrical region with enhanced pore fluid pressure (see Fig. 8.4) has been studied by Surkov (2000) to examine the electrokinetic effect in the above

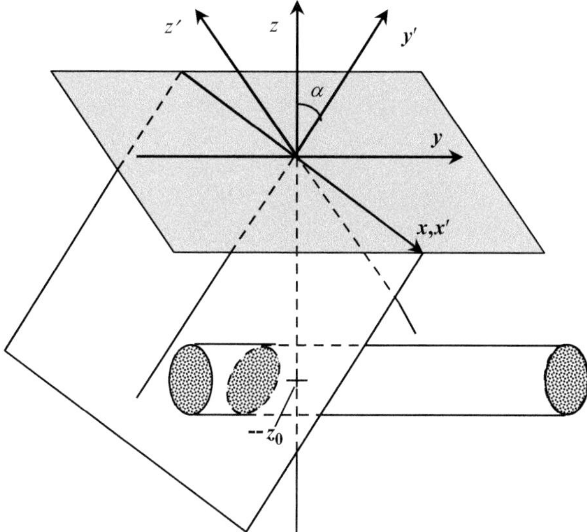

Fig. 8.4 A model of plane-stratified medium with inclined high-permeable layers. The *horizontal cylinder* represents a source of excess fluid pressure in pores

considered model of the plane-stratified medium. Despite the constant value σ, the magnetic variations have shown to appear both in the atmosphere and in the conducting ground. This result follows from the Maxwell equations and the tensor relationship between the electrokinetic current and the pore pressure gradient.

In the general case, considering the macro-anisotropical media, we can thus conclude that the scalar parameters C should be replaced by the streaming potential tensor $\hat{C}$ which depends on the structure of pore space and the presence of preferential directions for the underground fluid flow. Moreover the scalar Onsager relations (8.10) transform to the tensor ones in such a way that the scalar coefficients should be replaced by $\hat{L}_{12} = \hat{L}_{21} = \hat{\sigma}\hat{C}$, $\hat{L}_{22} = \hat{\sigma}$, and $\hat{L}_{11} = \hat{k}/\eta$ where $\hat{\sigma}$, $\hat{C}$, and $\hat{k}$ stand for the tensors of the rock conductivity, of streaming potential, and of rock permeability, respectively.

8.1.4   Electrokinetic Effect in Fractal Media

Taking into account that the dry rock conductivity is much smaller than that of the groundwater, we consider the model of porous rock which contains the non-conductive solid matrix and the channels and pores filled by the conducting underground water. As we have noted above, the isolated pores and cracks cannot be the conductor for the fluid flow as well as for the electrokinetic current. Only those cracks and channels which create a connected system or cluster are capable of conducting the electrokinetic current. It should be emphasized that a variety of the crack sizes can be described only in the framework of rather complicated percolation theory. Below we restrict our analysis to a simple percolation theory (e.g., see Snante and Kirkpatrick 1971; Staüffer 1979) in which the percolation cluster is able to appear under the requirement that the average rock porosity n is greater than the percolation threshold n_c. In this approximation the average rock conductivity is described by Eq. (8.9).

In reality the dry rock conductivity is never equal to zero because of the presence of ion conductivity in the solid matrix. Actually, as $n < n_c$ then the average rock conductivity falls off abruptly but it is not equal to zero. However, the parameter n_c can serve as a percolation threshold for both the fluid flow and the electrokinetic current.

The fractal properties near the threshold are determined by the correlation length

$$\xi \sim \frac{\xi_0}{(p - p_c)^\nu} \sim \frac{\xi_0}{(n - n_c)^\nu}, \qquad (8.17)$$

where ν is the correlation length critical exponent, p is the probability that an infinite cluster will appear in porous rocks while p_c and n_c denote the critical probability and porosity related to the percolation threshold, and ξ_0 is a constant of dimension of length. The infinite cluster has a fractal structure above the percolation threshold within spatial scale, which does not exceed the correlation length (Feder 1988; Staüffer 1979).

The upper crust contains a great deal of fluid-filled cracks, fracture zones, the sealed underground compartments with high pore pressure and etc. Some of such formations with high pore pressure may become unstable due to the variations of tectonic stresses (Bernard 1992; Fenoglio et al. 1995). The focal zone of a forthcoming earthquake is frequently associated with such unstable inhomogeneities which are capable of sustaining both the outward fluid migration and the electrokinetic current. In our model the pore space of the inhomogeneity forms the infinite cluster which has the fractal structure. This implies that the characteristic scale L of this inhomogeneity is of the order of the correlation length (8.17) whence it follows that $n - n_c \sim (\xi_0/L)^{1/\nu}$. Combining this relationship with Eqs. (8.7) and (8.9) one can find the rough estimation of the electrokinetic current density in fractal pore space (Surkov et al. 2002; Surkov and Tanaka 2005)

$$\mathbf{j}_{ek} \sim -\sigma_0 C \left(\frac{\xi_0}{L} \right)^{\tau/\nu} \nabla P. \qquad (8.18)$$

Using the critical exponents $\tau = 1.6$ and $\nu = 0.88$ obtained by numerical simulation on 3D (tree dimensional) grids (Staüffer 1979; Feder 1988) gives $\tau/\nu \approx 1.82$. We also suppose that $|\nabla P| \sim \Delta P/L$, where ΔP is the pore fluid pressure difference between the inhomogeneity and surrounding rock. In the case of large-scaled inhomogeneities such as earthquake hypocenter, ΔP is supposed to be proportional to the shear stress drop caused by rock fracture before the main shock. The shear stress drop is of the order of crushing/shear strength which is independent of the size L (Scholz 1990). In the same approximation ΔP is not a function of L whence it follows that $|\mathbf{j}_{\mathrm{ek}}| \propto L^{-\tau/\nu-1}$. Far away from the electrokinetic current source/inhomogeneity the electric field E is proportional to the current moment $d \sim |\mathbf{j}_{\mathrm{ek}}|\, V$, where $V \sim L^3$ is the source volume. This means that

$$E \propto L^{2-\tau/\nu}, \tag{8.19}$$

where $2-\tau/\nu \approx 0.18$; that is, the electric field weekly depends on the inhomogeneity size. By contrast, for the non-fractal inhomogeneity $E \propto L^2$.

In the model proposed by Surkov et al. (2002) the rock permeability and porosity were assumed to decrease from the center of the inhomogeneous region towards the periphery so that the porosity becomes close to the percolation threshold n_c at the external boundary of the inhomogeneity. The inner region of the inhomogeneity consists of broken rocks with so high permeability and porosity that the correlation length tends to zero in this region. This high permeable rock is surrounded by the layer with fractal structure, where the porosity decreases down to the percolation threshold n_c. If the thickness H of this fractal layer is much smaller than the typical size L of the inhomogeneity, then the porosity in this layer can be approximated as follows:

$$n \approx n_c + \nabla n \cdot H. \tag{8.20}$$

The porosity gradient can be estimated as $\Delta n/L$, where Δn denotes the porosity change inside the inhomogeneity. Substituting Eq. (8.20) for n into Eq. (8.17) and taking into account that the thickness H of the fractal layer is of the order of the correlation length we obtain that $H \propto L^{\nu/(\nu+1)}$.

The peripheral fractal region can make the major contribution to the total current moment under the requirement that the inner region has approximately a spherically symmetrical distribution of the electrokinetic currents because the vector $\mathbf{j}_{\mathrm{ek}}$ averaged over the inner region becomes close to zero. In such a case, the contribution of the fractal zone to the current moment can be estimated as follows: $d \sim |\mathbf{j}_{\mathrm{ek}}|\, L^2 H$, where the electrokinetic current density is given by

$$|\mathbf{j}_{\mathrm{ek}}| \sim \sigma_0 C \left(\frac{\xi_0}{H}\right)^{\tau/\nu} \frac{\Delta P}{L}. \tag{8.21}$$

Since the electric field varies directly as the current moment, we come to

$$E \propto L^{1-(\tau-\nu)/(1+\nu)}. \tag{8.22}$$

This theory can be applied to the earthquake preparation process in order to estimate amplitude electric field variations possibly associated with the electrokinetic phenomena in the seismo-active region.

8.2 Seismoelectric Effect Due to Propagation of Seismic Waves

To all appearance Ivanov (1939, 1940) was the first who detected electromagnetic effect associated with the propagation of seismic waves in the ground. High explosive charges with mass 1.5 kg were detonated just under the surface of the moist soil (surface explosions) in order to examine the acoustic and electric properties of the soil. During the explosion the appearance of an electric potential drop between two grounded electrodes was recorded at the distance of 120 m from the explosion point. This phenomenon was called the seismoelectric effect of the second kind in contrast to seismoelectric effect of the first kind that lies in the fact that the seismic wave changes the electric current flowing in the moist soil between two grounded electrodes. Frenkel (1944) and Martner and Sparks (1959) have shown that this phenomenon can be explained by the electrokinetic effect in fluid-filled cracks and channels contained in the surface layer of the ground. In the previous section we have studied the MHD mechanism which is capable of explaining such a kind of phenomena whereas this section will focus on the seismoelectric effect as an alternative way to explain the observation.

The main cause of the seismoelectric effect is that the seismic wave creates the deformations of porous rock followed by the generation of groundwater pressure gradient in pores and cracks. As is seen from Eq. (8.7), this results in the generation of electrokinetic current in porous rocks. According to the linear theory of porous water-saturated medium (Frenkel 1944) the excess of pore pressure δP over the hydrostatic level and the volume deformation θ are related by the following equation:

$$\frac{\partial_t^2 \delta P}{K_f} + \frac{(\beta - 1)\, \partial_t^2 \theta}{\alpha} = \frac{\nabla^2 \delta P}{\alpha \rho} - \frac{\eta}{k\rho}\left(\frac{\partial_t \delta P}{K_f} + \frac{\beta \partial_t \theta}{\alpha}\right). \qquad (8.23)$$

Here the following abbreviations are made

$$\alpha = 1 + (\beta - 1)\frac{K_f}{K_s}, \quad \beta = \frac{1}{n}\left(1 - \frac{K}{K_s}\right), \qquad (8.24)$$

where K_f, K_s, and K are the compression modules of the fluid, solid matrix, and dry porous rock (rock skeleton without fluid), respectively, n is the porosity, k is the rock permeability, and η and ρ are the fluid viscosity and density, respectively.

Below we focus on the large-scale seismic waves with typical wavelengths about 1 km and frequencies of the order of several Hertz. In this low frequency limit one may neglect the second order temporal and spatial derivatives in Eq. (8.23). As a result, this equation is reduced to the following one

$$\delta P = -K_f \beta \theta / \alpha. \qquad (8.25)$$

As it follows from Eq. (8.25), the pore fluid pressure is directly proportional to the volume deformation of the rock. The implication here is that the moving solid matrix completely drags the pore fluid.

In the seismic frequency range the displacement current is negligible as compared to the conduction one so that the electric potential in the homogeneous ground is described by Eq. (8.13) whence it follows that $\mathbf{E} = -C\nabla P$. Supposing that the rock parameters are constant and taking into account Eq. (8.25), we obtain

$$\mathbf{E} = CK_f\beta\nabla\theta/\alpha. \qquad (8.26)$$

For example, consider the surface quasi-harmonic Rayleigh wave propagating along the horizontal axis x. Taking the mass velocity components V_x and V_z given by Eqs. (7.59) and (7.60) one can derive the volume deformation $\theta = \partial_x V_x + \partial_z V_z$. Substituting this value into Eq. (8.26) gives the electric field in the ground ($z \le 0$)

$$\mathbf{E} = -\frac{\beta V_0 K_f \omega C}{\alpha s_1 C_l^2}\left(i\hat{\mathbf{x}} + s_1\hat{\mathbf{z}}\right)\exp\left\{k_R s_1 z + i\left(k_R x - \omega t\right)\right\}, \qquad (8.27)$$

where $s_1 = \left(1 - C_R^2/C_l^2\right)^{1/2}$, $k_R = \omega/C_R$ is the wave number of Rayleigh wave, and V_0 stands for the amplitude of the vertical velocity component V_z.

So, the electric field oscillates in-phase with the mass velocity. The same is true for the longitudinal wave. However, the transverse seismic wave cannot excite the seismoelectric effect because this wave does not produce the volume deformations; that is $\theta = 0$.

To estimate the electric field amplitude, we choose the following numerical values of the parameters: $K_s = 2.5$ GPa, $K = 0.5K_s$, $K_f = 0.1K_s$, $C_R = 1$ km/s, $C_l = 3$ km/s, $n = 0.1$, $C = 10^{-8} - 10^{-6}$ V/Pa and $V_0\omega = 1$ mm. Substituting these values into (8.27) we obtain the estimate $E_x \approx 0.01$–$1\,\mu$V/m which is close to the co-seismic signal amplitude 1–$10\,\mu$V/m.

Now we will use the simpler way to drive one more estimate of the maximal amplitude of seismoelectric signals. The pore pressure gradient can be roughly estimated as follows: $|\nabla P| \sim \delta P/\lambda$, where δP is the excess of pore pressure over the hydrostatic level and λ is the seismic wavelength. In order to obtain the maximum effect we assume that the pressure variation δP in the fluid is the same order-of-magnitude as that in solid matrix, so that $\delta P/K_s \sim V_0/C_l$, where $K_s \sim \rho_s C_l^2$ is the solid compressibility and ρ_s is the solid matrix density. It follows from this that $\delta P \sim \rho_s C_l V_0$. Hence we obtain that $E_{\max} \sim C\delta P/\lambda \sim \rho_s C V_0\omega$. Substituting $\rho_s \approx 2 \times 10^3$ kg/m^3 and the above values of the parameters we come to the same estimate $E_{\max} \approx 0.02$–$2\,\mu$V/m.

In contrast to the electric field amplitude due to the GMPs, the last estimate essentially depends on the porosity and underground water content. At the moment we cannot state which mechanism (that is, the GMP or seismoelectric effect) makes the main contribution to the co-seismic electric signals observed during seismic wave propagation. As for the magnetic component of the co-seismic signals, it seems likely that the GMP dominates the seismoelectric effect.

References

Bernard P (1992) Plausibility of long electrotelluric precursors to earthquakes. J Geophys Res 97B:17531–17546

Brace WF, Orange AS, Madden TR (1965) The effect of pressure on the electrical resistivity of water-saturated crystalline rocks. J Geophys Res 70:5669–5678

Brace WF, Walsh JB, Frangos WT (1968) Permeability of granite under high pressure. J Geophys Res 73:2225–2236

Brace WF (1977) Permeability from resistivity and pore shape. J Geophys Res 82:3343–3349

De Groot SR, Mazur P (1962) Non-equilibrium thermodynamics. North-Holland, Amsterdam; Wiley, New York, 510 pp

Dukhin SS (1975) Conductive and electrokinetic properties of dispersive systems. Naukova Dumka, Kiev, 246 pp (in Russian)

Feder E (1988) Fractals. Plenum Press, New York, 238 pp

Fenoglio MA, Johnston MJS, Byerlee JD (1995) Magnetic and electric fields associated with changes in high pore pressure in fault zones: Application to the Loma Prieta ULF emissions. J Geophys Res 100B:12951–12958

Frenkel YaI (1944) On the theory of seismic and seismoelectric phenomena in moist soil. Proc USSR Acad Sci Ser Geography Geophys (Izvestiya Akad. Nauk SSSR. Seriya Geografiya i Geofizika) 8(4):133–150 (in Russian)

Fuerstenau MC, Elgillani DA, Miller JD (1970) Adsorption mechanisms in nonmetallic activation systems. Trans AIME 247:11–14

Hunter RJ (1981) Zeta potential in colloid science. Academic Press, New York

Ishido T, Mizutani H (1981) Experimental and theoretical basis of electrokinetic phenomena in rock-water systems and its applications to geophysics. J Geophys Res 86B:1763–1775

Ivanov AG (1939) Effect of electrization of ground layers during passage of elastic wave through them. Rep USSR Acad Sci (Dokl Akad Nauk SSSR) 24(1):41–43 (in Russian)

Ivanov AG (1940) Seismoelectric effect of the second kind. Proc USSR Acad Sci Ser Geography Geophys (Izvestiya Akad Nauk SSSR Seriya Geografiya i Geofizika) 4(5):699–726 (in Russian)

James RO, Healy TW (1972) Absorption of hydrolyzable metal ions at the oxide-water interface, I, II, III. J Colloid Interface Sci 40:42–81

Jouniaux L, Pozzi J-P (1999) Streaming potential measurements in laboratory: a precursory measurements of the rupture and anomalous 0.1–0.5 Hz measurements under geochemical changes. In: Hayakawa M (ed) Atmospheric and ionospheric phenomena associated with earthquakes. TERRAPUB, Tokyo, pp 873–880

Jouniaux L, Bordes C (2012) Frequency-dependent streaming potentials: A review. Int J Geophys 2012(Article ID 648781):1–11

Luong DT, Sprik R (2013) Streaming potential and electroosmosis measurements to characterize porous materials. ISRN Geophys 2013(Article ID 496352):1–8

Martner ST, Sparks NR (1959) The electroseismic effect. Geophysics 24(2):297–308

Mizutani H, Ishido T (1976) A new interpretation of magnetic field variation associated with the Matsushiro earthquakes. J Geomagn Geoelectr 28:179–188

Mizutani H, Ishido T, Yokokura T, Ohnishi S (1976) Electrokinetic phenomena associated with earthquakes. Geophys Res Lett 3:365–368

Overbeek JThG (1952) Electrochemistry of the double layer. In: Kruyt HR (ed) Colloid science, vol 1, Irreversible systems. Elsevier, New York, pp 115–193

Parks GA (1965) The isoelectric points of solid oxides, solid hydroxides, and aqueous hydroxo complex systems. Chem Rev 65:177–198

Pfannkuch HO (1972) On the correlation of electrical conductivity properties of porous systems with viscous flow transport coefficients. Fundamentals of transport phenomena in porous media. Elsevier, New York, pp 42–54

Pride S (1994) Governing equations for the coupled electromagnetics and acoustics of porous media. Phys Rev B 50:15678–15696

Scholz CH (1990) The mechanics of earthquakes and faulting. Cambridge University Press, Cambridge, 439 pp

Semenov AS (1974) Electrical survey technique on a basis of natural electric field. Nedra, Moscow, 200 pp (in Russian)

Snante VK, Kirkpatrick S (1971) An introduction to percolation theory. Adv Phys 20:325–357

Staüffer D (1979) Scaling theory of percolation clusters, Phys Rep 54:1–74

Surkov VV (2000) Electromagnetic effects caused by earthquakes and explosions. Moscow, MEPhI, 448 pp (in Russian)

Surkov VV, Uyeda S, Tanaka H, Hayakawa M (2002) Fractal properties of medium and seismoelectric phenomena. J Geodynamics 33:477–487

Surkov VV, Tanaka H (2005) Electrokinetic effect in fractal pore media as seismoelectric phenomena. In: Dimri VP (ed) Fractal behaviours of the earth system. Springer, Berlin, Heidelberg, New York, pp 83–96

Watillon A, de Backer R (1970) Potentiel d'écoulement, courant d'écoulement et conductance de surface a l'interface eau-verre. J Electroanal Chem 25:181–196

Wiese GR, James RO, Healy TW (1971) Discreteness of charge and solvation effects in caution adsorption at the oxide/water interface. Discuss Faraday Soc 52:302–311

Chapter 9
Laboratory Study of Rock Deformation and Fracture

Abstract Our primary interest lies in the physics of electromagnetic phenomena associated with deformation and fracture of rocks. We start with a brief review of laboratory study of electromagnetic fields resulted from acoustic waves and shock polarization and magnetization effects in different materials. The vibration of charged dislocations and piezo-galvanic effect are considered to explain this effect in metals whereas the production and mobility of point and linear defects of atomic lattice is treated as a possible cause for the shock polarization effect in dielectrics. We discuss briefly a variety of electromagnetic phenomena caused by rock fracture. Among them are radiowave, optical and γ-radiation, and electron and ion emissions from fracturing rocks. We study the theories explaining the generation of strong electric fields in cracks and collapsing pores.

Keywords Dislocation • Piezo-galvanic effect • Point defect • Rock fracture • Shock magnetization • Shock polarization

9.1 Electromagnetic Effects Caused by Dynamic Deformation of a Solid

It is common knowledge that the dynamic deformation and fracture of solids are accompanied by a great variety of electromagnetic effects. The basic characteristics of these phenomena depend on the scales of fracture, intensity and duration of stress and strain, and a number of other factors. Such effects as generation of low-frequency electromagnetic fields and radiowaves, emission of charged particles, light flashes, X-ray emission and micro-discharges inside of cracks have been observed in laboratory experiments (e.g., see Parrot 1995; Surkov 2000).

It has been found experimentally that shock compression of different solids gives rise to a jump of electric potential at the shock front in all kinds of materials: metals, semiconductors, and dielectrics (e.g., Mineev and Ivanov 1976; Freund

V. Surkov and M. Hayakawa, *Ultra and Extremely Low Frequency Electromagnetic Fields*, Springer Geophysics, DOI 10.1007/978-4-431-54367-1_9,
© Springer Japan 2014

and Pilorz 2012). Effects of shock magnetization and demagnetization have been observed in magnetic materials. Fragments of fractured solid, as a rule, carry electric charges. The rock deformation and fracture can explain, in principle, some features of electromagnetic perturbations arising from large-scale tectonic processes such as EQs, volcanic eruptions and deeply buried high-yield detonations. In this section we first review laboratory studies and then examine these electromagnetic phenomena by analyzing the basic physical processes at microscopic level. Sometimes we will need to extend our research field to the frequency range of several MHz to gain a better insight into underlying mechanisms of these phenomena.

9.1.1   Electromagnetic Fields Originated from Acoustic Waves Propagation in Samples

Pioneering investigations of strain-induced polarization and depolarization in a solid dielectric were provided by Stepanow (1933) who observed the appearance of an electric potential difference between opposite sides of an ionic crystal under slow strain. This effect cannot be explained solely by pyro-electricity or piezoelectricity because it was observed in quite different materials. Caffin and Goodfellow (1955) and Fishbach and Nowick (1958) have shown that this phenomenon in ionic crystals can result from the motion of charged dislocations under mechanical stress. The same effect has been observed in a variety of materials subjected to dynamic loadings (e.g., see the review by Mineev and Ivanov 1976).

Consider first a typical laboratory test of electric signals excited by acoustic waves in dielectric samples. As usual the electrical signals are recorded by a standard radio-antenna with ferrite core or a rod antenna, which is placed several centimeters from the samples (for example, see Khatiashvily 1981). For the typical frequencies of observed signals (1–7 MHz) these antennas are situated in the near zone. Meanwhile these data are frequently interpreted as a radio emission, i.e., the authors proceed from theoretical conceptions that are valid only within wave zone. Therefore in fact the estimates of source parameters found in such a way are inexact.

The experiments with monocrystals of LiF, NaCl, and KCl have shown that the acoustic wave and electromagnetic field stimulated by the acoustic one have practically the same frequencies. It was found that the effect has the acoustic intensity threshold. Above the threshold the magnitude of electric signals was 1 mV for the annealed crystals. This value increases up to 4 mV with the increase of dislocation concentration (Khatiashvily 1981). These features of the effect were supposed to be due to excitation of stress-induced vibrations of the charged edge dislocations (Molotskiy 1980) or vibrations and motion of the fluctuation-charged walls of the micro-cracks (Khatiashvili and Perel'man 1982, 1989).

The same effect stimulated by acoustic wave propagation has been observed in pure metals (Misra 1978). The same mechanism; that is the bend vibration of the charged segments of dislocations has been proposed by Molotskiy (1980) to explain the observations. However, the free electrons in metals can be considered as a possible candidate for explanation of this effect (e.g., Gurevich 1957; Alekseev et al. 1984). The strains caused by the acoustic wave can modulate distributions of electron and ion number densities in metals in such a way that they exceed the equilibrium density level in those regions which are compressed by the acoustic wave. Due to their high mobility the electrons diffuse from the compressed regions towards the depression areas thereby producing the potential difference between these regions. The diffusion flux of electrons is partly compensated by electric current flowing in the inverse direction.

The state of thermodynamics equilibrium of the electron subsystem of metal is characterized by the equality

$$\mu - e\varphi = \mathrm{const},   \tag{9.1}$$

where μ denotes chemical potential which is approximately equal to the Fermi energy, ε_F, of electrons in the metal under zero temperature. The strength of the extrinsic forces, $\mathbf{E}_e$, can thus be described by $\mathbf{E}_e = \nabla\mu/e$. Rearranging this equation in the following way we obtain

$$\mathbf{E}_e = \frac{d\mu}{dn_e}\nabla n_e \cong \frac{\chi\varepsilon_F}{en_e}\nabla n_e = \frac{\chi\varepsilon_F}{e\rho}\nabla\rho,   \tag{9.2}$$

where χ is a dimensionless coefficient of the order of unity, n_e is the electron number density, e is the elementary charge, and ρ is the density of medium. Assuming that the magnetic field in metals can be neglected, then the total current is equal to zero; that is,

$$\sigma(\mathbf{E} + \mathbf{E}_e) + \varepsilon_0\partial_t\mathbf{E} = 0.   \tag{9.3}$$

In the acoustic frequency range the displacement current density $\varepsilon_0\partial_t\mathbf{E}$ is much smaller than the conduction one due to high conductivity σ of the metals. In the first approximation we suppose that σ goes into infinity. Considering that the sum of conduction and extrinsic current densities; that is $\mathbf{j} = \sigma(\mathbf{E} + \mathbf{E}_e)$, has to be finite we conclude that the electric field in the metal $\mathbf{E} \approx -\mathbf{E}_e$. We now estimate the last term in Eq. (9.3) $\mathbf{j}_d \approx \varepsilon_0\partial_t\mathbf{E}_e$ which is related to the charge density through Maxwell's equations. Substituting Eq. (9.2) for $\mathbf{E}_e$ into this equation and taking into account that the density variations are small, we get

$$\mathbf{j}_d = \frac{\chi\varepsilon_0\varepsilon_F}{e}\partial_t\left(\frac{\nabla\rho}{\rho}\right) \approx \frac{\chi\varepsilon_0\varepsilon_F}{e\rho}\nabla(\partial_t\rho).   \tag{9.4}$$

Now let us apply the continuity equation in the form

$$\partial_t\rho = -\nabla\cdot(\rho\mathbf{V}),   \tag{9.5}$$

where $\mathbf{V}$ is the mass medium velocity. Substituting Eq. (9.5) for $\partial_t \rho$ into Eq. (9.4) we finally find the expression for the current density

$$\mathbf{j}_d \approx \frac{\chi \varepsilon_0 \varepsilon_F}{e} \nabla^2 \mathbf{V}. \tag{9.6}$$

This phenomenon is referred to as the piezo-galvanic effect (e.g., see Gurevich 1957). In order to estimate the amplitude of this effect we note that $\left| \nabla^2 \mathbf{V} \right| \sim V/\lambda^2 = V f^2 / C_l^2$, where λ and f are the wavelength and frequency, respectively. Then we get $j_d \sim \left(\varepsilon_0 \varepsilon_F V f^2 \right) / \left(e C_l^2 \right)$ while the potential difference within a wavelength is $\Delta \varphi \sim \varepsilon_F V / (e C_l)$. Taking the following numerical values of the parameters $\varepsilon_F = 5$ eV, $V = 5$ mm/s, $f = 100$ kHz and $C_l = 5$ km/s, we find that $j_d \sim 0.1$ nA/m^2 and $\Delta \varphi \sim 3$ μV. Notice that the potential difference between the edges of the metal sample can reach a much bigger value of about 10–100 mV under the shock compression of several metals (Mineev and Ivanov 1976). The relaxation time of this short-term effect is about 0.1–1 μs.

The field experiments on the ore deposit of poly-metals have shown that the presence of the conductive bodies or inhomogeneities in a rock could promote the amplification of radio-emission caused by acoustic wave propagation (Sobolev et al. 1980). In these experiments the acoustic waves were excited due to the detonation of HE with masses of 2–6 kg. The measurements of the electric field were performed at the distance 120 m from the explosion site and 130–140 m from the ore body/inclusion. The electromagnetic signals in the frequency band of 0.2–3 MHz were recorded for the time when the explosive wave crosses the ore body/inclusion.

9.1.2 Shock Polarization of a Dielectric

A similar effect arises in the course of propagation of a shock wave (SW) in a dielectric. The schematic plot of a typical laboratory experiment is shown in Fig. 9.1 (Mineev and Ivanov 1976). The investigated sample (1) was sandwiched between an electrode (2) and a metallic screen (4). The arrows indicate the direction of plane SW propagation. The electrode and the protective ring (3) were prepared from metals which have an acoustic impedance close to the impedance of the sample material. Input resistance of the oscillograph and the load resistance of the protective ring are shown by R and R_1. The parameters of the circuit were selected in such a way that it was equivalent to that of short-circuited capacitor, with plates consisting of the screen and electrode. The explosive device to produce the shock and the sizes of the sample was selected in such a way that the damping of a plane SW and the influence of lateral unloading can be neglected for the time of SW propagation through the sample.

An electric current is observed in the external circuit immediately upon initiation of the SW. This signal continues for the duration of SW propagation in the sample. The external circuit has no current supply; that means that the formation of the

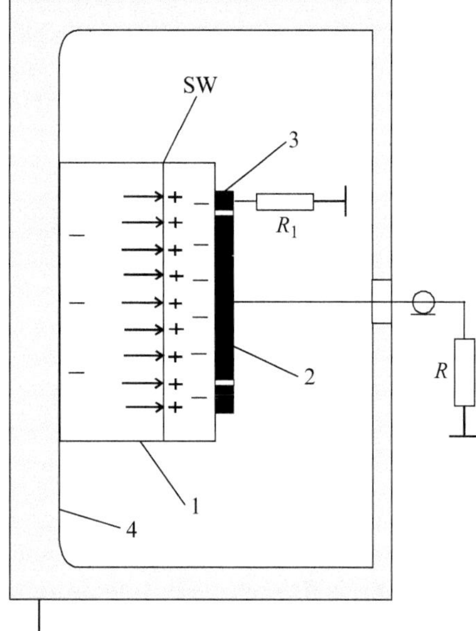

Fig. 9.1 A schematic arrangement of laboratory experiment devoted to the shock polarization effect. Adapted from Mineev and Ivanov (1976)

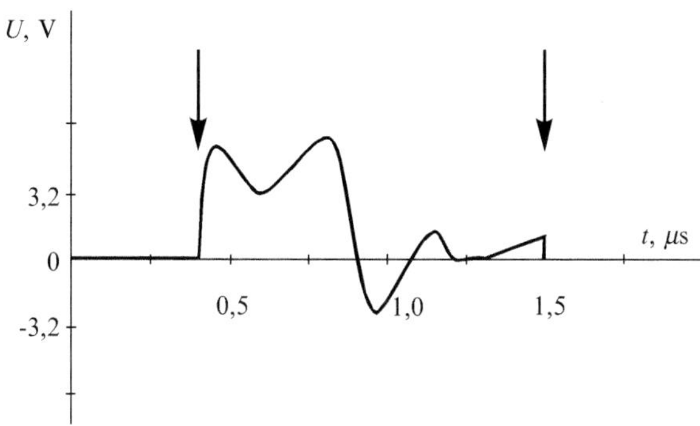

Fig. 9.2 A typical oscillogram of the polarization current caused by the shock loading of NaCl monocrystals. Adapted from Mineev and Ivanov (1976)

electromotive force (EMF) or a jump of electric potential most likely occurs at the shock front. The measured parameter is the voltage drop at the load resistance, R_1. The polarization current is directly proportional to the area, S, of the electrode. Therefore, the area of the protective ring, S_1, and the load resistance, R_1, are chosen so that $S_1 R_1 = SR$. In this case, the voltage drop at the resistances R and R_1 are equal. Under this arrangement the influence of lateral effects is avoided. The areas used were $S_1 = S = 1$–3 cm^2, the thickness of the sample was $l = 0.1$–3 cm, and $R_1 = R = 92$ Ohm are chosen to ensure the condition of a short-circuit. The latter means that the circuit relaxation time $t_r \ll l/V_s$ where V_s is the velocity of SW propagation.

A typical oscillogram of the polarization current is depicted in Fig. 9.2. The current results from shock loading of monocrystal NaCl with a pressure magnitude of 10 GPa (Mineev and Ivanov 1976). The vertical axis corresponds to the voltage drop at the load resistance, R. The current and voltage suddenly rise at the moment when the SW enters the sample and it abruptly decreases at the moment when the SW escapes the sample, as indicated by the vertical arrows.

The onset time of the SW front was 0.02–0.10 μs. Basically, this value is defined by non-simultaneity of the SW entrance into different parts of the sample. Tests with NaCl and KBr showed that the initial jump of the current is proportional to SV_s/l. The effective density of surface charges at the front of the SW amounted to 10^{-4}–10^{-3} C/m^2, and this value depends on the crystallographic direction of SW propagation. The relaxation time of polarization is 0.1–0.2 μs. These parameters are typical of all kinds of different ionic crystals that were tested (NaCl, LiF, KBr, RbCl, MgO, LiD, CsCl, and others). It was found that the features of the electric signals depend on the magnitude of the SW, characteristics of the atomic lattice of the tested crystal, concentration and type of alloy admixture, density of dislocations in the sample, and other parameters.

This shock polarization has been observed for many polar dielectrics: poly-methylmetacrylate, polyethylene, trinitrotoluene, polyamides pitch, dibutylftalate and water, as well as for a number of semi-conductors of p- and n-types (silicon, germanium and others). For these materials, the effective density of the surface charges at the SW varies between 10^{-7}–10^{-2} C/m^2 and the relaxation time is about 0.1–1 μs (Eichelberger and Hauver 1962).

The orientation of polar molecules along the direction of SW propagation is considered to be a plausible cause of shock polarization in polar dielectrics. Rotation of the molecules under the influence of mechanical stress could happen when the mass of one part of the molecule is greater than another. As a result, there occurs an induced dipole moment in the element of volume, i.e., the medium is polarized in the zone of SW compression. Thermal motion of the molecules eventually disorients the molecules and leads to relaxation of the shock polarization (Eichelberger and Hauver 1962). Some dielectrics acquire features of a conductor at the shock front that causes a decrease in the shock polarization. In this case, the polarization charges are shielded by the carriers of electric current.

A reverse effect, referred to as shock depolarization, has been observed in polar-ized seignette-electrics/ferroelectrics and piezoelectrics (Neilson 1957; Neilson and Benedick 1960). It was found that this phenomenon arises due to partial or total loss of the seignette-electrics/ferroelectric properties of a solid under shock compression.

9.1.3 Theory of Shock Polarization of Ionic Monocrystals

One possible cause of the shock polarization could be connected with the presence of charged dislocations (aged) and compensating clouds of point defects in the crystal (Linde et al. 1966; Wong et al. 1969). Mineev et al. (1967) assumed that the shock polarization in the ionic crystal could result from the diffusion of point defects through the SW front. Under the low pressure (up to 40–50 GPa) the key role had to be played by positively charged vacancies, because they are the carriers of electric current in the ionic crystal under normal conditions. It was also established that the amplitude of the electric signals depends on the initial concentration of alloy admixture (Ca^{2+}, Mn^{2+}, I^- and others) (Tyunyaev and Mineev 1973). Recently Freund (2000, 2002) have assumed that the shock polarization effect in different kinds of rocks can be due to a short-term increase in mobility of positive hole charge carriers, i.e., defect electrons/holes. The temperature increase at the shock front can result in transition from an insulator state to a conduction one similar to that in p-type semiconductors (Freund and Pilorz 2012).

It should be noticed that the charge of a moving edge dislocation can have a different sign for the cases of fast and slow deformation. The value of this charge depends on the different bonding energies of vacancies of negative and positive ions within a dislocation (Klein and Gager 1966; Tyapunina and Belozerova 1988). Taking into account this circumstance, we shall consider further both point defects and dislocations for the theoretical analysis of the shock polarization phenomenon. We start with the kinetics of point and linear lattice defects in the region of a SW front (Sirotkin and Surkov 1986). This approach has an advantage because it succeeds in accounting for the dependence of diffusion coefficients and the multiplication velocity of lattice defects on parameters of the SW and on the structure of the crystal lattice.

Let us consider 1D compression, along the x-axis, of a dielectric influenced by a plane SW. We suppose that the SW front has a finite spatial width. At first we consider the processes associated with electric charge transfer by point defects. An intensive multiplication of point defects gives rise to the generation of the so-called Frenkel defects, which represent the pairs composed of a vacancy and an interionic/interstitial ion. Compression of the crystal lattice causes distortion of the equilibrium configuration of ions in the vicinity of defects, which in turn can result in the defects motion.

Let $n_{i1}(x - a/2, t)$ and $n_{i2}(x + a/2, t)$ be the number density of defects in the sections with coordinates $(x - a/2)$ and $(x + a/2)$, where a denotes a lattice constant and i is a number of the defect species. Let $v_i^{\pm}$ be the frequencies of the defect jumps from an equilibrium place to another one, i.e., of one interatomic distance, a. The superscripts $\pm$ refer to displacement of the defects along and opposite to the direction of SW propagation (Malkovich 1982). The projection of particle flux density, f_{ix}, on the x-axis is assumed to be $f_{ix}(x, t) = \left(v_{i1}^+ n_{i1} - v_{i2}^- n_{i2}\right) a$, where $v_{i1}^+ = v_i^+ (x - a/2, t)$ and $v_{i2}^- = v_i^- (x + a/2, t)$. We expand the right-hand side of this expression in a power series of parameter a, and also take into account that motion of the material that results in transfer of the defects coupled with it. This can also be considered to be the transfer caused by an electric field. Thus, in the first approximation, adding to the total flux these additional fluxes, we get

$$f_{ix} = n_i a \left[v_i^+ - v_i^- - \frac{a}{2} \partial_x \left(v_i^+ + v_i^- \right) \right]$$

$$- \frac{a^2}{2} \left(v_i^+ + v_i^- \right) \partial_x n_i + n_i V_x + \frac{\sigma_i E_x}{q_i}. \tag{9.7}$$

Here V_x is the x-component of the material velocity in the SW, so that the term $n_i V_x$ describes defects transfer due to the material motion. The flux caused by the electric field is determined by the last term in Eq. (9.7). Here q_i is the particle charge for the given type, σ_i is ion conductivity, and E_x is the x-component of the electric field. Notice that all functions in the right-hand side of Eq. (9.7) are taken at the point x and at the time t, besides the functions σ_i and $v_i^{\pm}$ depend on temperature and mechanical stresses in the SW. The analysis shows that the difference between the functions v_i^{+} and v_i^{-} does not qualitatively change the resulting effect. Therefore in what follows we assume that $v_i^{+} = v_i^{-} = v_i$.

Generalizing Eq. (9.7) to 3D case we obtain the vector of the flux density

$$\mathbf{f}_i = -a^2 \nabla \left(v_i n_i \right) + n_i \mathbf{V} + \sigma_i \mathbf{E}/q_i. \tag{9.8}$$

The first term of Eq. (9.8) describes, as before, the motion of the defects with respect to atomic lattice. This term consists of two summands: $-a^2 v_i \nabla n_i$ and $-a^2 n_i \nabla v_i$. The first summand is usual diffusion flux due to the gradient of the number density of the defects and the value $a^2 v_i$ plays the role of diffusion coefficient. The next term can occur at the constant value of n_i. This flux arises from the deformation and heating of the lattice, which affect the jump frequency, v_i, of the defects. Below such effects are discussed in more detail.

The multiplication/reproduction rate of the defects under plastic deformation is proportional to the time-derivative of the strain deviator, $d_t \gamma$. As it follows from the observations, the number density of the point defects amounts to 10^{16}–10^{17} cm^{-3} per each percent of plastic strain in a SW (e.g., Klein 1965). The multiplication and recombination of the defects can be accounted for in the continuity equation and is given by

$$\partial_t n_i + \nabla \cdot \mathbf{f}_i = M_i d_t \gamma - \mu_{ki} n_k n_i. \tag{9.9}$$

Here M_i is the coefficient of defect multiplication, μ_{ki} is a recombination coefficient of vacancies and interionic/interstitial ions. The first term on the right-hand side of Eq. (9.9) describes the multiplication rate of defects whereas the second one describes their recombination. It follows from the charge conservation law that $M_k = M_i$ and $\mu_{ki} = \mu_{ik}$. Maxwell's equation for this case is

$$\varepsilon_0 \nabla \cdot (\varepsilon \mathbf{E}) = \rho_e, \qquad \rho_e = \sum_i q_i \left(n_i - n_{i0} \right), \tag{9.10}$$

where ε is dielectric permittivity, n_{i0} is initial number density of the defects, and ρ_e is electric charge density. The coefficients μ_{ki} are proportional to both the particle scattering cross-section (of the order of $4a^2$) and the particle velocity (of the order of va), whence it follows that $\mu_{ki} \sim va^3$.

At the SW the source function $M_i d_t \gamma$ plays a significant role due to the high value of strain rate $d_t \gamma \cong 10^7$ s^{-1}. Taking the following typical parameters of SW: $M_i \cong 10^{25}$ m^{-3}, $n_i \cong 10^{24}$ m^{-3}, $a = 0.281$ nm (for a NaCl lattice), we obtain that the recombination term in Eq. (9.9) is small in comparison with the source function up to a frequency $\nu_i = 10^{12}$ Hz. Therefore, we can neglect this term in further calculations.

Now we estimate the different terms in Eq. (9.8) for the flux density of the defects. If the condition $ea^2 |\nabla v_i n_i| > \sigma_i E$ is valid at the SW, then the conductivity of the crystal can also be neglected. Using the expression $\sigma_i = e^2 a^2 n_i \left(v_i^+ + v_i^- \right) / (2k_B T)$, where T is the temperature and k_B denotes Boltzmann constant, then the above condition reduces to: $E < k_B T / (e\lambda)$, where λ is typical size/width of the SW in the crystal. Thus, we obtain the estimation $E < 10^5$–10^6 V/m for the following parameters: $T \cong 10^3$ K and $\lambda \cong 10^{-7}$–10^{-6} m.

In this approach we shall ignore both the relaxation processes associated with the particle recombination and the influence of the conductivity at the front of the SW, that is at the length λ where the gradients of all parameters are large. Although both these processes can be very important just behind the SW. Substituting Eq. (9.8) for $\mathbf{f}_i$ into Eq. (9.9), omitting subscript i, and considering the plane SW propagating along the x-axis, we come to

$$\partial_t n + \partial_x (nv) - a^2 \partial_x^2 (nv) = M d_t \gamma. \tag{9.11}$$

Here $d_t \gamma$ denotes the absolute value of the rate of shear plastic strain. Equation (9.11) belongs to Fokker–Planck type equation with the source term on the right. For the case of constant v the last term on the left-hand side of Eq. (9.11) reduces to the form which is typical for diffusion type equation with the diffusion coefficient va^2.

Notice that the same equation can be applied to the number density of defect electrons/holes whose jump frequency v can be very sensitive to the temperature at the SW (Freund 2000, 2002). For the case of charge transfer by edge dislocations, the parameters n and M stand for the number of dislocations per unit area and multiplication coefficient, respectively. The charge q now has the meaning of electric charge per unit of dislocation length and $v = c_d / a$ where c_d is the velocity of dislocation slip. The charge q is considered to be independent of dislocation velocity. Using these re-designations, Eq. (9.11) can be applied to the description of dislocation kinetics as well.

In what follows we deal with a stationary SW. The parameters of such a wave depend solely on the variable $\xi = (x - V_s t) / \lambda$, where V_s is the speed of the SW propagation and λ stands for the characteristic scale of the SW front. The solution of Eq. (9.11) can be found as a power series of small parameter $\alpha = v_0 a^2 / (\lambda V_s)$. In the first approximation we obtain (Sirotkin and Surkov 1986)

$$n - n_0 = M\gamma - \alpha \frac{d}{d\xi} \frac{v}{v_0} (n_0 + M\gamma). \tag{9.12}$$

Here n_0 and v_0 denote the number density and frequency of defect jumps ahead of the SW front, i.e., when $\xi \to \infty$. In the zero-order approximation ($\alpha = 0$), the variations of number density $n - n_0$ arise only from the defect multiplication. In the next approximation, the variations of the defect number density due to displacement of defects with respect to the lattice with frequency v are accounted for by adding further terms of the power series of α.

Combining Eqs. (9.10) and (9.12) for the case of only one type of defect we find the electric charge density $\rho_e\,(\xi)$ and the x component of the electric field $E_x\,(\xi)$ in the vicinity of the SW front

$$\rho_e = -\frac{qa^2}{\lambda V_s}\frac{d}{d\xi}v\,(\xi)\,[n_0 + M\gamma\,(\xi)],\tag{9.13}$$

$$E_x = \frac{qa^2}{\varepsilon\varepsilon_0 V_s}\,\{v_0 n_0 - v\,(\xi)\,[n_0 + M\gamma(\xi)]\},\tag{9.14}$$

If the point defects and dislocations move by means of a thermofluctuational mechanism, the frequency of defect jumps numbered by subscript i can be written in the form

$$v_i = v_{i*}\exp\left[-u_i/\,(k_B T)\right],\tag{9.15}$$

where k_B is Boltzmann constant and T is the temperature. Because the inharmonicity of lattice oscillations is insignificant ($\gamma \ll 1$), then the linear dependence of the activation energy u_i on the strain γ can be applied, that is $u_i = u_{i0} - \beta_i\gamma$, where u_{i0} and β_i are constant values.

Let the subscripts c and v denote the cations and vacancies, respectively. The estimates of the parameters entering these equation for a crystal lattice of NaCl have shown that the behavior of cations and vacancies of Na$^+$ in the SW are essentially different (Sirotkin and Surkov 1986). The rise of shock pressure and shear strain results in the increase of u_c and the decrease of u_v due to the opposite signs of the parameters β_c and β_v. This implies that the shock polarization effect in ionic crystals is rather due to vacancies than the cations. On the other hand the estimate of electric field amplitude based on the thermofluctuational mechanism of the defect displacement is not in agreement with observation at least for the case of ionic crystals.

Another scenario can be realized if the shear strain exceeds a threshold value of $\gamma_* = u_{c0}/\beta_c$ followed by above-barrier movement of cations. Taking into account that the activation energy is equal to zero while the frequency v reaches the maximal value v_* we come to the following relationships:

$$\rho_e = -\frac{qa^2 v_* M}{\lambda V_c}\frac{d\gamma}{d\xi},\tag{9.16}$$

$$E_x \approx -\frac{qa^2 v_*}{\varepsilon \varepsilon_0 V_c} (n_0 + M\gamma) \tag{9.17}$$

which are valid for $\gamma > \gamma_*$. The sign of ρ in Eq. (9.16) is positive because $d\gamma/d\xi > 0$ so that the positive charge is situated at the front of the SW. The charges with opposite sign are either situated in the volume of the sample behind the SW at the relaxation length or they are located on the sides of sample.

To estimate the charge density and electric field in the SW for the case of above-barrier defect motion, we use the numerical values: $\varepsilon = 5.3$, $\lambda = 0.1\ \mu$m, $v_* = 5 \times 10^{12}$ Hz, $M = 10^{25}$ m^{-3}, $n_0 = 10^{22}$ m^{-3} and $a = 0.3$ nm as well as the empirical parameters of shock-compressed NaCl at 10 GPa (Baum et al. 1975): $V_c = 4.4$ km/s, and the strain amplitude $\gamma_m = 0.22$. Then we find that $\rho \approx 3.6 \times 10^2$ C/m^3 and $E \approx 7.7 \times 10^5$ V/m.

For the case of dislocations, the charge, q_d, per unit length can be different for the thermofluctuational and the sliding stages. Taking the parameters of dislocations $q_d \approx 1.7 \cdot 10^{-11}$ C/m (Tyapunina and Belozerova 1988) and $M \approx 10^{17}$ m^{-2}, we obtain $\rho \approx 3.8 \times 10^2$ C/m^3 and $E \approx 8.2 \times 10^5$ V/m which is close to the above estimates for the vacancies.

The strain threshold for dislocations is equal to $\gamma_* = Y/K$, where Y is yield strength and K is the modulus of bulk compression. For instance, in case of NaCl the strain threshold of dislocations $\gamma_* \approx (3 - 6) \times 10^{-3}$ which is smaller by 2–3 orders of magnitude than the threshold of cations (Sirotkin and Surkov 1986). This means that the dislocation mechanism of shock polarization effect in ionic crystals can prevail over cation one due to the smaller threshold.

The effective charge density, Σ, per unit area of the SW front surface can be estimated through the magnitude E_m of the SW-induced electric field as follows: $\Sigma = 2\varepsilon\varepsilon_0 E_m$. Taking E_m from Eq. (9.17) gives

$$\Sigma \approx 2qa^2 v_* M \gamma_m / V_c. \tag{9.18}$$

Results of laboratory tests with shock-compressed NaCl samples are displayed in Fig. 9.3 (Mineev and Ivanov 1976). Figures a, b, and c correspond to the SW propagation along different crystallographic directions. The surface charge density on the SW is shown with circles while the theoretical dependences of Σ on the strain amplitude given by Eq. (9.18) are shown with solid lines. It is evident that the linear character of the plotted functions $\Sigma (\gamma_m)$ follows the linear dependence of the multiplication rate of dislocations or point defects on the plastic strain rate. The numerical values of the parameter M are chosen in such a way to fit the experimental data. The results are shown in the table.

Direction	M_d, 10^{17} m^{-2}	M_p, 10^{25} m^{-3}	E_*, 10^6 V/m	γ_*
$[1,0,0]$	4.5	4.7	4.7	0.29
$[1,1,0]$	8.4	8.5	9.5	0.32
$[1,1,1]$	2.3	2.4	2.6	0.31

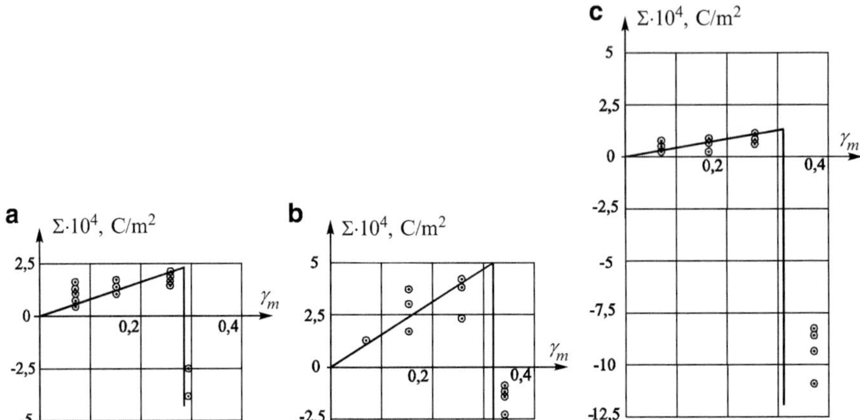

Fig. 9.3 Effective surface charge density at the SW front versus strain amplitude in a monocrystal NaCl. The plane SW propagated along the following crystallographic axes: (**a**) $[1, 0, 0]$, (**b**) $[1, 1, 0]$, and (**c**) $[1, 1, 1]$. The *circles* refer to experimental data (Mineev and Ivanov 1976) while the theory (Sirotkin and Surkov 1986) is shown with *solid lines*

Notice that the calculated multiplication coefficients of the dislocations, M_d, and of the point defects, M_p, are close to the experimental data (Pierce 1961) both for the case of a dislocation mechanism of charge transfer and for the case of a point defect. The strain threshold γ_* for the point defects is found in the last column of the table. As we have noted above, the mechanism of dislocations is more preferable because their strain threshold is smaller by 2–3 orders of magnitude. However, it should be noted that the parameters M_d and M_p are related as $q_d M_d \approx e M_p$. Based on this fact one can suppose that point defects are captured and carried away by dislocations; this means that such processes can define a value of dislocation charge.

Table shows maximum values of electric field, E_*, under the strain corresponding to the points of sharp changes in the trends of Fig. 9.3. The above analysis is not valid in the region of high strains, when a change in polarity of the electric signal is observed. What draws first attention is that the values of E_* are lower by 1–2 orders of magnitude than the critical field, E_c, of dielectric breakdown (e.g., $E_c \approx 1.3 \times 10^8$ V/m for NaCl). However, we should take into account that the increase in defect number density in the SW and substantial deformation of the lattice under $\gamma \approx 0.3$ can lead to the formation of additional local levels in the forbidden zone of the crystal, i.e., to the decrease of E_c. Thus one cannot exclude the effect of electron breakdown of the crystal to explain the change of signal polarity at high pressure. However this abrupt change of charge transfer mechanism in shock-compressed ionic crystals has been something of a mystery.

9.1.4   Phenomenological Models of Shock Polarization in Dielectrics

In phenomenological models of shock polarization effect, the structure of SW front is not considered and a jump/discontinuity of polarization at the SW front is usually assumed. The relaxation of shock polarization behind the front was supposed to be due to the thermal and mechanical relaxation (Allison 1965). The relaxation can also be a by-product of conductivity arising in the dielectric behind the SW front (e.g., Zeldovich 1967).

Now we consider a simple phenomenological model of the polarization current in the circuit of a shorted capacitor shown in Fig. 9.1 and try to explain the pattern of oscillogram displayed in Fig. 9.2. Following Zeldovich (1967) we first ignore the relaxation process due to the conductivity induced by the SW in the compressed medium. Then the electric charge has to be situated on the SW front and on the plates of a capacitor as schematically shown in Fig. 9.4. Let $-\Sigma_1$ and $-\Sigma_2$ be the surface charge densities on the right and left plates, respectively, while ε_1 and ε_2 are the dielectric permittivities of the medium in front of and behind the SW front, respectively. The plane SW is considered to be an extremely narrow layer or discontinuity of the medium parameters. If a load resistance of the external circuit is

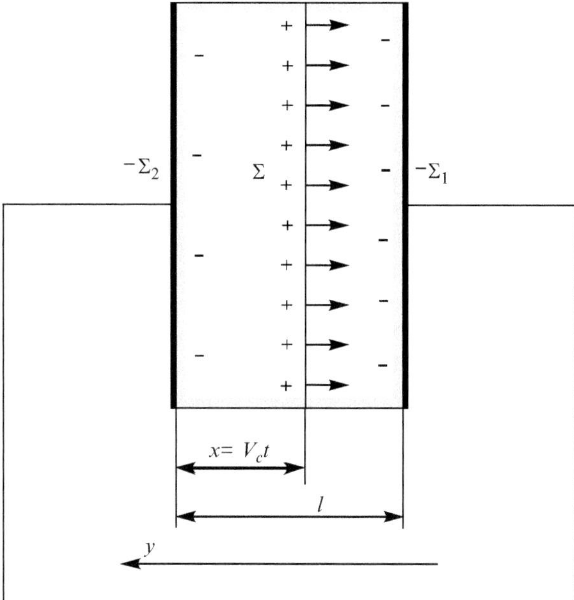

Fig. 9.4 A schematic plot of charge distribution in the shock-compressed sample and on plates of shorted capacitor. Here $-\Sigma_1$, $-\Sigma_2$, and Σ are the surface charge densities on the plates and SW front, respectively. The *arrows* show the direction of SW propagation in the sample

absent, then the potential difference between plates of the capacitor is equal to zero. In this case we have

$$\frac{\Sigma - \Sigma_1 + \Sigma_2}{2\varepsilon_2} x + \frac{\Sigma_2 - \Sigma_1 - \Sigma}{2\varepsilon_1} (l - x) = 0. \tag{9.19}$$

Here Σ is the surface charge density on the SW, l is a thickness of the dielectric, $x = V_c t$ is a distance traversed by the SW and $s = 1 + \gamma_m$ is a coefficient of shock compression of the material. Taking into account that $\Sigma = \Sigma_1 + \Sigma_2$ one can find the current density in the external circuit:

$$j = \frac{d\Sigma_1}{dt} = \frac{\Theta t_0 \Sigma}{[\Theta t_0 + t (1 - \Theta)]^2}, \tag{9.20}$$

where $\Theta = \varepsilon_2 s/\varepsilon_1$ and $t_0 = l/V_c$ is time of the SW propagation through the dielectric with initial thickness l.

It follows from Eq. (9.20) that the electric current density shows a maximum $j_m = j(0) = \Sigma/(\Theta t_0)$; that is, the current is not equal to zero at the initial moment. This result is due to the above approximation whereas, in reality, the current in the external circuit is formed during a finite time $t_r \approx RC$, where C is the capacitance of elements of the external circuit and R is the load resistance. In other respects, Eq. (9.20) gives a satisfactory approximation of the initial part of the signal shown in Fig. 9.2, at least of the interval between the first and second spikes.

Taking account of Eq. (9.18) for Σ we get an estimate of the current density magnitude:

$$j_m = \frac{\varepsilon_1 V_c \Sigma}{\varepsilon_2 s l} = \frac{2q a^2 v_* M \gamma_m}{\varepsilon_2 l (1 + \gamma_m)}. \tag{9.21}$$

Using the above parameters as well as $\gamma_m = 0.2$ and $l = 0.1$–1 cm, we find that $j_m \approx (0.2 - 2) \times 10^2$ A/m^2 which agrees with experimental data within an order of magnitude.

In the inverse case, if the relaxation time of shock polarization $\tau_r \ll t_0$, the qualitative distinctions of the current behavior will occur. Suppose that the matter behind the front of the SW becomes conductive with a constant conductivity σ_2. In such a case the charge relaxation time due to the conductivity of matter is of the order of $\tau_r = \varepsilon_2 \varepsilon_0/\sigma_2$ so that the above condition can be written as $\varepsilon_2 \varepsilon_0/\sigma_2 \ll t_0$. As before the plane SW front carries a surface charge with density Σ. To be specific in Fig. 9.4 this charge is chosen as positive. The rise of conductivity results in the formation of compensated charges with opposite sign behind the SW front. These charges are not shown in Fig. 9.4 because this figure displays the case of medium keeping of dielectric property behind of the SW.

Now we shall estimate a thickness of this oppositely charged layer and a potential difference between the capacitor sides. We use a reference frame and coordinate system connected with the moving SW front. The y-axis is directed oppositely to the SW velocity (Fig. 9.4). In this reference frame, the mass conservation law reads $V = V_s/s$, where V is the velocity of matter just behind the SW front and s is the shock compression, i.e., the ratio of the densities of the material behind and in front of the SW. The bulk charge motion at the velocity V together with the matter gives rise to convective current. The total current density caused both by the convective and conductive current is given by

$$j_y = \sigma_2 E_y + V_s \rho_e/s, \tag{9.22}$$

where ρ_e is the bulk charge density. The requirement of the current stationarity in the form: $j_y = 0$ allows us to find ρ_e

$$\rho_e = -s\sigma_2 E_y/V_s. \tag{9.23}$$

Substituting Eq. (9.23) into Poisson's equation (9.10) we obtain

$$\frac{dE_y}{dy} = -\frac{sE_y}{V_s\tau_r}. \tag{9.24}$$

Integration of Eq. (9.24) under the condition $E(0) = E_m = \Sigma/(\varepsilon_0\varepsilon_2)$ gives (Zeldovich 1967)

$$E = E_m \exp(-y/y_0), \quad y_0 = V_s\tau_r/s. \tag{9.25}$$

Laboratory tests of the NaCl samples of 0.1–1 cm thickness under a pressure of 10 GPa have shown that the shock-compressed matter exhibits a conductivity of the order of $\sigma_2 \approx 10^{-3}$–10^{-5} S/m (Mineev and Ivanov 1976). Under these parameters the relaxation length/typical size of the charged electric layer y_0 is of the order or much smaller than the thickness of the sample. This means that the effect of the stress-induced conductivity on the amplitude and shape of signals can be significant.

Since the load resistance of the external circuit is neglected, the potentials of the capacitor plates could be considered to be equal. The potential difference, $\Delta\varphi$, and effective capacity, C_e, formed by the front of the SW and the right plate of the capacitor are of the order of

$$\Delta\varphi = \int_0^\infty E_y dy = E_m y_0 = \frac{V_c\Sigma}{s\sigma_2}, \qquad C_e = \frac{\varepsilon_1\varepsilon_0 S}{l - V_c t}, \tag{9.26}$$

where S denotes the area of the plates. The current density in the external circuit is then

$$j = \frac{\Delta\varphi}{S}\frac{dC_e}{dt} = \frac{\varepsilon_1\varepsilon_0 V_c^2\Sigma}{s\sigma_2(l - V_c t)^2}. \tag{9.27}$$

It follows from Eqs. (9.26) and (9.27) that formally j, C_e and Σ_1 tend to infinity as $t \to l/V_c$. These results follow from the above approximations but, in fact, the charge density Σ_1 on the right plate of the capacitor cannot exceed the value of Σ. In this notation, the maximum value of j is not above the value of j_m in Eq. (9.21) that was found without accounting for the conductivity behind the SW.

Actually, the current $j = 0$ at the initial moment $t = 0$ since the external circuit contains both a non-nil load resistance R and capacity C. The shape of initial spike seen in Fig. 9.2 depends on the relation between the relaxation time of the external circuit, $t_r = RC$, and relaxation time τ_r due to stress-induced conductivity. The minimal parameter among these two ones determines the current rise time while the spike width is of the order of the maximal time parameter; that is, the current decreases down to the value of (9.27) for time on the order of maximal of these parameters. However, at least the initial portion of the signal shown in Fig. 9.2 is compatible with the predicted shape of the signal in both cases.

9.1.5 Shock Magnetization and Demagnetization of Magnetic Materials

In this section we consider ferromagnetic materials or rock bearing ferromagnetic inclusions which are immersed in an external magnetic field under normal conditions. Laboratory tests have shown that a shock compression of these materials gives rise to short-term shock magnetization followed by post-shock remanent magnetization of the samples (e.g., Nagata 1971; Pohl et al. 1975; Novikov and Mineev 1983). This effect is different from the usual magneto-elastic effect when the magnetization of the material is reversible (e.g., Goodenough 1954; Kondorsky 1959). The impact experiments on samples of iron, ferro- and ferrimagnets have shown that excess magnetization of the order of $(1 - 2) \times 10^6$ A/m arises at the shock front under shock pressures of 11 GPa during about 1–10 μs. Since the magnetization disappears in unloading wave, the effect is assumed to be due to reversible reorganization of domains into the SW along the direction of the external field (Novikov and Mineev 1983).

If the shock pressure in iron exceeds 11 GPa, then the inverse effect of abrupt decrease of magnetization at the SW front begins to prevail (e.g., Anderson and Neilson 1957; Royce 1966). Possibly the cause of the effect lies in reconstruction of the atomic lattice of iron in the SW from an α-phase (magnetic) to an ε-phase (non-magnetic) under pressures of 13 GPa. It should be noted that the partial transformation of iron from the ferromagnetic phase into a paramagnetic one was observed under a lower pressure of 5 GPa (Keller and Mitchell 1969). Substantial demagnetization (about 90 % under the pressure of 4 GPa) of nickel ferrites has been observed by Royce (1966). The most likely cause of the shock demagnetization is the decrease in the Curie temperature due to the pressure and temperature increase in the SW (e.g., Wayne 1969).

Brovkin et al. (1990) used the electromagnetic technique based on the shock demagnetization effect in order to distinguish individual detonations during a group shot. The detonator bodies were magnetized prior to detonation which results in a significant increase of the pulse amplitude. This phenomenon is caused by shock demagnetization of the ferromagnetic components during the destruction of the detonator body.

9.1.6 Remanent Magnetic Effects

Modification of crustal magnetization by strong explosions, EQs and by giant impact events can be due to both the phenomena; that is, the shock-induced remanent rock magnetization and partial or complete shock demagnetization of pre-existing rock remanence. The irreversible changes in the local geomagnetic field have been observed around the detonation point several minutes or hours after the detonation moment (Stacey 1964; Barsukov and Skovorodkin 1969; Hasbrouk and Allen 1972; Erzhanov et al. 1985). The effect of shock on the remanent magnetization of rocks and other magnetic anomalies resulted from impact cratering of planetary surface on the Mars (e.g., Hood et al. 2003; Louzada et al. 2011) and the Moon (e.g., Halekas et al. 2003; Hood and Artemieva 2008) have been extensively studied.

As one example, consider the irreversible magnetization of the granite samples studied in laboratory conditions (Shapiro and Ivanov 1969). The cubic samples with volume of 27 cm^3 were struck by the free falling weight with mass of 0.68 kg. The height of the fall was 7 cm. The primary natural magnetization was 1.5–2.6 A/m. A series of several tens strikes gives rise to additional irreversible magnetization of the order of 2–5 A/m which was aligned with the vector of geomagnetic field. Then the phenomenon of magnetic saturation was observed. Further tests with different magnetically isotropic rocks immersed in a weak magnetic field (< 1 mT) have shown that the shock remanent magnetization is linearly proportional to the amplitude of the ambient magnetic field while the magnetization vector is parallel to the magnetic field independently of the angle between the vectors of the shock wave velocity and of the magnetic field (Nagata 1971; Pohl et al. 1975; Pohl and Eckstaller 1981; Nagata et al. 1983; Gattacceca et al. 2008; Funaki and Syono 2008). The shock remanent magnetization is different from thermoremanent magnetization by larger magnitude and by shift of a coercivity spectrum towards lower values (Gattacceca et al. 2008).

Macroscopic theory of this effect is based on the empirical parameters of rocks. Let $\mathbf{J}$ be the vector of primary magnetization of a medium. It is generally accepted to suppose that the increment of the magnetization, $\Delta\mathbf{J}$, caused by elastic stress in an anisotropic crystal can be expressed through the deviator of the stress tensor, s_{ij}, in the following way (e.g., Dobrovolskiy 1991)

$$\Delta J_i = \frac{3}{2} C_m \sum_{i=1}^{3} J_j s_{ij}, \tag{9.28}$$

where C_m denotes the piezomagnetic coefficient which is varied within 0.5–1.7 GPa^{-1} (e.g., Stacey 1964). More usually we deal with the isotropic magnet so that the directly proportional dependence is used to analyze the laboratory tests

$$\Delta \mathbf{J} = C_m \mathbf{J} s_n, \qquad (9.29)$$

where s_n is normal component of the stress tensor or pressure.

In Chap. 11 we use this equation in order to estimate the effect of irreversible rock magnetization under large-scaled surface and underground detonations.

9.2 Electromagnetic Effects Caused by Fracture of a Solid Dielectric

Here we dwell on the laboratory observations of HF and VHF electromagnetic signals, particle emissions microdischarges, and other electromagnetic effect caused by the fracture of samples. These experiments have served such purposes as understanding of the solid fracture process, the development of nondestructive testing of inelastic deformations during solid loading, examination of anomalous electromagnetic phenomena possibly associated with EQs, volcano eruptions and etc.

9.2.1 Electrical Charges on the Surface of Fractured Solid

Fracture and intensive deformation of a solid results in the formation of static and transient electromagnetic fields in a wide band of frequencies. As usual the surfaces of fractured samples contain chaotically distributed charged areas with different signs. For example, the surface charge density of the order of 10^{-7}–10^{-8} C/m^2 have been measured upon the fracture of basalt, peridotite, fine-grained marble, polymethylmetacrylate/plexiglas, and other materials (Parkhomenko 1968; Shevtsov et al. 1975; Balbachan and Parkhomenko 1983). We notice that the results of these experiments depend on moisture of the rock. The local charge density on the surface of fractured monocrystals of alkaline-halogen compound reaches the value $(5–7) \cdot 10^{-7}$ C/m^2 (Urusovskaya 1968; Kornfeld 1975, 1978).

Below we examine the dynamic laboratory tests with dry rocks. Heterogeneous electric structure of the surface is typical for both fractured and undamaged crystals except for extra pure crystals prepared in the vacuum. In the undamaged crystals the linear sizes of negatively and positively charged macro-areas are about 50–500 μm. The rock fracture results in appearance of the new charges occurring on the newly formed fresh surfaces of the fractured material. These fresh charges are distributed in the form of fluctuation mosaic areas with positive and negative signs. Finkel et al. (1979) have used the electrometer with electric probes to measure the surface charge density. It was found that in some areas the local density reaches the value about 10^{-4}–10^{-2} C/m^2 which is much greater than that on the surface of undamaged materials.

9.2.2 Radiowave Emission Resulted from Fracture of Dielectric Solids

The rock fracture is frequently accompanied by intensive electrical discharge processes (Deryagin et al. 1973; Klyuev et al. 1984; Lipson et al. 1986; Martelli et al. 1989). The estimations show that characteristic times of the discharges are about 0.1 μs and their typical linear sizes $\sim 10^{-2}$–10^{-3} cm. The discharge processes as well as a vibration of gas-discharge micro-plasma arising between crack walls produce a wide-band electromagnetic emission from acoustic frequencies up to UHF/EHF band. Impulsive radio-emissions have been observed under loading of minerals, rocks, and ionic crystals (Gol'd et al. 1975; Nitsan 1977; Warwick et al. 1982; Bivin et al. 1982; Brady and Rowell 1986; Cress et al. 1987; Yamada et al. 1989; O'Keefe and Thiel 1995; Frid et al. 2000, 2003; Goldbaum et al. 2003; Hadjicontis et al. 2004). Notice that the same effect has been recorded during the process of water crystallization possibly due to the microcracks formation in solid phase (Kachurin et al. 1982).

One of possible causes of the phenomena could be piezoelectric effect in the samples of quartz-bearing rocks (Nitsan 1977; Warwick et al. 1982). This electric field generation is assumed to be at the moment of sharp stress drop occurring under the fracture of sample. The experiments have been made on the samples with volume from 1 to 10^4 cm^3. Typical signals caused by fracture of samples look like a burst of damped vibration with total duration from 5–10 μs (tourmaline and quartz) to 200 μs (sandstone, average size of grains is 1 mm).

The piezoelectric effect is not a non-unique one which is able to explain observations. For example, the crack generation in basalt that does not include the quartz is also accompanied by occurrences of the electric signals and of the light flashes but these intensities are much smaller than those in the piezoelectric samples (Brady and Rowell 1986; Cress et al. 1987). This effect has been observed by Bivin et al. (1982) under destruction of a variety of materials such as vinyl-plastic, ebonite, wood, sand, loam, plasticine and so on. In this experiment a striker throwing has been employed to destroy the samples/targets. The velocity of the striker varies in the interval 5–500 m/s. The electromagnetic field caused by the striker hit on target was recorded by antennas placed about 1 m from the target. There are three kinds of the antennas: pole, frame, and ferrite rod. The signals of the order of 10–50 mV have been observed in all cases. Duration of the signals was 1–5 ms that approximately corresponds to the time of the striker braking into target.

One peculiarity of these tests is that the size L of the wave zone was much greater than characteristic wavelength of the radiation. Even if the characteristic frequency of the fracture-induced signals reaches the value of the order of 1 MHz, then $L \gg$ 300 m. Since the antennas were located at the shorter distances, only near-field components can be measured in these experiments. The complexity of the problem is that the electric field decreases inversely proportional to the distance cubed; that is, more rapidly than that in the wave zone. On these conditions the antenna operates as an opened capacity rather than the usual radiowave antenna. Moreover, the size of antenna is much smaller than a typical wavelength so that one should calibrate the antenna before performing a measurement in order to give a correct interpretation of these data. One more problem is that the standard formulas for energy radiated by the source cannot be applied in this area.

Warwick et al. (1982) have noted that the electromagnetic signals appear synchronously with acoustic emission of the fractured sample. The frequencies of about 50 kHz dominate in the recording of electric field and the magnetic field is maximized in the range of 1–2 MHz. The similar experiments reported by Yamada et al. (1989) have shown that the spectrum of electromagnetic signals caused by failure had a maximum within 0.5–1 MHz. Brady and Rowell (1986) and Cress et al. (1987) have found that the maximum of spectral density of the electric signals has lower frequency and lies in the range 0.9–5 kHz. Ogawa et al. (1985) have studied fracture of granite samples under uniaxial loading and bending moment of external forces. In both cases wide-band electric fields (0.01–100 kHz) were detected.

Oscillations in the radio-band lasted about 15 ms have been observed during compression of the basalt and granite samples (Martelli et al. 1989). The fractured samples were surrounded by seven inductance coils, which measure the signals within a frequency band from 500 Hz to 830 kHz. Martelli et al. (1989) have assumed that the observed effect could be resulted from the low-frequency vibrations of plasma arising under the fracture.

Simultaneous recording of the acoustic and electric signals before and during destruction of the samples has been made by Yamada et al. (1989). Cylindrical granite samples with length of 62.5 mm and diameter of 25 mm were undergone by uniaxial compression with slow strain rate (about 10^{-6} s^{-1}). In this case the complete failure of the samples occurred in tens minutes. The acoustic emission sensors placed in different points of the sample surface recorded the impulses due to the process of microcracking. To measure electromagnetic impulses the inductance coils were put on the cylindrical sample with small gap, and their resonance frequency is changed from 80 kHz to 1.2 MHz. It was found that 15 events of the electromagnetic emission fall on 211 events of acoustic emission in one of the experiments and 31 events against 135 ones were detected in the next experiment. However, these results could be different from those in the case of higher sensitivity of the electromagnetic sensors.

Yamada et al. (1989) have found that the acoustic emission increases up to the moment of complete failure of the sample whereas the electromagnetic one is more intensive at the early stage of loading. It was supposed that the acoustic and electromagnetic impulses are simultaneously emitted because they are controlled by the same cause; that is, by the fast growth (during 1–100 μs) of separate microcracks. There are a number of indirect evidences that the sources of signals were tension cracks but not shear ones. This conclusion is consistent with the fact that the preparatory stage of failure is usually accompanied by accumulation of the tension microcracks around the tip of the main shear macrocrack.

Occurrence of low-frequency electric signals prior to the failure has been observed under the uniaxial compression of samples placed in a hydraulic machine (Hadjicontis and Mavromatou 1994). The cube-shaped samples with size of 3 cm made from natural materials (quartz, granite and limestone) were loaded with constant velocity. The dislocation model has been proposed by the authors to explain the dependence of the signal magnitude on the loading velocity. At normal condition the dislocation segments are surrounded by clouds of the electrically charged point defects. It was hypothesized that the dislocations bend and shift relative to the clouds of the point defects. This effect,which in turn gives rise the medium polarization, can be resulted from the stress increase.

The electromagnetic effects possibly related to the rock fracture in the field experiments have been reported by O'Keefe and Thiel (1995). The electromagnetic signals with the frequencies over 5 kHz have been detected during quarry detonations at the distance of 60 m from the quarry. There are a few reasonable mechanisms of this phenomenon: (1) the failure of the rock caused by the detonation, (2) the electric discharges arising during the impact of the ground particles upon the shaft bottom, and (3) micro-fracture at the fresh walls of the quarry due to the rock unloading and stress relaxation. The last mechanism is supported by the fact that the separate pulses were observed one minute after the detonation.

9.2.3   Optical Emissions

Observational evidence for optical glow under the brittle failure of certain dielectrics (Belyaev et al. 1962; Deryagin et al. 1973; Brooks 1965; Thiessen and Meyer 1970; Altier et al. 1979) and seignette-electrics/ferroelectrics (Chandra and Shrivastava 1978) have been reported although this phenomenon is not typical for all matters. Brady and Rowell (1986) have studied the fracture of samples immersed in different gases. The spectroscopic observations have shown that the glow of the fractured samples only contains the typical emission lines of gases, which the samples were placed in. Using the helium atmosphere they have found intensive emissions with wavelength of 0.7065 μm corresponding to emission transition between atomic levels with energies 22.76 and 20.96 eV. The continuous spectrum due to emission from the ionized region, which might be formed in the fracture zones, was not revealed. Possibly, this part of the emission spectrum was outside the sensitivity area of the sensors because the ionized atoms mostly radiate in the UV region. It was proposed the atoms excitation is caused by bombardment of exoelectrons emitted from the fracturing sample.

The formation of the ionized areas under destruction of granite and basalt samples placed in vacuum or air have been observed by Martelli et al. (1989). The photomultiplier recorded the light flashes with duration of 20 μs for all used samples. The flashes had the shape of luminous vertical lines and spots appeared on the fracture surface but not on the outward sides of the sample. This effect could be assigned to known triboluminescence/mechanoluminescence phenomenon, which results from the mechanical excitation of the F-centers in solids (electrons connected with vacancies of negative ions). However, in this case the triboluminescence does not play a significant role in this effect because the basalt does not belong to triboluminescence materials in contrast to the granite. In all cases the flash core was surrounded by luminous fluxes of flying away dust particles. Since the jet glow arises in both the atmosphere and vacuum, the source of this emission is not only the atmospheric gases but also the ionized and excited atoms of the fractured solid itself as well as the gases originally contained in pores and cracks.

Likewise, the optical phenomena have been observed under failure and plastic deformation of metals (Abramova et al. 1971; Tupik and Valuev 1980). Intensity of the glow has a maximum in the optical and infrared range and its duration is about 0.1–1 μs. There have been proposed a number of possible mechanisms of this effect such as the excitation of photoluminescence by the electric discharges into rising cracks (Belyaev et al. 1962; Deryagin et al. 1973), the energy release due to dislocations coming at the metal surface (Molotskiy 1978), and excitation of the electron states resulted from rupturing of the atomic bonds (Tupik and Valuev 1980).

9.2.4   *Roentgen and γ-Radiation*

Fracture of both dielectrics and metals is frequently accompanied by Roentgen radiation. The peculiarity of this effect is that the radiation spectrum has no clear boundaries (Gorazdovskiy 1967; Klyuev et al. 1984; Lipson et al. 1986). It is conjectured that this radiation can be due to both the local plasma nucleations inside the cracks and the bremsstrahlung caused by deceleration of the particles emitted during the fracture. Moreover, Slabkiy et al. (1973) and Lipson et al. (1986) have observed the lines of the Roentgen characteristic radiation due to the excitation of the inner shells of atoms under the destruction of crystal lattice.

Gamma-radiation with energy of 4 MeV has been recorded under splitting of crystals of radioactive materials (Klyuev et al. 1986). Abnormal increase of the γ-radiation background has been observed in in-situ measurements after a detonation of high explosive with mass of 2 kg in ore body which was located in a polymetal deposit (Sobolev et al. 1980). The sensors of γ-radiation were placed at the distance 12–15 m from the detonation point. The measurement in the quantum energy range of 0.4–0.7 MeV has shown that the rate of γ-radiation impulses increases up to $(5 - 8) \times 10^5$ impulses per second during 1.5–2 ms after acoustic wave arrival at the sensors. The excess rate of γ-radiation impulses over typical background was more than 10 %.

9.2.5 Electron and Ion Emissions Under Solid Failure

In most cases the total charge of fragments of fractured samples has the positive sign (e.g., Kornfeld 1975). It is generally believed that this phenomenon is due to the short-term emission of the electrons during the fracture of samples. The electrons emitted in vacuum under dynamical fracture of the samples have been observed by Deryagin et al. (1973), Krotova et al. (1975), Wollbrandt et al. (1975), Vladikina et al. (1980), Dickinson et al. (1985), Bykova et al. (1987). The rock fracture in the atmosphere is also accompanied by the electron emission (Dickinson et al. 1981; Enomoto and Hashimoto 1990, 1992; Enomoto et al. 1993).

Under the quasi-static loading of ionic crystals the electron emission takes place when the strain reaches the threshold value of 0.02–0.1. The mechanoemission begins still before the occurrence of large cracks and splitting off (Zakrevskiy et al. 1979). This effect was assumed to be associated with intersection of the slipbands followed by the large local strain. Generation of the excited electron states, like excitons (pair of an excited electron and an associated hole) in these regions could result in electrons output into vacuum. The triboluminescence and the sharp increase of conductivity have been observed simultaneously with particle emissions.

The high energy electron emission has been observed under the dynamical fracture of alkali halide crystals in vacuum (Krotova et al. 1975). The photomultiplier with threshold response of 20 keV has detected about 30 impulses under the fracture of a single crystal of LiF. The average energy of the emitted electron was about 30 keV though sometimes the electron energy reached the value of 100 keV. During the splitting of mica the energy of mechanoelectrons varies from 10 to 100 keV. The emission of electrons with energy about 100 keV can be due to the excitation and ionization of inner electron shells of atoms, which are located in the surface layer of the crack. This assumption can be supported by the fact that the emission of mechanoelectrons is often accompanied by not only Roentgen radiation but also the bremsstrahlung (Slabkiy et al. 1973; Lipson et al. 1986).

The simultaneous emission of electrons and positive ions has been observed under both brittle failure of solid dielectrics (Lipson et al. 1986; Klyuev et al. 1986; Martelli et al. 1989) and plastic deformation of metals (Tupik and Valuev 1985). The peak of ion emission intensity has been shown to fall on the moment of the cracks formation and growth. Klyuev et al. (1986) have observed the emission of atomic hydrogen under the fracture of hydrogenous material. About 10 neutrons per one act of the rupture have been detected during the impact rupture of cubic monocrystals of LiD with sizes of 3–4 mm.

9.2.6 Theory of Electric Field Formation in a Crack

The laboratory tests have shown that the electrical charges on the fresh surfaces of fractured samples decrease in time due to both the relaxation processes and the screening of surface charges by ions absorbed from air. So one can expect that the charge density on the surfaces of growing main crack is much greater than that measured several minutes after the destruction of sample. The indirect hints towards the existence of high charge density and strong electric field during the crack growth are the electric discharges between crack sides, radio and optical emissions and other phenomena observed during the fracture of solid.

The electrons that were captured by surface traps can release from these traps due to thermal fluctuations thereby producing a population of free electrons near the crack surface. These electrons can be accelerated by the strong electric field resulted from the fluctuation-mosaic charges distributed on the crack sides. In the model by Molotskiy and Malyugin (1983) the crack surface is considered as a plane x, y. The surface charges density is supposed to be a twice periodic function of coordinates x and y given by: $\Sigma_c = \Sigma_m \cos(k_x x) \cos(k_y y)$, where k_x and k_y are the parameters of periodic structure of the charge distribution, and Σ_m stands for maximal charge density. Then the electric potential in the whole space can be derived from Laplace's equation with corresponding boundary conditions at the crack surface and infinity. In what follows we use the simplest way to estimate the electron energy in the framework of this approach.

Let l be the typical size of the charged cells so that $k_x = k_y = \pi/l$. At small distances from the cell and far away from its boundaries the electric field is similar to that of infinite charged plane. The field amplitude in the vicinity of the center of the charged cell can thus be estimated as $E \sim \Sigma_m / (\varepsilon \varepsilon_0)$. This upward-directed electric field decreases at the short distances from the surface with attenuation distance on the order of l, because the charge density distribution is an alternating-sign function with spatial period of l. Consequently the electron emitted from the crack surface and accelerated by this field could accumulate the maximal energy $w \sim eEl \sim el\Sigma_m / (\varepsilon \varepsilon_0)$. This estimation agrees with the results by Molotskiy and Malyugin (1983) on the order of magnitude.

The transverse electric field directed perpendicular to the crack sides is a short-range one. To explain how the electrons can accumulate the observed energy of 10–100 keV one needs to introduce a great local density Σ of the surface charges. During the dynamical stage of fracture the value of Σ_m should be 2–3 order of magnitude greater than the residual charge density on the surfaces of the fractured sample.

Other mechanism of the high-energy electron production during the fracture of dielectrics has been proposed by Surkov (1986) and Gershenzon et al. (1986). It is very likely that the electron releasing from the surface traps can be accelerated by the longitudinal electric field which is parallel to the crack sides. This long-range electric field is caused by opposite electric charges situated at the tip and sides of the growing crack.

Now we examine this mechanism for charge separation between crack tip and its sides during quasi-brittle fracture of a crystal dielectric that results in the generation of the longitudinal electric field inside the crack (Surkov 1986). The large surface curvature at the crack tip results in an enhancement of mechanical stress near the tip up to the level which is several orders of magnitude greater than the average stress in a sample. Therefore a plastic/craze zone is usually formed in the vicinity of the crack tip. The high value of strain rate in the plastic zone gives rise to intensive production/multiplication of dislocations and point defects. Taking the notice of different mobility of the lattice defects one may expect that the charge separation can occur in the plastic zone. A close analogy exists with SW, in which, as we have noted, the intensive deformation results in both the production of the lattice defects and the charge pile up at the SW front.

Suppose that there are two kinds of the defects with opposite charges $\pm q$. Let us multiply Eq. (9.9) for the first ($i = 1$) and the second ($i = 2$) kinds of defects by $q_1 = q$ and by $q_2 = -q$, respectively. Then taking the sum of these two equations, we come to the continuity equation for the electric current density $\mathbf{j}$:

$$\partial_t \rho + \nabla \cdot \mathbf{j} = 0, \tag{9.30}$$

where

$$\mathbf{j} = \rho \mathbf{V} - q a^2 \nabla (v_1 n_1 - v_2 n_2) + (\sigma_1 + \sigma_2)\, \mathbf{E}. \tag{9.31}$$

Suppose that the mobility and conductivity of the defects of the first kind are greater than those of the second kind, i.e. $|\nabla (v_1 n_1)| \gg |\nabla (v_2 n_2)|$ and $\sigma_1 \gg \sigma_2$. Besides we assume that v_1 is a constant value and neglect the term $\rho \mathbf{V}$ which describes the current due to the medium displacement with the velocity $\mathbf{V}$. For convenience, we shall omit the subscript 1 everywhere. Then combining Eqs. (9.30), (9.31), and (9.10), we get

$$\partial_t \rho + \rho/\tau + \nabla \cdot \mathbf{j}_d = 0, \qquad \mathbf{j}_d = -q D \nabla n, \tag{9.32}$$

where $\tau = \varepsilon \varepsilon_0 / \sigma$ is the charge relaxation time, $D = v a^2$ is the coefficient of diffusion, and $\mathbf{j}_d$ is the diffusion current density. In the case of the dislocation one should replace the above value of D by the following one: $D = c_d a$.

Consider a thin plane crack growing at constant velocity $\mathbf{V}_c$ along the x-axis in a sample, which is infinite in the direction of z-axis as shown in Fig. 9.5. Let y-axis be perpendicular to the crack surface. The normal component of total current density must be continuous at the interfaces ($y = 0$) between crack sides and air/vacuum. It follows from the symmetry of problem that $E_y = 0$ in the space between crack sides. Then the boundary conditions at $y = 0$ read

$$\partial_t E_y + \sigma E_y - q D \partial_y n = 0. \tag{9.33}$$

At the initial moment $t = 0$ when the crack begins to grow the charge density $\rho = 0$. Given the function $\mathbf{j}_d$ and n, Eq. (9.32) can be solved for ρ then

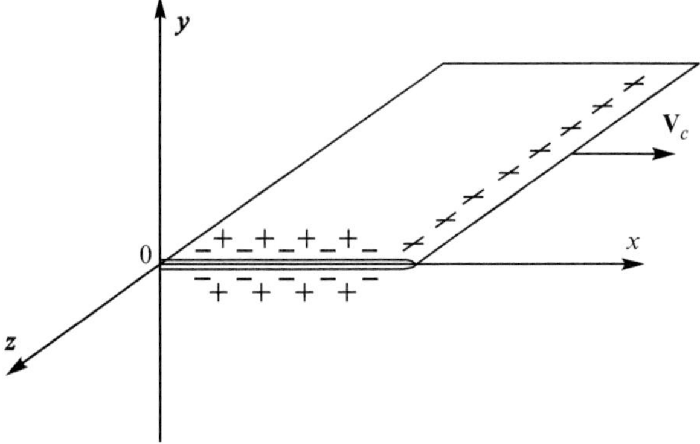

Fig. 9.5 A model of electric charge distribution near the tip and sides of thin tension crack growing in a sample (Surkov 1986)

$$\rho = qD \exp\left(-t/\tau\right) \int_0^t \exp\left(t'/\tau\right) \nabla^2 n \left(\mathbf{r}, t'\right) dt'. \tag{9.34}$$

The point defects/dislocations distribution in the plastic zone is assumed to be quasi-stationary. This implies that the defect number density is a function of variables $\zeta = (x - V_c t)/\lambda_\parallel$ and z, where $\lambda_\parallel$ is the characteristic length of the plastic zone. To be specific, consider the following approximation

$$n = n_m \left(1 - \exp\left(\zeta\right)\right) \exp\left(-|y|/\lambda_\perp\right) \eta\left(-\zeta\right), \qquad \zeta = (x - v_c t)/\lambda_\parallel, \tag{9.35}$$

where n_m is maximum of the defect number density, $\lambda_\perp$ is characteristic transverse size of the plastic zone, and η is the step-function. This implies that $n = 0$ as $\zeta > 0$; that is, in front of the crack. In the region $\zeta < 0$ the function n decreases with the increase of distance from the crack tip.

Consider first the extreme case when the characteristic length of the charge relaxation due to conductivity is much greater than the typical scales of the plastic zones, that is, $l_r = V_c \tau \gg \lambda_\parallel, \lambda_\perp$ and $\tau \gg t$. Solution of the problem shows that the electric charges predominantly pile up around the crack tip in the inner region of plastic zone in such a way that the charge per unit length of the crack front is $Q = 2qDn_m t \lambda_\perp/\lambda_\parallel$. Additionally, the charges are distributed along the crack sides in the DEL. This surface layer with width of the order of $\lambda_\perp$ has the total charge $-Q$. Both these charges vary proportional to time. In the inverse case when $t \gg \tau$, the charge per unit length of the crack front becomes approximately constant: $Q = 2qDn_m \tau \lambda_\perp/\lambda_\parallel$. The schematic charge distribution for the case of $q > 0$ is displayed in Fig. 9.5, and in this case the crack tip carries the positive charge Q. The mobile positive defects diffuse from the crack surface into the sample thereby producing the shortage of positive charges in the surface layer.

The low-frequency electric field is derivable from a potential function Φ through $\mathbf{E} = -\nabla\Phi$ while Φ satisfies the Poisson equation

$$\nabla^2 \Phi = -\frac{\rho}{\varepsilon\varepsilon_0}, \tag{9.36}$$

where the charge density ρ is given by Eqs. (9.34) and (9.35). Solution of Eq. (9.36) can be found in terms of the Green-function defining the potential of a uniformly charged infinite filament which is directed along y-axis. For example, if $t \ll \tau$ and $V_c t \gg \lambda_\parallel, \lambda_\perp$, then the difference of potentials between the sample surface with coordinate $x = 0$ and the crack tip with coordinate $x = V_c t$ is given by (Surkov 1986)

$$\Delta\varphi = -\frac{qDn_m t\lambda_\perp}{\pi\lambda_\parallel \varepsilon\varepsilon_0} \ln \frac{V_c t}{\lambda_\perp}. \tag{9.37}$$

The potential difference reaches a peak value at $t \sim \tau$ and then remains practically constant.

The implication of these results is that Eq. (9.37) determines the potential difference between two infinite parallel oppositely charged filaments with the above calculated charges Q and $-Q$. If $t \ll \tau$, the distance between these filaments is equal to the crack length $V_c t$ or, if $t \gg \tau$, this distance is of the order of relaxation length $l_r = V_c\tau$.

Gershenzon et al. (1986) reported that the charge concentrated at the crack tip can move together with the crack. The measurement of a stub antenna performed at short distance from fractured LiF samples has shown that the output electric signal increases approximately linearly with the crack length. Thus the dependence given by Eq. (9.37) is in qualitative agreement with the measurements. Substituting the following parameters of the point defects and the sample material: $n_m = 10^{23} - 10^{24}$ m^{-3}, $\nu = 5 \times 10^{12}$ Hz, $a = 3 \times 10^{-10}$ m, $\lambda_\perp = 10\lambda_\parallel = 1$ mm, $V_c t = 1$ cm, $\sigma = 10^{-5}$ S/m and $\varepsilon = 5$ into Eq. (9.37), we obtain the theoretical estimate $\Delta\varphi \approx$ 10–100 kV. This means that the electrons emitted from the crack surfaces can be accelerated in the longitudinal field up to the energies of 10–100 keV. The surface charge density at the tip of growing crack can be estimated as $\Sigma \sim Q / \left(2\lambda_\parallel\right) \sim 3 \times 10^{-3}$ C/m^2, which is by one order of magnitude smaller than that used by Molotskiy and Malyugin (1983) in the crack model with transverse electric field. Note that the possibility for the dislocation mechanism of charge separation cannot be excluded. For instance, if $q_d = 1.7 \cdot 10^{-11}$ C/m, $n_d = 10^{16}$ m^{-2} and $c_d = 1.5$ km/s, then we get the values of $\Delta\varphi$ and Σ of the same order of magnitude as above.

Thus, the observation of mechanoelectrons with energies 10–100 keV during the fracture of samples can be explained both by the ionization of inner electron shells of atoms and by the generation of the strong electric field predominantly directed along the main crack. Perhaps these two mechanisms can be distinguished by measuring the angular distributions of both the emitted mechanoelectrons and the Roentgen bremsstrahlung and characteristic radiation.

9.2.7 Electric Field in Collapsing Pores

Deformation of heterogeneous materials is accompanied by an enhancement of strain near pores and inclusions followed by the generation of local plastic and craze zones around the inhomogeneities. For example, the fast pore compression arises while the SW propagates in porous matter. Large shear strain near the pore surface results in the intensive heating of matter in plastic zones. Sometimes the shock compression is even accompanied by the formation of local melting zone around the pores that results in both material strength degradation and collapse of pores and micro-voids (Dunin and Surkov 1982). The fast deformation of the dielectric material around the inhomogeneities can impact the electric effects caused by the presence of space charges near the inhomogeneities. Below we show that the accumulation of electric charges in the vicinities of micro-inhomogeneities, grains and crazings leads to the generation of a strong electric field which may result in local disruption of dielectrics.

Suppose that both the width of the SW front and typical distances between pores are much greater than a typical size of pores. First of all we study the compression of individual pores in the SW (Surkov 1991). The high compression of the matter results in intensive production of the dislocations and point defects in plastic zones which arise around the pores. As before we also suppose that there are only two kinds of defects with charges of opposite signs. Considering the pore collapse one should take into account primarily the fast displacement of the material around the collapsing pore. In this notation the flux density of the defects is basically determined by the velocity $\mathbf{V}$ of the lattice displacement. As a first approximation, one can ignore other terms in Eq. (9.8) and substitute the flux density $\mathbf{f}_i = n_i \mathbf{V}$ $(i = 1, 2)$ into Eq. (9.9). If the matter is incompressible, i.e., $\nabla \cdot \mathbf{f}_i = 0$, then the solution of Eq. (9.9) is given by: $n_i = n_{i0} + M\gamma$. Here we have neglected the recombination of the defects. As the next approximation, we will search a small correction δn_i to the defect number density n_i, i.e., $\delta n_i \ll n_i$. Substituting the density in the form of $n_i = n_{i0} + M\gamma + \delta n_i$ into Eqs. (9.8) and (9.9) we get

$$\partial_t \delta n_i + \nabla \cdot (\delta n_i \mathbf{V}) - a^2 \nabla^2 (n_{i0} + M\gamma) v_i + \frac{1}{q_i} \nabla \cdot (\sigma_i \mathbf{E}) = 0, \qquad (9.38)$$

where the conductivities σ_i depend on the defect densities. Combining Eq. (9.38) for $i = 1$ and $i = 2$ we come to the continuity equation given by Eqs. (9.30) and (9.31) in which one should substitute $n_1 = n_2 = n_0 + M\gamma$, where $n_0 = n_{01} = n_{02}$.

For simplicity, all the pores are assumed to be ball-shaped with the same radius b. As the porosity is small so that the pore interaction can be neglected, then the distributions of the strain, electric charges, and fields around the pore are approximately spherically symmetric. Combining Eqs. (9.10), (9.30), and (9.31), we obtain

$$\partial_t E_r + \frac{V_r}{r^2} \partial_r \left(r^2 E_r \right) + \alpha_0 \partial_r \left[n \left(v_2 - v_1 \right) \right] + \frac{E_r}{\tau_0} = 0, \qquad (9.39)$$

where

$$n = n_0 + M\gamma, \qquad \alpha_0 = \frac{qa^2}{\varepsilon\varepsilon_0}, \qquad \tau_0 = \frac{\varepsilon\varepsilon_0}{\sigma_1 + \sigma_2}. \tag{9.40}$$

Here r is the distance from the center of the pore, E_r and V_r are radial components of the electric field and medium velocity, respectively. Now we introduce the flux, Φ, of the vector $\mathbf{E}$ through the sphere with radius r, i.e., $\Phi = 4\pi r^2 E_r$. Then we replace Eulerian variables r and t by Lagrangian ones, i.e., by r_0 and t, where r_0 is initial radius of small element of the medium with current radius $r = r(r_0, t)$. On rearrangement, Eq. (9.39) reduces to

$$\partial_t \Phi + \frac{\Phi}{\tau} + 4\pi\alpha_0 r^2 \partial_r \left[n(v_2 - v_1) \right] = 0. \tag{9.41}$$

The total current which includes the displacement one must be equal to zero at the initial pore radius $r_0 = b_0$ whence it follows that

$$\partial_t E_r + \alpha_0 \partial_r \left[n(v_2 - v_1) \right] + E_r / \tau_0 = 0, \qquad (r_0 = b_0 + 0). \tag{9.42}$$

Notice that owing to symmetry of the problem $E_r = 0$ inside the pore.

The solution of Eq. (9.41) with zero initial condition and the boundary requirement given by Eq. (9.42) can be written in the form

$$\Phi(r_0, t) = 4\pi\alpha_0 \int_0^t \exp\left(\frac{t' - t}{\tau_0}\right) r^2 \partial_r \left[n(v_1 - v_2) \right] dt', \tag{9.43}$$

where $r = r(r_0, t)$.

In the case of incompressible solid the relationship between Lagrangian and Eulerian variables is the following: $r^3(t) - r_0^3 = b^3(t) - b_0^3$. Taking into account that the shear strain $\gamma = \partial_r \xi - \xi/r$ where $\xi = r - r_0$ is the radial displacement, we obtain

$$\gamma(r, t) = \left(b_0^3 - b^3(t) \right) / r^3(t). \tag{9.44}$$

The frequencies of the defect displacement, v_i, depend on distributions of the stress and temperature around the pore. However, under the pressure of the order of 1 GPa, when collapse of the voids into the SW occurs, the heating of the solid is small. According to the model of pore compression by Dunin and Surkov (1979a,b) the stress weakly depends on distance (by logarithmic law) in the plastic zone which surrounds the pore. Therefore the value of $v_1 - v_2$ is considered to be approximately independent of r. Taking into account this approximation, assuming that $n \approx M\gamma$ and substituting Eq. (9.44) for γ into the integral given by (9.43), yields

$$\Phi\left(r_0, t\right) = 12\pi\alpha_0 M\left(\nu_1 - \nu_2\right) \int_0^t \exp\left(\frac{t'-t}{\tau_0}\right) \frac{\left(b^3\left(t'\right) - b_0^3\right) dt'}{\left(r_0^3 + b^3\left(t'\right) - b_0^3\right)^{2/3}}. \tag{9.45}$$

Most of solids exhibit a great viscosity as the plastic flow occurs. This means that during the shock compression the radius of pore fluently decreases without any appreciable vibrations. Therefore the time-dependence of the pore radius can be approximated by a smooth function, for example,

$$b^3\left(t\right) = b_*^3 + \left(b_0^3 - b_*^3\right) \exp\left(-\frac{t}{t_s}\right), \qquad t \geq 0, \tag{9.46}$$

where the characteristic time, t_s, and final pore radius, b_*, depend on SW magnitude. In fact, the parameter t_s defines the time scale of SW or duration of the compression stage.

Consider first the case $t \ll \tau_0$ when the exponent in the integral (9.45) is approximately equal to unity. Then performing integration in Eq. (9.45), and transforming Lagrangian variables to Eulerian ones one can find the flux of electric field as a function of r and t. Taking the notice of $E = \Phi/\left(4\pi r^2\right)$ we come eventually to the following expression for the electric field

$$E_r\left(r, t\right) = 9\alpha_0\left(\nu_1 - \nu_2\right) M t_s \frac{g\left(r, b\right)}{r^2}, \tag{9.47}$$

Here we have used the following auxiliary function

$$g = r_0 - r - \frac{\left(b_0^3 - b_*^3\right)}{r_*^2}\left[\frac{1}{6}\ln\frac{\left(r_0 - r_*\right)^2\left(r^2 + rr_* + r_*^2\right)}{\left(r - r_*\right)^2\left(r_0^2 + r_0 r_* + r_*^2\right)}\right.$$

$$\left. + \frac{1}{\sqrt{3}}\left(\arctan\frac{2r + r_*}{r_*\sqrt{3}} - \arctan\frac{2r_0 + r_*}{r_*\sqrt{3}}\right)\right], \tag{9.48}$$

where $r_*^3 = r^3 - b^3\left(t\right) + b_*^3$ and $r_0^3 = r^3 - b^3\left(t\right) + b_0^3$ while the value of $b\left(t\right)$ can be found from Eq. (9.46).

In the case of $t \gg \tau_0$ the integrand in Eq. (9.45) has a maximum in the vicinity of the point $t' = t$. So, we can substitute $t' = t$ into the integrand except for the exponential factor $\exp\left\{\left(t' - t\right)/\tau_0\right\}$. Performing integration in Eq. (9.45) we come to the approximate expression

$$E_r\left(r, t\right) = 3\alpha_0\left(\nu_1 - \nu_2\right) M \tau_0\left(b_*^3 - b_0^3\right)\left(1 - \exp\left(-t/\tau_0\right)\right)/r^4. \tag{9.49}$$

This result directly follows from Eq. (9.31) if one supposes that $\mathbf{j} = 0$ and $\mathbf{V} = 0$ when time tends to infinity.

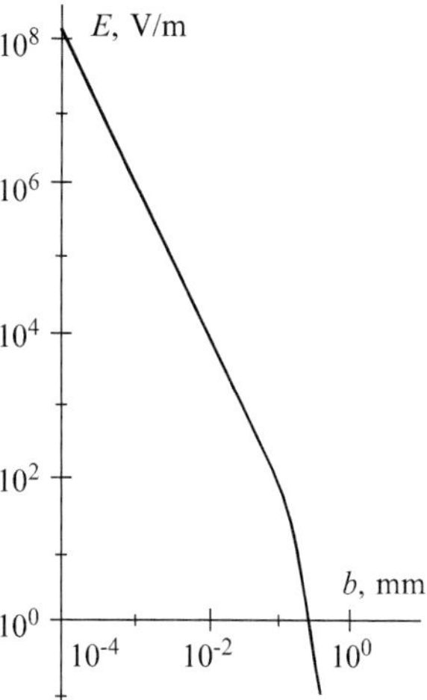

Fig. 9.6 Model calculation of the radial component of electric field at the surface of the collapsing pore versus pore radius (Surkov 1991)

The linear dependence $n\,(\gamma)$, which we have used above, holds true as the defect number density n is smaller than the density n_* of atoms in the crystal lattice. For example, taking $r = b$ and the numerical values $n_* \approx 10^{28}$–10^{29} m^{-3} and $M = 10^{25}$ m^{-3} we obtain that Eqs. (9.47) and (9.49) are valid if $b > b_c$ where the critical pore radius $b_c = b_0\,(M/n_*)^{1/3} \approx 0.1b_0$. In order to estimate the electric field for smaller pore radii one should consider that $n = n_*$. In the pore radius range of $b_* \ll b \ll b_0$ the electric field at the pore surface ($r = b$) can be approximated by the following expression (Surkov 1991)

$$E_r\,(b,t) = 3\left(15 + \sqrt{3}\right)\alpha_0\,(\nu_2 - \nu_1)\,b_0 t_s\,M^{1/3} n_*^{2/3}/\left(4b^2\right). \qquad (9.50)$$

The numerical value $\nu_2 - \nu_1 = 1$ s^{-1}, which is typical for ionic crystal under normal condition, can serve as the smallest estimate of this parameter. Then taking $t_s = 1$ ms and substituting these values into Eq. (9.50) we obtain that the electric field E_r begins to exceed the level of electrical breakdown for ionic crystals, which is approximately equal to 10^8 V/m, as the pore becomes smaller than $b \sim 0.5$ µm. A model calculation of E_r at the pore surface on the pore radius is displayed in Fig. 9.6, in which we need the above parameters and $b_* = 0.1$ µm. Notice that the electric discharge processes along with the recombination of the defects of crystal lattice can lead to the stabilization of electric field during pore collapse.

Light flashes and electron emissions during SW output from the sample into vacuum have been observed by Lyamkin et al. (1983) during high-pressure shock compression of powder materials. These findings are evidence in favor of the presence of strong electric fields in shock-compressed porous medium as it was predicted by the above theory.

9.3 Conclusions

The laboratory tests have shown that the effect of shock polarization is practically observed in all materials: dielectrics, metals, semiconductors and etc. The medium polarization occurs at different structural levels ranging from defects of crystal lattice to grain boundaries, cracks, pores, and other microscopic inhomogeneities. In monocrystals with ionic bond the shock polarization effect is due to the spatial selection of charged edge dislocations and point defects of lattice. This effect has the strain threshold associated with transition from the thermofluctuational mechanism of defect displacement to the over-barrier one. It appears that the strain threshold for dislocations is lower than that for the point defects. As the threshold is exceeded, the diffusion coefficient and mobility of the defects increase abruptly that result in the accumulation of electric charges at the SW front while the opposite charges leave at the sample surfaces.

Experimental evidence for the linear dependence of surface charge density on the SW front on strain amplitude has been observed under the compression of ionic crystals sandwiched between plates of short-circuited condenser. According to the theory this tendency follows the linear dependence between the rates of defect production and plastic strain. Drastic changes in signal polarity, as the shock pressure exceeds the certain threshold, are supposed to be due to the electron breakdown of ionic crystals.

A variety of electromagnetic effects has been observed under the fracture of rocks and other natural nonuniform materials. The polarization nuclei are mainly concentrated in plastic/craze zones which are located at the tips of growing cracks and in the vicinity of pores and inclusions. High strain rate inside these zones gives rise to intensive production of mobile charged linear and point defects of lattice that results in the generation of electric current. The electromagnetic variations caused by the rock fracture have a wide band spectrum in the range from 10 Hz to 1 MHz depending on the spatial scale of the fractured area. In laboratory conditions the maximum of intensity lies in the range 1–50 Hz. The typical signal generated by an individual microcrack looks as a burst of damped oscillations with total duration about 1–10 μs.

The fresh crack surfaces contain the fluctuation mosaic areas with positive and negative charges. The electrometric probe measurements have shown that the charge density can reach a value about 10^{-4}–10^{-2} C/m^2 at certain sites while the mean value on the cleavage surface is about 10^{-7}–10^{-8} C/m^2. However, the surface charge density can be much greater for the dynamical stage of crack growth. The indirect measurements are evidence for the generation of transient electric field $\sim 10^8$–10^9 V/m inside the cracks and pores. This is quite consistent with the theory which predicts that the cumulation of electric field in the vicinity of collapsing pore can result in local breakdown of dielectrics. Moreover, the short-term electron emissions with energies 10–100 keV have been observed during the fracture of dielectrics. This phenomenon is usually accompanied by Bremsstrahlung and even by characteristic X-ray radiation. The simultaneous emissions of electrons, positive ions, and X-ray radiation have been observed under fast plastic deformation of metals and alloys. It is conceivable that so high energy of electrons is due to the electric field originated from the electric charges accumulated at the crack tip and from the opposite charges distributed on the crack sides. This field is able to accelerate the free electrons in parallel with the crack surface. Besides, one may suppose that the intensive deformation of matter near the crack tip leads to the excitation and ionization of inner atomic shells, which in turn results the emission of both 100 keV electrons and characteristic X-ray radiation. The experimental evidence points to the presence of gas-discharge micro-plasma and electrical discharges between sides of the growing cracks. The duration of the discharges was estimated to be 0.1 μs and their linear sizes were of the order of 10^{-2}–10^{-3} cm. Optical measurements have shown that we observe the light flashes frequently during the destruction of different materials in vacuum and air. The flash duration varies from 0.1–1 μs for metals to 20 ms for the samples of granite and basalt. The light is radiated by at least two sites which are the fracture zone and the dusty streams flowing from the fracture zone. The sources of emissions are assumed to be the ionized and excited atoms of the fractured matter as well as the atmospheric gases.

The electromagnetic perturbations have been shown to connect with acoustic emissions of the fractured sample. The laboratory tests using a hydraulic machine have demonstrated that there arise the electric pulses only during the stress increase and the amplitude of the signals enhances with the increase of the loading rate. There is a few indirect evidence that the source of electromagnetic pulses is the tension microcracks rather than the shear ones. This is consistent with the observation that the buildup of the tension microcracks around the tip of main shear crack is typical for the preliminary stage of sample fracture. It is interesting to note further that the intensity of electromagnetic signals was higher at the early stage of the loading whereas the acoustic emission increases just before the moment of total sample destruction. This suggests that the same effects can be observed during large-scale tectonic processes associated with EQs, volcano eruptions and etc. However, in contrast to the laboratory tests, the range of typical frequencies is found to shift to the low frequency region; firstly because of the large sizes of the fractured zones located in the EQ focal area; and secondly due to the strong absorption of high frequency electromagnetic emission by the conducting layer of the ground.

References

Abramova KB, Valitskiy VP, Zlatin IA, Peregud BP, Pukhonto IYa (1971) Radiation caused by fast deformation and fracture of metals. Rep USSR Acad Sci (Doklady Akademii Nauk SSSR) 201(6):1322–1325 (in Russian)
Alekseev OG, Lazarev SG, Priemskiy DG (1984) On the theory of electromagnetic effects followed dynamical deformation of metals. J Appl Mech Tech Phys (Zhurnal Prikladnoi Mekhaniki i Tekhnicheskoi Fiziki) 4:145–147 (in Russian)
Allison FE (1965) Shock induced polarization in plastics. 1. Theory. J Appl Phys 36:2111–2113

Altier J, Camassel J, Mathieu H (1979) Effects of uniaxial stress on luminescence of GaP(S). Phys Rev Lett 72A:239–241

Anderson GM, Neilson FW (1957) Effects of strong shocks in ferromagnetic materials. Bull Am Phys Soc Ser II 2(6):302

Andreev SG, Babkin AV, Baum FA, Imkhovik NA, Kobylkin IF, Kolpakov VI, Ladov SV, Odintsov VA, Orlenko LP, Okhitin VN, Selivanov VV, Soloviev VS, Stanyukovich KP, Chelyshev VP, Shekhter BI (2002) Physics of explosion. Vol 1 and 2, Third edition, revised edition. In: Orlenko LP (ed). Fizmatlit, Moscow, 832 pp (in Russian)

Balbachan MYa, Parkhomenko EI (1983) Electret effect under rock fracture. Proc USSR Acad Sci Phys Earth (Izvestiya Akad Nauk SSSR Fizika Zemli) 8:104–108 (in Russian)

Barsukov OM, Skovorodkin YuP (1969) Magnetic observations in detonation region Medeo. Proc USSR Acad Sci Phys Earth (Izvestiya Akad Nauk SSSR Fizika Zemli) 5:68–69 (in Russian)

Baum FA, Orlenko LP, Stanyukovich KP, Chelyshev VP, Shekhter BI (1975) Physics of explosion. In: Stanyukovich KP (ed). Nauka, Moscow, 704 pp (in Russian)

Belyaev LM, Nabatov VV, Martyshev YuN (1962) On glow duration under tribo- and crystal-luminescence. Crystallography (Kristallografiya) 7(4):576–580 (in Russian)

Bivin YuK, Viktorov VV, Kulinich YuV, Chursin AS (1982) Electromagnetic radiation under dynamical deformation of different materials. Proc USSR Acad Sci Mech Solid (Izvestiya Akad Nauk SSSR Mekhanika Tverdogo Tela) 1:183–186 (in Russian)

Brady BT, Rowell GA (1986) Laboratory investigation of the electrodynamics of rock fracture. Nature 321(6069):488–492

Brooks W (1965) Shock-induced luminescence in quartz. J Appl Phys 36:2788–2789

Brovkin YuV, Dunin SZ, Popryadukhin AP, Surkov VV, Tochkin IN (1990) Recording of the number of explosions by electromagnetic methods. J Mining Sci 26(6):522–526

Bykova VV, Stakhovskiy IR, Fedorova TS, Khrustalev YaA, Deryagin BV, Toporov YuP (1987) Electron emission under fracture of rocks. Proc USSR Acad Sci Phys Earth (Izvestiya Akad Nauk SSSR Fizika Zemli) 8:87–90 (in Russian)

Caffin JE, Goodfellow TL (1955) Electrical effects associated with the mechanical deformation of single crystals of alkali halides. Nature 176(4488):878–879

Chandra BP, Shrivastava KK (1978) Dependence of mechanoluminescence in Rochelle-salt crystals on the charge-produced during their fracture. J Phys Chem Solids 39:939–940

Cress GO, Brady BT, Rowell GA (1987) Sources of electromagnetic radiation fromfracture of rock samples in the laboratory. Geophys Res Lett 14:331–334

Deryagin BV, Krotova NA, Smilga VG (1973) Adherence of solid. Nauka, Moscow, 279 pp (in Russian)

Dickinson JT, Donaldson EE, Park MK (1981) The emission of electrons and positive ions from fracture of materials. J Mater Sci 16:2897–2908

Dickinson JT, Jensen LC, Williams WD (1985) Fractoemission from lead zirconate-titanate. J Am Ceramic Soc 68(5):235–240

Dobrovolskiy IP (1991) Theory of preparation of tectonic earthquake. Inst. Phys. Solid Earth, USSR Acad. Sci., Moscow, 224 pp (in Russian)

Dunin SZ, Surkov VV (1979a) Dynamics of pore closing at shock front. Appl Math Mech 43(3):511–518

Dunin SZ, Surkov VV (1979b) Structure of a shock wave front in a porous solid. J Appl Mech Tech Phys 20(5):612–618

Dunin SZ, Surkov VV (1982) Effects of energy dissipation and melting on a shock compression of porous bodies. J Appl Mech Tech Phys 23(1):123–134

Eichelberger RJ, Hauver GE (1962) Solid state transducers for recording of intense pressure pulses. Les ondes de detonation, Paris, pp 363–381

Enomoto Y, Hashimoto H (1990) Emission of charged particles from indentation fracture of rocks. Nature 346(6285):641–643

Enomoto Y, Hashimoto H (1992) Transient electrical activity accompanying rock under indentation loadings. Tectonophysics 211:337

Enomoto Y, Akai M, Hashimoto H, Mori S, Asabe Y (1993) Exoelectron emission possibly related to geo-electromagnetic activities – microscopic aspect in geotribology. Wear 168:135–145

Erzhanov ZhS, Kurskeev AK, Nysanbaev TE, Bushuev AV (1985) Geomagnetic observations during MASSA experiment. Proc USSR Acad Sci Phys Earth (Izvestiya Akad Nauk SSSR Fizika Zemli) 11:80–82 (in Russian)

Finkel VM, Tyalin YuI, Golovin YuI, Muratova LN, Gorshenev MV (1979) Electrization of alkali-halide crystals during crystal split. Phys Solid (Fizika Tverdogo Tela) 21(7):1943–1947 (in Russian)

Fishbach DB, Nowick AC (1958) Some transient electrical effects of plastic deformation in NaCl crystals. J Phys Chem Solids 5(4):302–315

Freund F (2000) Time-resolved study of charge generation and propagation in igneous rocks. J Geophys Res 105:11001–11019

Freund F (2002) Charge generation and propagation in rocks. J Geodynam 33:545–572

Freund F, Pilorz S (2012) Electric currents in the Earth Crust and the generation of pre-earthquake ULF signals. In: Hayakawa M (ed) The frontier of earthquake prediction studies. Nihon-Senmontosho-Shuppan, Tokyo, pp 464–508

Frid V, Bahat D, Goldbaum J, Rabinovitch A (2000) Experimental and theoretical investigation of electromagnetic radiation induced by rock fracture. Isr J Earth Sci 49:9–19

Frid V, Rabinovitch A, Bahat D (2003) Fracture induced electromagnetic radiation. J Phys D Appl Phys 36:1620–1628. Doi:10.1088/0022-3727/36/13/330

Funaki M, Syono Y (2008) Acquisition of shock remanent magnetization for demagnetized samples in a weak magnetic field (7 μT) by shock pressures 5–20 GPa without plasma-induced magnetization. Meteorit Planet Sci 43:1–13

Gattacceca J, Berthe L, Boustie M, Vadeboin F, Rochette P, de Resseguier T (2008) On the efficiency of shock magnetization processes. Phys Earth Planet Inter 166:1–10

Gershenzon NI, Zilpimiani DO, Mandzhgaladze PV, Pokhotelov OA, Chelidze ZT (1986) Electromagnetic emission of crack tip under fracture of ionic crystalls. Rep USSR Acad Sci (Doklady Akademii Nauk SSSR) 288(1):75–78 (in Russian)

Gol'd RV, Markov GP, Mogila PG, Samokhvalov MA (1975) Impulsive electromagnetic radiation of minerals and rocks under mechanical loading. Proc USSR Acad Sci Phys Earth (Izvestiya Akad Nauk SSSR Fizika Zemli) 2:109–111 (in Russian)

Goldbaum J, Frid V, Bahat D, Rabinovitch A (2003) An analysis of complex electromagnetic radiation signals induced by fracture. Meas Sci Technol 14:1839–1844. Doi:10.1088/0957-0233/14/10/314

Goodenough JB (1954) A theory of domain creation and coercive force in polycrystalline ferromagnetic. Phys Rev Ser II 95(4):917–932

Gorazdovskiy TYa (1967) Effect of hard radiation under shear fracture of solids. Lett J Tech Phys (Pisma v Zhurnal Tehnicheskoi Fiziki) 5(3):78–82 (in Russian)

Gurevich LE (1957) On some electroacoustic effects. Proc USSR Acad Sci Ser Phys (Izvestiya Akad Nauk SSSR Seriya Fizicheskaya) 21(1):112–119 (in Russian)

Hadjicontis V, Mavromatou C (1994) Transient electric signals prior to rock failure under uniaxial compression. Geophys Res Lett 21:1687–1690

Hadjicontis V, Mavromatou C, Ninos D (2004) Stress induced polarization currents and electromagnetic emission from rocks and ionic crystals, accompanying their deformation. Nat Hazards Earth Syst Sci 4:633–639. Doi:10.5194/nhess-4-633-2004

Halekas JS, Lin RP, Mitchell DL (2003) Magnetic fields of lunar multi-ring impact basins. Meteorit Planet Sci 38:565–578

Hasbrouk WP, Allen JH (1972) Quasi-static magnetic field changes associated with CANNIKIN nuclear explosions. Bull Seism Soc Am 62:1479–1480

Hood L, Richmond NC, Pierazzo E, Rochette P (2003) Distribution of crustal magnetic fields on Mars: shock effects of basin-forming impacts. Geophys Res Lett 30. Doi:10.1029/2002GL016657

Hood LL, Artemieva NA (2008) Antipodal effects of lunar basin-forming impacts: initial 3D simulations and comparisons with observations. Icarus 193(2):485–502. Doi:10.1016/j.icarus.2007.08.023

Kachurin LG, Kolev S, Psalomschikov VF (1982) Impulsive radioemission caused by water crystallization and some dielectrics. Rep USSR Acad Sci (Doklady Akademii Nauk SSSR) 267(2):347–500 (in Russian)

Keller RN, Mitchell AC (1969) Electrical conductivity, demagnetization and high-pressure phase transition in shock-compressed iron. Solid State Comm 7:271–274

Khatiashvily NG (1981) Electromagnetic emission of ionic crystalls stimulated by acoustic wave. Lett J Tech Phys (Pisma v Zhurnal Tehnicheskoi Fiziki) 7(18):1128–1132 (in Russian)

Khatiashvili NG, Perel'man ME (1982) Generation of electromagnetic radiation during acoustic wave propagation in crystal dielectrics and several rocks. Rep USSR Acad Sci (Doklady Akademii Nauk SSSR) 263(4):839–842 (in Russian)

Khatiashvili NG, Perel'man ME (1989) On the mechanism of seismo-electromagnetic phenomena and their possible role in the electromagnetic radiation during periods of earthquakes, foreshocks and aftershocks. Phys Earth Planet Inter 57:169–177

Klein MJ (1965) The structure of explosively shocked MgO crystals. Phil Mag 12:735–739

Klein MJ, Gager WB (1966) Generation of vacancies in MgO by deformation. J Appl Phys 37:4112

Klyuev VA, Lipson AG, Toporov YuP, Aliev AD, Chalykh AE, Deryagin BV (1984) Characteristic radiation due to solid fracture and destruction of adhesive bonds in vacuum. Rep USSR Acad Sci (Dokl Akad Nauk SSSR) 279(2):415–419 (in Russian)

Klyuev VA, Lipson AG, Toporov YuP, Deryagin BV, Luschikov VI, Strelkov AV, Shabalin EP (1986) On high energy processes under fracture of solids. Lett J Tech Phys (Pisma v Zhurnal Tehnicheskoi Fiziki) 12(21):1333–1337 (in Russian)

Kondorsky EI (1959) On theory of magnetic state steadiness of ferromagnetic matter during magnetization. J Exp Theor Phys (Zhurnal Eksperimental'noy i Teoreticheskoy Fiziki) 37(4)(10):1110–1115 (in Russian)

Kornfeld MI (1975) Electrization of ionic crystal under plastic deformation and splitting. Adv Phys Sci (Uspekhi Fizicheskikh Nauk) 116(2):327–340 (in Russian)

Kornfeld MI (1978) Electrification of crystals by cleavage. J Phys D Appl Phys 11(9):1295–1301

Krotova NA, Wollbrandt Y, Khrustalev YaA, Linke E, Deryagin BV (1975) Generation of high energy electrons under fracture of solids. Rep USSR Acad Sci (Dokl Akad Nauk SSSR) 225(2):342–344 (in Russian)

Linde RK, Murri WJ, Doran DG (1966) Shock-induced electrical polarization of alkali halides. J Appl Phys 37(7):2527–2532

Lipson AG, Berkov VI, Klyuev VA, Toporov YuP (1986) On generation of vacuum spark during facture of solid. Lett J Tech Phys (Pisma v Zhurnal Tehnicheskoi Fiziki) 12(21):1297–1300 (in Russian)

Louzada KL, Stewart ST, Weiss BP, Gattacceca J, Lillis RJ, Halekas JS (2011) Impact demagnetization of the Martian crust: Current knowledge and future directions. Earth Planet Sci Lett 305:257–269. Doi:10.1016/j.epsl.2011.03.013

Lyamkin AI, Matytsin AI, Staver AM (1983) Electron emission during shock wave output from powder into vacuum. J Appl Mech Tech Phys (Zhurnal Prikladnoi Mekhaniki i Tekhnicheskoi Fiziki) 3:123–127 (in Russian)

Malkovich RSh (1982) Generalized diffusion equations in a crystal. Phys Solid (Fizika Tverdogo Tela) 24(2):463–465 (in Russian)

Martelli G, Smith PN, Woodward AJ (1989) Light, radiofrequency emission and ionization effects associated with rock fracture. Geophys J Int 98:397–401

Mineev VN, Tyunyaev YuN, Ivanov AG, Novitsky EZ, Lisitsyn YuV (1967) Polarization of alkali-haloid crystals under shock loading. J Exp Theor Phys (Zhurnal eksperimental'noy i teoreticheskoy Fiziki) 53(4):1242–1248 (in Russian)

Mineev VN, Ivanov AG (1976) Electromotive force produced by shock compression of a substance. Adv Phys Sci (Physics-Uspekhi) 19(5):400–419

Misra A (1978) A physical model for the stress-induced electromagnetic effect in metals. J Appl Phys 16:195–199

Molotskiy MI (1978) Dislocation mechanism of luminescence under fracture of metals. Phys Solid (Fizika Tverdogo Tela) 20(6):1651–1656 (in Russian)

Molotskiy MI (1980) Dislocation mechanism of Misra effect. Lett J Tech Phys (Pisma v Zhurnal Tehnicheskoi Fiziki) 6(1):52–55 (in Russian)

Molotskiy MI, Malyugin VB (1983) Energy spectrum of mechanoelectrons. Phys Solid (Fizika Tverdogo Tela) 25(10):2892–2895 (in Russian)

Nagata T (1971) Introductory notes on shock remanent magnetization and shock demagnetization of igneous rocks. Pure Appl Geophys 89:159–177

Nagata T, Funaki M, Dunn JR (1983) Shock remanent magnetization of meteorites. Mem Natl Inst Polar Res 30:435–446

Neilson FW (1957) Effects of strong chocks in ferroelectric materials. Bull Am Phys Soc Ser II 2(6):302

Neilson FW, Benedick WB (1960) Piezoelectric response of quarts beyond its Hugoniot elastic limit. Bull Am Phys Soc Ser II 5(7):511

Nitsan U (1977) Electromagnetic emission accompanying fracture of quartz-bearing rocks. Geophys Res Lett 4:333–336

Novikov VV, Mineev VN (1983) Magnetic effects under shock loading of magnetized ferro- and ferrimagnets. Phys Combust Explosion (Fizika goreniya i vzryva) 19(3):97–103 (in Russian)

O'Keefe SG, Thiel DV (1995) A mechanism for the production of electromagnetic radiation during fracture of brittle materials. Phys Earth Planet Inter 89:127–135

Ogawa T, Oike K, Miura T (1985) Electromagnetic radiations from rocks. J Geophys Res 90:6245–6249

Parkhomenko EI (1968) Electrization phenomenon in rocks. Nauka, Moscow, 210 pp (in Russian)

Parrot M (1995) Electromagnetic noise due to earthquake. In: Volland H (ed) Handbook of atmospheric electrodynamics. CRC Press, Boca Raton, London, Tokyo, pp 94–116

Pohl J, Bleil U, Hornemann U (1975) Shock magnetization and demagnetization of basalt by transient stress up to 10 kbar. J Geophys 41:23–41

Pohl J, Eckstaller A (1981) The effect of shock on remanence in multi-domain iron grains and implications for palaeointensity measurements. Lunar Planet Sci XII:851–853

Pierce CB (1961) Effect of pressure on the ionic conductivity in doped crystals of sodium chloride, potassium chloride and rubidium chloride. Phys Rev 123:744–754

Royce EB (1966) Anomalous shock-induced demagnetization of nickel ferrite. J Appl Phys 37(11):4066–4070

Shapiro VA, Ivanov NA (1969) Dynamical characteristics of impact magnetization of natural ferromagnetic samples. Proc USSR Acad Sci Phys Earth (Izvestiya Akad Nauk SSSR Fizika Zemli) 5:50–60

Shevtsov GI, Migunov NI, Sobolev GA, Kozlov EV (1975) Electrization of feldspar under deformation and destruction. Rep USSR Acad Sci (Dokl Akad Nauk SSSR) 225(2):313–315 (in Russian)

Sirotkin VK, Surkov VV (1986) Mechanism of space charge generation in shock compression of ionic crystals. J Appl Mech Tech Phys 27(4):495–500

Slabkiy LI, Odnovol LA, Kozenko VP (1973) On Roentgen radiation caused by collision between solids. Rep USSR Acad Sci (Dokl Akad Nauk SSSR) 210(2):319–320 (in Russian)

Sobolev GA, Demin VM, Los' VF, Maybuk YuYa (1980) Mechano-electrical radiation of ore bodies. Rep USSR Acad Sci (Dokl Akad Nauk SSSR) 252(6):1353–1355 (in Russian)

Stacey FD (1964) The seismomagnetic effect. Pure Appl Geophys 58:5–23

Stepanow AW (1933) Uber den mechanismus der plastisen deformation. Zs f Phys Bd 61(S):560–564

Surkov VV (1986) Emission of electrons during fracture of crystal dielectric. J Tech Phys 56(9):1818–1820

Surkov VV (1991) Electric field buildup in pore collapse. J Appl Mech Tech Phys 32(4):481–484

Surkov VV (2000) Electromagnetic effects caused by earthquakes and explosions. MEPhI, Moscow, 448 pp (in Russian)

Thiessen PA, Meyer K (1970) Tribolumineszenz bei verformungsvorgangen fester korper. Naturwiss Bd 57:423–427

Tyapunina NA, Belozerova EP (1988) Charged dislocations and properties of alkali-haloid crystals. Adv Phys Sci (Physics-Uspekhi) 156(4):683–717 (in Russian)

Tupik AA, Valuev NP (1980) Electromagnetic radiation under fracture of metals. Lett J Tech Phys (Pisma v Zhurnal Tehnicheskoi Fiziki) 6(2):82–85 (in Russian)

Tupik AA, Valuev NP (1985) Ion emission under deformation and fracture of metals and alloys. Rep USSR Acad Sci (Dokl Akad Nauk SSSR) 281(4):852–853 (in Russian)

Tyunyaev YuN, Mineev VN (1973) Polarization mechanism of aloyed alkali-haloid crystals in shock waves. Phys Solid (Fizika Tverdogo Tela) 15(5):1901–1904 (in Russian)

Vladikina TN, Derjaguin BV, Kluev VA, Toporov YuP, Khrustalev YuA (1980) Emission of high-energy electrons during friction and wear of dielectrics. J Lubric Tech 102(4):552–555. doi:10.1115/1.3251594

Urusovskaya AA (1968) Electrical effects under deformation of ionic crystals. Adv Phys Sci (Physics-Uspekhi) 96(1):39–60

Warwick JW, Stoker C, Meyer TR (1982) Radio emission associated wit rock fracture: Possible application to the great Chilean earthquake on May 22, 1960. J Geophys Res 87:2851–2859

Wayne RC (1969) Effect of hydrostatic and shock-wave compression's on the magnetization. J Appl Phys 40:15–20

Wollbrandt J, Linke E, Meyer K (1975) Emission of high energy electrons during mechanical treatment of alkali halides. Phys Status Solid A Appl Res A 27(2):K53–K55

Wong JW, Linde RK, White RM (1969) Electrical signals in dynamically stressed ionic crystals: A dislocation model. J Appl Phys 40(10):4137–4145

Yamada I, Masuda K, Mizutani H (1989) Electromagnetic and acoustic emission associated with rock fracture. Phys Earth Planet Inter 57:157–168

Zakrevskiy VA, Pakhotin VA, Vaytkevich SK (1979) Electron emission under uniaxial compression of ionic crystals. Phys Solid (Fizika Tverdogo Tela) 21(3):723–729 (in Russian)

Zeldovich YaB (1967) EMF arising during shock wave propagation in dielectric. J Exp Theor Phys (Zhurnal Eksperimental'noy i Teoreticheskoy Fiziki) 53(1):237–243 (in Russian)

Chapter 10
Electromagnetic Effects Resulted from Natural Disasters

Abstract The short-term EQ prediction on the basis of non-seismic technique is an intriguing problem since conventional seismic techniques provide us with long-term EQ forecast but cannot predict an impending EQ a few days or hours before main seismic event. Our prime interest is in the theories which are capable of explaining different electromagnetic phenomena possibly related to the EQs. In this chapter we consider the theories of ULF electromagnetic noise produced by rock fracture and crack formation. Here we explore GMPs, electrokinetic effect, variations of the rock basement conductivity, ionospheric perturbations, and other physical mechanisms which have been studied in previous chapters. Particular emphasis has been placed on the problem of direction finding for the ULF electromagnetic source. In the remainder of this chapter we discuss electromagnetic phenomena associated with large-scale natural disasters such as volcano eruptions, tsunamis, and hurricanes.

Keywords Crack formation • Direction finding problem • Earthquake (EQ) prediction • Tsunamis • Volcano eruptions

A destructive effect of natural catastrophes such as an EQ and a tsunami brings about great damage and loss of lives. One of the challenges of short-term EQ prediction is to know enough about physical mechanisms triggering a main shock. Since the discovery of tectonic plate motion several decades ago, the short-term prediction problem can be seemingly solved for a short time. At the time being it is clear that the tectonic plate dynamics can predict only a possibility of EQ occurrence on the geological time scale (10–100 years) that may offer long-term prediction, but nothing tells about short-term EQ prediction. Foreshocks and other seismic precursors of EQs are very sporadic in nature, sometimes they appear before the quakes and sometimes they are absent at all. This demands the search for an alternative, that is, non-seismic techniques for predicting an impending

V. Surkov and M. Hayakawa, *Ultra and Extremely Low Frequency Electromagnetic Fields*, Springer Geophysics, DOI 10.1007/978-4-431-54367-1_10,
© Springer Japan 2014

EQ. In this picture the study of seismo-electromagnetic phenomena and search of electromagnetic precursors of EQs seems to be a promising direction of researches.

10.1   ULF Electromagnetic Variations Possibly Associated with Earthquakes (EQs)

10.1.1   Observations of ULF Electromagnetic Noise Before and After EQs

We start our study with observations of ULF electromagnetic noise occasionally measured before strong crust EQs. There are known at least three authentic seismic events, which were accompanied by the ULF electromagnetic noise several hours or days before and after the main shock. First of all, it is worth mentioning practically simultaneously published observations of the ULF noise before and after the large Spitak (Spitak, Armenia, December 8, 1988, with magnitude $M_s = 6.9$, hypocenter depth $h = 6$ km, Kopytenko et al. 1990) and the large Loma Prieta (Loma Prieta, California state, USA, October 17, 1989, $M_s = 7.1$, $h = 15$ km, Fraser-Smith et al. 1990) EQs. The electromagnetic activity lasted for several weeks for both events and they exhibit similar characteristics. For example, the noise intensity began to be enhanced 3–5 days before the Spitak EQ and 12 days before the Loma Prieta EQ. A maximum of the noise intensity occurred 3 h before the Spitak event and 4 h before the Loma Prieta event (Bernardi et al. 1991; Molchanov et al. 1992; Kopytenko et al. 1993).

The power spectrum of electromagnetic noise observed during October 1989 at Corralatios station, Central California is shown in Fig. 10.1 (Fraser-Smith et al. 1990). Taking the notice of the frequency of measured channel, one can estimate the amplitude of electromagnetic noise around the moment of Loma Prieta event as much as 4–6 nT. The typical noise amplitude associated with Spitak EQ was about 0.05–0.2 nT. To explain the lower amplitude of the signals during Spitak event as compared to Loma Prieta event, one should take into account the difference in the epicentral distances; that is, 129 km for Spitak and 7 km for Loma Prieta.

The third authentic seismic event occurred at Guam (50 km south of the island of Guam, August 8, 1993, $M_s = 8.1$, $h = 60$ km). An increase in ULF magnetic activity has been observed one month before the great EQ at Guam (Hayakawa et al. 1996; Kawate et al. 1998). The amplitude of electromagnetic noise around the moment of main shock reached a peak value about 0.1 nT at the epicentral distance 65 km.

This phenomenon can be considered as a possible candidate for ULF electromagnetic EQ precursor and thus it has been studied intensively for the last decade. Wide-band ULF perturbations possibly associated with two EQs happened at Southern California on April, 17, 1990, $M = 4.6$, and at the Northridge on January 17, 1994, $M = 6.7$ were observed before and during these EQs

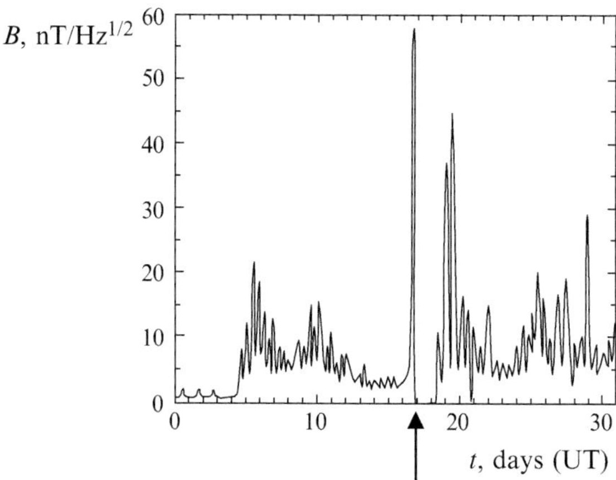

Fig. 10.1 Power spectrum of electromagnetic noise observed in October 1989 at Corralatios station, Central California. The frequency of measured channel is 7.32 mHz. The *arrow* shows the moment of Loma Prieta occurrence. Adapted from Fraser-Smith et al. (1990)

(Dea et al. 1991; Dea and Boerner 1999). An increased ULF magnetic activity has been observed before the great EQ in Indonesia, Biak, $M = 8.1$ at the distance about 80 km (Hayakawa et al. 2000). On the other hand Fraser-Smith et al. (1994) have not observed electromagnetic noise exceeding background or any unusual phenomena associated with either the 1992, $M = 7.4$ Landers EQ or the same Northridge earthquake. Based on this fact and taking into account that the receivers were located far away from the EQ epicenter, one can suppose that the zone of ULF precursor reception does not exceed about 100 km (Fraser-Smith et al. 1994). We will return to this point later after theoretical analysis.

However, some skepticism always persists as to the plausibility of ULF noise variations in possible association to impending EQs (Park et al. 1993; Geller 1996, 1997; Bakun et al. 2005; Campbell 2009; Thomas et al. 2009). Though this controversy is concentrated both on the measurement reliability and also on the mechanisms which are able to explain the observed effects.

In spite of this skepticism the majority of researchers believed that the ULF electromagnetic noise can be considered as a promising candidate for short-term EQ prediction. The networks of sensitive three-component ULF magnetometers and electric sensors have been installed in Kamchatka peninsula, Russia (Uyeda et al. 2002a; Gladychev et al. 2002), Japan (Hobara et al. 2002; Hattori 2004; Hayakawa et al. 2011), Taiwan (Hattori et al. 2002a), Greece (Makris et al. 2003), and India (Arora et al. 2012).

The recent studies of the ULF perturbations have shown that the signal-to-noise ratio is small (Fraser-Smith et al. 1990, 1994; Kopytenko et al. 1990, 1993; Johnston et al. 1994; Yepez et al. 1995; Hayakawa et al. 1996, 2000, 2007a, 2011; Akinaga et al. 2001; Ismaguilov et al. 2001; Gotoh et al. 2002; Hattori et al. 2002b, 2004; Hattori 2004, 2013; Ohta et al. 2005; Hobara et al. 2012) so that the problem is how to distinguish EQ-induced electromagnetic signatures from the background noise produced by ionospheric and magnetospheric currents, man-made interference and so on. The interested reader is referred to the books and reviews by Johnston (1989), Park et al. (1993), Hayakawa and Fujinawa (Eds., 1994), Parrot (1995), Hayakawa (Ed., 1999), Surkov (2000a), Hayakawa and Molchanov (Eds., 2002), Molchanov and Hayakawa (2008), Hayakawa (Ed., 2009) and Hayakawa (Ed., 2013) for a history and recent studies of electromagnetic phenomena which can be associated with the impending EQs.

10.1.2 Theory of Transient Electromagnetic Field Generated by Electric Charges on Crack Surfaces

In an early study of the EQ precursors, it was hypothesized by Gokhberg et al. (1979, 1982), Sadovsky et al. (1979), Warwick et al. (1982), Parrot et al. (1985) and Oike and Ogawa (1986) that the increase in natural radiowave emission occasionally observed before the EQ occurrence can serve as a possible candidate for the EQ precursor. The narrow-band receivers at frequencies from $f = 81$ kHz (Gokhberg et al. 1979, 1982) to 10 MHz (Warwick et al. 1982) have been used to detect the radiowave emission. It was generally accepted that the possible cause of the observed emission is oscillations of fluctuating charges on the sides of underground cracks (e.g., see Gershenzon et al. 1989).

First of all it should be noted that the skin-depth in the conducting ground at this frequency range is about $\delta \sim (\mu_0 \sigma f)^{-1/2} \approx 1$–30 m. Actually this means that the radiowave can come from the depth whose value is no more than 1–30 m because of strong damping of such a kind of emissions in the conducting ground (Surkov 2000a; Molchanov and Hayakawa 2008). Thus, the radiowaves coming from the ground surface cannot provide us with any information about tectonic processes in the EQ focal zone.

Moreover, the above frequency range seems to be typical for emission of microcracks. The electric charges formed during the microcrack growth can radiate electromagnetic waves at the frequencies which can be roughly estimated as $f \sim V_c / l_c$, where V_c is the velocity of crack growth and l_c stands for the crack length. Taking the mean values: $f = 1$ MHz and $V_c = 1$ km/s, we obtain the estimate of the crack size $l_c = 1$ mm. Surkov (2000a) has shown that any reasonable number density of such microcracks situated in the surface layer of the ground cannot explain the observed intensity of radiowave emission.

In the theory by Molchanov and Hayakawa (1994, 1995) the main emphasis is on the electric currents generated in the conducting rock due to the formation of microcracks. The ULF electromagnetic field was assumed to be excited by all the currents resulted from the development of microcrack ensemble around the focal zone. Below we do not follow the papers by Molchanov and Hayakawa (1994, 1995, 1998) and choose the simplest but not rigorous way to reproduce the main results of these papers.

Fig. 10.2 A schematic plot of macroscopic crack surrounding by the cloud of microcracks. A craze/plastic zone is restricted by *dashed line*. Adapted from Surkov (2000a)

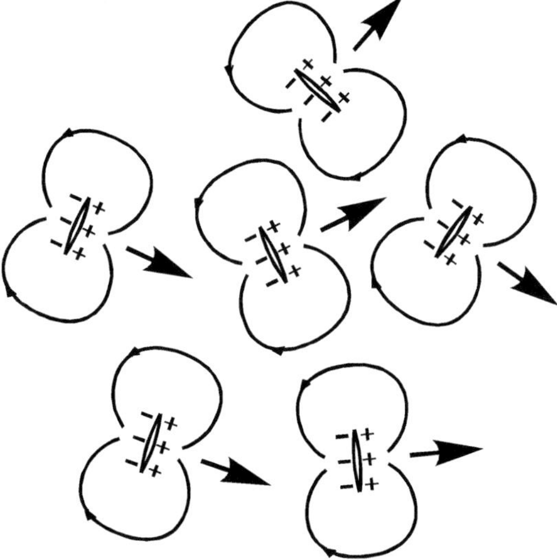

Fig. 10.3 Electric charges at the surfaces of microcracks. The *lines* show the directions of conduction currents that lead to the charge relaxation. The electric dipole moments of the microcracks are shown with *arrows*. Adapted from Surkov (2000a)

Laboratory tests have shown that the microcracks are formed in the so-called craze/plastic zones at the crack tip or in the vicinity of crack walls where the micro-failure remains trapped (Fig. 10.2). The formation and relaxation of electric charges on the sides of microcracks result in the excitation of wide-band electromagnetic noise and currents which are dissipated in the conducting media thereby transforming into ULF vibrations bounded above by the frequency $\sim 1\,\mathrm{Hz}$. The theory assumes that the sides of microcracks are homogeneously charged: one of them positively while the other negatively. The charges on the crack walls are situated in such a way that the vectors of electric dipole moments of all the microcracks are approximately parallel to each other as schematically shown in Fig. 10.3. Here a mechanism of the charge generation is not specific while the charge relaxation is assumed to be due to electric currents.

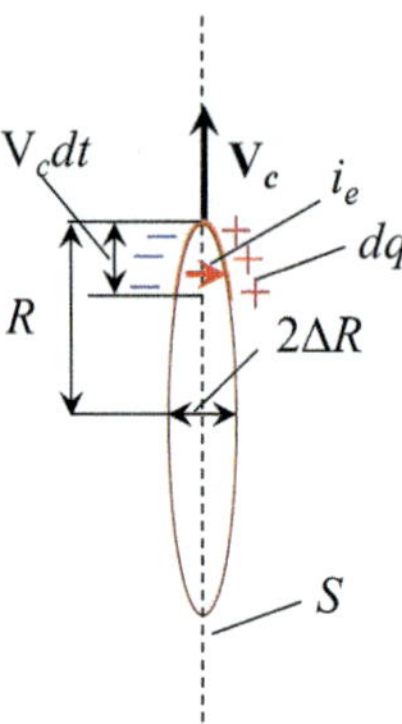

Fig. 10.4 Schematic plot of a disk-shaped crack expanding at velocity V_c. The extrinsic current flowing through the surface S is shown with *red arrow*

Consider first an individual disk-shaped microcrack increasing in size at constant velocity V_c. A schematic plot of the microcrack with radius R and half-width ΔR is shown in Fig. 10.4. Let dq be a small charge generated at the fresh crack sides forming for a short time dt due to the microcrack opening. As a result the extrinsic current i_e flows through the surface S as shown in Fig. 10.4 with red arrow. Taking into account that the fresh microcrack surface equals $2\pi R V_c dt$, the charge increment for the time dt is equal to $dq = 2\pi R V_c \Sigma_c dt$, where Σ_c denotes the surface charge density. For simplicity, we have ignored the rock conductivity that may lead to the charge relaxation due to conduction currents. Whence we find the extrinsic current produced by a single crack

$$i_e = 2\pi R V_c \Sigma_c. \tag{10.1}$$

Now we estimate the macroscopic current density J_e caused by the ensemble of expanding microcracks

$$J_e \sim i_e \Delta n \Delta R = 2\pi R \Delta R V_c \Sigma_c \Delta n, \tag{10.2}$$

where Δn is the number density of mobile cracks. Let n be the total number density of all the cracks, that is, mobile and stationary ones. Here we ignore the distribution in the microcrack sizes. The mobile crack number density can be thus estimated as $\Delta n \sim \tau \partial_t n$ where $\tau \sim R/V_c$. The rate of crack number density can be estimated as follows: $\partial_t n \sim n_*/t_*$, where $n_* \sim R^{-3}$ denotes the extreme crack concentration at which the multiple crack intersection happens that results in the complete rock disruption. Here the parameter t_* stands for the time scale of disruption process.

Now we leave out of account the field attenuation due to the skin effect in order to estimate the upper bound of the magnetic effect caused by microfracturing in the earthquake focal zone. According to *Biot–Savart–Laplace* law and Eqs. (7.2) and (7.3), the magnetic field due to the current J_e is on the order of

$$E \sim \frac{JV}{4\pi\sigma r^3}, \quad \delta B \sim \frac{\mu_0 J V}{4\pi r^2}, \tag{10.3}$$

where V denotes the volume of EQ focus. Substituting Eq. (10.2) for $J = J_e$ into Eq. (10.3), we obtain

$$E \sim \frac{\delta B}{\mu_0 \sigma r}, \quad \delta B \sim \frac{\mu_0 \Sigma_c \Delta R V}{2 r^2 R t_*}. \tag{10.4}$$

Taking the numerical values of the parameters $t_* = 10^3$ s, $R = 1$ mm, $\Delta R / R = 0.01$, $\Sigma_c = 10^{-3}$ C/m^2, $V = 10^3$ km^3, $\sigma = 10^{-3}$ S/m, and distance $r = 50$ km we come to the following estimates: $E \sim 30$ nV/m and $\delta B \sim 2$ pT.

The charges are actually randomly distributed on the crack sides forming the so-called fluctuating alternating-sign mosaic. Thus the mean value of the surface charge density at the crack sides can be much smaller than the maximal value of Σ_c which is typical for separate points of fresh crack sides. In this picture the used value $\Sigma_c = 10^{-3}$ C/m^2 appears as being overestimated.

However the main difficulty of this theory is incoherence of electromagnetic microfields produced by individual dipoles/charged cracks because the space orientation of dipole vectors is random rather than ordered. This means that the amplitude of the total field is proportional to the square root of the crack number but not to the crack number N. Taking the notice of a very large value of N, the above estimate of the microfracturing effect can decrease by many orders of magnitude.

In the other model (Vallianatos and Tzanis 1998; Tzanis and Vallianatos 2002), the main emphasis is on the motion of charged edge dislocations associated with the microfacturing process in the EQ focus. The majority of the edge dislocation is assumed to slip parallel to the applied shear stresses thereby producing a maximal value of the current. Since there may be several types of the charged dislocations with opposite charges, we suppose the excess of certain type of dislocation. Consider, for simplicity, only one type of such dislocations, the current density due to edge dislocation motion reads

$$J_d = q_d N_d V_d, \tag{10.5}$$

where q_d is the charge per unit dislocation length, and N_d is the number of dislocation per unit area. The mean dislocation velocity V_d is related to the strain rate $\dot{\varepsilon}$ through

$$\dot{\varepsilon} = b N_d V_d, \tag{10.6}$$

where b stands for the absolute value of Burgers vector. Combining Eqs. (10.5) and (10.6) leads to

$$J_d = q_d \dot{\varepsilon}/b. \tag{10.7}$$

Within a factor of the order of unity this equation coincides with the result obtained by Tzanis and Vallianatos (2002).

For the estimate we choose the following values for numerical parameters: $\dot{\varepsilon} = 10^{-4}\,\mathrm{s}^{-1}$, $q_d = 10^{-11}\,\mathrm{C/m}$, and $b = 0.5\,\mathrm{nm}$ (Tzanis and Vallianatos 2002). Substituting Eq. (10.7) for $J = J_d$ into Eq. (10.3) and taking above values of the parameters, we obtain the estimates $E \sim 0.6\,\mu\mathrm{V/m}$ and $\delta B \sim 40\,\mathrm{pT}$. In making this estimate we have assumed that the dislocation motion occupies the whole volume V of the EQ focal zone.

10.1.3   Theory of ULF GMPs Due to Acoustic Noise Produced by Rock Fracture and Crack Formation

In what follows we focus alone on GMPs from the rock fracture and energization of crack formation in the rock surrounding the fault zone. As we have noted in Sect. 7.3, the acoustic emission of the cracks results in the excitation of electric currents due to the motion of conductive ground in the geomagnetic field. In the case of tension cracks the effective magnetic moment of the electric currents must be pointed oppositely to the vector of geomagnetic induction. The magnetic moments of all the cracks have been shown to be co-directed independently of the crack plane orientation that gives rise to effective coherent amplification of the ULF GMPs, whereas the acoustic emission of the cracks is incoherent in nature and thus it cannot bring the same effect (Surkov 1997, 1999, 2000a,b; Surkov et al. 2003). This model can be extended for the shear cracks (Surkov 2000a, 2001; Molchanov et al. 2002) but in this case a certain ordering in the crack orientation is required in order to produce the coherent effect.

We recall that at far distance the electromagnetic precursor of acoustic wave (see Sect. 7.2.4) has the same shape and polarization for all the cracks while the next stage of the signal which associates with the acoustic wave arrival, consists of co-seismic oscillations, whose frequency and phase depend on the inclination of geomagnetic field, the crack size and the crack plane orientation. In an early study of the crack-generated GMPs, Surkov et al. (2003) took into account only the initial part of the signals, that is the coherent part, in order to avoid some mathematical complexities.

Here we follow the more accurate model (Surkov and Hayakawa 2006), which allows for an accidental character in the moments of the crack growth and formation. When calculating the net electromagnetic signal produced by all the cracks, we take into consideration both coherent and incoherent/co-seismic parts of the signals. The model includes such important details as random crack orientation, distribution of the crack sizes, and the attenuation of the acoustic waves.

According to Scholz (1990), the region of a preparing EQ can occupy the area with size of about several hundred kilometers. Accumulation of tectonic energy before an EQ results in the generation of a system of cracked zones with sizes of about 0.1–1 km at higher depth around the EQ focus. An enhancement of the crack formation activity inside this zone may give rise to both acoustic and electromagnetic noises, which can be detected on the ground surface. For example, the intense acoustic emission in the frequency band of 0.03–1 kHz along with noticeable ULF magnetic noise has been detected by Gorbatikov et al. (2002) at Matsushiro Observatory, Japan before and after seismic events occurred within distance range of about 150 km.

As the cracked zone is situated at higher depth, the rock conductivity will result in strong damping of the electromagnetic noise. Considering the ULF band, the cracked zone is supposed to be located at the depth which is smaller than the corresponding skin-depth, that is, no more than several kilometers from the ground surface. The rock fracture inside the cracked zone brings about the crack growth and formation of the fresh crack followed by radiation of acoustic waves. The sequence of acoustic impulses due to the crack growth is supposed to be a stationary random process which can be described by Poisson distribution given by Eq. (4.64). Magnitude of acoustic impulses radiated by a single crack depends not only on distance to the sensor but also on both the crack size and orientation of the crack plane with respect to the sensor.

The electric currents in the conducting ground and GMPs resulted from the acoustic emission of the crack ensemble form a random process as well. Thus, the net GMP, $\delta\mathbf{B}_t(\mathbf{r}, t)$, is the sum of random impulses. Additionally, we assume that the distance from the cracked zone to the ground-recording station is much greater than the typical cracked zone size. We obtain the following,

$$\delta\mathbf{B}_t(\mathbf{r}, t) = \sum_k \delta\mathbf{B}_k(\mathbf{r}, t - t_k, \mathbf{n}_k, l_k), \tag{10.8}$$

where the subscript k is the impulse number occurred during the interval $(0, t)$, and the unit vector $\mathbf{n}_k$ normal to the crack plane defines a random orientation of the crack. Likewise, the moment of radiation onset, t_k, and the crack size l_k are random values. Suppose that all these random values, t_k, $\mathbf{n}_k$ and l_k, are statistically independent of each other and their probability distributions are independent of the impulse number.

The mean value and other characteristics of Poisson random process are studied in more detail in Appendix B. In particular, if all the cracks are identical in size and orientation, the mean value of magnetic field is given by an equation similar to Eq. (4.71), that is

$$\langle\delta\mathbf{B}_t(\mathbf{r}, t)\rangle = \dot{N} \int_{-\infty}^{\infty} \delta\mathbf{B}_1(\mathbf{r}, t')dt', \tag{10.9}$$

where $\dot{N}$ is the number of cracks generated in unit time and $\delta\mathbf{B}_1$ denotes the magnetic field excited by a single crack. Notice that $\langle\delta\mathbf{B}_t(\mathbf{r}, t)\rangle$ is not a function of time since the Poisson process is a stationary random process.

To take into account the distribution of cracks over their sizes, we replace $\dot{N}$ by the function $\dot{N}(l)$ which denotes the number of cracks arising per unit time with a length greater than l occurring in a specified area. Instead of Eq. (10.9), we come to the following:

$$\langle \delta \mathbf{B}_t \rangle = - \int_0^{l_{\max}} \frac{d\dot{N}(l)}{dl} \mathbf{b}(\mathbf{r}, l)\, dl, \qquad (10.10)$$

where

$$\mathbf{b}(\mathbf{r}, l) = \int_0^{\infty} \langle \delta \mathbf{B}_1(\mathbf{r}, t, \mathbf{n}, l)\rangle\, dt, \qquad (10.11)$$

and $l_{\max}$ is the maximal crack size. The angular brackets in Eq. (10.11) denote the statistical averaging over the angles which determine the random orientation of the vectors $\mathbf{n}$. Below we show that the small cracks make a little contribution to the integral in Eq. (10.10) due to strong damping of the acoustic waves radiated by the small cracks. In this picture a minimal crack size in Eq. (10.10) is unimportant. Because of this statement the considered theory differs from that by Molchanov and Hayakawa (1994, 1995) where the main emphasis was on microcracks.

As before, the ground is supposed to be a uniform conductor immersed in the constant geomagnetic field $\mathbf{B}_0$. Consider first the electromagnetic perturbations, $\delta \mathbf{B}_1$ and $\mathbf{E}_1$, caused by acoustic emission of a single crack. The magnetic perturbations ($\delta B_1 \ll B_0$) satisfy the quasi-stationary Maxwell equation (7.12). For convenience, we introduce the vectorial and scalar potentials by Eqs. (5.73) and (5.74); that is, $\delta \mathbf{B}_1 = \nabla \times \mathbf{A}_1$ and $\mathbf{E}_1 = -\nabla \Phi_1 - \partial_t \mathbf{A}_1$. These potentials satisfy the standard gauge for a conductive medium (e.g., see Molchanov et al. 2002)

$$\nabla \cdot \mathbf{A}_1 + \mu_0 \sigma \Phi_1 = \text{const.} \qquad (10.12)$$

Substituting the above representations of the electromagnetic field through the potentials into Eq. (7.12), taking into account Eq. (10.12) and rearranging, we obtain

$$\partial_t \mathbf{A}_1 = v_m \nabla^2 \mathbf{A}_1 + \mathbf{V} \times \mathbf{B}_0, \qquad (10.13)$$

where $v_m = (\mu_0 \sigma)^{-1}$ is the coefficient of magnetic diffusion. The mass medium velocity, $\mathbf{V}(\mathbf{r}, t)$, can be expressed through the vector of medium displacement, $\mathbf{u}(\mathbf{r}, t)$, via $\mathbf{V} = \partial_t \mathbf{u}$.

In order to obtain the time-integrated magnetic perturbation $\mathbf{b}(\mathbf{r}, l)$ in Eq. (10.11) one should integrate Eq. (10.13) with respect to time from 0 to infinity under the condition that $\mathbf{A}_1(0) = \mathbf{A}_1(\infty) = 0$ and then take the mean value over the crack orientation. Thus, we get

$$v_m \nabla^2 \mathbf{a} + \langle \mathbf{u}_s \rangle \times \mathbf{B}_0 = 0, \qquad (10.14)$$

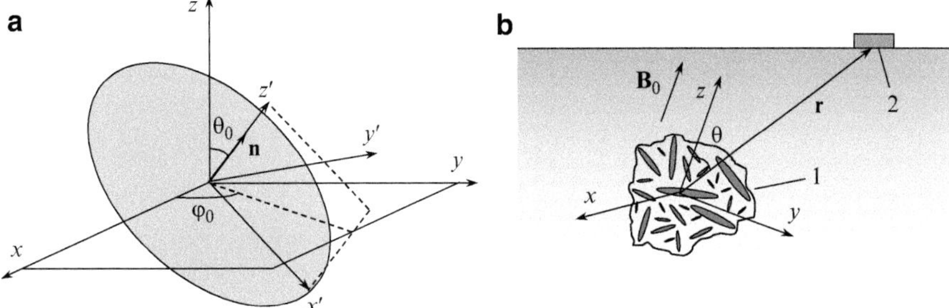

Fig. 10.5 Schematic plot of a cracked zone model. (**a**) The general reference frame and a local frame associated with the crack. (**b**) General scenario. *1*—the cracked zone, *2*—a ground-based recording station. Taken from Surkov and Hayakawa (2006)

where **a** is the mean value of $\mathbf{A}_1$, $\mathbf{b} = \nabla \times \mathbf{a}$ and the vector $\langle \mathbf{u}_s \rangle = \langle \mathbf{u}(\mathbf{r}, \infty) \rangle$ is defined as the static/residual displacements in the medium. These displacements are assumed to be maximal inside the cracked zone and they should gradually decline in the surrounding rock.

As one example, the displacement field for a tension crack is found in Appendix H. A coordinate system and random orientation of the crack plane are shown in Fig. 10.5a. The displacement components at the observation point depend on the random angles θ_0 and φ_0. For simplicity, an equal probability for the crack plane orientation inside the cracked zone is assumed. As follows from Eq. (10.75), the averaging of the displacement vector over the crack orientation gives only the radial component. This means that the mean displacement field of the crack ensemble is spherically symmetric at far distance from the cracked zone, and it is not surprising since the equal probability for the vector **n** orientation is assumed. Conversely, if the probability distribution for the crack plane orientation is non-spherically symmetric, there must occur certain declination from Eq. (10.75).

For reasons of convenience, all the cracks are considered to have the same disk-shaped form with different radius R. The displacement discontinuity/jump, $[u_z]$, normal to the crack surface is considered as a given function of time. The static value of the discontinuity (at $t \to \infty$) is supposed to be proportional to the crack length $l = 2R$, so that $[u_z] = kl$ where $k = 0.001$–0.01.

As has already been stated, the attenuation of acoustic waves due to dissipation and absorption of the acoustic energy in actual rocks may greatly affect the magnitude of both the acoustic waves and the GMPs. In order to estimate this effect we introduce the acoustic damping factor $T_a(r, R) = \exp(-r/L(R))$, which depends on the distance r and the crack radius. Multiplying Eq. (10.75) by this factor, taking into account Eq. (10.58) and above expression for $[u_z]$, we obtain

$$\langle u_r \rangle = \frac{u_0}{r^2} \exp\left(-\frac{r}{L(R)}\right) \quad \text{and} \quad u_0 = \frac{kl^3}{4\pi}\left(1 - \frac{4w^2}{3}\right), \qquad (10.15)$$

where $w = C_t/C_l$ and $\langle u_r \rangle$ stands for the mean radial displacement caused by the acoustic emission of stochastic crack ensemble. The characteristic length of the acoustic wave attenuation, L, is estimated as follows (Surkov et al. 2003)

$$L = \alpha R, \quad \alpha \sim \frac{QC_l}{2\pi V_c}, \tag{10.16}$$

where V_c is the velocity of crack growth and Q is the quality/energy-factor, which is approximately constant in the frequency range from 10^{-4} to $100\,\mathrm{Hz}$ depending on the variety of materials (Aki and Richards 2002). Taking the typical parameters $C_l = 5\,\mathrm{km/s}$, $V_c = 1.5\,\mathrm{km/s}$, $Q = 100$ we can estimate the coefficient of proportionality in Eq. (10.16) as $\alpha \approx 50$. As it follows from Eqs. (10.15) and (10.16), the small cracks are of little importance in the sense that their contribution to the net displacement field appears to be exponentially small, while the large cracks make a main contribution to the mean displacement and therefore to the crack-generated magnetic perturbations.

In the analysis that follows, we first seek for the solution of Eq. (10.14) in the case of cracks with fixed radius R and then extend the solution to the case of crack size distribution. Since the mean displacement has been obtained to be spherically symmetric, the spherical coordinates r, θ, φ are needed. We shall use a coordinate system in which the z axis is positive parallel to the vector of geomagnetic field, $\mathbf{B}_0$. For illustrative purposes, the reference frame and the cracked zone are sketched in Fig. 10.5b. In the case study all the values are independent of the azimuthal angle φ, and only the azimuthal component of $\mathbf{a}$ is nonzero. Equation (10.14) is thus reduced to the following form

$$\frac{v_m}{r^2}\left[\partial_r\left(r^2\partial_r a_\varphi\right) + \partial_\theta\left\{\frac{\partial_\theta\left(a_\theta \sin\theta\right)}{\sin\theta}\right\}\right] = B_0 \langle u_r\left(r\right)\rangle \sin\theta, \tag{10.17}$$

where ∂_r and ∂_θ denote the partial derivatives with respect to r and θ, accordingly, and θ is the polar angle between the vectors $\mathbf{B}_0$ and $\mathbf{r}$. We seek for the solution of Eq. (10.17) in the form $a_\varphi = g\left(r\right)\sin\theta$. Substituting this function and Eq. (10.15) for $\langle u_r \rangle$ into Eq. (10.17), we obtain

$$d_r\left(r^2 d_r g\right) - 2g = \frac{u_0 B_0}{v_m}\exp\left(-\frac{r}{L}\right), \tag{10.18}$$

where $d_r = d/dr$. This differential equation has a straightforward analytical solution

$$g\left(x\right) = c_1 x + \frac{c_2}{x^2} + \frac{u_0 B_0}{3 v_m}\left[\left(\frac{1}{x^2} + \frac{1}{x} - 1\right)\exp\left(-x\right) + xE_1\left(x\right)\right], \tag{10.19}$$

where c_1 and c_2 are arbitrary constants, $x = r/L$ and $E_1\left(x\right)$ denotes the exponential integral, that is

$$E_1(x) = \int_x^\infty \frac{\exp(-x')}{x'} dx'.$$ (10.20)

One makes sure of validity of the solution by direct substitution of Eq. (10.19) for $g(r)$ into Eq. (10.18). The sought function $\mathbf{b} = \nabla \times \mathbf{a}$ can be expressed through $g(x)$ in the following manner

$$\mathbf{b} = \frac{1}{r}\left[2\hat{\mathbf{r}}g(x)\cos\theta - \hat{\boldsymbol{\theta}}\partial_r\{rg(x)\}\sin\theta\right].$$ (10.21)

In order to find c_1 and c_2 these equations should be supplemented by the proper boundary conditions. Since the function $\mathbf{b}$ in Eq. (10.21) must go to zero at infinity, we get that $c_1 = 0$. Moreover, $\mathbf{b}$ must to be limited when $r \to 0$. Eventually, we obtain that

$$g(x) = \frac{u_0 B_0}{3v_m}\left[\left(\frac{1}{x^2} + \frac{1}{x} - 1\right)\exp(-x) - \frac{1}{x^2} + xE_1(x)\right].$$ (10.22)

The function $g(x)$ can be simplified in two extreme cases corresponding to large and small values of argument x. Consider first the case of large distances/small cracks, that is $r \gg L$ ($x \gg 1$), when the function $g(x)$ simplifies to

$$g \approx -\frac{u_0 B_0}{3v_m x^2}.v$$ (10.23)

It should be noted that if the maximal crack size satisfies the inequality $l_{\max} \ll r/\alpha$ where α is given by (10.16), this approach is valid for all the cracks.

For now, we shall be interested in the averaging over the crack size. We suppose that the number of cracks generated per unit time with length greater than l occurring in a specified area can be estimated from the known empirical law obtained by Gutenberg and Richter (1954) for the number of EQs. According to Turcotte (1997)

$$\dot{N}(l) = \frac{\dot{\beta}}{l^{2b}},$$ (10.24)

where b is the dimensionless empirical constant whose value varies from region to region but is generally in the range $0.8 < b < 1.2$. The constant $\dot{\beta}$ is a measure of the regional level of seismicity. This value is measured in units of m^{2b}/s. The worldwide data correlate with (10.24) taking $b = 1.11$ and $\dot{\beta} \approx 2 \times 10^3$ m^{2b}/s. Assuming for the moment that the dependence (10.24) can be extrapolated down to the crack sizes of about several meters and combining Eq. (10.24) and Eq. (10.10), yields

$$\langle\delta\mathbf{B}_t(\mathbf{r})\rangle = 2b\dot{\beta}\int_0^{l_{\max}} \frac{\mathbf{b}(\mathbf{r}, l)}{l^{2b+1}} dl.$$ (10.25)

Substituting Eqs. (10.21) and (10.23) for $\mathbf{b}$ and g into Eq. (10.25) and performing integration over l gives the sought value of the mean magnetic perturbations. The result of integration can be written in the form of a quasi-steady field of magnetic dipole (Eq. (7.55)) whose effective magnetic moment, $\mathbf{M}$, is given by

$$\mathbf{M} = -\frac{\sigma \dot{\beta} k b \alpha^2 l_{\max}^{5-2b}}{6\,(5-2b)} \left(1 - \frac{4w^2}{3}\right) \mathbf{B}_0. \tag{10.26}$$

The vector $\mathbf{M}$ is directed oppositely to the unperturbed geomagnetic field $\mathbf{B}_0$. It is not surprising since, as we have noted above, the conductor motion in external magnetic fields must result in the usual diamagnetic effect, which makes for formation of the effective magnetic moment with negative sign.

As is seen from Eq. (7.5) which describes the solution of problem, the average level of the electromagnetic noise at far distances decreases as r^{-3}. This case that is $r \gg L$, corresponds to strong attenuation of acoustic emissions followed by strong attenuation of GMPs. It follows from the numerical estimation (Surkov and Hayakawa 2006) that at such far distances the crack-generated electromagnetic noise is practically undetectable and thus this case-study is of little importance.

At short distances of interest here, that is, as $r \ll \alpha l_{\max} \sim 50 l_{\max}$, the contribution of the small and large cracks to the integral (10.25) should be estimated separately. As it follows from Eq. (10.15) and Eq. (10.23), which is valid for the small cracks with length $l \ll r/\alpha$, the function $\mathbf{b}\,(\mathbf{r}, l)$ in Eq. (10.21) is proportional to l^5 and hence the integrand in Eq. (10.25) is proportional to $l^{4-2b} \approx l^{1.8}$, where b is the fractal dimension in Eq. (10.24). In the case of the large cracks when $x \ll 1$, Eq. (10.22) for g is transformed to

$$g \approx -\frac{u_0 B_0}{2 v_m}. \tag{10.27}$$

In this case g is not a function of distance. This implies that the attenuation of the acoustic waves radiated by large cracks is nearly unimportant.

It follows from Eqs. (10.21) and (10.27) that as long as $l \gg r/\alpha$ the function $\mathbf{b}\,(\mathbf{r}, l) \propto l^3$ and therefore the integrand in Eq. (10.25) is proportional to $l^{2-2b} \approx l^{-0.2}$. The rough estimate of the contributions to the integral (10.25) due to the small and large cracks gives the ratio $[r/(\alpha l_{\max})]^{3-2b}$. This means that at the distance $r \ll \alpha l_{\max}$ the small cracks make a little contribution to the integral sum in Eq. (10.25) compared to that due to the large cracks. Taking the notice of this fact, substituting Eqs. (10.21) and (10.27) for $\mathbf{b}$ and g into Eq. (10.25), and performing integration, we obtain

$$\langle \delta \mathbf{B}_t\,(\mathbf{r}) \rangle = -\frac{\mu_0 \sigma B_0 \dot{\beta} k b l_{\max}^{3-2b}}{4\pi r\,(3-2b)} \left(1 - \frac{4w^2}{3}\right) \left(2\hat{\mathbf{r}}\cos\theta - \hat{\boldsymbol{\theta}}\sin\theta\right). \tag{10.28}$$

It should be emphasized that Eq. (10.28) determines, as a matter of fact, only a statistical average, which indicates the mean level of the crack-induced electromagnetic noise. This rough estimate does not depend on the damping constant α since it holds as the ground-recording station is located not far from the cracked zone.

For the numerical estimation of the magnetic effect caused by rock fracture we take typical parameters of regional seismicity $\dot{\beta} = 2 \times 10^3 \, \mathrm{m}^{2b}/\mathrm{s}$ and $b = 1.11$ (Turcotte 1997) and the following constants $k = 0.01$, $l_{\max} = 1 \, \mathrm{km}$, $\sigma = 10^{-2} \, \mathrm{S/m}$, $B_0 = 5 \times 10^{-5} \, \mathrm{T}$, $C_t/C_l = 0.5$ and $\theta = \pi/4$. Suppose also that the magnetometer is situated at the distance $r \approx 1 \, \mathrm{km}$ from the cracked zone. Substituting these parameters into Eq. (10.28) gives the rough assessment of the mean level of the ULF electromagnetic noise $|\langle \delta \mathbf{B}_t \rangle| \approx 0.3 \, \mathrm{pT}$. It should be noted that we have considered the case of equal probability for the crack orientation. A certain order of the crack orientation may enhance our estimate of the mean noise level. Likewise, due to the fluctuations the noise magnitude is rather large as compared to the mean value $|\langle \delta \mathbf{B}_t \rangle|$.

The ULF magnetic noise occasionally observed prior to the strong crust EQs (e.g., see Fraser-Smith et al. 1990; Kopytenko et al. 1990; Hayakawa et al. 1996, 2000) lies in the frequency range of 0.01–1 Hz. The average level $|\langle \delta \mathbf{B}_t \rangle|$ related to the square root of these characteristic frequencies gives the value of about 0.3–$3 \, \mathrm{pT/Hz}^{1/2}$ that can serve as a rough estimate of the power spectral intensity. This assessment is consistent in magnitude with the observation considering the uncertainties in the parameters. Moreover, it is worth mentioning that the seismic activity before and after the EQ occurrence may be greatly enhanced that leads to an increase in the actual value of regional seismicity parameter, $\dot{\beta}$, and eventually to an increase in the ULF electromagnetic noise level.

In this study we have dealt only with the tension cracks since the consideration of the shear cracks requires very complicated expressions. Actually all types of cracks including the tension and shear cracks are formed during rock fracture and probably the majority will tend to be shear ones (Scholz 1990). As alluded earlier in Sect. 7.3, the shear crack can radiate the acoustic waves in such a way that the effective magnetic moment induced in the conductive medium is non-parallel to the vector $\mathbf{B}_0$ depending on the crack orientation. If there is an equal probability for the shear crack orientation, then the mean magnetic moment of the crack ensemble is equal to zero. At the same time one may expect that the shear cracks will predominantly grow along the axis of maximal shear stresses or, more precisely, along the directions that make the angle $0.5 \arctan k_f$ with this axis (Scholz 1990). Here k_f denotes the coefficient of internal friction of the rock. Likewise, most of the shear cracks will possibly tend to be parallel to the fault plane. Taken together, this means that the mean magnetic moment of the shear crack ensemble can be nonzero. On account of the fact that the displacement discontinuity $[u]$ along the shear crack surface is proportional to the crack surface, we come to an estimate similar to that given by Eq. (10.28).

In summary, we describe two aspects of the problem.

(1) It follows from our calculations that the large cracks make a main contribution to the ULF electromagnetic noise. The small cracks appear to have no effect

due to the strong damping of their signals. Inside the region where the acoustic damping is of minor importance the magnitude of the electromagnetic perturbations is found to decrease with distance as r^{-1}, while far from this region the mean level of the noise falls off more rapidly with distance, that is as r^{-3}. According to our estimations the magnitude of the electromagnetic noise amounts to the value which is greater than or of the order of $1\,\mathrm{pT}/\mathrm{Hz}^{1/2}$.

(2) In the case of tension cracks the effect treated here can arise independently of the distribution of the crack plane orientation since the mean effective magnetic moment of the cracks is always directed in opposition to the geomagnetic field. In order to obtain the same effect, as the shear crack ensemble is considered, a certain order of the crack orientation is necessary. For example, the predominant directions for the shear crack growth can concentrate around the axis of maximal shear stresses.

10.1.4   Theory of ULF Electromagnetic Noise Due to Electrokinetic Effect

Perhaps Terada (1931) was the first that hypothesized a possible connection between the EQ preparation process and electrokinetic phenomena in the ground. It is in favor of this hypothesis that during Matsushiro EQ swarm (Japan, in August 1965) the simultaneous perturbations of the Earth's electromagnetic field and local changes in subterranean water level have been occasionally observed. The fluid lifting resulted from diffusion of the subterranean water through the porous rock can be accompanied by the generation of electrokinetic currents followed by perturbations of the Earth's electromagnetic field (Mizutani and Ishido 1976; Mizutani et al. 1976). Many researches have considered the electrokinetic effect as a promising candidate for explaining the electromagnetic anomalies possibly associated with EQs (e.g., Fitterman 1978; Miyakoshi 1986; Dobrovolsky et al. 1989; Gershenzon et al. 1993; Surkov 2000a; Molchanov and Hayakawa 2008).

In Chap. 8 we have shown that the electrokinetic phenomenon developed in a homogeneous conductive ground cannot produce the magnetic field in the atmosphere. The anomalies of both electric and magnetic fields can be generated only due to any kind of underground inhomogeneities where the rock conductivity and streaming potential coefficient undergo strong changes. As one example, consider the earlier model of the media (Fitterman 1978, 1979a,b, 1981; Dobrovolsky et al. 1989) as schematically shown in Fig. 10.6. In the region $x < 0$ the conductivity σ_1 and streaming potential coefficients C_1 are assumed to be constant and in the next region $x > 0$ they are characterized by σ_2 and C_2 constant as well. The EQ focus is modeled as an ellipsoidal inclusion shown in Fig. 10.6 with filled area. The changes in tectonic stress may result in the enhanced/lowered pore pressure in the focal region. Since a typical time scale of the underground fluid filtration process is supposed to be much greater than that of EQ preparation, we treat the quasi-steady regime of the fluid filtration.

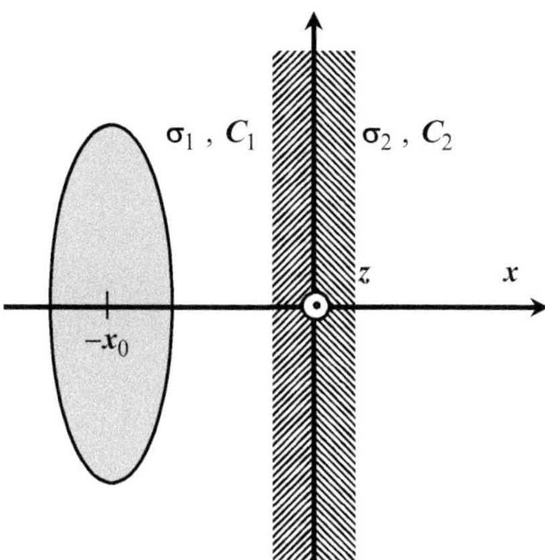

Fig. 10.6 A model of conductive medium containing two half-spaces with different conductivity σ and streaming potential C coefficients. A pore pressure inside the *fill area* differs significantly from that in the surrounding media

Below we do not follow the cited papers in any detail since they consider only a plane medium model. For simplicity, we assume that the pore pressure is a given function and the pressure gradient is not equal to zero at the boundary between two half-spaces. The pore pressure P has an axially symmetric distribution around x-axis which is directed perpendicular to the boundary surface. The center of this distribution is situated inside the focal region on the x-axis at the point with coordinate $x = -x_0$.

For the axially symmetric problem, Eq. (8.12) in cylindrical coordinates x and r reads

$$\partial_x^2 \Psi_{1,2} + r^{-1}\partial_r \left(r\partial_r \Psi_{1,2} \right) = 0, \tag{10.29}$$

where $\Psi_{1,2} = \Phi_{1,2} + C_{1,2}P$ and Φ stands for the electric potential. The subscripts 1 and 2 are related to the first and second regions, respectively. We seek for the solution of Eq. (10.29) in the form of Bessel transform

$$\Psi_{1,2} = \int_0^\infty \psi_{1,2}(x,k) J_0 \left(kr \right) k\,dk, \tag{10.30}$$

where J_0 denotes Bessel function of the first kind and zero order, $\psi_{1,2} = \varphi_{1,2}\left(x,k \right) + C_{1,2}p\left(x,k \right)$ while φ and p are Bessel transforms of the potential and pore pressure, respectively. Substituting Eq. (10.30) for $\Psi_{1,2}$ into Eq. (10.29) yields

$$\psi''_{1,2} - k^2 \psi_{1,2} = 0, \tag{10.31}$$

where the prime denotes a derivative with respect to x. The solution of the problem must be limited as $x \longrightarrow \pm\infty$. Thus, we obtain

$$\psi_1 = A_1 \exp(kx), \quad \psi_2 = A_2 \exp(-kx). \tag{10.32}$$

The undetermined constants A_1 and A_2 can be found by applying the boundary conditions at $x = 0$ where the potential φ and the normal component of current density should be continuous. Substituting the result of calculations into Eq. (10.30), we get

$$\Psi_2(x,r) = \frac{C_2 - C_1}{1 + \sigma_2/\sigma_1} \int_0^\infty p(0,k) \, J_0(kr) \exp(-kx) \, k \, dk, \tag{10.33}$$

where

$$p(0,k) = \int_0^\infty P(0,r') \, J_0(kr') \, r' \, dr'. \tag{10.34}$$

Permuting the indices 1 and 2 provides us with the solution for $\Psi_1(x,r)$. Assuming for the moment the absolute convergence of integrals, we change the order of integration in Eq. (10.33) with respect to variables k and r':

$$\Psi_2(x,r) = \frac{C_2 - C_1}{1 + \sigma_2/\sigma_1} \int_0^\infty P(0,r') \, D(x,r,r') \, r' \, dr'. \tag{10.35}$$

Here we have introduced the following function

$$D(x,r,r') = \int_0^\infty J_0(kr) \, J_0(kr') \exp(-kx) \, k \, dk. \tag{10.36}$$

The function D can be expressed through associated Legendre functions of the second order. However, in order to derive an asymptotic form of this function at great values of x, it will be enough to perform integration several times by parts. As a result, Eq. (10.35) reduces to

$$\Phi_2(x,r) = \frac{C_2 - C_1}{(1 + \sigma_2/\sigma_1) \, x^2} \int_0^\infty P(0,r') \, r' \, dr' - C_2 P(x,r). \tag{10.37}$$

As is seen from Eq. (10.37) the electric potential may have a maximum in the vicinity of the fault plane. It should be noted that the similar effect can take place at the contact area of two conductors with different conductivities. Assuming for the moment that far away from the EQ focus the last term on the right-hand side of Eq. (10.37) can be neglected, we conclude that the electric potential decreases with the distance x from the fault plane as x^{-2}. Consequently, the electric field strength falls off with distance as x^{-3}. Notice that the quasi-static elastic deformations obey the similar law but they fall off inversely cubed distance from the EQ hypocenter.

One more approach is based on the assumption that the Earth's crust may contain the unstable inhomogeneities such as groundwater reservoirs and local failure zones filled by fluid (Bernard 1992). Some of these inhomogeneities are close to an instability threshold and thus they are very sensitive to small changes in rock deformation. On the other hand the deformation threshold has to exceed the threshold for tidal deformation $\left(\sim 10^{-7}\right)$. The EQ focus as a source of seismic activity may trigger the rock fracture in the vicinity of the unstable homogeneity followed by sporadic fluid fluxes from the groundwater reservoirs into fresh cracks and pores originated from the rock failure. This may result in the generation of local electrokinetic effects far away from the epicenter of a forthcoming EQ. In this picture the effect can be observed by chance if the sensors are located near such an unstable inhomogeneity. It was hypothesized that this or similar effect can explain several of a series of successful EQ predictions in Greece on the basis of the so-called seismoelectric signals (e.g., see Varotsos and Alexopoulos 1984a,b; 1986; Varotsos et al. 1996; Varotsos 2005).

Fenoglio et al. (1994, 1995) have suggested that the sealed high pore pressure compartments can occur in the zones of intense shear deformation and around the fault. The compartment size is supposed to vary from 100 m to 1 km. The bridging of pore space around the compartment can be due to the deposition of silica, fracturing and other processes. Under the rock shift or subduction the sealings between the compartments can be crushed that gives rise to rapid changes of pore pressure followed by the fluid stream along the cracks and underground channels. The volume and velocity of the underground fluid can vary significantly because of the generation of fresh pores, cracks, and channels due to the rock decompaction or dilatation effect in the zone of intense shear deformation (Scholz 1990; Nikolaevskiy 1996).

In the model by Fenoglio et al. (1994, 1995), the sealed compartment with volume V has a form of narrow rectangular parallelepiped similar to a plane crack as shown in Fig. 10.7. The sealing at the end of compartments breaks down due to a weak seismic event, which makes for the fluid filtration through the surface S towards the low pressure region. Suppose that the fluid has advanced a distance of Δl thereby producing the electrokinetic current in the volume ΔV. Taking into account Eq. (8.7) for the electrokinetic current density j_{ek}, the electrokinetic current moment is estimated as

$$d = j_{ek}\,\Delta V = \sigma C\,|\nabla P|\,\Delta V, \qquad (10.38)$$

where $|\nabla P|$ is absolute value of the pore pressure gradient. Substituting Eq. (10.38) for d into Eq. (7.3) gives a rough estimate of magnetic variation resulted from the electrokinetic effect

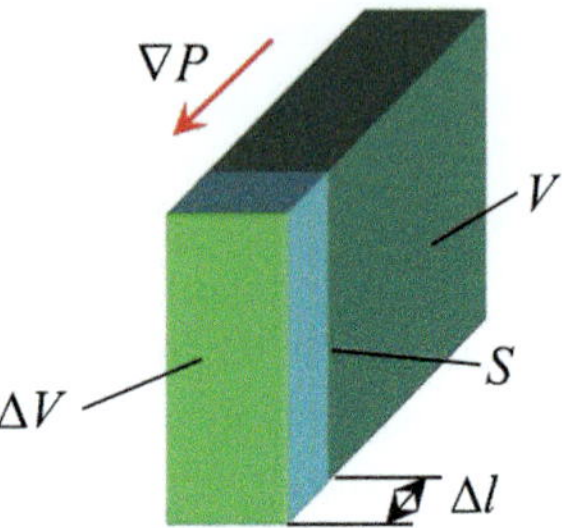

Fig. 10.7 A model of sealed underground compartments V with high pore pressure P. The fluid filtration through the surface S towards the low pressure region ΔV may be triggered by weak seismic events to generate electrokinetic effects (Bernard 1992; Fenoglio et al. 1995)

$$\delta B \sim \frac{\mu_0 \sigma C \, |\nabla P| \, \Delta V}{4\pi r^2}. \tag{10.39}$$

Taking the notice of the numerical values of the parameters: $|\nabla P| = 10^2$–10^4 Pa/m, $\Delta V = 10^3$–10^5 m^3, $C = 10^{-6}$–10^{-8} V/Pa, $\sigma = 10^{-2}$–10^{-3} S/m and distance $r = 10$ km, we come to the following upper estimate $\delta B_{\max} \approx 10^{-2}$ pT. Certainly, it is only a rough estimate which depends on the chosen values of the parameters.

In the model, the duration of the electrokinetic phenomena is determined by the time scale of the underground fluid diffusion through the volume ΔV of the crashed rock. The typical time of the fluid diffusion is of the order of $\tau \sim \Delta l^2 / D$, where $D \sim k K_f / \eta$ is the fluid diffusion coefficient (Frenkel 1944). Unfortunately, at higher depth the fluid viscosity coefficient η, the compressibility modulus of the fluid K_f and especially the rock permeability k are not well known to provide reliable estimates of both the time scale of the process and typical frequencies.

One more model is based on the assumption that the electromagnetic perturbations observed at the ground recording station can be due to the variations of streaming potential coefficient in the vicinity or inside the EQ focus (Surkov 2000a; Fedorov et al. 2001). This large-scale effect can be observable around the EQ epicenter but not only near the water-bearing zones or near the faults coming to the ground surface.

In this case the electrokinetic current moment (7.1) can be estimated as follows:

$$|\mathbf{d}| = \left| \int_V \mathbf{j}_e dV \right| \sim \sigma \Delta C \, |\nabla P| \, V, \tag{10.40}$$

where ΔC is the variation of the streaming potential within the volume V of the EQ focal zone. For simplicity, we have ignored the changes in pore pressure gradient and conductivity. Substituting Eq. (10.40) for $|\mathbf{d}|$ into Eq. (7.3), we obtain the estimate for magnetic variations

$$\delta B \sim \frac{\mu_0 \sigma \Delta C \, |\nabla P| \, V}{4\pi r^2}. \tag{10.41}$$

This estimate differs from that given Eq. (10.39) by replacement of C and ΔV by ΔC and V, respectively. Taking the volume of the EQ focal zone $V = 10^3 \, \mathrm{km}^3$, distance $r = 50 \, \mathrm{km}$, $\Delta C \sim C$ and values of other parameters we obtain $\delta B_{\max} \approx 4$ pT; that is, several orders of magnitude greater than the previous value obtained in the framework of the model by Fenoglio et al. (1994, 1995). Certainly this result directly follows from the large value of the focus volume V.

Notice that the numerical estimates of the ULF signal cover a wide range of amplitudes because of the lack of information on the rock structure at higher depths, i.e. rock permeability, underground water content, crack distribution over size and orientation, and so on.

In Chap. 8 we have demonstrated that the electrokinetic effect in a fractal porous medium results in power law dependence of electric field variations on the source size L. According to Eqs. (8.19) and (8.22) this dependence can vary depending on the structure of fractal pore space, transport critical exponents τ and correlation length critical exponent ν. Now we consider the model of an earthquake hypocenter zone in which the electrokinetic current is considered as a by-product of the fluid filtration in the fractal pore network above percolation threshold (Surkov et al. 2002a; Surkov and Tanaka 2005). The typical size of EQ focal zone L can be related with the EQ magnitude M by the known empirical rule (Kanamori and Anderson 1975):

$$\log L = 0.5M - 1.9 \tag{10.42}$$

where L is measured in kilometers. Combining Eqs. (8.19) and (10.42) yields

$$\log E = aM + b \tag{10.43}$$

where $a = 1 - \tau/(2\nu) \approx 0.09$. Combining Eqs. (8.22) and (10.42), we come to the same dependence where $a = 0.5\{1 - (\tau - \nu)/(1 + \nu)\} \approx 0.31$. Here we have used the following critical exponents : $\tau = 1.6$ and $\nu = 0.88$ derived from numerical simulations on 3D grids (Staüffer 1979; Feder 1988). By contrast, in the case of non-fractal focal zone we obtain that $a = 1$. Although the value $a = 0.31$ is very close to the observational one ($a \approx 0.34$–0.37) reported by Varotsos et al. (1996), we cannot relate the electrokinetic phenomena and pre-seismic activity with confidence primarily due to a paucity of actual observations and the complexity of the ULF electric field variations in the near-surface atmospheric layer.

The physical mechanisms treated above, i.e. the geomagnetic field perturbation and the electrokinetic effect, seem to be the most promising and creditable to explain, in principle, the co-seismic electromagnetic phenomena and pre-seismic ULF electromagnetic noise. Both mechanisms have to be stimulated by pre-seismic activity, which is accompanied by an enhancement of underground fluid migration, by rock fracture and an increase of the crack number in the vicinity of fault zone.

It should be emphasized that both theoretical models predict that the ULF seismo-electromagnetic signals related to the EQ focus become undetectable at the distance greater than approximately 100 km far from the EQ epicenter. According to our estimations the magnitude of the electromagnetic noise inside this region can be of the order of approximately $1\,\mu V/m \cdot Hz^{1/2}$ and $0.1\,nT/Hz^{1/2}$.

10.1.5 Variations of the Rock Basement Conductivity and of Telluric Voltage Possibly Associated with EQs

Long-term observations in seismo-active regions have shown that the variations of the rock basement conductivity can be generated synchronously with an enhancement of seismic activity several months before and after the EQ occurrence (Myachkin et al. 1972; Sobolev et al. 1972; Sobolev 1975; Rikitake and Yamazaki 1978; Honkura 1981; Rikitake 1987; Meyer and Teisseyre 1989; Park and Fitterman 1990; Park 1991; Bragin et al. 1992). For example, Rikitake and Yamazaki (1978) observed the gradual decrease in the ground conductivity down to several percent several hours before the main shock in 21 cases among 30 ones when the considered effect was observed. In-situ measurements, both passive and active techniques have been used for measurements of rock conductivity and telluric voltage. In an early study, the potential difference between a pair of grounded electrodes was measured to reveal abnormal behavior of telluric currents prior to and after an EQ. A net of ground-based stations located 100–200 km apart at Kamchatka shore, Russia have been operated to detect this effect (Myachkin et al. 1972; Sobolev et al. 1972; Sobolev 1975). In some case-studies the amplitude of telluric voltage variations reached the value about 100 mV while the signal-to-noise ratio was 10–100. Myachkin et al. (1972) have detected the abnormal variation of the telluric electric field about 100–300 $\mu V/m$ 3–16 days before a seismic event.

The variations of transient electric current ($\sim$ 1–100 Hz) flowing between two grounded electrodes were considered to be indicative of the rock conductivity variations (Rikitake and Yamazaki 1978; Park 1991). The active techniques such as the dipole DC probing, vertical electric probing, and field formation method can be sensitive to the rock conductivity variations up to the depth 15–20 km (Bragin et al. 1992).

Many researchers have discussed whether the conductivity and telluric current variations can serve as an EQ precursor (Sobolev 1975; Allen et al. 1975; Smith and Johnston 1976; Honkura et al. 1976; Fedotov et al. 1977; Corwin and Morrison 1977; Koyama and Honkura 1978; Murakami et al. 1984; Varotsos and Alexopoulos 1984a,b, 1986; Miyakoshi 1986; Ralchovsky and Komarov 1988; Jackson and Kagan 1998; Uyeda 1998; Uyeda et al. 2000; Varotsos 2005). A few of workers believed that the correlation between the abovementioned phenomena and EQs is very weak (e.g., Burton 1985; Johnston 1989).

It is generally accepted that there are two candidate mechanisms, which can be responsible for variations of the rock basement conductivity: (1) The content variations of the conducting underground fluid (Mazzella and Morrison 1974; Takahashi and Fujinawa 1993) and (2) The changes of dry rock conductivity itself associated with variations of tectonic stresses and strains at higher depths (Varotsos and Alexopoulos 1986; Slifkin 1993; Freund 2000, 2002; Freund and Pilorz 2012).

The study in the last decades was indicative of the presence of the underground fluid at higher depths down to tens kilometers (Nikolaevskiy 1996). At the moment there are no reliable data about the physical state, mineral composition, and other parameters of the underground fluid. However, the underground fluid flow plays a key role in the preliminary stage of the EQ preparation in many models of the EQ preparation process such as the dilatancy-diffusion model (Nur 1972; Scholz et al. 1973) or crack migration model (Surkov et al. 2002b). On the other hand, the water conductivity is much greater than that of dry rocks at least at low pressure. In this picture the EQ preparation can affect the mean rock conductivity due to changes of the underground fluid content.

As has already been stated in Chap. 9, the laboratory tests (e.g., see reviews by Urusovskaya (1969) and by Mineev and Ivanov (1976)) have shown that the solid conductivity is enhanced under the applied stress basically due to the increase in number density of charged dislocations and point defects of atomic lattice. Though the mobility of interstitial ions, vacancies, and especially dislocations are not so high because of their large effective masses. Recently Freund (2002) has suggested an alternative mechanism of the stress-induced rock conductivity based on the time-resolved impact experiments. The increase in the rock conductivity is supposed to be due to the clouds of highly mobile positive hole charge carriers, i.e. defect electrons/holes. Below we consider this mechanism in a little more detail since it may be of interest in understanding the rock conductivity at higher depths.

Majority of minerals in the rock basement are silicates which contain $O_3Si–O–SiO_3$ and $O_3Si–O–AlO_3$ structural units or bonds, in which all oxygen anions have the valence $2-$. When traces of H_2O structurally dissolve in the matrix of such silicates, the following reactions can take place

$$H_2O + O_3Si–O–SiO_3 \rightarrow O_3Si–O–O–SiO_3 + H_2, \qquad (10.44)$$

thereby producing the so-called peroxy bond. If the applied shear stresses exceed the elastic limit that results in the generation of large amount of dislocations, the peroxy link can be destroyed due to intersection of the mobile dislocation and the peroxy link. When the broken peroxy link meets O^{2-}, it acts as an electron receptor, which can hold the electron for a long time. The O^{2-}, which has donated the electron, turns into O^-. This new anion plays a role of positive hole because it has one electron less as compared to all other oxygen anions. The lifetime of the holes is assumed to vary from several seconds to several weeks. In this picture, this process is similar to the conventional transition from an insulator state to a semiconductor state. It is conjectured that this p-type conductivity can take place in the middle and partly in the lower crust in the wide temperature interval from about $400°C$ to $500–550°C$ (Freund 2000, 2002; Freund and Pilorz 2012).

It is worth mentioning that the similar effect has been observed in the ionic crystal samples during the shock loading (Mineev and Ivanov 1976). In Chap. 9 we have discussed that the increase in the shock amplitude can result in the sharp changes in the sign of potential difference between sides of the sample. This effect can be due to the sharp changes in the character of the sample conductivity, that is, due to the transition from ionic conductivity to electron/hole conductivity. However, the above effect was not observed at constant pressure so that the validity of the above mechanism for the rock at higher depth where the static loading prevails is questionable.

In conclusion, we note that many other models have been developed to explain the electromagnetic phenomena in seismo-active regions. Among them are the piezoelectric (Cutolo 1988; Yoshino and Tomizawa 1988; Kingsley 1989; Sornette and Sornette 1990) and tectonomagnetic (Stacey 1964; Stacey and Johnston 1972; Sasai 1991; Gershenzon and Bambakidis 2001) effects caused by the influence of tectonic stress on piezoelectric and piezomagnetic minerals in the rocks. In the model by Draganov et al. (1991) the underground water motion along a water-bearing stratum is treated as a possible source mechanism for ULF noise preceding an EQ. However, Surkov and Pilipenko (1999) noted that the rock permeability which has been used by Draganov et al. (1991) for their numerical calculations was overestimated at least by four orders of magnitude. In a very exotic model by Lockner et al. (1983) and by Lockner and Byerlee (1985), a number of electrical and optical effects are assumed to be due to the water vaporation inside underground cavities and cracks situated near the fault.

10.1.6 Ionospheric Effects Observed Around the Time of EQs

Considerable recent attention has been focussed on experimental evidences for ionospheric perturbations associated with seismic activity (Leonard and Barnes 1965; Davies and Baker 1965; Wolcott et al. 1984; Tanaka et al. 1984; Kelley et al. 1985a; Le Pichon et al. 2002; Komjathy et al. 2012). The energy transfer from an EQ to the atmosphere and ionosphere is essentially due to acoustic waves and internal/atmospheric gravity waves (IGW) generated in the atmosphere after strong earthquakes. Four to five minutes after the main shock the atmospheric air waves arrive at the ionosphere thereby exciting the ionospheric plasma motion that, in turn, results in the generation of GMPs. The electromagnetic channel of the energy transfer from the EQ to the ionosphere seems to be improbable because of the low level of electromagnetic perturbations on the ground surface.

The ground surface vibrations bring about atmospheric air waves not only in the epicentral region but also at far distances from the EQ. After strong EQs such as a great Alaska EQ on March 28, 1964 the atmospheric waves with horizontal phase velocities about 3 km/s and periods about tens seconds have been detected (e.g., Bolt (1964)). The source of atmospheric air waves is believed to be the EQ-produced Rayleigh surface wave (the detail about Rayleigh wave is found in

Sect. 7.2) propagating along the ground surface just at the same velocity and with the same periods. For example, the first Rayleigh wave originated from the great Alaska EQ propagated at the velocity 3.0–3.3 km/s with period 23 s. The maximal vertical ground displacement was about 4.2 cm that results in the atmospheric pressure excess in the near surface layer about 4.0 Pa (Bolt 1964). Considering the Mach number $M = C_R/C_a$ where C_a is the sound speed in the air, we should note that its value is about $M \approx 9$–10. In this case the wave vector of atmospheric acoustic waves is nearly vertically directed. According to the calculations by Golitsyn and Klyatskin (1967) the angle included between the acoustic wave vector and the normal must be about $6°$. Moreover, the Rayleigh waves can excite only acoustic waves but not IGW in the atmosphere.

The amplitude of mass velocity of an upgoing acoustic wave increases with altitude because of an exponential decrease in air density and pressure. At the E-layer of the ionosphere (90–100 km) the mass velocity of the acoustic wave V_a can reach a value about tens m/s that may greatly affect the ionospheric plasma. The relative variations of the ionospheric plasma density is of the order of $\Delta n/n \sim V_a/C_a$, that is about several percent or more. We consider this effect in more detail in the next section.

Much emphasis has been paid in the past on the studies of EQ precursors in the ionosphere. Space-borne observations by OGO-6, Intercosmos-19, Intercosmos-24, Aureol-3 and DEMETER satellites over seismo active regions have shown the ionospheric perturbations which can be associated with impending EQs. Among them ULF/ELF/VLF electromagnetic variations and noses occasionally observed several hours before an EQ (Migulin et al. 1982; Gokhberg et al. 1983; Larkina et al. 1985; Parrot and Mogilevsky 1989; Chmyrev et al. 1989; Parrot 1990, 2012; Serebryakova et al. 1992; Molchanov et al. 1993), changes in ion composition (Boskova et al. 1993; Pulinets et al. 1994), small scale plasma inhomogeneities (Chmyrev et al. 1997), weak variations of particle precipitations (Voronov et al. 1989; Galper et al. 1989; Serebryakova et al. 1992) and etc. However, analyses of space-borne data collected by GEOS-2 (Matthews and Leberton 1985), DE-2 (Henderson et al. 1993) and by ISIS-2 (Rodger et al. 1996) satellites have not shown any correlation between seismic activity and low-frequency electromagnetic variations in the ionosphere.

The IGW is frequently considered as a possible mechanism/source for the pre-seismic phenomena in the ionosphere (e.g., see Molchanov and Hayakawa 2008; Sorokin et al. 2003). However, the acoustic mechanism for pre-EQ perturbations in the ionosphere is very questionable since the ground surface vibrations due to seismic wave propagation can hardly be applied to the phenomena under consideration because of the weak amplitude of the acoustic waves caused by such vibrations (Hayakawa et al. 2007b). Indeed, the pressure variations in the near-surface layer of the atmosphere can be estimated as follows:

$$\Delta P = \rho C_a \partial_t u \sim \rho C_a \frac{u}{T}, \tag{10.45}$$

where ρ is the air mass density, C_a the acoustic wave velocity, and $\partial_t u$ denotes the time-derivative of the ground surface displacement u. Considering the typical microseism measurement with displacement amplitude about 10^{-4} cm and with the period $T = 6$ s and taking $\rho = 1.2$ kg/m^3 and $C_a = 340$ m/s brings the pressure variation in the atmosphere $\Delta P \sim 0.1$ mPa, which is found to be 2–3 order-of-magnitude smaller than the level of the atmospheric acoustic noise at the ground surface.

10.1.7   A Problem of Direction Finding for the ULF Electromagnetic Source

In spite of much progress towards the search of electromagnetic EQ precursors, a major question of whether the ULF electromagnetic signals detected prior to a seismic event are really associated with the tectonic activity is still an open question. In this notation, the problem of direction finding of the underground source as well as the problem of identifying a weak electromagnetic signal in the background of natural ionospheric and magnetospheric noises and man-made interference are of special interest in geophysical studies.

In order to solve this problem one comes across a number of serious difficulties and complexities. First, the characteristic wavelength (in vacuum) in the ULF frequency range is so large that an observer is always situated in the near zone, i.e. in such a case the traditional radiowave methods, such as the wave time lag measurement or miscellaneous interference schemes, are hardly applicable. Second, a primary ULF source which is of main interest here, is seemingly located under the ground, maybe at higher depths. In such a case the electromagnetic signals undergo a strong dissipation and dispersion since the ULF field spreading in conductive layers of the ground is governed by the diffusion law. The typical time scale of the perturbations that can be related to the EQ precursor is as large as several tens minutes or hours so that the front of the perturbations considerably extends in time. The application of traditional technique to the ULF source is based on a network of the ground-recording stations equipped with magnetometers in order to detect the time lag or the phase difference between signals recorded at different points (Kopytenko et al. 2002; Ismaguilov et al. 2003; Hayakawa et al. 2007a, 2011).

One of the challenges of EQ prediction is how to distinguish useful signals from various electromagnetic noises originated from ionospheric and magnetospheric sources. In this notation, one should pay attention to the polarization of the field generated by ionospheric and magnetospheric sources. Considering the ground-based observation, we note that the vertical component B_z of this field should be smaller than the horizontal one B_h due to the conventional impedance boundary conditions at the conducting ground surface. Here $B_h^2 = B_H^2 + B_D^2$, where B_H and B_D are both orthogonal components of the horizontal magnetic field. The ratio of spectral densities of vertical and horizontal components of magnetic field perturbations is believed to be indicative of the presence of seismo-electromagnetic signal (Hayakawa et al. 1996). According to this criterion, one may expect that the value of B_z/B_h on the ground surface is on the order of, or even more than unity, as the underground source is operative. In order to discriminate the global and local variations, (1) the data obtained at one ground-based station is compared with data of other remote stations; and (2) the temporal evolutions of B_z/B_h and of averaged power spectra are compared with the trend of K_p index of global geomagnetic activity. This technique has been successfully applied to two shallow EQs (the hypocenter depth was 20 km in both cases) happened at Kagoshima, Japan on March 23, 1997 ($M_s = 6.5$) and on May 16, 1997 ($M_s = 6.3$) (Hattori et al. 2002b, 2004). The observation at ground-based station 60 km far from the EQ epicenters has shown (1) the increase in polarization ratio B_z/B_h before the seismic events, and (2) a distinct correlation between the changes of field polarization and of local seismic activity.

According to the simplest theory, the ULF underground sources can be split into two types: current element and magnetic dipole. As is seen from Eqs. (7.1)–(7.5) and Figs. 7.1–7.3, in all cases the ratio B_z/B_h is a function of the angle θ between the vector of dipole moment and the position vector. This implies that the above technique is sensitive to the epicentral distance since the value B_z/B_h can vary with distance (through θ) especially as the ground-based station is situated near the EQ epicenter.

Schekotov et al. (2007, 2008) have developed the polarization technique by analyzing the components of polarization matrix. A possibility to locate an EQ epicenter has been demonstrated on the basis of the data collected at Karimshino station (52.83° N, 158.13° E, Kamchatka, Russia) for seven or more years. However, the validity of this technique to find direction of the underground ULF source has not been discussed.

One more polarization technique for locating the source of pre-EQ electromagnetic activity has been proposed by Dudkin et al. (2010) and Dudkin and Korepanov (2012). The theoretical basis of this technique is based on the assumption that this source is equivalent to an effective magnetic dipole which has axially symmetrical magnetic field (that is, only two components, B_r and B_θ). It should be noted that this assumption is valid if only the vector of magnetic dipole moment is perpendicular to the boundary between the earth and the atmosphere, otherwise there must be all the components of the magnetic field including azimuthal one.

However, in the model by Dudkin et al. (2010, 2012), the magnetic field polarization ellipse and the vector of magnetic moment are assumed to be in the same plane. Since the polarization ellipses are measured at two different ground-based stations S_1 and S_2, the intersection line of the polarization ellipse planes contains the magnetic dipole vector $\mathbf{M}$, as shown in Fig. 10.8. Thus the measurement of the polarization ellipses parameters makes it possible to calculate a position of $\mathbf{M}$ which is assumed to be located at the EQ focus. A basic limitation of this method is that it can be applied only for magnetic-dipole sources (poloidal-type magnetic field) whereas the actual underground source may be an electric-dipole source (toroidal-type magnetic field).

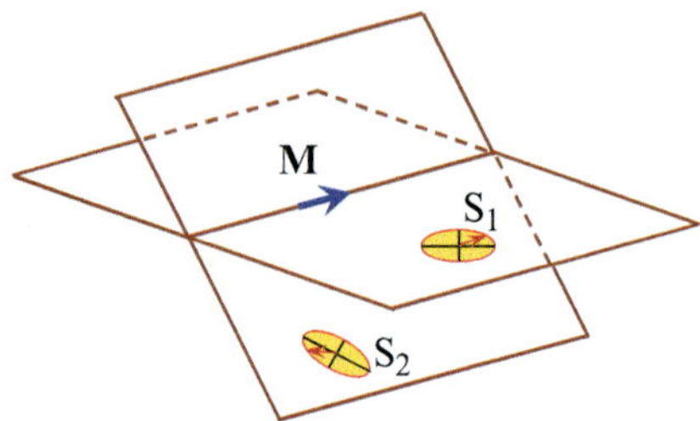

Fig. 10.8 According to the model by Dudkin et al. (2010), the magnetic dipole vector **M** is on the intersection line of polarization ellipse planes measured at two different ground-based station, S_1 and S_2

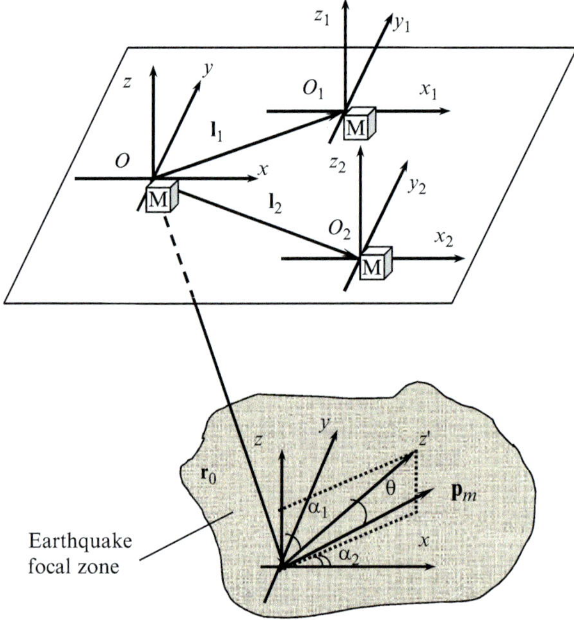

Fig. 10.9 A sketch of equipment arrangement and the magnetic-dipole source location. The reference magnetometer is located at the origin of coordinate system O, and the places of the second and third magnetometers are shown by the vectors $\mathbf{l}_1$ and $\mathbf{l}_2$

Semi-empirical and semi-analytic techniques for finding the ULF source in the background of natural noise have been reported by Surkov et al. (2004). To distinguish a weak ULF source signal from the natural noise, a network of multicomponent magnetometers is supposed to be used. A sketch of equipment arrangement is shown in Fig. 10.9. The distances between magnetometers are designed such that $|\mathbf{l}_1|$, $|\mathbf{l}_2| \ll L$, where L is the spatial scale of the variations originated from the ionosphere-magnetosphere origin. The influence of these large-scale variations can be eliminated by subtraction of data measured by a set of magnetometers from those measured by a reference magnetometer. The data processing is based on theoretical formulas describing the quasi-steady field of the magnetic dipole or of the current element. More precisely, these formulas give an evaluation of the partial derivatives of the magnetic perturbations. In principle, this procedure allows us to extract all the dipole parameters including its depth orientations and so on, as was illustrated by the example of magnetic dipole. This promising technique has not been used yet in practice possibly because of complicated mathematical supplements.

The problem of direction finding for a ULF source is still far from completion, although this information can be extremely useful for the interpretation of experimental data and for understanding the origin of the ULF signals and, in particular, for finding the electromagnetic signals possibly related to an impending EQ.

10.1.8   *Other Electromagnetic Phenomena Possibly Associated with EQs*

A special credit has been paid in the last decade to monitoring of the seismo-active regions from space in visible and infrared (IR) ranges (e.g., Gorny et al. 1988; Qiang et al. 1991, 1999; Salman et al. 1992; Tronin 1996, 1999, 2002; Tronin et al. 2002; Liu et al. 2000; Tramutoli et al. 2001; Ouzounov and Freund 2004). The analysis of the outgoing IR radiation detected over sparsely populated territories at nighttime conditions has shown the presence of anomalies of the outgoing Earth's radiation flux associated with large structures and fault systems of the Earth's crust. The nighttime conditions are the most favorable for the IR monitoring of the ground-surface temperature since there are no solar heating. The minimal absorption of IR by the atmosphere corresponds to wavelength intervals 3–5 μm and 8–13 μm. The second interval termed as a second transparency window is preferable to the first one because the outgoing radiation peak which corresponds to the ground surface temperature 273–373 K lies within 7.7–10.5 μm.

Because of high spatial (0.5–1 km) and temperature (0.1–0.5 K) resolutions this method has been used for the analyses of lineaments, morphological structures, and tectonic movements. The data processing of NOAA-series satellite survey has shown an increase in the average ground temperature of several Kelvin at the foot of Kopetdag in Central Asia (Tronin 2002; Tronin et al. 2002). The anomaly area was 25–30 km wide and about 500 km in length. Another type of anomaly of about 50 km wide and 300 km long was found at the foot of the Karatau Range that roughly coincides spatially with the Karatau fault. This procedure promotes the elimination of average seasonal temperature trend, the reduction of the climatological and other effects (Tramutoli et al. 2001). The capability of the refined noise-removal approach has been demonstrated for the case of the Irpina–Basilicata EQ of November 23, 1980 ($M_s = 6.9$).

Interestingly enough, the size variations of the anomaly areas in seismo-active regions have been observed occasionally several days before and after strong EQs in Central Asia (Gorny et al. 1988; Tronin et al. 2002), China (Qiang et al. 1991, 1999), European Union (Tramutoli et al. 2001), and Japan (Hayakawa et al. 2001). Whether these effects are really associated with seismic activity and what is the origin of such stable and non-stationary IR anomalies is not well understood. Perhaps, the abnormal behavior of such an area is due to an output of optically active gases such as CO_2, CH_4, and water vapor which results in the local green house effect (Tronin 1999).

In the theory by Surkov et al. (2006) the stable ground temperature anomalies can be due to the geothermal convective heat flow induced by slow filtration of the underground fluid. This heat flux is much smaller in magnitude than that arising from the solar radiation and other thermal fluxes of natural origin. Nevertheless this mechanism can play a substantial role in the presence of groundwater that can filtrate along the system of interconnected cracks in the upper crust. The tectonic stresses in the fault zone result in gradual squeezing-out of the groundwater from higher depth towards the ground surface thereby heating the rocks since the groundwater temperature is higher than that of the upper layers. The theory predicts a surface temperature increase of a few K if the mean filtration velocity amounts to 10^{-6} m/s and if the mean rock porosity near the fault exceeds the value $n = 0.01$–0.1. The estimations have shown that there happen the anomalous regions with enhanced surface temperature of the order of a few K on the geological time scales. However, the convection mechanism appears to have no effect during the short time domain and thus cannot provide any explanation of the nonstationary IR anomalies that occasionally appear several days before EQs. Perhaps, one can concede that a sudden enhancement of rock porosity may occur prior to an EQ followed by rapid groundwater lifting that in turn results in an increase of the ground surface temperature. In principle, this picture correlates with the known effect of water level changes occasionally observed prior to seismic events (Wakita 1975; Barsukov et al. 1985; Roeloffs 1988; King et al. 1999; Sobolev and Ponomarev 2003)

This effect can be connected with other phenomena probably associated with impending EQs, that is, the propagation of over-the-horizon FM signals (Kushida and Kushida 1998, 2002; Fukumoto et al. 2001, 2002; Fujiwara et al. 2004). Though we have not received any signal from a VHF transmitter out of the light-of-sight, we sometime receive the signals with small incident angle from the transmitter and we define this as being abnormal. Kushida and Kushida (1998) have detected the signals from an over-the-horizon transmitter in Central Japan several days or weeks prior to the Kobe EQ. Some correlation was found between the abnormal VHF wave propagation and the EQs which happened at certain sensitive regions. This phenomenon has been studied intensively for a number of EQs in Central Japan (Fukumoto et al. 2001, 2002). In spite of the FM transmitter (77.1 MHz) in Sendai, that is 312 km far from the receiver in Chofu, the FM signals have been occasionally received in Chofu although the distance of line-of-sight was 80 km. Fukumoto et al. (2001, 2002) have found that the cross-correlation between the abnormal over-the-horizon FM signals and EQs exhibits a significant peak around 7 days before an EQ.

Hayakawa et al. (2007b) have suggested that the over-the-horizon VHF wave propagation in seismo active regions can be due to the tropospheric ducting via the wave reflection below the tropopause. In this case the VHF waves are guided in the same way as in a metallic waveguide. The ducting in the troposphere followed by the ray distortion builds up as a result of refraction index changes with altitude. In the geometric optics approach the curvature radius, R, of the ray can be expressed via the refraction index n as follows (e.g., see Landau and Lifshits 1982):

$$\frac{1}{R} = \mathbf{N} \cdot \frac{\nabla n}{n},$$

(10.46)

where $\mathbf{N}$ denotes the unit vector of a principal normal to the curve/ray. The refraction index of air depends on meteorological conditions including the atmospheric pressure, temperature and the partial pressure of water vapor, which is contained in the air. Under a standard meteorological condition the refractive index of air decreases with height z, and the typical value of its derivative is $dn/dz = -4 \times 10^{-8} \, \mathrm{m}^{-1}$. The ducting takes place under the requirement that the refractive index falls off more rapidly with height so that

$$\frac{dn}{dz} < -\frac{1}{R_e},$$

(10.47)

where R_e is the Earth's radius. This implies that the rays will follow the curvature of the Earth. The major cause of radio ducting is humidity and temperature inversion, which occasionally occurs in the coastal region during anticyclone conditions. The tropospheric ducting requires low-angle entry and exit of the ray into the duct that is consistent with the elevation angle (smaller than $10°$) observed by Fukumoto et al. (2002).

Hayakawa et al. (2007) have pointed a possible connection between the abnormal variations of refractive index and the increase in the ground surface temperature in seismo-active regions. As it follows from simple estimates, the weak heating of the ground surface ($\Delta T = 1$–3 K) for several days or weeks can produce a large amount of water vapor. If the favorable meteorological conditions occur in the near-surface layer over seismo-active regions, it can produce the air humidity and temperature inversion followed by the abnormal variations of the air refractive index.

The enhancement of radon concentration in the air due to its penetration through soil into the atmosphere has been occasionally observed before an EQ (e.g., see Pulinets et al. 1997; Yasuoka et al. 2012). The radioactive radon is a source of α-particles with a mean kinetic energy about 6 MeV. This effect can result in additional ionization of narrow near-surface layer of 1–3 m thickness. Every particle can produce about 2×10^5 electron–ion pairs. The electrons are captured by oxygen and nitrogen molecules for a short time and then the ions are absorbed by dusty particles and by aerosols, which in turn climb at the upper layer of the air. The amplitude of electric field originated from the radon input is hardly estimated due to the deficiency of information about the distribution of radon excess over the seismo active area.

In conclusion we note that all the effects treated here seem to be sporadic and case sensitive, since they depend on meteorological conditions, diurnal and seasonal variation of the air humidity, proximity of the coastline, etc.

10.2   Other Large-Scale Disasters

10.2.1   *Electromagnetic Phenomena Associated with Volcano Eruption*

Among a variety of electromagnetic phenomena associated with volcano eruption the most significant effect is the strong electric field generated by volcanic ash clouds. Short lightning discharges have frequently been observed inside volcanic ash clouds (Arabadji 1951; Anderson et al. 1965; MacDonald 1972; Brook and Moore 1974; Uman 1987). The majority of intracloud (IC) lightning propagates horizontally. The mean dipole moment of IC lightning is estimated as $100\,C{\cdot}m$ (Rulenko 1985), which is approximately 10^3 lower than that of CG lightning discharges. The energy of IC discharges in volcano clouds is estimated as $10^6\,J$ (Anderson et al. 1965), that is 10^3 smaller than that of typical CG lightning. The typical lightning length is about 8–$10\,m$ while the CG lightning channel length varies within several km. This implies that the spatial separation of electric charges in the volcanic ash cloud is much smaller than that in conventional thunderclouds. The charge density in the volcanic cloud is assumed to be of the order of 10^{-11}–$10^{-12}\,C/cm^3$. The experimental evidence of the volcanic lightning is displayed in Fig. 10.10, in which the image was obtained by NASA during Sakurajima volcano in southern Japan on March 11, 2013. As is seen from this figure, the lightning occurs between the hot magma bubbles and the volcanic ash clouds above the volcano's tip.

The generation of the electric charges occurring in the volcanic ash cloud is believed due to the following mechanisms: (1) Electrification of the smallest lava particles (with sizes up to $10^{-7}\,cm$) during explosive fragmentation of the lava at the initial stage of explosion. Above a volcano crater the number density of such particles reach a value $10^{-9}\,m^{-3}$ (Zemtsov et al. 1976); (2) A friction between ash particles and walls of volcano vent during the upward motion of ash–gas flux; (3) Electrification of ash particles due to their collisions and friction as well as due to the friction between the particles and air fluxes (Lenchenko 1988).

Interestingly enough, the strong electric field can take place near the point where the lave comes in contact with sea water. Brook and Moore (1974) have measured the electric field amplitude about $7\,kV/m$ at the distance 100 far from this point. Weak electric discharges and jumps of electric field following 0.6–$1.4\,s$ have been observed during a sudden release/outburst of vapor-drop mixture from the crater

Fig. 10.10 A volcanic lightning occurring during the Sakurajima volcano eruption in southern Japan (NASA—March 11, 2013)

of Karymsky volcano (Rulenko 1979). The measurements have shown that the hot vapor contained positive charges. The electric field variations associated with steam plumes from the geothermal well have been observed by Björnsson and Vonnegut (1965).

Enhancement of ULF electromagnetic noise in the frequency bands of 0.01–0.6 Hz and 1.0–3.0 Hz has been detected several days before and after the volcano eruption (Fujinawa et al. 1992). It was hypothesized that this enhancement can be due to the magma motion along the volcano vent that results in electrokinetic effects in the rocks followed by the electromagnetic noise. It is also believed that in the volcano regions there are a lot of underground cameras filled by the fluid with pressure from the hydrostatic up to the lithostatic level (Johnston 1989). The destruction of such cameras can generate electromagnetic signals in the range of 0.01–100 Hz. We have estimated the electrokinetic effect when we studied electromagnetic phenomena related to an EQ. Although this mechanism can explain the observed anomalies, only rough estimates are available because such important parameters as rock permeability, pore fluid pressure and etc. vary in an extremely wide band. Additionally, tectonomagnetic, tectonoelectric, magnetohydrodynamical, and other mechanisms are considered as possible causes of the observed phenomena (Johnston 1997; Uyeda et al. 2002; Zlotnicki and Nishida 2003). The history of investigation for field volcanology including the applied electromagnetic method such as resistivity sounding or self-potential and geomagnetic field measurements are found in a review by Hashimoto (2005).

In the model by Kopytenko and Nikitina (2004a,b), the GMP caused by a conducting magma upwelling along the volcano vent is treated as a possible source of ULF noise observed prior to and during the volcano eruption. Below we choose our own way to reproduce their result. In order to model the magma motion along the volcano vent, we first consider an ellipsoid inclusion in the rock as a model of magma reservoir. The origin of the coordinate system is chosen to be in the center of the ellipsoid. The coordinate axes point from the center and coincide with the symmetry axes, so that the ellipsoid equation is $x^2/a^2+y^2/b^2+z^2/c^2 = 1$, where a, b, and c are the ellipsoid semi-axes. This is only a mathematical technique because in what follows we consider the extreme case $c \rightarrow \infty$, which corresponds to an infinite cylindrical channel.

As the magma can move inside the inclusion, the electric current density inside and outside this inclusion is then

$$\mathbf{j} = \sigma_m \left(\mathbf{E} + \mathbf{V} \times \mathbf{B}_0 \right), \text{ and } \mathbf{j} = \sigma_r \mathbf{E}, \tag{10.48}$$

where $\mathbf{V}$ denotes the local magma velocity, $\mathbf{B}_0$ is the induction of undisturbed geomagnetic field, and $\mathbf{E}$ is the electric field. Since the magma conductivity σ_m is of the order of 10^{-2}–10 S/m (e.g., Gaillard and Marziano 2005), the conductivity σ_r of the surrounding rock is assumed to be much smaller than σ_m. Note that the Hall/imposed current, $\mathbf{j}_H = \sigma_m \left(\mathbf{V} \times \mathbf{B}_0 \right)$ in Eq. (10.48) plays a role of a source of GMPs.

For simplicity we assume that inside the ellipsoid the vector $\mathbf{V}$ is positive parallel to z axis and approximately constant. Notice that this assumption can be applied to the unclosed space such as an infinite cylinder, which will be considered below. However the problem under consideration is similar to that for a dielectric ellipsoid immersed in the uniform electric field. Likewise, one can find a certain similarity of the problem to that for a magnetized ellipsoid immersed in the uniform magnetic field. According to Landau and Lifshits (1982), the electromagnetic field outside of the ellipsoid is thus determined through the effective current moment

$$\mathbf{d} = \frac{\sigma_r m_x \hat{\mathbf{x}}}{\sigma_m n^{(x)} + \sigma_r \left(1 - n^{(x)} \right)} + \frac{\sigma_r m_y \hat{\mathbf{y}}}{\sigma_m n^{(y)} + \sigma_r \left(1 - n^{(y)} \right)}, \tag{10.49}$$

where $\hat{\mathbf{x}}$ and $\hat{\mathbf{y}}$ are unit vectors and m_x and m_y are projections of the Hall current moment $\mathbf{m} = \mathbf{j}_H V$ on the x and y axes. Here $V = 4\pi abc/3$ is the ellipsoid volume and $n^{(x)}$ and $n^{(y)}$ denote the so-called geometrical depolarization factors depending on the parameters a, b, and c. It should be noted that the vectors $\mathbf{d}$ and $\mathbf{m}$ are not parallel.

Following a simple model by Kopytenko and Nikitina (2004a,b), we consider the magma motion along the infinite cylindrical vent with constant radius, that corresponds to the case when $c \rightarrow \infty$ and $a = b$. On account that in this limit $n^{(x)} = n^{(y)} = 0.5$ we get

$$\mathbf{d}_{\text{eff}} = \frac{\mathbf{d}}{c} = \frac{2\pi a^2 \sigma_m \sigma_r \left(\mathbf{V} \times \mathbf{B}_0 \right)}{\sigma_m + \sigma_r}, \tag{10.50}$$

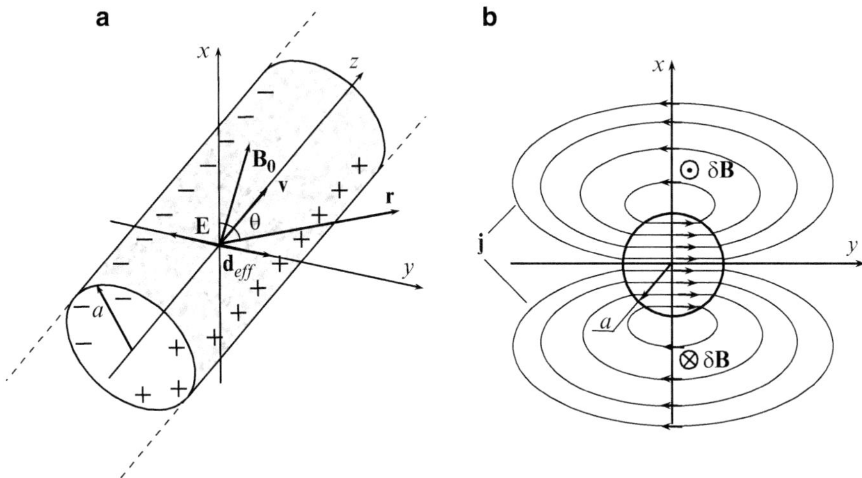

Fig. 10.11 A model of conducting magma upwelling along the volcano vent. (**a**) The magma velocity vector **V** is positive parallel to z axis while the effective magnetic moment **d**$_{\text{eff}}$ is perpendicular to both **V** and the Earth's magnetic field **B**$_0$. (**b**) A configuration of total current **j** flowing across the vent. The *circles with point and cross* inside indicate the directions of magnetic field perturbation δ**B**

where **d**$_{\text{eff}}$ is the effective magnetic moment per unit length and a is the radius of the cylinder. Let y axis be positive parallel to the vector **d**$_{\text{eff}}$. Thus, the vectors **V** and **B**$_0$ must be in xz plane as shown in Fig. 10.11a. We use a polar coordinate system in which r is the distance from the z-axis and θ ($0 \leq \theta < \pi$) is the polar angle between the x-axis and the vector **r**. Taking into account the symmetry of the problem and the fact that the moment (10.50) is normal to the cylinder axis, one can find that the GMPs have only z-projection

$$\delta B_z = -\frac{\mu_0 d_{\text{eff}} r \cos\theta}{2\pi a^2}, \quad (r < a); \tag{10.51}$$

and

$$\delta B_z = -\frac{\mu_0 d_{\text{eff}} \cos\theta}{2\pi r}, \quad (r > a), \tag{10.52}$$

where $d_{\text{eff}} = |\mathbf{d}_{\text{eff}}|$. On account of $\sigma_m \gg \sigma_r$, Eqs. (10.51)–(10.52) can be simplified. For example, in the region $r > a$ we obtain

$$\delta B_z = -\mu_0 \sigma_r a^2 B_{0x} V \frac{\cos\theta}{r}. \tag{10.53}$$

This result differs from that obtained by Kopytenko and Nikitina (2004a,b) by the factor σ_r/σ_m because they left out of account the conductivity of the rock surrounding the vent and hence they overestimated the magnitude of GMPs.

Applying Eq. (10.53) to the case of magma upwelling, one may estimate the amplitude of magnetic perturbations in the vicinity of the volcano vent as follows: $\delta B_{\max} \approx \mu_0 \sigma_r a^2 B_0 V_{\max}/r$.

The back conduction current is in turn driven by the transverse electric field resulted from the charges at the vent walls as shown in Fig. 10.11a. A schematic plot of the total current density lines is shown in Fig. 10.11b. It should be noted that inside the vent ($r < a$) the total current $\mathbf{j}$ is directed in opposition to the $\mathbf{E}$, that is

$$\mathbf{j} = \frac{\sigma_r \mathbf{j}_H}{\sigma_m + \sigma_r}, \quad \mathbf{E} = -\frac{\mathbf{j}_H}{\sigma_m + \sigma_r}. \tag{10.54}$$

In the framework of the model both fields, $\mathbf{j}$ and $\mathbf{E}$, are constant inside the vent. In the surrounding rock they have a power-law decrease with distance ($r > a$)

$$E_r = \frac{|\mathbf{j}_H|\, a^2 \sin\theta}{(\sigma_m + \sigma_r)\, r^2}, \quad \text{and } E_\theta = -E_r \cot\theta. \tag{10.55}$$

In order to estimate the magnitude of GMPs in the vicinity of a volcano we choose the typical parameters $\sigma_r = 10^{-3}$–10^{-2} S/m, $B_0 = 5\times 10^{-5}$ T and $r = 5$ km. Taking the following values of the magma parameters $V_{\max} = 5$ m/s and $a = 0.1$–1 km (Kopytenko and Nikitina 2004a,b), we obtain the estimate of maximal effect related to the active period of volcano activity: $\delta B_{\max} = 0.6 \times 10^{-3}$–0.6 nT and $E_{\max} \approx B_0 a^2 V_{\max}/r^2 = 0.1$–10 μV/m.

The ULF electromagnetic noise caused by the crack generation due to volcano activity can be as much as that due to an EQ since the spatial scale of the fractured zones appears to be comparable. The above estimates derived for the microfracturing and GMPs can be applied to this case. However, there may be expected the enhancement of the effect in the vicinity of volcano crater because in this region the underground cracked zones can be closer to the ground surface. In order to estimate this effect, one can substitute the shorter distances in the above equations.

The effect of the volcano on the ionosphere is believed to be due to IGWs and acoustic waves generated during the volcano eruption. Notice that the IGW can be excited not only by the volcanic explosion but also because of air temperature and density gradients in the vicinity of volcano crater. These waves can disturb the ionospheric plasma density and generate electric currents thereby exciting low-frequency electromagnetic perturbations. The variations of total electron content (TEC) in the ionosphere with period from 16 to 30 min have been observed during the Mount Pinatubo volcano eruption on June 15, 1991 in Taiwan (Cheng and Huang 1992) and around Japan (Igarashi et al. 1994). Heki (2006) has recorded the TEC variations 12 min after the Asama volcano eruption, central Japan in September 2004. The period of the ionospheric variations was about 1.3 min, which is close to the typical period of acoustic waves. The TEC variations after a major volcanic explosion at the Soufrière Hills volcano in July 2003 have been measured by Dautermann et al. (2009). These measurements exhibit the ionospheric variations with periods of 17 and 4 min, which is typical for IGWs and acoustic waves. The influence of volcanic eruption on the ionosphere has been recently analyzed on the basis of date obtained by DEMETER satellite over active volcanoes during the period 2004 August–2007 December (Zlotnicki et al. 2010).

10.2.2 Electromagnetic Phenomena Associated with Tsunami

The major sources of natural electromagnetic fields in the seas and oceans are the magnetospheric and ionospheric electric current and perturbation of the Earth's magnetic field due to the motion of seawater. This latter effect is similar to that observed during the seismic wave propagation in the continent. However, ocean-induced electromagnetic fields are poorly understood, primarily due to the paucity of actual observations and the complexity of the ocean velocity field (e.g., Sanford 1971; Chave 1984). The conductivity of seawater near the ocean surface can vary within interval 3–6 S/m depending on the temperature and water saltiness, which in turn is a function of ion contents in the seawater. The ocean-induced electromagnetic field is dependent on the type of hydrodynamical source that is surface or internal waves, sea current, ocean tides, long waves such as tsunami and etc.

The tsunami wave is usually generated by a seaquake followed by sudden displacements in the ocean floor or by underwater volcanic activity. Though the tsunami wave arrives at the coast for many hours after a major EQ, it brings about a significant danger. The largest EQ of the past 40 years ($M = 9.1$–9.3) in the Indian Ocean on December 26, 2004 was followed by a devastating tsunami, which killed more that 280,000 people in the south-east Asian region (Lay et al. 2005; Manoj et al. 2010). Although the spatial scale of tsunami generated waves in the open ocean can be larger than 100 km, the monitoring of tsunami wave propagation is difficult because in the deep ocean the tsunami wave is smaller than a few cm high (Artru et al. 2005). The tsunami waves become dangerous near the coastal line basically because they may increase in height to become a fast moving wall of turbulent water several meters high.

It has long been hypothesized that tsunamis produce a detectable value of GMPs (Larsen 1968; Gershenzon and Gokhberg 1992; Tyler 2005). To estimate the GMP caused by tsunami waves, one should take into account that this effect is sensitive to the movement of the entire water column. The simplest model of tsunami waves in a deep ocean is sketched in Fig. 10.12. The wave has a shape of vertical column with height h and horizontal sizes λ and l. The seawater column as a whole is assumed to move horizontally along x axis parallel to the ocean floor at a constant speed. The mass velocity $\mathbf{V} = V\hat{\mathbf{x}}$ of the seawater inside the wave is assumed to be constant as well whereas outside this region the medium is at rest. Certainly according to hydrodynamical equations, the mass velocity must gradually increase

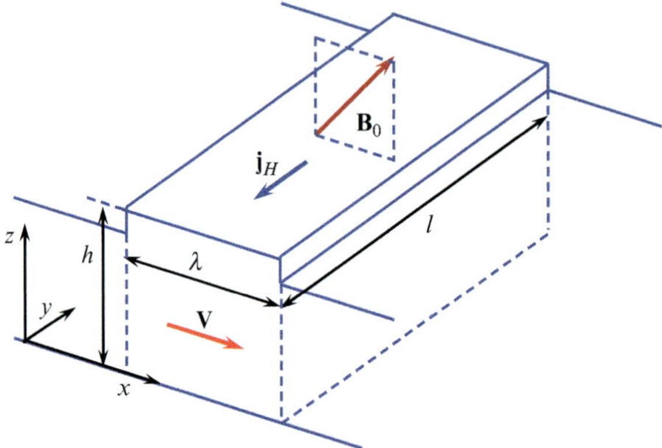

Fig. 10.12 A schematic plot of tsunami waves in an open ocean. The mass velocity **V** and Hall current density $\mathbf{j}_H$ are shown with *red and blue arrows*, respectively

at the front of wave and gradually decrease at its end. It is unimportant as far as the estimate of electromagnetic effect since only the total volume of moving water will enter this estimate. Assuming that both the unperturbed magnetic field $\mathbf{B}_0$ and the mass velocity $\mathbf{V}$ are in zx plane, the Hall current $\mathbf{j}_H = \sigma_w \mathbf{V} \times \mathbf{B}_0$ is parallel to y axis as shown in Fig. 10.12. Moreover we shall ignore the variations of seawater conductivity σ_w with depth. As the distance r to the tsunami wave is smaller than its lateral size l, the linear infinite current should be considered as the most appropriate model for the calculation of GMPs $\delta\mathbf{B}$. Taking into account that $\delta B = \mu_0 I_H / (2\pi r)$ where the total Hall current is equal to $I_H = j_H \lambda h$, we obtain the simple estimate

$$\delta B \sim \frac{\mu_0 \sigma_w V B_0 \lambda h}{2\pi r}. \tag{10.56}$$

Using the following numerical values of the parameters: $V = 5\,\text{cm/s}$, $\sigma_w = 5\,\text{S/m}$, $B_0 = 5 \times 10^{-5}\,\text{T}$, $\lambda = 100\,\text{km}$, $h = 2\,\text{km}$ and the distance $r = 500\,\text{km}$, we get $\delta B = 1\,\text{nT}$. This value is close to other estimates (Tyler 2005) in which the magnetic field variations generated by the Indian ocean tsunami could have reached an amplitude of 4 nT. Manoj et al. (2010) have reported that tsunamis could also induce an observable electric potential difference in deep submarine cables. It should be emphasized that the electromagnetic effect of tsunamis is essentially due to the large volume of the moving water although the mass velocity is as small as several centimeters per second.

The hypothesis that tsunamis can generate detectable magnetic variations has been recently examined for the strong Chilean EQ ($M = 8.8$) on February 27, 2010 (Nair et al. 2010) followed by a moderate tsunami that crossed the Pacific Ocean westward. The measurements provided by the magnetic observatory on Easter Island, 3,500 km west of the epicenter have shown a periodic signal of 1 nT in the vertical magnetic component, coincident with the arrival time of the tsunami. These weak signals were identified because the exceptionally deep solar minimum on this day provided ideal conditions to identify these signals in the background noise. However, it appears that the practical application of electromagnetic monitoring for tsunami wave propagation is extremely limited except for periods of low solar activity.

Alternative methods in tsunami wave monitoring have been the subject of special interest during recent years. It was hypothesized that tsunami waves in the open ocean can generate IGWs which in turn may result in the ionospheric perturbations (Hines 1972; Peltier and Hines 1976; Pavlov and Sukhorukov 1987). In the period range of 10–20 min the horizontal velocity of IGW is close to that of tsunami wave whereas the vertical IGW velocity is about 50 m/s, so that the IGW-induced perturbations in the ionosphere have a time lag about several hours. The short-scale ionospheric TEC perturbations with the same characteristic have been observed in Japan after the $M_w = 8.2$ EQ in Peru on June 23, 2001 (Artru et al. 2005). The measurements were made around the time of tsunami wave arrival. The ionospheric perturbations observed during the giant tsunami following the Sumatra–Andaman event on December 26, 2004 ($M_w = 9.3$), the 2009 Samoa and 2010 Chile tsunamis have been reported by DasGupta et al. (2006), Liu et al. (2006), Lognonné et al. (2006), Occhipinti et al. (2006, 2013), Rolland et al. (2010), and Galvan et al. (2011). As these perturbations are really associated with tsunami wave propagation, the atmospheric and ionospheric remote sensing can provide a new tool for oceanic monitoring and tsunami detection.

10.2.3   *Space-Borne Measurements over Hurricanes and Typhoons*

The effect of intense meteorological processes in the bottom atmosphere on the ionospheric parameters have been examined by Kelley et al. (1985b), Holzworth et al. (1985), and Korepanov et al. (2009). Space-borne measurements over an active typhoon have shown variations of the plasma conductivity and the ionospheric electric field at the frequencies below 100 kHz. Mikhailova et al. (2000, 2002) have reported the satellite observations of the ULF/ELF ionospheric perturbations over powerful typhoons in the Pacific Ocean. The electric field perturbations in the ionosphere over a tropical storm region have been measured onboard the COSMOS-1809 satellite (Isaev et al. 2002; Sorokin et al. 2005). The amplitude of electric perturbations reached an abnormally high value about 25 mV/m, which is not typical for the equatorial ionosphere. This effect is supposed to be due to the vertical electric current associated with the amplification of vertical air convective motion or turbulent transfer in the lower atmosphere.

Some of electromagnetic phenomena observed over hurricanes and typhoons are believed to be due to the CG and IC lightning or due to high altitude discharges such as Red sprites and Blue jets (BJs), which are developed over a thunderstorm.

An observational hint towards the lightning effect on the ionosphere was provided by Burke et al. (1992) using DE-2 satellite measurements over the hurricane Debbie. The burst of electric field with amplitude about 40 mV/m and the bursts of electron flux intensity with energies 854 eV and 35 keV have been detected by DE-2 space-borne sensors. At the same time, the variation of quasi-static electric field is not found to exceed 0.5 mV/m while the electron number density varies within 10 %. One may speculate that an Alfvén pulse generated by the mesospheric discharge can be a cause of the observed effects. Additionally, the bursts of 35 keV electron flux intensity following 1–2 s agreed satisfactorily with the Trimpi-effect, caused by the interaction of electrons trapped in the overlying radiation belt with whistlers radiated by the lightning that in turn result in the scattering and precipitation of electrons (e.g., Rodger 1999).

Final remarks we wish to emphasize are as follows. All the ULF/ELF electromagnetic effects treated here seem to be weak at the background level of natural and man-made electromagnetic noises. A majority of the theories predict that the EQs, volcano eruption, tsunami, and other disasters can provide a small but detectable value of the electromagnetic effects only at short distance from the sources. One of the challenges of experimental study is how to increase the signal-to-noise ratio in order to distinguish the effect of our interest from other noises and how to find the location of the ULF/ELF source. At the moment we have a few competitive theories, which can explain, in principle, the ULF phenomena associated with natural catastrophes. The problem is that the most theories describing the tectonic and electromagnetic processes at higher depth are based on such parameters, whose real values cannot be extracted with confidence from the ground-based data. Further experiments are required to sort out this interesting problem in seismically active regions in order to elaborate the theoretical conceptions studied in this section. Despite the recent studies have shown that at the moment the short-term prediction of impending EQs is impossible, we must extend our knowledge on electromagnetic phenomena associated with natural disasters. The understanding of the origin of the ULF signals can be extremely useful for the interpretation of experimental data possibly related to impending EQs.

Appendix H: Averaging over the Crack Orientation

In order to study the displacements due to the formation and evolution of the crack ensemble we first consider the dynamic displacement field due to single crack growth (Surkov and Hayakawa 2006). Acoustic waves generated by the growing cracks in a uniform elastic medium are defined by the wave equation (Aki and Richards 2002)

$$\rho \partial_t^2 \mathbf{u} = (\lambda + 2\mu)\, \nabla\,(\nabla \cdot \mathbf{u}) - \mu \nabla \times (\nabla \times \mathbf{u}). \qquad (10.57)$$

where λ and μ are the Lame coefficients and ρ is the rock density. If the distance between the crack and sensor is much greater than the crack size, the solution of Eq. (10.57) can be expressed through the so-called seismic moment tensor, $\hat{\mathbf{M}}$, of the crack. For example, let us consider a tension flat crack lying in the x', y' plane of the local coordinate system fixed to the crack. Let $[u_z(t)] = u_z(t, z' = 0+) - u_z(t, z' = 0-)$ be a given function, which defines the displacement discontinuity/jump normal to the crack surface and parallel to the z'-axis. In this case only three diagonal components of the seismic moment tensor are nonzero (Aki and Richards 2002)

$$M_{xx} = M_{yy} = \lambda S[u_z(t)], \quad M_{zz} = (\lambda + 2\mu) S[u_z(t)], \tag{10.58}$$

where S is the area of the crack surface.

The crack growth gives rise to changes in the seismic moment, which in turn results in the radiation of acoustic waves. The components of medium displacement, $u_i(\mathbf{r}', t)$, is derivable from the wave equation (10.57) via the seismic moment in the following way

$$u_i = \frac{\gamma_i'}{4\pi\rho r'} \left\{ \frac{[6f_1(t_1) - f_2(t_1) - 2M_{ii}(t_1)]}{r'C_l^2} - \frac{[6f_1(t_2) - f_2(t_2) - 3M_{ii}(t_2)]}{r'C_t^2} \right.$$
$$\left. + \partial_t \left(\frac{f_1(t_1)}{C_l^3} - \frac{[f_1(t_2) - M_{ii}(t_2)]}{C_t^3} \right) + \frac{3}{r'^3}\hat{G}(5f_1 - f_2 - 2M_{ii}) \right\}, \tag{10.59}$$

where r' is the distance from the crack, and $C_l = [(\lambda + 2\mu)/\rho]^{1/2}$ and $C_t = [\mu/\rho]^{1/2}$ are the velocities of longitudinal and transverse acoustic waves, the subscript $i = x', y', z'$ and

$$f_1 = \left(\gamma_x'^2 + \gamma_z'^2\right) M_{xx} + \gamma_z'^2 M_{zz}, \quad f_2 = 2M_{xx} + M_{zz}, \tag{10.60}$$

$$t_1 = t - \frac{r'}{C_l}, \quad t_2 = t - \frac{r'}{C_t}.$$

The directional cosines, γ_i', can be written as $\gamma_x' = x'/r'$, $\gamma_y' = y'/r'$ and $\gamma_z' = z'/r'$, and the integral operator $\hat{G}$ is defined in the following manner

$$\hat{G}M_{ii} = \int_{r'/C_l}^{r'/C_t} \tau M_{ii}(t - \tau)\, d\tau, \tag{10.61}$$

The set of Eqs. (10.58)–(10.61) contains both the near- and far-field displacement components. As it follows from the general solution (10.59), the crack opening produces both the longitudinal and transverse acoustic waves, which propagate with different velocities. The corresponding terms in Eq. (10.59) depend on different arguments, t_1 and t_2, accordingly. In the static limit when $t \to \infty$ the time-derivatives $\partial_t M_{ii}$ in Eq. (10.59) vanish, whereas the terms $\hat{\mathbf{G}} M_{ii}$ tend to the value

$$\hat{\mathbf{G}} M_{ii} = \frac{M_{ii(s)} r'^2}{2C_t^2} \left(1 - w^2\right), \tag{10.62}$$

where $w = C_t / C_l$ and $M_{ii(s)} = M_{ii}(\infty)$ is the static/residual value of the seismic moment of the crack. In what follows we omit the subscript s in order to simplify the expressions for the residual seismic moment. Combining Eqs. (10.59), (10.60), and (10.62) gives the residual displacement field in the form

$$u_i = \frac{\gamma_i'}{8\pi\rho r'^2 C_t^2} \left[(3f_1 - f_2)\left(1 - w^2\right) + 2w^2 M_{ii}\right], \tag{10.63}$$

where the functions f_1, f_2 are also taken in a static limit, i.e. at $t \to \infty$.

As we have noted above, the rock fracture due to the crack development is predominantly concentrated inside the cracked zone. In what follows we use the general reference frame, x, y, z, fixed to a certain point/"center" of the cracked zone. Since we are interested in the distances large compared to the size of the cracked zone, the spatial crack distribution inside the cracked zone is ignored. This implies that the origin of the general reference frame approximately coincides with the those of the local reference frames associated with separate cracks. The displacement components of each crack should be transformed from the local reference frame, x', y', z', to the general one. For now, the size and plane crack orientation are considered as accidental values. Let θ_0 and φ_0 be the Euler angles, which define the random orientation of the unit vector $\mathbf{n}$ normal to the crack plane. In addition, we introduce the spherical coordinates, r, θ, φ of the general reference frame. For convenience, both the general and local reference frames are sketched in Fig. 7.9. The crack plane is shown in Fig. 7.9 with gray ellipse. The displacement field of the separate crack in the spherical coordinates can then be expressed through the displacement components given by Eq. (10.63) in the following way

$$u_r = \gamma_x' u_{x'} + \gamma_y' u_{y'} + \gamma_z' u_{z'}, \tag{10.64}$$

$$u_\theta = u_{x'}\left(\cos\theta_0 \cos\theta \cos(\varphi - \varphi_0) + \sin\theta_0 \sin\theta\right) + u_{y'}\cos\theta \sin(\varphi - \varphi_0)$$
$$+ u_{z'}\left(\sin\theta_0 \cos\theta \cos(\varphi - \varphi_0) - \cos\theta_0 \sin\theta\right). \tag{10.65}$$

$$u_\varphi = -u_{x'}\cos\theta_0 \sin(\varphi - \varphi_0) + u_{y'}\cos(\varphi - \varphi_0) - u_{z'}\sin\theta_0 \sin(\varphi - \varphi_0). \tag{10.66}$$

The directional cosines are transformed to the view

$$\gamma_x' = \sin\theta\cos\theta_0\cos(\varphi - \varphi_0) - \cos\theta\sin\theta_0, \tag{10.67}$$

$$\gamma_y' = \sin\theta\sin(\varphi - \varphi_0), \tag{10.68}$$

$$\gamma_z' = \sin\theta_0\sin\theta\cos(\varphi - \varphi_0) + \cos\theta_0\cos\theta. \tag{10.69}$$

Substituting Eq. (10.63) for u_i and Eqs. (10.67)–(10.69) for γ_i' into the set of Eqs. (10.64)–(10.66), and rearranging, we obtain

$$u_r = \frac{1}{8\pi\rho r^2 C_t^2}\left[\left(3 - w^2\right)f_1 - \left(1 - w^2\right)f_2\right], \tag{10.70}$$

$$u_\theta = \frac{\sin 2\theta\,(M_{zz} - M_{xx})}{8\pi\rho r^2 C_l^2}\left[\sin^2\theta_0\cos^2(\varphi - \varphi_0)\right.$$
$$\left. + \frac{\sin 2\theta_0}{2}\cos 2\theta\cos(\varphi - \varphi_0) - \cos^2\theta_0\right], \tag{10.71}$$

$$u_\varphi = \frac{\sin\theta_0\sin(\varphi - \varphi_0)\,(M_{xx} - M_{zz})\,\gamma_z'}{4\pi\rho r^2 C_l^2}. \tag{10.72}$$

As is seen from Eqs. (10.70)–(10.72), the displacement components depend on the random angles θ_0 and φ_0 as well as on the fixed/deterministic angles θ and φ, which define the direction from the cracked zone to the ground-recording station. Assuming for the moment that there is an equal probability for the crack plane orientation, the probability density distribution does not depend on azimuthal angle φ_0. In such a case the averaging of Eqs. (10.70)–(10.72) with respect to φ_0 yields

$$\langle u_r \rangle = \frac{1}{8\pi\rho r^2 C_t^2}\left\{(M_{xx} - M_{zz})\left[\left(3 - w^2\right)\left(\frac{\sin^2\theta\sin^2\theta_0}{2}\right.\right.\right.$$
$$\left.\left.\left. + \cos^2\theta\cos^2\theta_0\right) - 1\right] + w^2\,(M_{xx} + M_{zz})\right\}, \tag{10.73}$$

$$\langle u_\theta \rangle = \frac{\sin 2\theta\,(M_{zz} - M_{xx})}{8\pi\rho r^2 C_l^2}\left(\frac{\sin^2\theta_0}{2} - \cos^2\theta_0\right), \quad \langle u_\varphi \rangle = 0. \tag{10.74}$$

In a similar fashion we can average Eqs. (10.73) and (10.74) with respect to the polar angle θ_0. Finally, taking into account that $\langle \cos^2 \theta_0 \rangle = 1/3$ and $\langle \sin^2 \theta_0 \rangle = 2/3$, we get

$$\langle u_r \rangle = \frac{u_0}{r^2} \quad \text{and} \quad \langle u_\theta \rangle = 0, \quad \text{where} \quad u_0 = \frac{(2M_{xx} + M_{zz})}{12\pi\rho C_l^2}. \qquad (10.75)$$

References

Aki K, Richards P (2002) Quantitative seismology, 2nd edn. Univ. Sci. Books, Sausalito, California, 932 p

Akinaga Y, Hayakawa M, Liu JY, Yumoto K, Hattori K (2001) A precursory signature for Chi-Chi earthquake in Taiwan. Nat Hazards Earth Syst Sci 1:33–36

Allen CR, Bonilla MG, Brace WF, Bullock M, Clough RW, Hamilton RM, Holfheinz R, Kisslinger C Jr, Knopoff L, Park M, Press F, Raleigh CB, Sykes LR (1975) Earthquake research in China. EOS Trans Am Geophys Union 56:838–881

Anderson R, Björnsson S, Blanchard DC, Gathman S, Hughes J, Jonasson S, Moore CB, Survilas HJ, Vonnegut B (1965) Electricity in volcanic clouds. Science 148(3674):1179–1189

Arabadji VI (1951) Electrical phenomena associated with volcano eruption. Meteorol Hydrol 7:38–48 (in Russian)

Arora BR, Rawat G, Mishra SS (2012) Indexing of ULF electromagnetic emission to search earthquake precursors. In: Hayakawa M (ed) The frontier of earthquake prediction studies. Nihon-senmontosho-Shuppan, Tokyo, pp 346–362

Artru J, Ducic V, Kanamori H, Lognonné P, Murakami M (2005) Ionospheric detection of gravity waves induced by tsunamis. Geophys J Int 160:840–848. Doi:10.1111/j.1365-246X.2005.02552.x

Bakun WH, Aagaard B, Dost B, Ellsworth WL, Hardebeck JL, Harris RA, Ji C, Johnston MJ, Langbein J, Lienkaemper JJ, Michael AJ, Murray JR, Nadeau RM, Reasenberg PA, Reichle MS, Roeloffs EA, Shakal A, Simpson RW, Waldhauser F (2005) Implications for prediction and hazard assessment from the 2004 Parkfield earthquake. Nature 437(7061):969–974

Barsukov VL, Varshal GM, Zamolina NS (1985) Recent results of hydrogeochemical studies for earthquake prediction in the USSR. PAGEOPH 122:143–156

Bernard P (1992) Plausibility of long electrotelluric precursors to earthquakes. J Geophys Res 97B:17531–17546

Bernardi A, Fraser-Smith AC, McGill PR, Villard OG Jr (1991) ULF magnetic field measurements near the epicenter of the Ms7.1 Loma Prieta earthquake. Phys Earth Planet Inter 68:45–63

Bolt BA (1964) Seismic air waves from the great 1964 Alaskan earthquake. Nature 202(4937):1095–1096

Björnsson S, Vonnegut B (1965) A note on electric potential gradient measurements near Icelandic geyser and steam plume from geothermal well. J Appl Meteorol 4:295–296

Boskova J, Shmilauer J, Jiricek F, Triska P (1993) Is the ion composition of the outer ionosphere related to seismic activity? J Atmos Terr Phys 55:1689–1695

Bragin VD, Volykhin AM, Trapeznikov YA (1992) Electrical-resistivity variations and moderate earthquakes. Tectonophysics 202:233–238

Brook M, Moore CB (1974) Lightning in volcanic clouds. J Geophys Res 79:472–475

Burke WJ, Aggson TL, Maynard NC, Hoegy WR, Hoffman RA, Candy RM, Liebrecht C, Rodgers E (1992) Effects of a lightning discharge detected by the DE 2 satellite over hurricane Debbie. J Geophys Res 97:6359–6367

Burton PW (1985) Electric earthquake prediction. Nature 315(6018):370–371

Campbell WH (2009) Natural magnetic disturbance fields, not precursors, preceding the Loma Prieta earthquake. J Geophys Res 114:A05307. Doi:10.1029/2008JA013932

Chave AD (1984) On the electromagnetic fields induced by oceanic internal waves. J Geophys Res 89(C6):10519–10528

Cheng K, Huang YH (1992) Ionospheric disturbances observed during the period of Mount Pinatubo eruptions in June 1991. J Geophys Res 97:16995–17004. Doi:10.1029/92JA01462

Chmyrev VM, Isaev NV, Bilichenko SV, Stanev G (1989) Observation by space-borne detectors of electric fields and hydromagnetic waves in the ionosphere over an earthquake centre. Phys Earth Planet Inter 57:110–114

Chmyrev VM, Isaev NV, Serebryakova ON, Sorokin VM, Sobolev YaP (1997) Small scale plasma inhomogeneities and correlated ELF emissions in the ionosphere over an earthquake region. J Atmos Terr Phys 59:967–974

Corwin RF, Morrison HF (1977) Self-potential variations preceding earthquakes in Central California. Geophys Res Lett 4:171–174

Cutolo M (1988) On a new general theory of earthquake. Il Nuovo Cimento 11:209–217

DasGupta A, Das A, Hui D, Bandyopadhyay KK, Sivaraman MR (2006) Ionospheric perturbation observed by the GPS following the December 26th, 2004 Sumatra-Andaman earthquake. Earth Planets Space 35:929–959

Davies K, Baker DM (1965) Ionospheric effects observed around the time of the Alaskan Earthquake on March 28, 1964. J Geophys Res 70:2251–2253

Dautermann T, Calais E, Mattioli GS (2009) GPS detection and energy estimation of the ionospheric wave caused by the 13 July 2003 explosion of the Soufrière Hills Volcano, Montserrat. J Geophys Res 114:B02202. Doi:10.1029/2008JB005722

Dea JY, Richman CI, Boerner WM (1991) Observations of seismo-electromagnetic earthquake precursor radiation signatures along Southern California fault zones: Evidence of long-distance precursor ultra-low frequency signals observed before a moderate Southern California earthquake episode. Can J Phys **69**, 1138–1145

Dea JY, Boerner WM (1999), Observations of anomalous ULF signals preceding the Northridge earthquake on January 17, 1994. In: Hayakawa M (ed) Atmospheric and ionospheric phenomena associated with earthquakes. Terra Scientific Publishing (TERRAPUB), Tokyo, pp 137–146

Dobrovolsky IP, Gershenzon NI, Gokhberg MB (1989) Theory of electrokinetic effects occurring at the final stage in the preparation of a tectonic earthquake. Phys Earth Planet Inter 57:144–156

Draganov AB, Inan US, Taranenko YuN (1991) ULF magnetic signature at the Earth surface due to ground water flow: a possible precursor to earthquakes. Geophys Res Lett 18:1127–1130

Dudkin F, Rawat G, Arora BR, Korepanov V, Leontyeva O, Sharma AK (2010) Application of polarization ellipse technique for analysis of ULF magnetic fields from two distant stations in Koyna-Warna seismoactive region, West India. Nat Hazards Earth Syst Sci 10:1513–1522. Doi:10.5194/nhess-10-1513-2010

Dudkin F, Korepanov V (2012) Magnetic field polarization ellipse: approach for detection of pre-earthquake lithosphere activity. In: Hayakawa M (ed) The frontier of earthquake prediction studies. Nihon-Senmontosho-Shuppan, Tokyo, pp 212–244

Feder E (1988) Fractals. Plenum Press, New York, 238 pp

Fedorov E, Pilipenko V, Uyeda S (2001) Electric and magnetic fields generated by electrokinetic processes in a conductive crust. Phys Chem Earth 26:793–799

Fedotov SA, Sobolev GA, Boldyrev SA, Gusev AA, Kondratenko AM, Potapova OV, Slavina LV, Theophylaktov VD, Khramov AA, Shirokov VA (1977) Long- and short-term earthquake prediction in Kamchatka. Tectonophysics 37:305–321

Fenoglio M, Johnston M, Byerlee J (1994) Magnetic and electric fields associated with changes in high pore pressure in fault zones – application to the Loma Prieta ULF emissions. In: Proceedings of workshop LXIII the mechanical involvement of fluids in faulting, Menlo Park, 1994

Fenoglio MA, Johnston MJS, Byerlee JD (1995) Magnetic and electric fields associated with changes in high pore pressure in fault zones: application to the Loma Prieta ULF emissions. J Geophys Res 100B:12951–12958

Fitterman DV (1978) Electrokinetic and magnetic anomalies associated with dilatant regions in a layered earth. J Geophys Res 83B:5923–5928

Fitterman DV (1979a) Theory of electrokinetic-magnetic anomalies in a faulted half-space. J Geophys Res 84B:6031–6040

Fitterman DV (1979b) Calculations of self-potential anomalies near vertical contacts. Geophysics 44:195–205

Fitterman DV (1981) Correction to "Theory of electrokinetic-magnetic anomalies in a faulted half-space". J Geophys Res 86B:9585–9588

Fraser-Smith AC, Bernardi A, McGill PR, Ladd ME, Helliwell RA, Villard OG Jr (1990) Low-frequency magnetic field measurements near the epicenter of the Ms 7.1 Loma Prieta earthquake. Geophys Res Lett 17:1465–1468

Fraser-Smith AC, McGill PR, Helliwell RA, Villard OG Jr (1994) Ultra-low frequency magnetic-field measurements in Southern California during the Northridge earthquake of 17 January 1994. Geophys Res Lett 21:2195–2198

Frenkel YaI (1944) On the theory of seismic and seismoelectric phenomena in moist soil. Proc USSR Acad Sci Ser Geography Geophys (Izvestiya Akad Nauk SSSR Seriya Geografiya i Geofizika) 8(4):133–150 (in Russian)

Freund F (2000) Time-resolved study of charge generation and propagation in igneous rocks. J Geophys Res 105:11001–11019

Freund F (2002) Charge generation and propagation in rocks. J Geodynamics 33:545–572

Freund F, Pilorz S (2012) Electric currents in the Earth Crust and the generation of pre-earthquake ULF signals. In: Hayakawa M (ed) The frontier of earthquake prediction studies. Nihon-Senmontosho-Shuppan, Tokyo, pp 464–508

Fujinawa Y, Kumagai T, Takahashi K (1992) A study of anomalous underground electric-field variations associated with a volcanic-eruption. Geophys Res Lett 19:9–12

Fujiwara H, Kamogawa M, Ikeda M, Liu JY, Sakata H, Chen YI, Ofuruton H, Muramatsu S, Chuo YJ, Ohtsuki YH (2004) Atmospheric anomalies observed during earthquake occurrences. Geophys Res Lett 31:L17110

Fukumoto Y, Hayakawa M, Yasuda H (2001) Investigation of over-horizon VHF radio signals associated with earthquakes. Nat Hazards Earth Syst Sci 1:107–112

Fukumoto Y, Hayakawa M, Yasuda H (2002) Reception of over-horizon signals associated with earthquakes. In: Hayakawa M, Molchanov OA (eds) Seismo electromagnetics: lithosphere-atmosphere-ionosphere coupling. TERRAPUB, Tokyo, pp 263–266

Gaillard F, Marziano GI (2005) Electrical conductivity of magma in the course of crystallization controlled by their residual liquid composition. J Geophys Res 110:B06204. doi:10.1029/2004JB003282

Galper AM, Dmitrienko VV, Nikitina NV, Grachev VM, Ulin SE (1989) Interconnection of high-energy particle fluxes in radiation belt and the Earth seismicity. Cosmic Res 27:789–792 (in Russian)

Galvan DA, Komjathy A, Hickey MP, Mannucci AJ (2011) The 2009 Samoa and 2010 Chile tsunamis as observed in the ionosphere using GPS total electron content. J Geophys Res 116:A06318. doi:10.1029/2010JA016204

Geller RJ (1996) Debate on evaluation of the VAN method: editor's introduction. Geophys Res Lett 23:1291–1293

Geller RJ (1997) Earthquake prediction: a critical review. Geophys J Int 131:425–450. doi:10.1111/j.1365-246X.1997.tb06588.x

Gershenzon NI, Gokhberg MB, Karyakin AV, Petviashvili NV, Rykunov AI (1989) Modeling the connection between earthquake preparation processes and crustal electromagnetic emission. Phys Earth Planet. Inter 57:129–138

Gershenzon NI, Gokhberg MB (1992) Electromagnetic forecasting of tsunami. Phys Solid Earth (Izvestiya Akad Nauk SSSR Fizika Zemli) 28(2):38–42

Gershenzon NI, Gokhberg MB, Yunga SL (1993) On electromagnetic field of an earthquake focus. Phys Earth Planet. Inter 77:13–19

Gershenzon NI, Bambakidis G (2001) Modeling of seismo-electromagnetic phenomena. Russ J Earth Sci 3(4):247–275

Gladychev V, Baransky L, Schekotov A, Fedorov E, Pokhotelov O, Andreevsky S, Rozhnoi A, Khabazin Y, Belyaev G, Gorbatikov A, Gordeev E, Chebrov V, Sinitsin V, Lutikov A, Yunga S, Kosarev G, Surkov V, Molchanov O, Hayakawa M, Uyeda S, Nagao T, Hattori K, Noda Y (2002) Some preliminary results of seismo-electromagnetic research at Complex Geophysical Observatory, Kamchatka. In: Hayakawa M, Molchanov OA (eds) Seismo electromagnetics: lithosphere-atmosphere-ionosphere coupling. TERRAPUB, Tokyo, pp 421–432

Gokhberg MB, Morgunov VA, Aronov EL (1979) On high frequency electromagnetic emission under seismic activity. Rep USSR Acad Sci (Dokl Akad Nauk SSSR) 248(5), 1077–1081 (in Russian)

Gokhberg MB, Morgunov VA, Yoshino T, Tomizawa I (1982) Experimental measurement of electromagnetic emissions possibly related to earthquakes in Japan. J Geophys Res 87B:7824–7828

Gokhberg MB, Pilipenko VA, Pokhotelov OA (1983) Satellite observation of electromagnetic radiation above the epicentral region of incipient earthquake. Dokl Acad Sci USSR Earth Sci Ser (Engl. Transl.) 268(1):5–7

Golitsyn GS, Klyatskin VI (1967) Atmospheric oscillations caused by motion of the ground surface. Proc USSR Acad Sci Phys Atmos Ocean (Izvestiya Akad Nauk SSSR Fizika atmosfery i okeana) 3(10):1044–1052 (in Russian)

Gorbatikov AV, Molchanov OA, Hayakawa M, Uyeda S, Hattori K, Nagao T, Tanaka H, Nikolaev AV, Maltsev P (2002) Acoustic emission possibly related to earthquakes, observed at Matsushiro, Japan and its implications. In: Hayakawa M, Molchanov O (eds) Seismo electromagnetics: lithosphere-atmosphere-ionosphere coupling. TERRAPUB, Tokyo, pp 1–10

Gorny VI, Salman AG, Tronin AA, Shilin BB (1988) The Earth outgoing IR radiation as an indicator of seismic activity. Proc USSR Acad Sci 301:67–69

Gotoh K, Akinaga Y, Hayakawa M, Hattori K (2002) Principal component analysis of ULF geomagnetic data for Izu islands earthquake in July 2000. J Atmos Electr 22:1–12

Gutenberg B, Richter CF (1954) Seismicity of the earth and associated phenomenon, 2nd edn. Princeton University Press, Princeton

Hashimoto T (2005) Electromagnetism for volcanology – history, significance, and prospect. Kazan (Volcano) 50(Special issue):115–138 (in Japanese)

Hattori K, Takahashi I, Yoshino C, Nagao T, Liu JY, Shieh CF (2002a) ULF geomagnetic and geopotential measurement at Chia-Yi, Taiwan. J Atmos Electr 22:217–222

Hattori K, Akinaga Y, Hayakawa M, Yumoto K, Nagao T, Uyeda S (2002b) ULF magnetic anomaly preceding the 1997 Kagoshima earthquakes. In: Hayakawa M, Molchanov OA (eds) Seismo electromagnetics: lithosphere - atmosphere - ionosphere coupling. TERRAPUB, Tokyo, pp 19–28

Hattori K, Takahashi I, Yoshino C, Iisezaki N, Iwasaki H, Harada M, Kawabata K, Kopytenko E, Kopytenko Y, Maltsev P, Korepanov V, Molchanov OA, Hayakawa M, Noda Y, Nagao T, Uyeda S (2004) ULF geomagnetic field measurements in Japan and some recent results associated with Iwataken Nairiku Hokubu earthquake in 1998. Phys Chem Earth 29:481–494

Hattori K (2004) ULF geomagnetic changes associated with large earthquakes. Terr Atmos Ocean Sci 15:329–360

Hattori K (2013) ULF geomagnetic changes associated with major earthquakes. In: Hayakawa M (ed) Earthquake prediction studies: seismo electromagnetics. TERRAPUB, Tokyo, pp 129–152

Hayakawa M, Fujinawa Y (eds) (1994) Electromagnetic phenomena related to earthquake prediction. Terra Scientific Pub. Comp., Tokyo, 677 pp

Hayakawa M, Kawate R, Molchanov OA, Yumoto K (1996) Results of ultra-lowfrequency magnetic field measurements during the Guam earthquake of 8 August 1993. Geophys Res Lett 23:241–244

Hayakawa M (ed) (1999) Atmospheric and ionospheric electromagnetic phenomena associated with earthquakes. Terra Scientific Pub. Comp., Tokyo, 996 pp

Hayakawa M, Itoh T, Hattori K, Yumoto K (2000) ULF electromagnetic precursors for an earthquake at Biak, Indonesia on February 17, 1996. Geophys Res Lett 27:1531–1534

Hayakawa M, Molchanov O, Tronin A, Hobara Y, Kodama T (2001) NASDA's earthquake remote sensing frontier research seismo-electromagnetic phenomena in the lithosphere, atmosphere and ionosphere. NASDA report, Final report, 229 pp

Hayakawa M, Molchanov OA (eds) (2002) Seismo electromagnetics: lithosphere - atmosphere - ionosphere coupling. TERRAPUB, Tokyo, 477 pp

Hayakawa M, Hattori K, Ohta K (2007a) Monitoring of ULF (ultra-low-frequency) geomagnetic variations associated with earthquakes. Sensors 7:1108–1122

Hayakawa M, Surkov VV, Fukumoto Y, Yonaiguchi N (2007b) Characteristics of VHF over-horizon signals possibly related to impending earthquakes and a mechanism of seismo-atmospheric perturbations. J Atmos Solar-Terr Phys 69:1057–1062

Hayakawa M (ed) (2009) Electromagnetic phenomena associated with earthquakes. Transworld Research Network, Trivandrum, 279 pp

Hayakawa M, Hobara Y, Ohta K, Hattori K (2011) The ultra-low-frequency magnetic disturbances associated with earthquakes. Earthquake Sci 24:523–534

Hayakawa M (ed) (2013) Earthquake prediction studies: seismo electromagnetics. TERRAPUB, Tokyo, 794 pp

Henderson TR, Sonwalkar VS, Helliwell RA, Inan US, Fraser-Smith AC (1993) A search for ELF/VLF emissions induced by earthquakes as observed in the ionosphere by the DE-2 satellite. J Geophys Res 98:9503–9509

Heki K (2006) Explosion energy of 2004 eruption of the Asama Volcano, central Japan, inferred from ionospheric disturbances. Geophys Res Lett 33:L14303. doi:10.1029/2006GL026249

Hines CO (1972) Gravity waves in the atmosphere. Nature 239:73–78

Hobara Y, Koons HC, Roeder JL, Yamaguchi H, Hayakawa M (2002) New ULF/ELF observation in Seikoshi, Izu, Japan, and the precursory signal in relation with recent large seismic events near Izu. In: Hayakawa M, Molchanov OA (eds) Seismo electromagnetics: lithosphere-atmosphere-ionosphere coupling. TERRAPUB, Tokyo, pp 41–44

Hobara Y, Yamaguchi H, Hayakawa M (2012) A statistical analysis of ULF magnetic data associated with seismic activity. In: Hayakawa M (ed) The frontier of earthquake prediction studies. Nihon-Senmontosho-Shuppan, Tokyo, pp 306–320 (in Japanese)

Holzworth RH, Kelley MC, Siefring CL, Hale LC, Mitchell JD (1985) Electrical measurements in the atmosphere and the ionosphere over an active thunderstorm, 2. Direct current electric fields and conductivity. J Geophys Res 90:9824–9830

Honkura Y, Niblett ER, Kurtz RD (1976) Changes in magnetic and telluric fields in a seismically active region of eastern Canada: preliminary results of earthquake prediction studies. Tectonophysics 34:219–230

Honkura Y (1981) Electric and magnetic approach to earthquake prediction. In: Rikitake T (ed) Current research in earthquake prediction. D. Reidel Publishing, Dordrecht, pp 301–383

Igarashi K, Kainuma S, Nishimura I, Okamoto S, Kuroiwa H, Tanaka T, Ogawa T (1994) Ionospheric and atmospheric disturbances around Japan caused by the eruption of Mount Pinatubo on June 15, 1991. J Atmos Terr Phys 56:1227–1234

Isaev NV, Sorokin VM, Chmyrev VM, Serebryakova ON, Ovcharenko Ya (2002) Electric field enhancement in the ionosphere above tropical storm region. In: Hayakawa M, Molchanov OA (eds) Seismo electromagnetics: lithosphere-atmosphere-ionosphere coupling. TERRAPUB, Tokyo, pp 313–315

Ismaguilov V, Kopytenko Y, Hattori K, Hayakawa M (2001) ULF magnetic emission connected with under sea bottom earthquake. Nat Hazards Earth Syst Sci 1:23–31

Ismaguilov VS, Kopytenko YuA, Hattori K, Hayakawa M (2003) Variations of phase velocity and gradient of ULF geomagnetic disturbances connected with the Izu strong earthquakes. Nat Hazards Earth Syst Sci 3:211–215

Jackson D, Kagan J (1998) VAN method lacks validity. EOS Trans Am Geophys Union 79:573–579

Johnston MJS (1989) Review of magnetic and electric field effects near active faults and volcanoes in the USA. Phys Earth Planet Inter 57:47–63

Johnston MJS, Mueller RJ, Ware RH, Davis PM (1994) Precision of geomagnetic field measurements in a tectonically active region. J Geomagn Geoelectr 36:83–95

Johnston MJS (1997) Review of electric and magnetic fields accompanying seismic and volcanic activity. Surv Geophys 18:441–475

Kanamori H, Anderson DL (1975) Theoretical basis of some empirical relations in seismology. Bull Seism Soc Am 65:1073–1095

Kawate R, Molchanov OA, Hayakawa M (1998) Ultra-low-frequency magnetic fields during the Guam earthquake of 8 August 1993 and their interpretation. Phys Earth Planet Inter 105:229–238

Kelley MC, Livingston R, McCready M (1985 a) Large amplitude thermospheric oscillations induced by earthquakes. J Geophys Res 12:577–580

Kelley MC, Siefring CL, Pfaff RF, Kintner PM, Larsen M, Green M, Holzworth RH, Hale LC, Mitchell JD, Vine DI (1985b) Electrical measurements in the atmosphere and the ionosphere over an active thunderstorm, 1. Campaign overview and initial ionospheric results. J Geophys Res 90:9815–9823

King C-Y, Azuma S, Igarahi G, Ohno G, Saito H, Wakita H (1999) Earthquake-related water-level changes at 16 clustered wells in Tono, central Japan. J Geophys Res 104:13073–13082

Kingsley SP (1989) On the possibilities for detecting radio emissions from earthquakes. Il Nuovo Cimento 12C:117–120

Komjathy A, Galvan DA, Stephens P, Butala MD, Akopian V, Wilson B, Verkhoglyadova O, Mannucci AJ, Hickey M (2012) Detecting ionospheric TEC perturbations caused by natural hazards using a global network of GPS receivers: the Tohoku case study Earth Planets Space 64:1287–1294

Kopytenko YuA, Matiashvili TG, Voronov PM, Kopytenko EA, Molchanov OA (1990) Detection of ULF emissions connected with the Spitak earthquake and its aftershock activity, based on geomagnetic pulsation data at Dusheti and Vardzia observations. Preprint of IZMIRAN, 25 pp (in Russian)

Kopytenko YuA, Matiashvili TG, Voronov PM, Kopytenko EA, Molchanov OA (1993), Detection of ULF emissions connected with the Spitak earthquake and its aftershock activity, based on geomagnetic pulsation data at Dusheti and Vardzia observations. Phys Earth Planet Inter 77:85–95

Kopytenko YuA, Ismaguilov VS, Molchanov OA, Kopytenko EA, Voronov PM, Hattori K, Hayakawa M, Zaitsev DB (2002) Investigation of ULF magnetic disturbances in Japan during seismic active period. J Atmos Electr 22:207–215

Kopytenko YuA, Nikitina LV (2004a) ULF oscillations in magma in the period of seismic vent preparation. Phys Chem Earth 29:459–462

Kopytenko YuA, Nikitina LV (2004b) A possible model for initiation of ULF oscillation in magma. Ann Geophys 47:101–105

Korepanov V, Hayakawa M, Yampolski Y, Lizunov G (2009) AGW as a seismo-ionospheric coupling responsible agent. In: Hayakawa M, Liu JY, Hattori K, Telesca L (eds) Electromagnetic phenomena associated with earthquakes and volcanoes. Elsevier, Amsterdam, pp 485–495

Koyama S, Honkura Y (1978) Observation of electric self-potential at Nakaizu (1). Bull Earthquake Res Inst Tokyo Univ 53(Pt. 3):939–942 (in Japanese)

Kushida Y, Kushida R (1998) On a possibility of earthquake forecast by radio observation in the VHF band. RIKEN Rev 19:152–160

Kushida Y, Kushida R (2002) Possibility of earthquake forecast by radio observations in the VHF band. J Atmos Electr 22:239–255

Landau LD, Lifshits EM (1982) Electrodynamics of continuous media. Nauka, Moscow (in Russian)

Larkina VI, Migulin VV, Mogilevsky MM, Molchanov OA, Galperin YuI, Jorjio NV, Gokhberg MV, Lefeuvre F (1985) Earthquake effects in the ionosphere according to Intercosmos-19 and Aureol-3 satellite data. In: Results of the ARCAD-3 Project and of the recent programs in magnetospheric and ionospheric physics, Toulouse, 1984, pp 685–699

Larsen JC (1968) Electric and magnetic fields induced by deep sea tides. Geophys J Roy Astron Soc 16:47–70

Lay T, Kanamori H, Ammon CJ, Nettles M, Ward SN, Aster RC, Beck SL, Bilek SL, Brudzinski MR, Butler R, DeShon HR, Ekstrom G, Satake K, Sipkin S (2005) The great Sumatra-Andaman earth- quake of 26 December 2004. Science 308:1127–1133. doi:10.1126/science.1112250

Lenchenko VM (1988) Electrization of solid by slip flow. J Tech Phys (Zhurnal Tekhnicheskoi Fiziki) 58:995–997 (in Russian)

Leonard RS, Barnes RA (1965) Observations of ionospheric disturbances following Alaska earthquake. J Geophys Res 70:1250–1253

Le Pichon A, Guilbert J, Vega A, Garce's M, Brachet N (2002) Ground-coupled air waves and diffracted infrasound from Arequipa earthquake on June23, 2001. Geophys Res Lett 29:1886–1889

Liu Qi-Qi, Ding J, Cui C (2000) Probable satellite thermal infrared anomaly before the Zhangbei $M = 6.2$ earthquake on Jan 10, 1998. Acta Seimologica Sin 13:203–209

Liu JY, Tsai YB, Chen SW, Lee CP, Chen YC, Yen HY, Chang WY, Liu C (2006) Giant ionospheric disturbances excited by the M9.3 Sumatra earthquake of 26 December 2004. Geophys Res Lett 33:L02103. doi:10.1029/2005GL023963

Lockner DA, Johnston MJS, Byerlee JD (1983) A mechanism to explain the generation of earthquake lights. Nature 302(3):28–32

Lockner DA, Byerlee JD (1985) Complex resistivity of fault gouge and its significance for earthquake lights and induced polarization. Geophys Res Lett 12:211–214

Lognonné P, Artru J, Garcia R, Crespon F, Ducic V, Jeansou E, Occhipinti G, Helbert J, Moreaux G, Godet PE (2006) Ground based GPS imaging of ionospheric post-seismic signal. Planet Space Sci 54:528–540

MacDonald GA (1972) Volcanoes. Prentice-Hall, Englewood Cliffs, 510 pp

Makris JP, Valliantos F, Kopytenko Y, Smirnova N, Korepanov V, Kopytenko E, Soupios P, Vardiambasis I, Stampolidis A, Mavromatidis A (2003) ULF geomagnetic observations and seismic activity on the southern part of hellenic Arc. Geophys Res Abstracts 5 (EGS-AGU-EUC Joint Assembly, 06-11 April 2003, Nice, France) 12338 (CDROM)

Manoj C, Kuvshinov A, Neetu S, Harinarayana T (2010) Can undersea voltage measurements detect tsunamis? Earth Planets Space 62:353–358

Matthews JP, Leberton JP (1985) A search for seismic related wave activity in the micropulsation and ULF frequency ranges using GEOS-2 data. Ann Geophys 3:749–754

Mazzella A, Morrison FH (1974) Electrical resistivity variations associated with earthquake on the San Andreas fault. Science 185(4154):855–857

Meyer K, Teisseyre R (1989) Observation and qualitative modeling of some electrotelluric earthquake precursors. Phys Earth Planet Inter 57:45–46

Migulin VV, Larkina VI, Molchanov OA, Gokhberg MB, Nalivaiko AV, Pilipenko VA (1982) Detection of earthquake influence on the ELF/VLF emissions at the upper ionosphere. Preprint IZMIRAN, 25 (2390), Moscow, 22 pp (in Russian)

Mikhailova G, Mikhailov Yu, Kapustina O (2000) ULF-VLF electric field in the external ionosphere over powerful typhoons in Pacific ocean. Int J Geomagn Aeronomy 2:153–158

Mikhailova GA, Mikhailov YuM, Kapustina OV (2002) Variations of ULF-VLF electric fields in the external ionosphere over powerful typhoons in Pacific Ocean. Adv Space Res 30:2613–2618

Mineev VN, Ivanov AG (1976) Electromotive force produced by shock compression of a substance. Phys Uspekhi (Adv Phys Sci) 19(5):400–419

Miyakoshi J (1986) Anomalous time variation of the self-potential in the fractured zone of an active fault preceding the earthquakes occurrence. J Geomagn Geoelectr 38:1015–1030

Mizutani H, Ishido T (1976) A new interpretation of magnetic field variation associated with the Matsushiro earthquakes. J Geomagn Geoelectr 28:179–188

Mizutani H, Ishido T, Yokokura T, Ohnishi S (1976) Electrokinetic phenomena associated with earthquakes. Geophys Res Lett 3:365–368

Molchanov OA, Kopytenko YuA, Voronov PM, Kopytenko EA, Matiashvili TG, Fraser-Smith AC, Bernardi A (1992) Results of ULF magnetic field measurements near the epicenters of the Spitak (Ms=6.9) and Loma Prieta (Ms=7.1) earthquakes: comparative analysis. Geophys Res Lett 19:1495–1498

Molchanov OA, Mazhaeva OA, Goliavin AN, Hayakawa M (1993) Observation by the Intercosmos-24 satellite of ELF-VLF electromagnetic emissions associated with earthquakes. Ann Geophys 11:431–440

Molchanov OA, Hayakawa M (1994) Generation of ULF seismogenic electromagnetic emission: a natural consequence of microfracturing process. In: Hayakawa M (ed) Electromagnetic phenomena related to earthquake prediction. TERRAPUB, Tokyo, pp 537–563

Molchanov OA, Hayakawa M (1995) Generation of ULF electromagnetic emissions by microfracturing. Geophys Res Lett 22:3091–3094

Molchanov OA, Hayakawa M (1998) On the generation mechanism of ULF seismogenic electromagnetic emissions. Phys Earth Planet Inter 105:201–210

Molchanov O, Kulchitsky A, Hayakawa M (2002) ULF emission due to inductive seismo-electromagnetic effect. In: Hayakawa M, Molchanov OA (eds) Seismo electromagnetics: lithosphere-atmosphere-ionosphere coupling. TERRAPUB, Tokyo, pp 153–162

Molchanov OA, Hayakawa M (2008) Seismo-electromagnetics and related phenomena: history and latest results. TERRAPUB, Tokyo, 189 pp

Murakami H, Mizutani H, Nabetani S (1984) Self-potential anomalies associated with an active fault. J Geomagn Geoelectr 36:351–376

Myachkin VI, Sobolev GA, Dolbilkina NA, Morozow VN, Preobrazensky VB (1972) The study of variations in geophysical fields near focal zones of Kamchatka. Tectonophysics 14:287–293

Nair MC, Maus S, Neetu S, Kuvshinov AV, Chulliat A (2010) The magnetic fields generated by the tsunami of February 27, 2010. American Geophysical Union, Fall Meeting 2010, abstract #GP41B-03

Nikolaevskiy VN (1996) Geomechanics and fluidodynamics. Kluwer Academic Publishers, Dodrecht/Boston/London

Nur A (1972) Dilatancy, pore fluids, and premonitory variations of t_s/t_p travel times. Bull Seismol Soc Am 62:1217–1222

Occhipinti G, Lognonné P, Alam Kherani E, Hebert H (2006) Three-dimensional waveform modeling of ionospheric signature induced by the 2004 Sumatra tsunami. Geophys Res Lett 33:L20104. doi:10. 1029/2006GL026865

Occhipinti G, Rolland L, Lognonné P, Watada S (2013) From Sumatra 2004 to Tohoku-Oki 2011: the systematic GPS detection of the ionospheric signature induced by tsunamigenic earthquakes. Geophys Res Phys 118(A6):3626–3636. doi:10:1002/jgra.50322

Ohta K, Watanabe N, Hayakawa M (2005) The observation of ULF emissions at Nakatsugawa in possible association with the 2004 Mid Niigata Prefecture earthquake. Earth Planets Space 57:1003–1008

Oike K, Ogawa T (1986) Electromagnetic radiations from shallow earthquakes observed in the LF range. J Geomagn Geoelectr 38:1031–1040

Ouzounov D, Freund F (2004) Mid-infrared emission prior to strong earthquakes analyzed by remote sensing data. Adv Space Res 33(3):268–273

Park SK, Fitterman DV (1990) Sensitivity of the telluric monitoring array in Parkfield, California, to changes of resistivity. J Geophys Res 95B:15557–15571

Park SK (1991) Monitoring resistivity changes prior to earthquake in Parkfield, California, with telluric arrays. J Geophys Res 96B:14211–14237

Park SK, Johnson M, Madden JS, Morgan FD, Morrison HF (1993) Electromagnetic precursors to earthquakes in the ULF band: a review of observations and mechanisms. Rev Geophys 31:117–132

Parrot M, Lefeuvre F, Corcuff Y, Godefroy P (1985) Observation of VLF emissions at the time of earthquakes in the Kerguelen Islands. Ann Geophysicae 3:731–736

Parrot M, Mogilevsky MM (1989) VLF emissions associated with earthquakes and observed in the ionosphere and the magnetosphere. Phys Earth Planet Inter 57:86–89

Parrot M (1990) Electromagnetic disturbances associated with earthquakes: an analysis of ground-based and satellite data. J Sci Exploration 4:203–211

Parrot M (1995) Electromagnetic noise due to earthquake. In: Volland H (ed) Handbook of atmospheric electrodynamics. CRC Press, Boca Raton/London/Tokyo, pp 94–116

Parrot M (2012) Satellite observations of electromagnetic noise related to seismic activity. In: Hayakawa M (ed) The frontier of earthquake prediction studies. Nihon-Senmontosho-Shuppan, Tokyo, pp 738–764

Pavlov VI, Sukhorukov AI (1987) About moving ionospheric perturbations caused by propagation of tsunami wave in ocean. Lett J Tech Phys (Pisma v Zhurnal tekhicheskoy Fiziki) 13(6):351–354 (in Russian)

Peltier WR, Hines CO (1976) On the possible detection of tsunamis by a monitoring of the ionosphere. J Geophys Res 81(12):1995–2000

Pulinets SA, Legenka AD, Alekseev VA (1994) Pre-earthquake ionospheric effects and their possible mechanisms. In: Kikuchi H (ed) Dirty plasmas, noise and chaos in space and in the laboratory. Plenum Press, New York, pp 545–556

Pulinets SA, Alekseev VA, Legenka AD, Khegai VV (1997) Radon and metallic aerosols emanation before strong earthquakes and their role in atmosphere and ionosphere modification. Adv Space Res 20:2173–2176

Qiang ZJ, Xu XD, Dian CG (1991) Thermal infrared anomaly - precursor of impending earthquakes. Chin Sci Bull 36:319747–319750, 323

Qiang ZJ, Dianv CG, Li LZ (1999) Satellite thermal infrared precursor of two moderate-strong earthquakes in Japan and impeding earthquake prediction. In: Hayakawa M (ed) Atmospheric and ionospheric phenomena associated with earthquakes. Terra Scientific Publishing, Tokyo, pp 747–750

Ralchovsky TM, Komarov LN (1988) Periodicity of the Earth electric precursors before strong earthquake. Tectonophysics 145:325–327

Rikitake T, Yamazaki Y (1978) Precursory and seismic changes in ground resistivity. In: Kisslinger C, Suzuki S (eds) Earthquake prediction. Center for Acad. Publications, Tokyo, pp 161–173

Rikitake T (1987) Magnetic and electric signals precursory to earthquakes: an analysis of Japanese data. J Geomagn Geoelectr 39:47–61

Rodger CJ, Thomson NR, Dowden RL (1996) A search for ELF/VLF activity associated with earthquakes using ISIS satellite data. J Geophys Res 101(A6):13369–13378

Rodger CJ (1999) Red sprites, upward lightning, and VLF perturbations. Rev Geophys 37:317–336

Roeloffs EA (1988) Hydrological precursors to earthquakes: a review. Pure Appl Geophys 126:177–209

Rolland L, Occhipinti G, Lognonné P, Loevenbruck A (2010) The 29 September 2009 Samoan tsunami in the ionosphere detected offshore Hawaii. Geophys Res Lett 37:L17191. doi:10.1029/2010GL044479

Rulenko OP (1979) Electrical processes in vapor-gas clouds of Karymsky volcano. Rep USSR Acad Sci (Dokl Akad Nauk SSSR) 245(5):1083–1086 (in Russian)

Rulenko OP (1985) Electrization of volcano clouds. Volcanology seismology 2:71–83 (in Russian)

Sadovsky MA, Sobolev GA, Migunov NI (1979) Changes in natural radiowave emission under strong earthquake at Karpaty. Rep USSR Acad Sci (Dokl Akad Nauk SSSR) 244(2):316–319 (in Russian)

Salman A, Egan WG, Tronin AA (1992) Infrared remote sensing of seismic disturbances. In: Polarization and remote sensing. SPIE, San Diego, pp 208–218

Sanford TB (1971) Motionally induced electric and magnetic fields in the sea. J Geophys Res 76:3476–3492

Sasai Y (1991) Tectonomagnetic modeling on the basis of linear piezomagnetic effect. Bull Earthquake Res Inst Tokyo Univ 66(Pt. 4):585–722

Schekotov AY, Molchanov OA, Hayakawa M, Fedorov EN, Chebrov VN, Sinitsin VI, Gordeev EE, Belyaev GG, Yagova NV (2007) ULF/ELF magnetic field variations from atmosphere induced by seismicity. Radio Sci 42:RS6S90. doi:10.1029/2005RS003441

Schekotov AY, Molchanov OA, Hayakawa M, Fedorov EN, Chebrov VN, Sinitsin VI, Gordeev EE, Andreevsky SE, Belyaev GG, Yagova NV, Gladishev VA, Baransky LN (2008) About possibility to locate an EQ epicenter using parameters of ELF/ULF preseismic emission. Nat Hazards Earth Syst Sci 8:1237–1242

Scholz CH, Sykes LR, Aggarwal YP (1973) Earthquake prediction: a physical basis. Science 181:803–810

Scholz CH (1990) The mechanics of earthquakes and faulting. Cambridge University Press, Cambridge, 439 pp

Serebryakova ON, Bilichenko SV, Chmyrev VM, Parrot M, Rauch JL, Lefeuvre F, Pokhotelov OA (1992) Electromagnetic ELF radiation from earthquakes regions as observed by low-altitude satellites. Geophys Res Lett 19:91–94

Slifkin L (1993) Seismic electric signals from displacement of charged dislocations. Tectonophysics 224:149–152

Smith BE, Johnston MJS (1976) A tectonomagnetic effect observed before a magnitude 5.2 earthquake near Hollister, California. J Geophys Res 81:3556–3560

Sobolev GA, Morozov VN, Migunov NI (1972) Electrotelluric field and strong earthquake at Kamchatka. Izvestiya Akad Nauk SSSR Fizika Zemli (Proc USSR Acad Sci Solid Earth Phys) 2:73–78 (in Russian)

Sobolev GA (1975) Application of electric method to the tentative short-term forecast of Kamchatka earthquakes. Pure Appl Geophys 113:229–235

Sobolev GA, Ponomarev AV (2003) Physics of earthquakes and precursors. Nauka, Moscow, 270 pp (in Russian)

Sornette A, Sornette S (1990) Earthquake rupture as a critical point: consequences for telluric precursors. Tectonophysics 179:327–334

Sorokin VM, Chmyrev VM, Yaschenko AK (2003) Ionospheric generation mechanism of geomagnetic pulsations observed on the Earth's surface before earthquake. J Atmos Solar-Terr Phys 65(1):21–29

Sorokin VM, Isaev NV, Yaschenko AK, Chmyrev VM, Hayakawa M (2005) Strong DC electric field in the low latitude ionosphere over typhoons. J Atmos Solar-Terr Phys 67:1269–1279

Stacey FD (1964) The seismomagnetic effect. Pure Appl Geophys 58:5–23

Stacey FD, Johnston MJS (1972) Theory of piezomagnetic effect in titanomagnetite-bearing rocks. Pure Appl Geophys 97:146–155

Staüffer D (1979) Scaling theory of percolation clusters. Phys Rep 54:1–74

Surkov VV (1997) The nature of electromagnetic forerunners of earthquakes. Trans (Doklady) Russ Acad Sci Earth Sci Sections 355:945–947

Surkov VV, Pilipenko VA (1999) The physics of pre-seismic electromagnetic ULF signals. In: Hayakawa M (ed) Atmospheric and ionospheric phenomena associated with earthquakes. TERRAPUB, Tokyo, pp 357–370

Surkov VV (1999) ULF electromagnetic perturbations resulting from the fracture and dilatancy in the earthquake preparation zone. In: Hayakawa M (ed) Atmospheric and ionospheric phenomena associated with earthquakes. TERRAPUB, Tokyo, pp 371–382

Surkov VV (2000a) Electromagnetic effects caused by earthquakes and explosions. MEPhI, Moscow, 448 pp (in Russian)

Surkov VV (2000b) On the nature of ULF electromagnetic noise anticipating some earthquakes. Phys Solid Earth 12:61–66

Surkov VV (2001) The role of shear cracks in the formation of electromagnetic noise preceding some earthquakes. Doklady Earth Sci (Translated from Doklady Akademii Nauk Russ Acad Sci) 377A:349–355

Surkov VV, Uyeda S, Tanaka H, Hayakawa M (2002a) Fractal properties of medium and seismoelectric phenomena. J Geodynamics 33:477–487

Surkov VV, Iudin DI, Molchanov OA, Korovkin NV, Hayakawa M (2002b) Thermofluctuational mechanism of cracks migration as a model of earthquake preparation. In: Hayakawa M, Molchanov OA (eds) Seismo electromagnetics: lithosphere-atmosphere-ionosphere coupling. TERRAPUB, Tokyo, pp. 195–201

Surkov VV, Molchanov OA, Hayakawa M (2003) Pre-earthquake ULF electromagnetic perturbations as a result of inductive seismomagnetic phenomena during microfracturing. J Atmos Solar-Terr Phys 65:31–46

Surkov VV, Molchanov OA, Hayakawa M (2004) A direction finding technique for the ULF electromagnetic source. Nat Hazards Earth Syst Sci 4:513–517

Surkov VV, Tanaka H (2005) Electrokinetic effect in fractal pore media as seismoelectric phenomena. In: Dimri VP (ed) Fractal behaviours of the earth system. Springer, Berlin/Heidelberg/New York, pp 83–96

Surkov VV, Pokhotelov OA, Parrot M, Hayakawa M (2006) On the origin of stable IR anomalies detected by satellite above seismo-active regions. Phys Chem Earth 31:164–171

Surkov VV, Hayakawa M (2006) ULF geomagnetic perturbations due to seismic noise produced by rock fracture and crack formation treated as a stochastic process. Phys Chem Earth 31:273–280

Takahashi K, Fujinawa Y (1993) Locating source regions of precursory seismo-electric fields and the mechanism generating electric-field variations. Phys Earth Planet Inter 77:33–38

Tanaka T, Ichinose T, Okuzawa T, Shibata T, Sato Y, Nagasawa C, Ogawa T (1984) Hf-Doppler observations of acoustic waves excited the Urakawa-Oki earthquake on 21 March 1982. J Atmos Solar-Terr Phys 46:233–245

Terada T (1931) On luminous phenomena accompanying earthquake. Bull Earthquake Res Inst Tokyo Univ 9:225–255

Thomas JN, Love JJ, Johnston MJS (2009) On the reported magnetic precursor of the 1989 Loma Prieta earthquake. Phys Earth Planet Inter 173:207–215

Tramutoli V, Bello D, Pergola GN, Piscitelli S (2001) Robust satellite technique for remote sensing of seismically active areas. Ann Di Geofisica 44:295–312

Tronin AA (1996) Satellite thermal survey - a new tool for the study of seismoactive regions. Intern J Remote Sensing 17:1439–1455

Tronin AA (1999) Satellite thermal survey application for earthquake prediction. In: Hayakawa M (eds) Atmospheric and ionospheric phenomena associated with earthquakes. Terra Scientific Publishing, Tokyo, pp 357–370

Tronin AA (2002) Atmosphere-lithosphere coupling. Thermal anomalies on the Earth surface in seismic processes. In: Hayakawa M, Molchanov OA (eds) Seismo electromagnetics. Lithosphere-atmosphere-ionosphere coupling. Terra Scientific Publishing, Tokyo, pp 173–176

Tronin AA, Hayakawa M, Molchanov O (2002) Thermal IR satellite data application for earthquake research in Japan and China. J Geodynamics 33:519–534

Turcotte DL (1997) Fractal and chaos in geology and geophysics, 2nd edn. Cambridge University Press, Cambridge

Tyler RH (2005) A simple formula for estimating the magnetic fields generated by tsunami flow. Geophys Res Lett 32:L09608. doi:10.1029/2005GL022429

Tzanis A, Vallianatos F (2002) A physical model of electrical earthquake precursors due to crack propagation and the motion of charged edge dislocations. In: Hayakawa M, Molchanov O (eds) Seismo electromagnetics: lithosphere - atmosphere - ionosphere coupling. TERRAPUB, Tokyo, pp 117–130

Uman MA (1987) The lightning discharge. Academic, New York

Urusovskaya AA (1969) Electric effects associated with plastic deformation of ionic crystals. Phys Uspekhi (Adv Phys Sci) 11(5):631–643

Uyeda S (1998) VAN method of short-term earthquake prediction shows promise. EOS Trans Am Geophys Union 79:573–580

Uyeda S, Nagao T, Orihara Y, Yamaguchi T, Takahashi I (2000) Geoelectric potential changes: possible precursors to earthquakes in Japan. Proc Nat Acad Sci 97:4561–4566

Uyeda S, Hayakawa M, Nagao T, Molchanov OA, Hattori K, Orihara Y, Gotoh K, Akinaga Y, Tanaka H (2002) Electric and magnetic phenomena observed before the volcano-seismic activity in 2000 in the Izu islands regions, Japan. Proc Natl Acad Sci 99:7352–7355

Vallianatos F, Tzanis A (1998) Electric current generation associated with the deformation rate of a solid : preseismic and co-seismic signals. Phys Chem Earth 23:933–938

Varotsos P, Alexopoulos K (1984a) Physical properties of the variations of the electric field of the earth preceding earthquake – I. Tectonophysics 110:73–98

Varotsos P, Alexopoulos K (1984b) Physical properties of the variations of the electric field of the earth preceding earthquake – II. Determination of epicenter and magnitude. Tectonophysics 110:99–125

Varotsos P, Alexopoulos K (1986) Stimulated current emission in the earth: piezostimulated currents and related geophysical aspects. In: Amelinckx S, Gevers R, Nihoul J (eds) Thermodynamics of point defects and their relation with bulk properties. North Holland, Amsterdam, pp 136–142, 403–406, 410–412, 417–420

Varotsos P, Lazaridou M, Eftaxias K, Antonopoulos G, Makris J, Kopanas J (1996) Short term earthquake prediction in Greece by Seismic Electric Signals. In: Lighthill J Sir (ed) A critical review of VAN. World Scientific, Singapore, pp 29–76

Varotsos A (2005) The physics of seismic electric signals. TERRAPUB, Tokyo, 338 pp

Voronov SA, Galper AM, Kirillov-Ugryumov VG, Koldashev SV, Mikhailov VV, Nikitina NV, Popov AV (1989) An increase in fluxes of high-energy particles in the region of Brazil geomagnetic anomaly on September 10, 1985. Cosmic Res 4:629–631 (in Russian)

Wakita H (1975) Water wells as possible indicators of tectonic strain. Science 189:553–555

Warwick JW, Stoker C, Meyer TR (1982) Radio emission associated with rock fracture: possible application to the Great Chilean earthquake of May 22, 1960. J Geophys Res 87B:2851–2859

Wolcott JH, Simons DJ, Lee DD, Nelson RA (1984) Observations of an ionospheric perturbations arising from the Coaglinga earthquake of May 2, 1983. J Geophys Res 89:6835–6839

Yasuoka Y, Ishikawa T, Nagahama H, Kawada Y, Omori Y, Tokomami S, Shinogi M (2012) Radon anomalies prior to earthquake. In: Hayakawa M (ed) The frontier of earthquake prediction studies. Nihon-Senmontosho-Shuppan, Tokyo, pp 410–427 (in Japanese)

Yepez E, Angulo-Brown F, Peralta JA, Pavia CG, Gonzalel-Santos G (1995) Electric field patterns as seismic precursors. Geophys Res Lett 22:3087–3090

Yoshino T, Tomizawa I (1988) LF seismogenic emissions and its application on the earthquake prediction. The technical report of institute of electronic information and communications. Technical report EMCJ pp 88–64

Zemtsov AN, Tron AA, Markhinin EK (1976) Electric charges in ash-gas clouds arising from volcano eruption. Bull Volcanological Stations 52:18–23 (in Russian)

Zlotnicki J, Nishida Y (2003) Review of morphological insights of self-potential anomalies on volcano. Surv Geophys 24:291–338

Zlotnicki J, Li F, Parrot M (2010) Signals recorded by DEMETER satellite over active volcanoes during the period 2004 August – 2007 December. Geophys J Int 183(3):1332–1347

Chapter 11
Electromagnetic Effects Resulted from Explosions

Abstract In this chapter we study man-made low-frequency electromagnetic fields resulted from high explosive or nuclear detonations. The main emphasis is on underground explosion effects and a variety of accompanying electromagnetic phenomena caused by rock deformation and perturbations of the Earth magnetic field. We start with basic mechanisms for a so-called electromagnetic pulse (EMP) occurring just after the detonation and belonging to ULF/ELF frequency range. It is usually the case that the EMP precedes the co-seismic phenomena caused by seismic wave arrival at the observation point. Then we examine the atmospheric effects caused by the generation of dusty clouds and propagation of aerial shock waves (SWs). In the remainder of this chapter we consider the perturbations of the ionospheric plasma caused by an upward propagating SW.

Keywords Aerial shock wave (SW) • Electromagnetic pulse (EMP) • Gas-dust cloud • Residual electromagnetic field • Underground explosion

11.1 Diamagnetic Plasma Effect of Explosions

11.1.1 Observations of EMP Resulted from Underground Explosions

The earliest detailed recordings of the EMP caused by underground tests have been published after the series of nuclear detonations referred as Hardtack II on the proving ground in Nevada in 1958 (Zablocki 1966). At first the examination of ground conductivity was planned in the vicinity of an underground explosion chamber. However the strong low-frequency electric field was unexpectedly detected at the moment of detonation. The electric pickup arising simultaneously with the detonation was so high that it prevented seriously the recording of conductivity changes. These findings lent impetus to a study of interrelationship between electromagnetic and seismic effects because of importance of this research for the

V. Surkov and M. Hayakawa, *Ultra and Extremely Low Frequency Electromagnetic Fields*, Springer Geophysics, DOI 10.1007/978-4-431-54367-1_11,
© Springer Japan 2014

treaty verification of nuclear undearground tests (e.g., Latter et al. 1961b; Gorbachev et al. 1999a,b). Much emphasis has been put on studies of the EMP in order to detect any underground nuclear testing especially in the case of the so-called decoupling of underground nuclear explosion (Zablocki 1966; Sweeney 1989). The decoupling means that the underground explosion is realized in an evacuated volume in the chamber of large size in order to diminish seismic effect of the explosion (Latter et al. 1961a; Herbst et al. 1961; Patterson 1966). Although the EMP magnitude of the explosion with decoupling can even be greater in comparison with that of the explosion conducting under the usual size of the explosion chamber (Gorbachev and Semenova 2000a,b).

Below we review experimental data and then focus on basic physical mechanisms of this phenomenon and estimate the amplitude of ULF electromagnetic variations. The observations have shown that the EMP of underground explosions decreases rapidly with distance so that it is practically undetectable at the distance over 10 km from the detonation point. For instance, the electric field amplitude was approximately $1\,\mu V/m$ at the epicentral distance of 6.72 km (Zablocki 1966). A typical scheme of the recording electrodes arrangement under Hardtack II series of nuclear testing is shown in Fig. 11.1. The non-screened isolated copper wire with length of 750 m was laid on the ground from the observation point towards to the explosion epicenter in the East–West direction. The wire ends were linked with the lead electrodes buried in the ground 1–3 m deep. The same length wire was laid in the perpendicular North–South direction. Next one was put into a hole at the depth of 30 m. This wire is ended by the lead electrode as well. The recording sensors measured the potential difference between grounded ends of each wire. The natural potential difference that always occurs while a pair of grounded electrodes is connected was compensated at the inputs of the recorder by means of a potentiometer circuit. The bandwidth of sensors was in an interval from 0 to 220 Hz. Such a system allows us to control all three components of the low-frequency electromagnetic field.

The magnetic field perturbations were also recorded during a series of nuclear tests in 1961. The magnetic coils with vertically directed axis were used to perform the measurements of vertical component of the magnetic perturbation. Eight turns of wire were winded round the frame of coils, whose size was from 7.5 to 18.6 m. The eigenfrequency of electromagnetic vibrations of the coil was within 10–20 kHz, and these values significantly exceed the typical frequencies of the EMP.

Magnetic antennas with horizontal axis measured the horizontal component of magnetic field variations. The coils have an area of $2\,m^2$ area and 3.2×10^4 turns that correspond to the eigenfrequency about 60 Hz. The other kind of coils are $1\,m^2$ area, 7×10^3-turn loop of wire, so that the eigenfrequency is 200 Hz. A variety of amplifiers and filtering schemes were used to give the maximum of signal-to-noise ratio.

The EMP signal recorded at a proving ground in Nevada in 1958 during one of five underground explosions of the Hardtek series is depicted in Fig. 11.2. The depth of this explosion was 254 m, and trinitrotoluene/TNT equivalent was 19 kt (kiloton) (Zablocki 1966). What draws first attention is almost complete polarization of the electric field in the direction of azimuthal component, E_φ, and this feature was practically observed in all the tests.

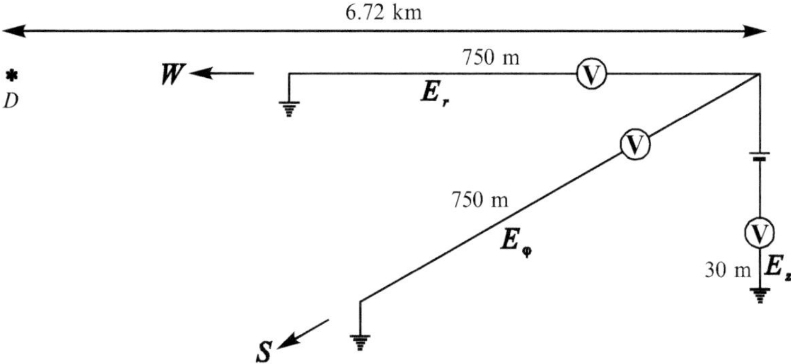

Fig. 11.1 A schematic plot of the equipment arrangement used during Hardtack II series of nuclear testing. The detonation hypocenter is marked by D. Adapted from Zablocki (1966)

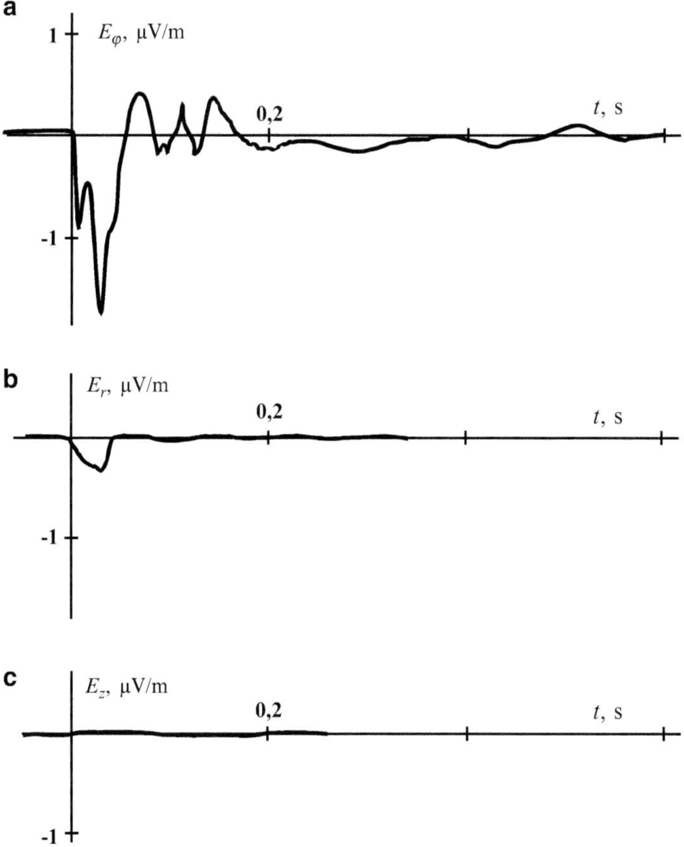

Fig. 11.2 A schematic plot of (**a**) tangential, (**b**) radial, and (**c**) vertical components of electric field variations measured at the proving ground in Nevada in 1958 during one of five underground explosions of the Hardtek series. The measurements were performed at 6.72 km distance east of explosion point and magnetic meridian. Adapted from Zablocki (1966)

The summary of the characteristics of majority of the signals could be made in the following way. The rise time of the initial spike is about 8–15 ms, and the polarization of perturbations is so that the vector of electric field is predominantly directed from South to North (Fig. 11.1). Within the time interval of 30–70 ms the field decreases approximately exponentially. Powerful detonations, as a rule, are accompanied by the shock excitation of the irregular vibrations lasted nearly 0.4 s. These vibrations reach 20 % of the signal magnitude (for example, see Fig. 11.2a). On numerous occasions the initial positive spike was followed by the negative half-wave with duration from 0.5 to 1 s as shown in Fig. 11.3 a, b with lines 1. The power spectrum of the EMP has spikes in the frequency ranges of 2–8 and 20–30 Hz. Besides the power spectrum tends to increase with decrease in frequency, which are lower than 2 Hz.

One more example is the nuclear test referred as "Bilbi," which was detonated at the depth of 714.5 m on September 1963. This contained underground explosion had a TNT equivalent $Y = 235$ kt, i.e. more than that considered above by one order of magnitude (Zablocki 1966). In this case the amplitude of electric field component, E_φ, was 3.6 µV/m at the distance 7.62 km to the South (along the magnetic meridian) of the detonation epicenter point. The rise time of the initial spike of the EMP did not exceed 15 ms and the time of the decrease down to zero level was approximately 150 ms. Notice that the rise time of the initial spike varies within 8–15 ms for all the tests. This value is larger than that of atmospherics, whose typical build-up time does not exceed 5 ms.

The relaxation time, τ_r, of the electric component of EMP as observed in the series of experiments is shown in Fig. 11.4 as a function of TNT equivalent, Y, of the detonation. This empirical dependence can be approximated by the following:

$$\tau_r = 30Y^{1/3} \text{ ms,} \tag{11.1}$$

where Y is measured in kt (1 kt is approximately equal to 4×10^{12} J).

The magnitude of electric field variations decreases approximately inversely proportional to the cubed distance, at least as the distance is smaller than 10 km. The empirical dependence of the horizontal component of electric field on the epicentral distance and TNT equivalent is given by (Zablocki 1966):

$$E = 2.2 \cdot 10^2 Y^{0.44}/R^3, \quad \text{µV/m.} \tag{11.2}$$

where the distance R is measured in kilometers. The typical magnitude of electric field can reach several tens mV/m under the detonation with TNT equivalent smaller than 150 kt (Malik et al. 1985; Sweeney 1989).

It is usually the case that the magnetic component of the EMP by nuclear underground explosions varies from several pT to several nT at the epicentral distances which are no more than 10 km. For example, the magnetic measurements during the detonation "Hardin" at the testing area in Nevada in 1987 have shown that

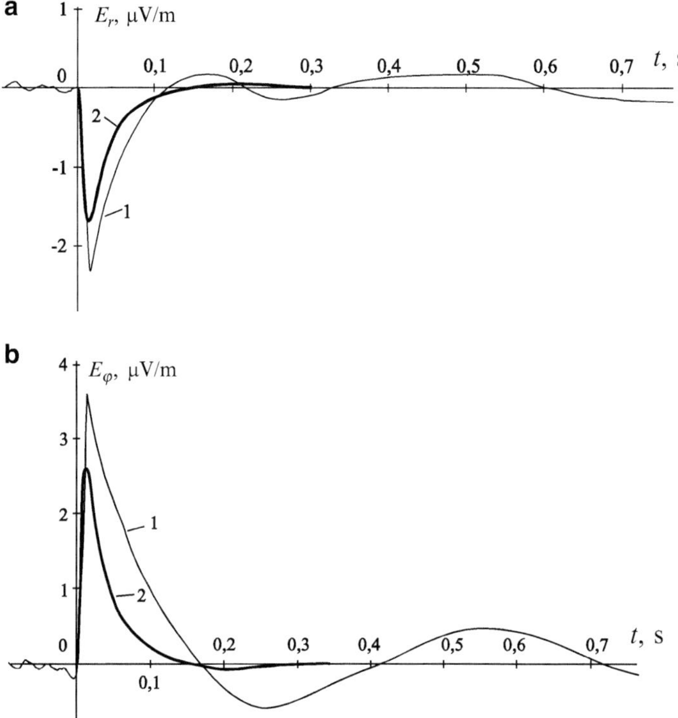

Fig. 11.3 Radial E_r (**a**) and azimuthal E_φ (**b**) components of electric field at the epicentral distance of 7.62 km from the explosion. (*1*) Experimental observations during the underground explosion "Bilbi" performed at the depth 714.5 m (Zablocki 1966); (*2*) Numerical calculations based on the model of expanding plasma ball (Ablyazov et al. 1988)

Fig. 11.4 Relaxation time of the electric component of EMP observed in a series of the experiments versus TNT equivalent of the detonation. Adapted from Zablocki (1966)

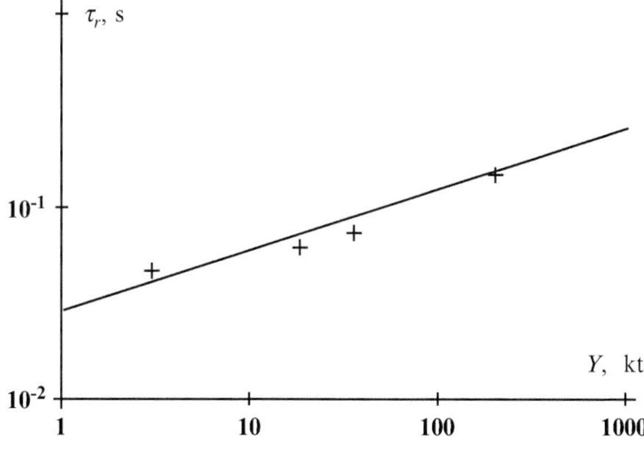

the magnitude of horizontal component of magnetic perturbations decreased from 40 pT at the distance of 5 km to 10 pT at the distance of 11 km from the detonation point (Sweeney 1989). At the same time the vertical component of the magnetic perturbations exceeded several hundreds pT.

The natural ULF electromagnetic background due to the atmospherics, the ionospheric and magnetospheric micropulsations and others restricts the possibility for detection of the EMP. As the distance is of the order of or much greater than 10 km, it appears that the EMP becomes undetectable because of small value of the signal-to-noise ratio. Moreover, the instability of the signal polarization produces the additional difficulty in the utilization of the cross-correlation technique to separate the signals from the background noise (Sweeney 1989, 1995, 1996). However the duration of initial part of the EMP can be utilized in order to estimate the energy of nuclear underground explosions (Gorbachev et al. 1999a,b).

11.1.2 Physical Mechanisms of EMP Caused by Atmospheric and Space Nuclear Explosions

At first let us consider possible physical mechanisms of the EMP under the atmospheric explosion. This effect can be due to the generation of radial currents of Compton recoil electrons originated from the short-term interaction ($\sim$ 0.25 μs) between the gamma-quantums of nuclear detonation and environment (Kompaneets 1958). It is known that approximately 0.03 % of the whole energy of the explosion is transformed into gamma-quantum radiation (Karzas and Latter 1962a,b). The average gamma-quantum energy is 1 MeV, and about 7.5×10^{21} gamma-quantums are generated per 1 kt of the TNT equivalent of the explosion. The gamma-quantums interact with the matter of nuclear device and with the molecules of air that causes the electron fluxes due to the Compton effect. On average the vectors of the electron velocities coincide with the directions of gamma-quantums motion. The radial electric current is also attributed to the photoelectrons resulted from X-rays emitted by the heated matter of the nuclear device.

Every Compton's electron ionizes the medium that leads to the generation of a great number of ion pairs. For example, the electron with kinetic energy of 2 MeV gives rise to approximately 3×10^4 pairs of ions in the air. In the air the free electrons are captured by molecules of O_2 having great chemical affinity with electrons. Under normal conditions the characteristic time of the electron attachment to molecules O_2 is about 0.01 μs. As a result, the reverse ionic current is developed thereby producing the relaxation of the dipole moment and radiation of electromagnetic waves in the frequency range around 10 kHz (Troitskaya 1960; Latter et al. 1961b; Gilinsky 1965; Gilinsky and Peebls 1968; Medvedev et al. 1980).

If the explosion gives rise to the spherically symmetrical system of currents and electric charges like the spherical condenser, then the electromagnetic field is strongly equal to zero outside this system. Actually always there are some causes for the non-symmetrical gamma-quantum spatial distribution. The irregularities of the gamma-quantum flux can be due to the construction features of the nuclear device or they could be excited by nonuniformity and anisotropy of the medium in which the gamma-quantums move. The asymmetry of gamma-quantum fluxes results in the asymmetrical distribution of the currents caused by Compton's electrons which in turn give rise to the generation of the dipole moment of the current system. The evolution of the dipole moment defines the temporal dependence of the EMP at far distance from the explosion site. However Latter et al. (1961) have noted that at the distances about several thousands kilometers from the detonation point the spectrum of the signals radiated by the atmospheric nuclear detonation is practically the same as that of typical atmospherics.

Leypunskiy (1960) has assumed that the EMP of the atmospheric nuclear detonation could be radiated because of the GMP caused by the fast extension of strongly heated plasma generated by the detonation. Since the plasma conductivity is so high as 10^3 S/m, the plasma motion in the geomagnetic field gives rise to the generation of electric currents, which screen the geomagnetic field. This results in the displacement of the geomagnetic field lines by the expanding plasma from the ionized area into the surrounding space. Such an effect which is often referred to as the "magnetic bubble effect," is followed by a subsequent current relaxation after the arrest and cooling of the plasma. These processes are accompanied by the radio-emission of the magneto-dipole type (Karzas and Latter 1962a; Kompaneets 1977). Additional effects can be due to the secondary gamma-quantums resulted from inelastic scattering and capture of the thermal neutrons by nuclei of atoms in the molecules of air and explosion products. A lot of aspects of the excitation of electromagnetic fields due to gamma and neutron radiation have been studied (e.g., see Sandmeier et al. 1972; Medvedev and Fedorovich 1975). The EMP of nuclear explosions in the outer space is different from that in the atmosphere in respect to the great value of mean free path of gamma-quantums and electrons. Johnson and Lippman (1960) and Karzas and Latter (1962b, 1965) have pointed out that the electrons in a so high-rarefied medium can produce cyclotron radiation in the Earth's magnetic field.

11.1.3   *GMP Due to a Strongly Heated Plasma Ball Produced by Underground Explosions*

The gamma radiation of the underground nuclear explosion is resulted from the fission of atomic nuclei, directly, as well as from the inelastic scattering of neutrons in the material of nuclear device and in the rock surrounding the underground chamber. Considering the Compton electrons mechanism of the EMP, we note that in the ground the free path lengths of the gamma-quantums and electrons are significantly shorter than those in the air. In the case of Compton interaction the mean free path of quantum with energy on the order of 1 MeV is $\lambda_\gamma = b_\gamma/\rho$ where $b_\gamma \approx 1.5 \times 10^2$ kg/m^2 and ρ is the medium density. For example, one can find that $\lambda_\gamma \approx 100$ m at the sea level in the atmosphere and $\lambda_\gamma \approx 0.1$ m in the ground with the density $\rho = 1.7 \times 10^3$ kg/m^3. The free path length of the Compton electrons in the ground is $\lambda_e = b_e/\rho \approx 1$ mm ($b_e \approx 2$ kg/m^2), i.e. this value is small too. Based on these simple estimates one may expect that the ratio of linear sizes of the electric dipoles caused by underground and atmospheric explosions with the same energy is inversely proportional to the ratio of densities of the corresponding media, i.e. 1–10^3. For the more accurate estimate we should take into account radiation-induced conductivity of the rock around the underground chamber.

One more significant factor which may greatly decrease the EMP of an underground explosion is the natural conductivity of the rock. The gamma-quantum pulse originated from the nuclear fission has a duration about 0.1 μs, which corresponds to the characteristic frequency $\omega \approx 10^7$ Hz. Taking a typical value of the rock conductivity $\sigma = 10^{-2}$–10^{-3} S/m, we obtain the estimate of the corresponding skin-depth in the ground $r_s \sim (\mu_0\sigma_e\omega)^{-1/2} \approx 3$–9 m. To illustrate this strong attenuation, we note that if the explosion point is situated at the depth of 500 m, then this short signal can attenuate 10^{24} times or larger.

One more effect can be associated with the neutrons produced by explosions and by secondary gamma-radiation. The deceleration of these neutrons down to thermal energy is basically due to the interaction of the neutrons with nuclei of the light elements such as hydrogen. This seemed entirely possible since the ground usually contains about 16 % of hydrogen, 57 % of oxygen, 19 % of silicon, and 8 % of aluminum (Straker 1971). The duration of the deceleration process is on the order of 10 ns, whereas the life time of the thermal neutrons in the ground is about 0.1–1 ms that is significantly greater than the duration of primary gamma-quantum pulse. The inelastic scattering and capture of the thermal neutrons by the nuclei of aluminum and silicon causes the secondary gamma-radiation followed by the generation of electric current. The characteristic frequencies of this process are $\omega = 10^3$–10^4 Hz. This means that this effect could be observed in principle since the corresponding skin-depth is about 90–900 m, that is compared with the explosion depth.

The high-temperature plasma in the underground explosion cavity is believed to be one of the main sources for the EMP generation during a nuclear detonation. In the nuclear device the fission reaction is completed for the times about 10^{-8}–10^{-7} s. By this moment the matter still occupies the volume of about several cubic centimeters. Since the temperature of fissioned matter reaches 10^7 K, the atoms of light elements are completely ionized. This indicates that the electron-ion collisions prevail in the plasma. In such a case the conductivity σ_p of two-component plasma can be written in the form (e.g., Lifshitz and Pitaevskii 1981):

$$\sigma_p = \frac{4\sqrt{2}}{\pi^{3/2}} \frac{T^{3/2}}{Ze^2 m_e^{1/2} L}. \tag{11.3}$$

Here T is the plasma temperature measured in energy units, Ze is the charge of ions, m_e is the electron mass, and L is the Coulomb logarithm (Gaussian system of units)

$$L = \begin{cases} \ln \frac{r_D T}{Z e^2}, & \frac{Z e^2}{u \hbar} \gg 1; \\ \ln \frac{r_D (m_e T)^{1/2}}{\hbar}, & \frac{Z e^2}{u \hbar} \ll 1; \end{cases} \tag{11.4}$$

where $r_D = \{T/(4\pi n_e e^2)\}^{1/2}$ is the Debye shielding radius, n_e is the electron number density, u is the average relative velocity of electrons and ions, and $\hbar$ is the Plank constant. Substituting the average charge of ions $Z = 2$, the initial temperature $T = 1\text{--}10\,\text{keV}$ and parameter $L = 4$ into Eqs. (11.3) and (11.4) we obtain the value $\sigma_p \approx 4 \times 10^7\text{--}1 \times 10^9\,\text{S/m}$ which is close to the conductivity of metals under normal conditions.

The perturbations of the Earth magnetic field can diffuse in the conducting plasma according to Eq. (7.7). Let R be the radius of the underground chamber filled with the plasma. Then the characteristic time of the diffusion of GMPs inside the plasma ball can be estimated as follows:

$$\tau_d \approx \mu_0 \sigma_p R^2/4. \tag{11.5}$$

Taking the above value of σ_p and $R = 1\,\text{m}$ one can find that $\tau_d \approx 10^8\text{--}10^9\,\text{s}$, while the characteristic time of the plasma extension is about $t_p \approx 10\text{--}100\,\text{ms}$ depending on the energy of explosion. Since $\tau_d \gg t_p$ the Earth magnetic field lines are completely frozen to the conducting plasma, so that the field lines move together with the plasma. Thus the plasma motion results in the local distortion of Earth's magnetic field. The equidistant lines of undisturbed magnetic field $\mathbf{B}_0$ are schematically shown in Fig. 11.5a while Fig. 11.5b displays a picture resulted from expansion of the conducting plasma ball.

Since the "frozen in" magnetic field is a uniform one in the plasma ball, the conservation of the magnetic field flux can be written as $B_0 \pi R_0^2 = B \pi R^2$ whence it follows that

$$B = B_0 R_0^2/R^2, \tag{11.6}$$

where B is the induction of uniform magnetic field into the ball with current radius R and R_0 is the initial radius of the ball. The perturbation, $\delta \mathbf{B}$, of the magnetic field in the ball is given by

$$\delta \mathbf{B} = \mathbf{B} - \mathbf{B}_0 = -\left(1 - R_0^2/R^2\right) \mathbf{B}_0. \tag{11.7}$$

If the ground conductivity around the plasma ball can be neglected, then the magnetic perturbations out of the plasma ball is described by the field of the effective magnetic dipole given by Eq. (7.5) where one should replace $\mathbf{B}$ by $\delta \mathbf{B}$. The magnetic dipole moment $\mathbf{M}$ is directed oppositely to the vector $\mathbf{B}_0$ and this absolute value

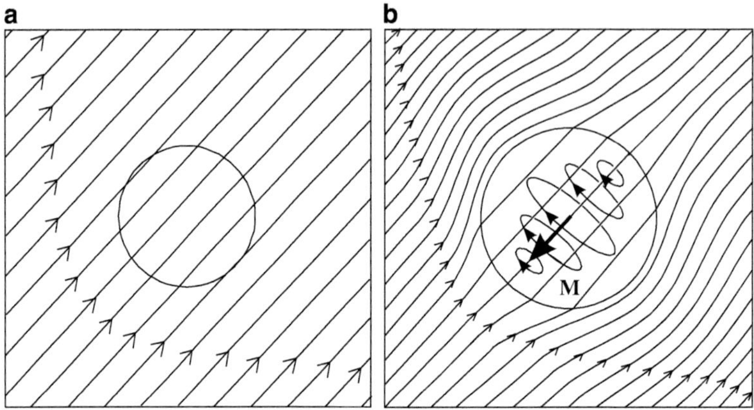

Fig. 11.5 Distortion of the geomagnetic field lines caused by the expansion of a highly heated plasma ball. (**a**) Equidistant field lines of undisturbed magnetic field $\mathbf{B}_0$; (**b**) Field line pattern resulted from the conducting plasma motion. The *closed "circular" lines* indicate the currents induced in the plasma ball. The effective magnetic moment is shown with vector $\mathbf{M}$

depends on the current system in the plasma. In order to find the value of this moment we use the boundary condition which requires the continuity of the normal component of $\delta\mathbf{B}$ at the plasma ball surface. Equating the normal component of magnetic perturbations in Eq. (11.7) with that given by Eq. (7.5) at $r = R$, we obtain

$$\mu_0 M / \left(2\pi R^3\right) = \left(1 - R_0^2/R^2\right) B, \tag{11.8}$$

hence it follows that the effective magnetic moment of the plasma ball is

$$\mathbf{M} = -\frac{2\pi \mathbf{B}_0 R^3}{\mu_0}\left(1 - \frac{R_0^2}{R^2}\right). \tag{11.9}$$

Substituting Eq. (11.9) for $\mathbf{M}$ into Eq. (7.7) one can find the radial δB_r and tangential δB_θ components of the GMPs

$$\delta B_r = -B_0\frac{R^3}{r^3}\left(1 - \frac{R_0^2}{R^2}\right)\cos\theta, \tag{11.10}$$

$$\delta B_\theta = -B_0\frac{R^3}{2r^3}\left(1 - \frac{R_0^2}{R^2}\right)\sin\theta. \tag{11.11}$$

Here θ is the polar angle measured from the direction of $\mathbf{B}_0$.

As is seen from Eqs. (11.10) and (11.11) the magnitude of EMP decreases with distance inversely proportional to the distance cubed. Substituting the following numerical parameters $B_0 = 5 \times 10^{-5}\,\text{T}$, $R = 30\,\text{m}$, $R_0 = 1\,\text{m}$, and $r = 7\,\text{km}$

into Eqs. (11.10) and (11.11) we get the estimate $\delta B \approx 4\,\mathrm{pT}$ which is consistent in magnitude with the signals observed during underground detonations. At the distance exceeding approximately $10\,\mathrm{km}$, the amplitude of the signals falls off below the level of background noise.

The maximal radius R of the explosion cavity is proportional to $Y^{1/3}$ where Y is the TNT equivalent of the explosion (e.g., see Chadwick et al. 1964; Rodionov et al. 1971). This implies that the magnitude of the GMPs in Eqs. (11.10) and (11.11) is proportional to Y. The similar relationship holds true for the electric field variations that contradicts the empirical dependence given by Eq. (11.2). This means that the simplified model considered above does not describe the EMP effect quite adequately.

To estimate the EMP relaxation time due to return diffusion of the magnetic field into the plasma, we now suppose that the cooling of the uniformly expanding plasma follows the adiabatic law. The adiabatic equation of a perfect gas reads

$$TR^{3(\gamma-1)} = T_0 R_0^{3(\gamma-1)}, \tag{11.12}$$

where γ stands for the adiabatic exponent and the subscript zero is related to the initial values of the plasma temperature and the radius of underground cavity. Considering the moment of the cavity stoppage and substituting the numerical values $R/R_0 = 30$ and $\gamma = 5/3$ into Eqs. (11.3), (11.4), and (11.12), we find that at this moment the plasma temperature and conductivity are $T \approx 1\text{--}10\,\mathrm{eV}$ and $\sigma_p \approx 1.5 \times 10^3\text{--}4 \times 10^4\,\mathrm{S/m}$. Substituting these values into Eq. (11.5) gives the rough estimate $\tau_d \approx 0.4\text{--}10\,\mathrm{s}$ which is compatible with the duration of the EMP signals shown in Figs. 11.2 and 11.3.

In the strict sense, the amplitude estimates given by Eqs. (11.10) and (11.11) are valid in the extreme case of a perfectly conducting plasma ball. To study the effect of finite plasma conductivity we consider the expanding uniform plasma ball situated in the rock at higher depth. The conductivity and radius of the plasma ball are assumed to be given functions of time; that is $\sigma_p = \sigma_p(t)$ and $R(t) = R_0\beta(t)$, where R_0 is the initial ball radius (Ablyazov et al. 1988). In this model the rock conductivity is much smaller than the plasma one. A detailed analysis of this problem presented in Appendix I has shown that the ULF GMPs outside the ball can be qualified as magnetic dipole field. The solution of the problem is represented as a series with respect of eigenfunctions of the problem. The effective magnetic moment of the plasma ball can be found from Eq. (11.80)

$$\mathbf{M}(t) = -\frac{12R_0^3\mathbf{B}_0\beta(t)}{\pi\mu_0}\sum_{n=1}^{\infty}\frac{1}{n^2}\int_0^t\frac{d\beta^2}{dt'}\exp\left(-\int_{t'}^t\frac{\pi^2 n^2}{\mu_0\sigma_p R_0^2\beta^2}dt''\right)dt'. \tag{11.13}$$

In the limit $\sigma_p \to \infty$ we get

$$\mathbf{M}(t) = -\frac{12\zeta(2)R_0^2 R(t)\mathbf{B}_0}{\pi\mu_0}\left(\frac{R^2(t)}{R_0^2} - 1\right), \tag{11.14}$$

where $\zeta(x)$ denotes the ζ-function of Riemann. Taking into account that $\zeta(2) = \pi^2/6$ we obtain that Eq. (11.14) for $\mathbf{M}$ coincides with Eq. (11.9) which was derived in the same extreme case. In the opposite case $\sigma_p \to 0$ we have a so apparent result $\mathbf{M} = 0$.

The low-frequency conductivity of the heated plasma is defined by Eqs. (11.3) and (11.4) because the electron-ion collisions prevail over other ones at high temperature. Suppose that the plasma is the perfect gas that expands according to the adiabatic law (11.12). Then the plasma temperature varies as $T = T_0/\beta^{3(\gamma-1)}$ where T_0 is the initial plasma temperature. Now we first examine the exponent function under the integral sign in Eq. (11.13). The expression standing in the index of the exponent function can be written in the form (Gaussian system of units)

$$\frac{\pi c^2 n^2}{4\sigma_p R_0^2 \beta^2} = \frac{n^2}{\tau_d}\left(\frac{\beta}{\beta_m}\right)^\mu, \tag{11.15}$$

where

$$\tau_d = \frac{16\sqrt{2}}{\pi^{5/2}}\frac{R_0^2 T_0^{3/2}}{Z e^2 L c^2 m_e^{1/2}\beta_m^\mu}, \qquad \mu = \frac{9\gamma - 13}{2}. \tag{11.16}$$

Here c is the light speed in the free space and β_m denotes maximum of the function $\beta(t)$; that is $\beta_m = R_m/R_0$, where R_m is the final radius of the plasma. The parameter τ_d determines the back diffusion time of the perturbed magnetic field into the plasma ball. This parameter has the same sense as the relaxation time given by Eq. (11.5). It can be shown that Eq. (11.16) coincides with Eq. (11.5) within a constant factor. Substituting the numerical parameters $R_0 = 1\,\mathrm{m}$, $T_0 = 1\,\mathrm{keV}$, $Z = 2$, $L = 4$, $\gamma = 5/3$ and $\beta_m = 30$ into Eq. (11.16) we obtain $\tau_d = 0.55\,\mathrm{s}$. This value is compatible with the relaxation time of EMP observed during underground explosions. However the dependence $\tau_d \propto Y^{2/3}$ which follows from Eq. (11.16) contradicts with the empirical dependence $\tau_d \propto Y^{1/3}$ displayed in Fig. 11.4.

There are a lot of factors which may affect the electromagnetic signals under the explosions and thus may concern this discrepancy. For example, the melting and evaporation of the surface of an underground chamber subjected to the radiation of nuclear explosion results in changing the plasma constituents due to the injection of evaporated particles. A fall in plasma temperature brings the decrease in the plasma ionization degree due to recombination process. These effects lead to the changes in the plasma conductivity, adiabatic exponent, and other plasma parameters that leave out of account in the above models.

The changes in the underground chamber size can be approximated by a smooth function, for example,

$$\beta(t) = 1 + (\beta_m - 1)\left[1 - \exp(-t/\tau_b)\right], \tag{11.17}$$

where τ_b is the characteristic time of chamber expansion. In Fig. 11.3 the data recorded during the containing underground explosion "Bilbi" (Zablocki 1966) are compared with the numerical calculations which are based on the above model and Eq. (11.17) (Ablyazov et al. 1988). As is seen from this figure, the theoretical dependencies shown with lines 2 are in qualitative agreement with the observations shown with lines 1. The vibrations followed by the initial spike can be explained by the hydrodynamical instability of the expanding plasma. Hydrodynamic waves excited in the plasma and products of detonation can be reflected from the walls and center of the chamber thereby producing the modulation of the EMP signals in amplitude and frequency (Gorbachev et al. 1999). It should be noted that in specific events the EMP exhibits the polarization corresponding to the field of a magnetic dipole whereas the polarization in other cases is rather close to the electric dipole one.

11.2   Electromagnetic Effects Due to Shock Wave (SW) and Rock Fracture

11.2.1   Electric Dipole Moment Due to Shock Polarization of Rocks

A SW generated by the contained underground explosion gives rise to rock polarization which in turn can serve as a possible source for the electric dipole (Surkov 1986). The shock polarization effect in laboratory conditions have been studied in any detail in Sect. 9.1. Here we deal with large-scale polarization phenomena under the natural situation. There are a few stages of the deformation and rock fracture caused by an underground explosion. At first the fast expansion of the underground chamber due to plasma impact and vaporation of the chamber walls results in the generation of the strong SW with pressure amplitude $\sim 10^{11}$–10^{12} Pa (e.g., Zeldovich and Raizer 1963; Chadwick et al. 1964; Rodionov et al. 1971; Baum et al. 1975). At this stage called as hydrodynamical one the rock strength can be neglected, and the pressure amplitude decreases with distance as r^{-3} over a length of several meters or tens meters. During this stage the pressure falls off by 3–4 order of magnitude, and then the amplitude attenuation obeys the law r^{-n}, where $1 < n < 2$. As before the shear stress will exceed the crushing strength of the rock so that the fracturing of rocks occurs behind the SW front. Then the crushing wave begins to decelerate and thus fails to keep up with the main shock so that the primary wave is split into two waves. At the moment of the crushing wave stop the radius of zone of complete fracturing reaches tens or hundreds meters depending on the energy of explosion. The tension stresses take place in the region between the fracturing zone and the SW front. Since these stresses exceed the ultimate tensile strength, there develop the radial cracks in this region. Typically the zone of intensive radial fracturing is as much as several hundreds meters in length. When the wave amplitude falls off below the rock strength, the medium behaves like elastic one. At this stage which is referred as the seismic one, the SW transforms into the elastic/seismic wave. If the dispersion-dissipative properties of the rock are not important, then the amplitude of the longitudinal seismic wave decreases with distance as r^{-1}. When this wave reflects from the day surface, it is split into the primary/longitudinal (P-wave), secondary/transverse (S-wave) and surface Rayleigh and Love waves.

In this brief overview we have omitted a few details of camouflet underground explosions such as the generation of the unloading wave, dilatancy of the fractured rock, dynamics of the camouflet chamber and etc.

The stress wave propagating in the ground and rocks is known to generate a variety of low-frequency electromagnetic phenomena so that the interpretation of the observation is often troublesome. One source of this variety is that the natural materials and rocks are very inhomogeneous as for their rheological structure and electrical parameters. For example, the ground conductivity strongly depends on the humidity and porosity which vary with depth. The rock fracturing and pore collapse caused by the SWs gives rise to the generation of the local electric fields and great charges near the cracks and pores that can be accompanied by the local electric breakdowns of the medium.

Now we discuss the existence of the shock polarization effect in nonuniform media at different structural levels (Surkov 2000). Firstly, the microscopic movements of the charged dislocations and point defects can result in the polarization of individual monocrystals and grains. This effect is enhanced essentially in the vicinity of the grain boundaries, microcracks, small inclusions, and pores. Secondly, considering the macroscopic scale we note that the polarization processes are localized in the regions of enhanced stresses; that is, near tips of large cracks and individual blocks of fractured rock. So one may expect that there exists certain hierarchy of relaxation times of the shock polarization. The largest values of the rise and decay times of the shock polarization can exceed by several orders of magnitude the same parameters observed under laboratory tests. Thus, we come to the conclusion that the SW generates its own, as a rule, low-frequency electromagnetic field due to the polarization of different structural units of nonuniform matter.

In what follows we assume the linear dependence between the rock polarization Π and the amplitude of pressure P_m (Allison 1965)

$$\Pi = \alpha P_m \left[1 - \exp\left(-\frac{t}{\tau_f} \right) \right] \exp\left(-\frac{t}{\tau_r} \right) \eta(t), \qquad (11.18)$$

where α is empirical coefficient of proportionality, τ_f is the rise time, τ_r is the decay time of the polarization, and $\eta(t)$ denotes the step function.

The shape of the SW resulted from an explosion is very similar to a spherical one. Suppose that behind of the wavefront the matter is polarized in the radial direction. Besides the medium polarization has a weak asymmetry that can be due to an irregular distribution of fracturing, asymmetry of the shock wavefront, influence of the gravity, presence of large-scale inhomogeneity of the medium and other causes. So, the actual distribution of the shock polarization is not entirely spherically symmetrical. For instance, we assume the cylindrical symmetry of such a distribution (Surkov 1986). The origin of coordinate system is placed in the center of symmetry of the spherical SW. The radial polarization of the medium is described by the following equation:

$$\mathbf{\Pi}_1 \left(r, \theta, t - t_a \right) = \left(1 + \beta \cos \theta \right) \Pi \left(r, t - t_a \right) \hat{\mathbf{r}}, \qquad (11.19)$$

where β is the small parameter of asymmetry, $\hat{\mathbf{r}}$ is the unit vector and the parameter t_a denotes the moment of the SW arrival at the point with spherical coordinates r and θ. Here the polar angle θ is measured from the axis of symmetry z.

As has already been stated in Sect. 9.1, the SW in a solid carries the electric charge. Let R_0 be the initial radius where the primary charge of the SW is formed. By assuming that the SW velocity, U, is constant, we obtain that $t_a = \left(r - R_0 \right) / U$. The pressure magnitude in the spherical SW changes with distance by a power law, i.e. $P_m = P_* \left(R_0 / r \right)^n$, where P_* denotes the pressure magnitude at the radius $r = R_0$. For the ground the value $n \approx 1.6$ is usually accepted though the exponent n can vary depending on distance (e.g., see Rodionov et al. 1971). The actual values of the characteristic times τ_f and τ_r can be of the order of the SW duration. Since the width of shock wavefront in the ground can reach several meters, these parameters can be as large as a few ms or even more. In the model by Grigor'iev et al. (1979) the SW rise time is inversely proportional to the pressure magnitude and hence it follows that $\tau_f = \tau_0 \left(r / R_0 \right)^n$ where τ_0 is the constant.

The vector of the dipole electric moment $\mathbf{d}$ of the polarized matter is directed along the axis of symmetry z. In order to find the absolute value of $\mathbf{d}$, one should integrate the projection of $\mathbf{\Pi}_1$ on z-axis over the volume V occupied by the SW, i.e. over the volume restricted by the radius $R_f = R_0 + Ut$

$$d \left(t \right) = \int_V \mathbf{\Pi}_1 \left(r, \theta, t - t_a \right) \cos \theta \, dV. \qquad (11.20)$$

Here we consider the time interval when the SW has not yet reached the ground surface.

The space charges due to the shock polarization of the medium are mainly situated in the vicinity of the shock wavefront in the layer with depth on the order of $U \left(\tau_r + \tau_f \right)$. These charges have a certain sign depending on the properties of the rock. The opposite charges are situated behind the SW front at the distance which is equal to the charge relaxation length or they are concentrated around the underground cavity. Certainly, in all cases the total electric charge contained in the rock is equal to zero.

Below we assume an inequality $U \left(\tau_r + \tau_f \right) \ll R_f$. Since the integral sum in Eq. (11.20) is mainly accumulated within a short length $U \left(\tau_r + \tau_f \right)$, the functions

$r^2 \Pi (r, t - t_a)$ and $\tau_f (r)$ under the integral sign can be considered as constant values. Taking these function at $r = R_f$ and performing integrations yields

$$d (t) = \frac{4\pi \alpha \beta P_* R_0^2 \tau_r}{3 (1 + t/\tau_s)^{n-2}} \left[1 - \exp\left(-\frac{t}{\tau_r}\right) - \gamma \left\{ 1 - \exp\left(-\frac{t}{\gamma \tau_r}\right) \right\} \right] \qquad (11.21)$$

Here we have introduced the auxiliary function

$$\gamma (t) = \frac{\tau_f (R_f)}{\tau_r + \tau_f (R_f)} = \frac{\tau_0 (1 + t/\tau_s)^n}{\tau_r + \tau_0 (1 + t/\tau_s)^n}, \qquad (11.22)$$

where $\tau_s = R_0/U$. We notice that the spherically symmetric portion of Eq. (11.19) which is independent of θ does not contribute to the dipole moment in Eq. (11.21) at all. This obvious result follows the fact that the spherically symmetric charge distribution confined by two concentric spheres does not create the electromagnetic field in the outer space.

The dipole moment given by Eq. (11.21) determines the electromagnetic field of SW far away from the explosion point. The analysis of the near field spectrum caused by the SW has shown that the spectral intensity is enhanced in the vicinity of typical frequencies $\omega \approx \tau_r^{-1}$, $\omega \approx \tau_s^{-1}$ and $\omega \approx \tau_*^{-1}$ where $\tau_* = \tau_0 \tau_r / (\tau_0 + \tau_r)$ (Surkov 1986). This frequency range lies within an interval from several Hz to one kHz that is in a reasonable agreement with typical spectra of the EMP observed during large-scale underground explosions.

In Sect. 3.1.4 we have noted that in the atmosphere the vertical dipole antenna is a more effective radiator of electromagnetic waves than the horizontal one (see Fig. 3.5). By contrast, it follows from the theory that the horizontal component of the underground dipole antenna plays more significant role than the vertical component (Wait 1961, 1970). To clarify this statement we note that the vertical dipole generates a symmetrical distribution of electric current in the surrounding conductive space as shown in Fig. 11.6a. On the ground surface these currents flow in the radial directions from the point O. The effective electric dipole **d** of such a system of surface radial currents is equal to zero, which means that the effective antenna produced by these surface currents does not radiate. On the other hand in the case of the horizontal dipole, shown in Fig. 11.6b, the surface electric currents flow approximately along the direction of dipole vector **d**. These current systems are equivalent to the nonzero electric dipole turned to the same direction.

Thus, as a first approximation, the electromagnetic field generated by the SW of underground explosion is equivalent to that of horizontal component of the effective current dipole, which is usually assumed to be located in a homogeneous conducting half-space. The shape of the signals detected on the ground surface depends on both the function $d(t)$ given by Eq. (11.21) and the conductivity of the half-space. A theoretical analysis of this problem has shown that the EMP has a bipolar shape similar to that shown with lines 2 in Fig. 11.3 (Surkov 1986). In this model the first narrow spike is due to the fast variation of the shock-induced dipole moment

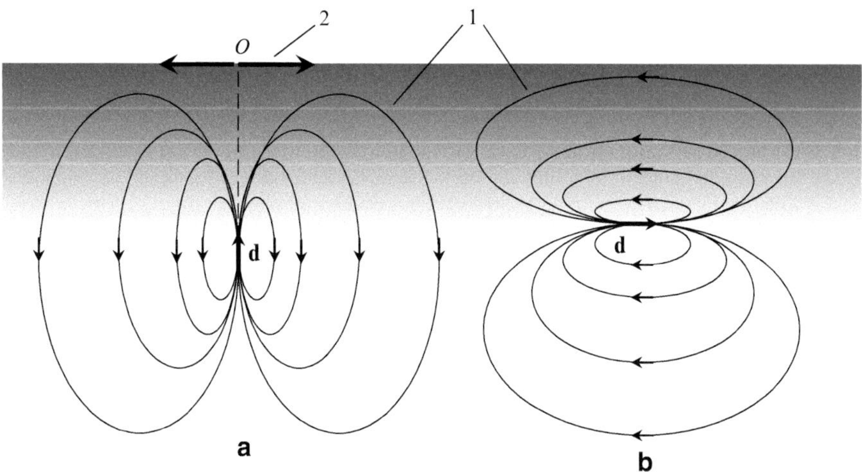

Fig. 11.6 Effective underground (**a**) vertical and (**b**) horizontal dipoles **d** produced by electric current systems in the homogeneous conducting ground. *1*—current field lines, *2*—radial currents on the ground surface

for the time of the order of $\tau_r + \tau_f$ while the negative half-wave is caused by the subsequent attenuation of the SW amplitude. The predicted amplitudes of the EMP at the distances 5–10 km are on the order of several μ V/m and 1–10 pT, which is also compatible with observations though the duration of the observed signals is greater than calculated values.

It appears that the EMP is a combined effect resulted from different sources such as the high-temperature plasma ball, SW polarization, and perhaps gamma and neutron radiations. There are a few difficulties in distinguishing these sources because the direction of the shock-induced dipole is unknown prior to the test; that is, the dipole direction can be accidental in character.

11.2.2   Residual Electromagnetic Field on the Ground Surface

In what follows we examine the phenomena arising after the abatement of the EMP. The residual quasi-static magnetic perturbations at the epicenter of surface and buried detonations have been observed by Stacey (1964) and by Undzenkov and Shapiro (1967). One of the first measurements of residual magnetic field has been made in the region Medeo (USSR) at the distance of 700 m from the detonation point (Barsukov and Skovorodkin 1969). The detonation was made in granite, whose natural magnetization was $J = 5\text{--}100$ mA/m. The ground-based observation showed that 1 h 50 min after the moment of detonation the local geomagnetic field changes by 8 nT and this perturbation is halved 5 h after it. The field relaxation back to the former level was lasted during 24 h. Hasbrouk and Allen (1972) have reported magnetic measurements during the underground nuclear explosion, which is referred to as CANNIKIN experiment. The explosion with TNT equivalent of 5 Mt was conducted on the Amchitka Island (Aleutian Islands) on January 6, 1971. The proton magnetometer, which was placed at the epicentral distance of 3 km, recorded the gradual increase of the magnetic field by 9 nT in 30 s after the detonation. The field variation about 2 nT was detected at the distance of 9 km. The magnetic survey around the epicenter of detonation revealed that the 10 nT changes of the geomagnetic field were kept approximately constant for 8 days.

Thus the magnetic perturbations due to the detonation in rock could be conventionally divided into three stages (Erzhanov et al. 1985): (1) the transient alternating-sign pulse (EMP) with duration smaller than 1 s and with magnitude 0.1–100 nT; (2) the 10–20 nT residual changes which can relax during several hours or days; (3) the long-term residual changes with magnitude of several nT that can last for several days or months.

It was hypothesized by Stacey (1964) and by Undzenkov and Shapiro (1967) that the residual magnetic perturbations near the detonation site are excited by means of changes in the natural rock magnetization which in turn are based on the occurrence of inelastic/plastic deformations in the rock. As noted in Sect. 9.1, the laboratory tests with magnetite-bearing rocks have shown that the sample magnetization can change by 1% under the stress of 10 MPa. Restoration of the local geomagnetic field back to the former level could be resulted from the relaxation of inelastic deformation in the rock. However this effect is likely if the rock contains sufficient amount of the ferromagnetic inclusions.

To estimate the above effect, we consider the model in which the SW and residual stresses around the detonation site are spherically symmetric (Surkov 1989). The SW magnitude exceeds the crushing strength of the rock in the vicinity of the powerful explosion. The crushed zone is assumed to have a spherical shape with radius of R_c. As a rule this radius is of the order of several tens meters. The residual magnetization in this zone is probably chaotic owing to the repacking of the broken rock. Therefore the contribution of this zone to magnetic perturbations is neglected as compared to the residual rock magnetization which occurs in the region $r > R_c$. In this region the SW magnitude is lower than the crushing strength but higher than the tensile strength of the rock. The medium is monolithic in character while there occur separate large cracks. The typical size of this region is of the order of several hundreds meters, and the rock deformation is elastic and reversible from outside of this zone.

The primary rock magnetization, $\mathbf{J}$, is assumed to be constant. In the region $r > R_c$ the magnetization increment, $\Delta\mathbf{J}$, due to the SW is described by Eq. (9.29), where $s_n = s_{rr}$ denotes the magnitude of radial component of the stress tensor. The stress magnitude depends only on the distance r from the explosion point. The effect of residual rock magnetization ceases at the certain radius $R_e > R_c$ when the stress magnitude falls short of certain threshold stress so that in the field $r > R_e$ the residual magnetization is absent. The magnetic permeability of the rock inside the zone $R_c < r < R_e$ can be slightly different from that of surrounding medium.

For simplicity we ignore this difference and set the magnetic permeability is equal to unity everywhere. The induction of remanent magnetic field, $\mathbf{B}$, within the zone $R_c < r < R_e$ is described by the following Maxwell equation:

$$\nabla \times \mathbf{B} = \mu_0 \nabla \times \Delta \mathbf{J}. \tag{11.23}$$

The detonation point is chosen to be the origin of coordinate system with z-axis parallel to the vector $\mathbf{J}$. Spherical coordinates, i.e. the radius r and the polar angle θ measured from z-axis are used. For the region $r > R_e$ the solution of Eq. (11.23) is given by Eq. (7.5) which describes the field of magnetic dipole. Taking Eq. (9.29) for $\Delta \mathbf{J}$ the effective magnetic moment of the magnetized rock can be written as follows:

$$\mathbf{M} = 4\pi C_m \mathbf{J} \int_{R_c}^{R_e} r'^2 s_{rr}\left(r'\right) dr', \tag{11.24}$$

where C_m denotes the piezomagnetic coefficient.

According to this model, far away from the detonation point the residual magnetic field decreases with distance as $B \sim r^{-3}$. However, this dependence contradicts the data obtained during the experiment referred to as MASSA (Erzhanov et al. 1985). The detonation of chemical high explosive (HE) with mass of 251 t was made on the sandstone surface. Survey of the changes of the geomagnetic field for this detonation was made at the different points in the distance range from 0.5 to 10 km. It was found that the decrease of the residual magnetic field is closer to the dependence $B \sim r^{-1}$ in character.

This discrepancy between the theory and experiment can be due to the fact that the observation point was located at the distances within $R_c < r < R_e$ where the remanent rock magnetization should occur (Surkov 1989). The solution of this problem is found in Appendix I. Since the elastic strain and stress are predominant in this region, the magnitude of the normal stress in the seismic wave satisfies the following law: $s_{rr}(r) = P_c R_c / r$, where the parameter P_c is of the order of the crushing strength or of tensile one. Substituting this expression into Eqs. (11.90) and (11.91) and performing integration, we obtain the solution of the problem $(R_c < r < R_e)$

$$B_r = \frac{\mu_0 C_m J P_c R_c \cos\theta}{r}\left(1 - \frac{R_c^2}{r^2}\right), \tag{11.25}$$

$$B_\theta = -\frac{\mu_0 C_m J P_c R_c \sin\theta}{2r}\left(1 + \frac{R_c^2}{r^2}\right). \tag{11.26}$$

As is seen from Eqs. (11.25) and (11.26), the magnetic field components decrease with distance slower than the rate expected from the dipole law. When $r^2 \gg R_c^2$, they decrease with distance approximately as r^{-1}. However, this solution provides

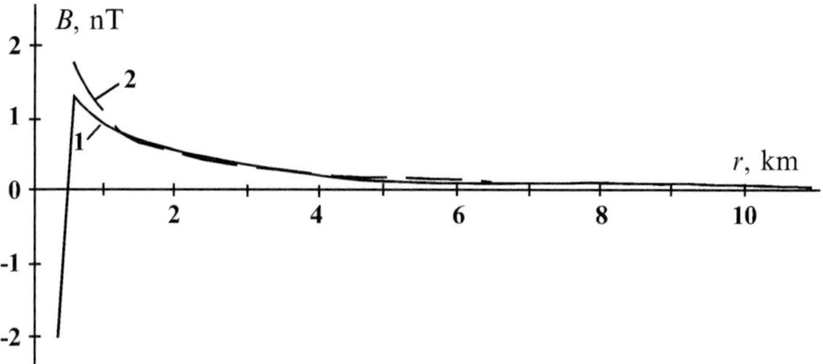

Fig. 11.7 Remanent changes in the Earth magnetic field resulted from the surface detonation of HE with mass 251 t as a function of distance from the detonation point. *1*—Experimental data (adapted from Erzhanov et al. 1985); *2*—model calculations (Surkov 1989)

only a rough estimate of the phenomenon because we do not take into account the effect of medium unloading near the free surface.

The remanent changes of the Earth magnetic field measured in the case of MASSA experiment are shown in Fig. 11.7 with line 1 (Erzhanov et al. 1985). The numerical calculation shown with the dotted line 2 was made under the following parameters: $P_c = 0.1\,\mathrm{GPa}$, $C_m = 1\,\mathrm{GPa}^{-1}$, $R_c = 100\,\mathrm{m}$, and $J = 0.12\,\mathrm{A/m}$. It is obvious from this figure that the theoretical and experimental dependencies are close except for the region of $r < 0.5\,\mathrm{km}$. Actually the surface detonation generates the non-spherically symmetric SW. The study of such a problem has shown that the asymmetry of the rock magnetization can lead to an increase of the above estimate (Surkov 1989).

The additional effect can be due to the impact action on technical constructions and installations which are magnetized under the shock and vibrations in the Earth magnetic field. The mechanism of this effect has been discussed more fully in Sect. 9.1. One of such construction is the steel casing/encasement pipe which is used in order to protect utility lines from getting damaged and for lowering the explosive device in the rockhole. Vibrations of the casing pipe due to SW propagation can result in the pipe magnetization thereby producing local changes of the geomagnetic field.

One more effect can be due to the redistribution of natural and man-made telluric currents flowing in the rock around the place of underground detonation. The major origin for the man-made current is believed to be the electrochemical processes at the interface between the casing pipe and surroundings. The occurrence of the contact potential difference at the surface of the metallic pipes can be associated with the difference between mechanisms of conductivity, namely electronic conductivity in the metal and ionic one in the rock surrounding the pipe. It should be noted that the similar effect is usually observed in the vicinity of ore deposit (e.g., see Semenov 1974). The contact electromotive force at the casing pipe surface depends on the depth because the mineral content of underground water varies with depth. As a result the currents develop in both the casing pipe and the rock around the pipe.

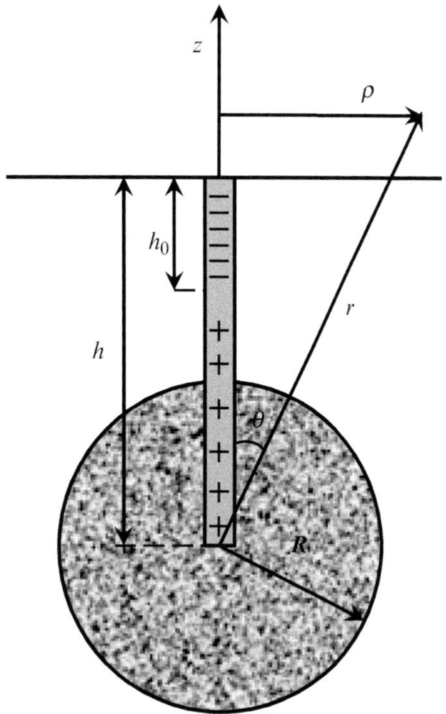

Fig. 11.8 Model distribution of contact EMF along the casing pipe. A *circle* indicates the boundary of fractured zone resulted from the explosion

The quasi-static electromagnetic perturbations originated from these currents could be detected just after the casing pipe installation and even before the underground detonation itself.

Considering the electrochemical processes around the metallic ore body, it is usually the case that the top of ore body plays a role of a negative electrode (Semenov 1974). In this notation, we suppose that the upper part of the casing pipe operates as a negative electrode while the lower part serves as a positive electrode. Let h be the length of casing pipe while h_0 be the length of the effective negative electrode as schematically shown in Fig. 11.8. The detonation at the depth h results in the formation of spherical region of the fractured rock with radius R, whose conductivity, σ_p, is different from the conductivity, σ_g, of the surrounding rock. Suppose that the current distribution is stationary and the total current, I, generated by the electrodes is a given value. The currents per unit length of negative electrode, I/h_0, and positive electrode, $I/(h - h_0)$, are constant values. Then the potential, φ, of the electric field is determined by Poisson equation which should be supplemented by the proper boundary conditions; that is, the potential and its derivative with respect to radius is continuous at the ball surface and the vertical component of the current density has to be zero at the free surface.

The solution of this problem for the potential φ can be represented in the form $\varphi = \varphi_0 + \varphi_p$, where φ_0 denotes the potential generated by the casing pipe in the uniform conducting half-space with constant conductivity σ_g while φ_p stands for the perturbations caused by the appearance of the fractured region with conductivity σ_p. In the atmosphere ($z \geq 0$) the potential φ_0 reads

$$\varphi_0 = \frac{I}{2\pi\sigma_g (h - h_0)} \left\{ \ln \frac{\sqrt{(z + h_0)^2 + \rho^2} - z - h_0}{\sqrt{(z + h)^2 + \rho^2} - z - h} \right.$$
$$\left. + \left(1 - \frac{h}{h_0}\right) \ln \frac{\sqrt{(z + h_0)^2 + \rho^2} + z + h_0}{\sqrt{z^2 + \rho^2} + z} \right\}, \tag{11.27}$$

where ρ denotes the polar radius shown in Fig. 11.8. The approximate relationship for the perturbations caused the rock fracture is given by (Surkov 1989)

$$\varphi_p = \frac{I \left(\sigma_g - \sigma_p\right)}{2\pi\sigma_g (h - h_0)} \sum_{k=1}^{\infty} \frac{\Lambda_k P_k (\cos \theta)}{k\sigma_p + (k + 1) \sigma_g} \left(\frac{R}{r}\right)^{k+1}, \tag{11.28}$$

where

$$\Lambda_k = \frac{2k + 1}{k + 1} - \frac{h R^k}{h_0 (h - h_0)^k} - \left(1 - \frac{h}{h_0}\right) \left(\frac{R}{h}\right)^k. \tag{11.29}$$

Here $P_k (\cos \theta)$ are Legendre's polynomials, $r = \left\{(z + h)^2 + \rho^2\right\}^{1/2}$, and the angle θ is shown in Fig. 11.8.

If $R \ll h$ and $R \ll h - h_0$, then Eq. (11.28) is simplified because only first term of the series can be taken into account. If, in addition, $\rho \gg h$,s then Eq. (11.27) is also simplified. For example, on the plane $z = 0$ the solution of the problem is reduced to

$$\varphi_0 = -\frac{I h (h + h_0)}{4\pi\sigma_g \rho^3}, \qquad \varphi_p = \frac{3I \left(\sigma_g - \sigma_p\right) h R^2}{4\pi (h - h_0) \sigma_g \left(\sigma_p + 2\sigma_g\right) \rho^3}. \tag{11.30}$$

It follows from Eq. (11.30) that both the casing pipe and fractured zone produce only a local effect since the electric field $\mathbf{E} = -\nabla\varphi$ falls off faster with distance, that is, as ρ^{-4}. The numerical estimates of this effect seem to be accident-sensitive because of the lack of information about actual values of the parameters I and h_0.

The magnetic perturbation in the ground has only the azimuthal component B_ϕ, owing to the symmetry of the given problem, though $B_\phi = 0$ in the atmosphere. Actually the magnetic perturbation is equal to a finite value near the ground surface because the symmetry of the problem gets broken due to asperity of the ground surface. Notice that the magnetic field associated with this effect and the field due to shock magnetization are different in symmetry that enables us to distinguish these phenomena.

The irreversible deformation and fracture of rocks may greatly affect natural terrestrial currents which in turn result in the similar effects. There are a few causes for the terrestrial current generation such as variations of magnetospheric and ionospheric fields, electrochemical processes in the ground, groundwater flow, local hydrological factors, precipitations and etc. The shock impact on the rock leads to the changes in permeability of capillaries, channels, and fluid-filled cracks followed by the changes in rock conductivity.

The heated gaseous products of detonation contained in the underground cavity can produce thermoelectric and thermo-galvanomagnetic phenomena in the surrounding space. The heating of the medium and temperature gradient can save for the long times if the decay of radioactive elements contained in the detonation products goes on. These processes could also provide for stable changes of the local electromagnetic Earth's field in the vicinity of the detonation point.

11.2.3   *Electric Field of Gas-Dust Clouds*

Surface and shallow buried detonations are accompanied by dustfall and the ejection of broken ground. The explosion products and air heated by an explosion are mixed with fragments of the broken ground, thereby producing the gas-dust cloud which can emerge in the atmosphere by the action of buoyancy force. There are a few stages of the cratering explosion. At first the ground dome is developed under the influence of the explosion for subseconds to be followed by the gas break through the dome and by the gas output in the atmosphere. Then the air-SW is created in the atmosphere. The coarse fragments of the flying rocks follow ballistic trajectories, while fine particles are pulled into the gas motion behind the shock front. The gas-dust cloud rises upward during several seconds or minutes depending on the scale of explosion. The conventional time scale is several minutes or hours for the dust deposition and for the dispersal of gas-dust clouds.

The generation of the gas-dust cloud caused by an explosion is accompanied by the appearance of low-frequency (up to $100\,\mathrm{Hz}$) electromagnetic field in the atmospheric surface layer (Holzer 1972). Adushkin and Soloviev (1988) have observed the variations of vertical electric field during the surface detonation of HE with mass $1\,\mathrm{t}$. The amplitude of electric field reached several tens kV/m at the distance $1\,\mathrm{km}$ from the explosion site. As the cratering explosion is performed in the medium-moisture rock, the electric signals, as a rule, have a bipolar shape and the polarization of the first phase is negative if z axis is downward directed (Adushkin and Soloviev 1996). This negative phase is usually observed during the course of ground dome development and over the period of ballistic flying of fractured rock fragments. The next more durable phase of the electric field evolution is due to the relaxation of electric charges and the motion of the gas-dust cloud. The duration of this phase is on the order of precipitation time of charged particles and aerosols.

The observed low frequency electric field is believed to be due to the ionization of gas and explosion products as well as from the electrification of fractured rock and soil fragments. In the theory by Adushkin et al. (1990) the gas-dust cloud is modeled by a uniformly charged column filled by detonation products. At first, the negatively charged ground fragments are assumed to be located at the top of the cloud. The kinematic characteristics of the model were chosen in such a way to fit the numerical calculation of the dust cloud evolution with the filming of surface explosion. The ground fragments fall with gravity acceleration, while the precipitation of the small dust particles is described by the following equation:

$$\frac{d\mathbf{v}}{dt} = \mathbf{g} - \frac{\mathbf{V} - \mathbf{V}_g}{\tau} + \frac{q\mathbf{E}}{m}, \tag{11.31}$$

where $\mathbf{g}$ is the free fall acceleration, $\mathbf{V}$ is the speed of descent of the dust particles, $\mathbf{V}_g$ is gas velocity, η is the air viscosity, and m, q, and a stand for average mass, charge, and size of particles, respectively. In the case of laminar flow the relaxation time is given by $\tau = m/(6\pi\eta a)$. Here $\mathbf{E}$ is the sum of geoelectric field and self-consistent field generated by the charged particles in the air and by induced charges on the ground surface. Assuming for the moment that the electric field is close to the breakdown threshold in the air, that is $E = 32\,\text{kV/m}$, and taking the numerical parameters $a = 20\text{–}50\,\mu\text{m}$, $q = 100e$ (e is elementary charge), we obtain that $qE/m \approx 0.7\text{–}1.8\,\text{m/s}^2 \ll g$. In practice this means that the electric field can be neglected as compared to the gravity.

Figure 11.9 displays the results of numerical calculation shown with dotted line 1 and the experimental observation of vertical electric field generated by a cratering explosion with HE mass 23.8 g (line 2). To fit the numerical and experimental data, the maximal charge of the dust cloud and the height of the cloud lift are estimated as $1.84\,\mu\text{C}$ and 4.2 m, respectively. In making the plot of E_z the following parameters were used $\eta = 1.7 \times 10^{-5}\,\text{Pa}\cdot\text{s}$, $a = 48\,\mu\text{m}$. It should be noted that the shape of initial half-wave essentially depends on the charge distribution in the ground dome and in the gas-dust cloud. The sharp spike in the beginning of the second half-wave (dotted line 1) is based on the assumption that all the rock fragments begin to fall simultaneously. As is seen from Fig. 11.9, the experimental data is qualitatively consistent with the simple model presented above. So, the quasi-static electric field caused by the excavating explosion is most likely to be due to the motion of electric charges located in the ground dome and gas-dust cloud.

One more example of transient electric fields detected during a powerful surface explosion with mass 500 t is displayed in Fig. 11.10 (Soloviev and Surkov 2000). The seismic wave arrival at the ground-based station has not a visible effect on the electric field, and the reason may be that the signal was below the sensor sensitivity. The vertical electric field at the distance 1.5 km (line a) reaches a peak value about 2.5 kV/m at the moment $t = 40\,\text{s}$. The numerical calculation shown with line c is based on the simple model of gas-dust cloud which consists of the spherically symmetric charge q_1 situated at the altitude h_1 over the ground and the uniformly charged column with total charge q_2 and altitude h_2. The best fit of the calculated

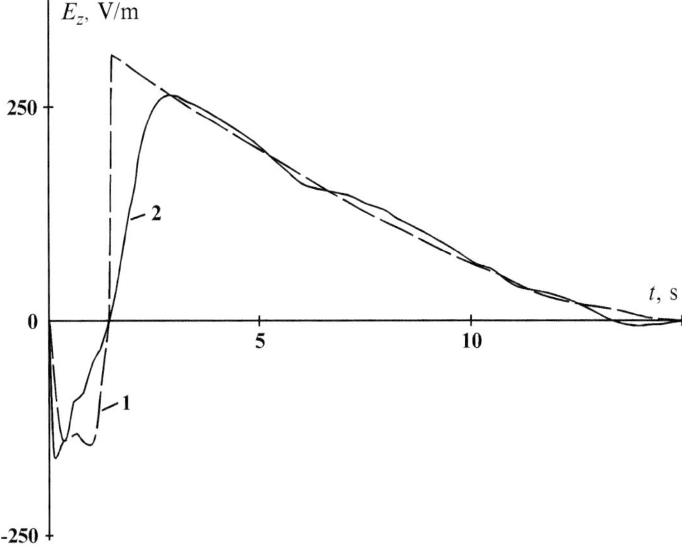

Fig. 11.9 Vertical electric field arising in the surface atmospheric layer under a cratering explosion of HE with mass 23.8 g. The explosion was performed at normalized depth 0.68 m/kg$^{1/3}$. *1*—theoretical calculations, *2*—measurements at distance 5.2 m from the explosion site. Adapted from Adushkin et al. (1990)

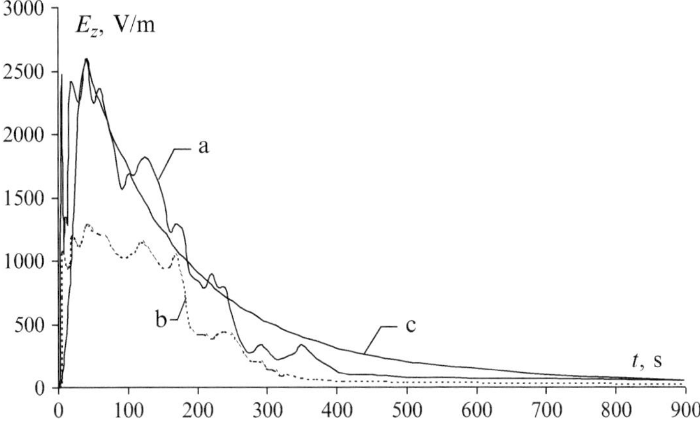

Fig. 11.10 Variations of vertical electric field caused by the HE explosion with mass 500 t as observed at different distances R from the explosion site (Soloviev and Surkov 2000). (**a**) $R = 1.5$ km; (**b**) $R = 2$ km; (**c**) $R = 1.5$ km (model calculation)

and experimental data at the moment 40 s corresponds to the following parameters: $q_1 = 1.3\pm0.1$ C, $h_1 = 1.39\pm0.06$ km, $q_2 = -0.4\pm0.1$C, and $h_2 = 0.30\pm0.06$ km.

A series of surface detonations of HE charges with mass of a few kilograms were carried out by Soloviev et al. (2002). The study showed that the amplitude of the

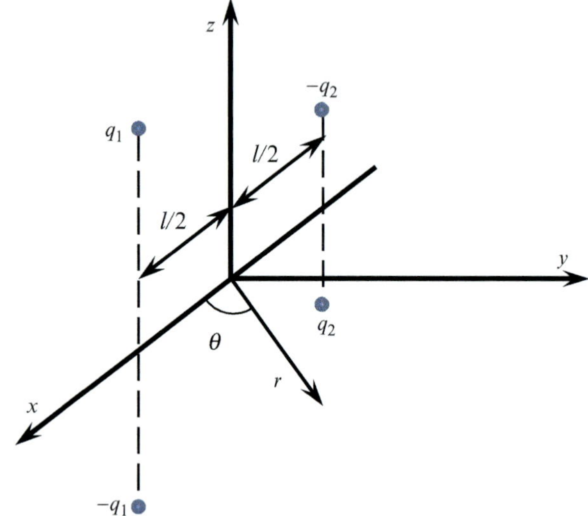

Fig. 11.11 A coordinate system and effective point charges which model the actual charge distribution

electric variations decreases with distance, r, approximately as r^{-4}, at least at the initial stage of the surface explosion. This is indicative of high symmetry of the electric charge distribution which results in a quadrupole character of the electric field in the initial stages of the explosion. In order to interpret the experimental data, the actual charge distribution was modeled by two effective point charges q_1 and $-q_2$ and their mirror images in the conducting ground. The origin of coordinate system is placed at the detonation site while the point charges are located on x, z plane as shown in Fig. 11.11. Far away from the charges the electric field might be expanded in a series of a small parameter l/r where l denotes the distance between the charges. The vertical electric field on the ground surface is given by

$$E_z\left(r, \theta, t\right) = -\frac{d_z\left(t\right)}{4\pi\varepsilon_0 r^3} - \frac{Q_{xz}\left(t\right)}{4\pi\varepsilon_0 r^4} + \dots \tag{11.32}$$

Here, $d_z = 2\left(q_1 h_1 - q_2 h_2\right)$ is the projection of the dipole moment onto the z-axis, and $Q_{xz} = 3l\left(q_1 h_1 + q_2 h_2\right)\cos\theta$ is the single nonzero component of the tensor of quadrupole moment of the charge system.

The experimental data suggest that the second term on the right-hand side of Eq. (11.32) is greater than the first one during the initial stage. For the detonation of HE with mass of 5 kg the numerical value $Q_{xz} = -(0.4–1.3) \times 10^{-4}\,\text{C}\cdot\text{m}^2$ brings the closest fit with experimental data. Taking the notice of empirical dependence, according to which the electric charge of products of the surface explosion varies as the 0.65 ± 0.05 power of the explosive mass (Adushkin and Soloviev 1996), the effective charges and the characteristic horizontal distance between them were estimated as $q_1 \approx q_2 \approx 2\,\mu\text{C}$ and 2–7 mm, respectively (Soloviev et al. 2002). It is plausible that the horizontal separation of positive and negative charges in space will behave randomly. Effective charges may rotate around the vertical axis because of vortices arising beyond the shock front.

The dipole and quadrupole terms in Eq. (11.32) were comparable in some experiments. In the case of explosives with mass of $50\,\mathrm{kg}$ the dipole term begins to be prevalent over the quadrupole one at the distances larger than $57\,\mathrm{m}$. It is not surprising because the dipole term decreases more slowly away from the epicenter. So the quadrupole law r^{-4} is applicable in the near zone of explosions which is limited by some critical radius $r_0 = |Q_{xz}/d_z| \sim l\,(q_1 h_1 + q_2 h_2)\,/\,(q_1 h_1 - q_2 h_2)$. In the area $r > r_0$ the dipole term predominates and the field amplitude decreases as r^{-3}. It is usually the case that the vertical separation of electric charges and dipole moment increases with time at least when the time is greater than 1–$10\,\mathrm{s}$ (Adushkin and Soloviev 1996). This implies that the critical distance r_0 decreases with time.

The nuclear explosions are frequently accompanied by the generation of lightning discharges. The five upward-propagating discharges were detected during a thermonuclear detonation "Mike" test with TNT equivalent $10.4\,\mathrm{Mt}$ (Uman et al. 1972). The detonation was in the large ground-based hall at Eniwetok Atoll in the Pacific on 31 October 1952. It appears that the lightning discharges were initiated from instrumentation stations slightly above sea level. The major cause of the electric field generation is believed to be the flux of Compton electrons produced by nuclear detonations. For the detonation with such a TNT equivalent the estimated initial number density of ionized particles reaches the value about 10^{15} pair/cm^3, which is sufficient for electrical breakdown in the ionized air (Uman et al. 1972). Laboratory tests and numerical simulations showed that the breakdown conditions, branching and configuration of the discharge channels are determined by the spatial charge distribution in the exposed atmosphere (Hill 1973; Grover 1981; Colvin et al. 1987; Williams et al. 1988).

For the powerful surface explosions the strong quasistatic electric fields can be interpreted in terms of a vortex ring of the heated gas and dust. The lifting of the vortex ring in the atmosphere is caused by Archimedian force which results in the electric charge separation between the vortex ring and dust column. Holzer (1972) has observed an enhancement of the Earth electric field by as much as $60\,\mathrm{V/m}$ for the time of the vortex lifting in the atmosphere (2–3 min). The numerical calculations have shown that the electric field at the top of the dust column can reach the breakdown level in the air as the column height increases up to 1–$2\,\mathrm{km}$ (Surkov 2000). It appears that the lightning can be initiated in the dust cloud of explosion similar to that occurring in the volcanic ash cloud. The IC lightning discharges can explain the sharp peaks which are occasionally observed at the background of quasistatic electric field produced by explosions (Soloviev and Surkov 2000).

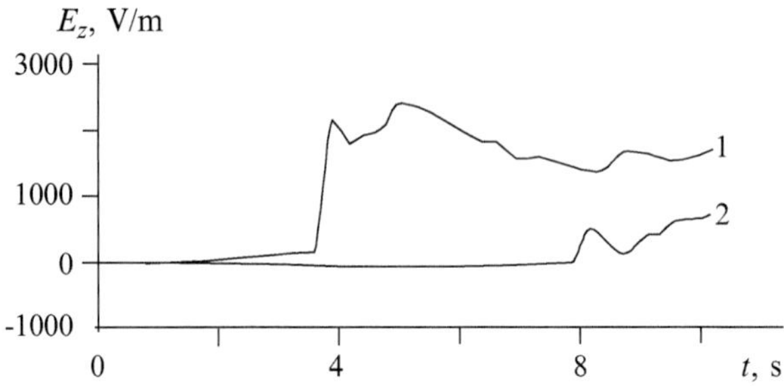

Fig. 11.12 Electric signals caused by aerial wave propagation in the surface atmospheric layer. The signals were measured at the distances (*1*) 1.5 km and (*2*) 2.8 km from the detonation of HE with mass 500 t (Soloviev and Surkov 1994)

11.2.4   Effect of Aerial SWs Propagating in a Surface Atmospheric Layer

A sharp narrow spike in the initial portion of signal shown in Fig. 11.10 took place several seconds after the detonation at the moment of aerial SW arrival at the ground-based station. The electric perturbations caused by the seismic wave propagating in a conductive ground are lower than this spike because the amplitude of seismo-induced effect at the distance $1.5 = 2$ km is about several μV/m (Chaps. 7 and 8). Figure 11.12 illustrates the initial portion of the signals measured at the distances (1) 1.5 km and (2) 2.8 km from the detonation site (Soloviev and Surkov 1994). As is seen from this figure, the front of geoelectric field perturbation approximately propagates at the velocity of aerial wave. So, one may expect that the source of electric variations is the local changes of pressure in the aerial SW.

Almost without exceptions the atmospheric air contains the heavy ions and aerosols which may be both the neutrals and also the charged particles (e.g., see Chalmers 1967; Wåhlin 1986; Sorokin 2007). It is common that the spatial electric charge in the surface atmospheric layer is about 10–500 pC/m^3. Taking into account that the total number density of the heavy ions is on the order of $5 \cdot 10^5 - 10$ m^{-3}, the heavy ion excess is estimated as $10^8 - 10^9$ m^{-3}. Hence the perturbations of the geoelectric field are induced by the changes in the spatial electric charge which is formed by the heavy ions and charged aerosols.

The heavy ions play a crucial role in the formation of the atmospheric electrode layer. However the aerosol particle may greatly affect the electrical parameters of the atmosphere such as the composition and number density of heavy ions and the structure of electrode layer. For the particles with size 0.01–0.2 μm the aerosol number density is about $10^9 - 10^{10}$ m^{-3} in rural areas and $10^{10} - 10^{11}$ m^{-3} near towns. The enhancement of the aerosol density gives rise to an increase of the electrode layer depth. Numerical simulations have shown that this depth can vary within 1–100 m by the action of turbulent stirring of the air near the ground surface (Hoppel 1967).

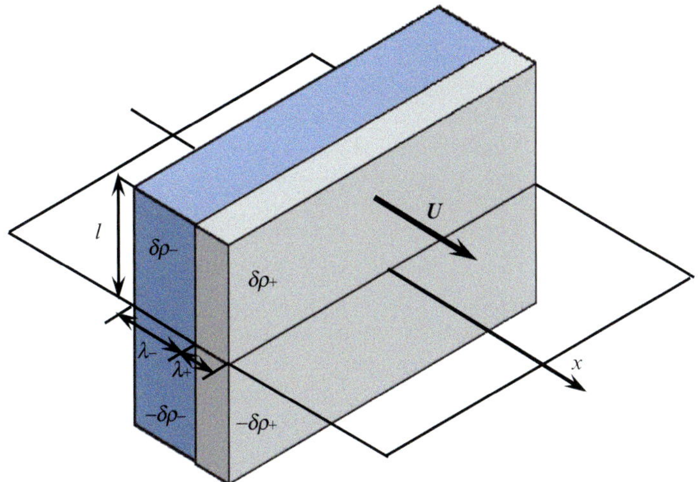

Fig. 11.13 A model of electrical effect caused by an aerial SW propagating in the sur-
face/electrode layer with width l. The charge density variations due to the presence of heavy
ions and aerosols are modeled by a step-function. Also shown are the electrical images of the
atmospheric charges in the conducting ground

To estimate the aerial SW effect in the electrode layer, we now consider a simple
model in which a plane steady SW propagates at constant velocity U along the x
axis parallel to the ground surface. The dynamics of the particles motion behind
the SW is described by Eq. (11.31). The numerical estimate has shown that the
relaxation time $\tau = m/(6\pi\eta a)$, which enter Eq. (11.31), is much smaller than the
wave duration. This implies that there will be complete entrainment of aerosols and
heavy ions with air flow. In the first approximation the changes in the spatial charge
density is supposed to follow the changes in the air density. We set the profile of the
gas mass velocity as a rectangular pulse with positive polarity followed by the next
one with negative polarity. The amplitudes, V_+ and V_-, and lengths, λ_+ and λ_-,
of these rectangular pulses are connected through $V_+\lambda_+ = V_-\lambda_-$. The continuity
equation for the electric charge flowing through the SW front is given by

$$\rho_0 U = \rho_+ (U - V_+), \tag{11.33}$$

where ρ_+ denotes the charge density in the air compressed by the SW while ρ_0 is
the undisturbed charge density. Hence the small perturbations of the charge density
can be written as $\delta\rho_+ = \rho_+ V_+/U$. The charge variations behind the SW front is
described by a step-function in such a way that $\delta\rho = \delta\rho_+$ within the length λ_+ of
"positive half-wave" and $\delta\rho = \delta\rho_-$ within the length λ_- of "negative half-wave."
The model distribution of the charges and their electrical images in the conducting
ground is sketched in Fig. 11.13.

If the typical length λ_+ of the aerial wave is much smaller than the width l of the electrode layer, then the model is reduced to the field of a charged thin lateral strip and its electrical mirror image in the conducting ground. In this case we get the following simple estimate of the electric field amplitude on the ground surface

$$E_{\max} \approx \frac{\rho_0 \lambda_+ V_+ \ln(l/\lambda_+)}{\varepsilon_0 U}. \qquad (11.34)$$

In the inverse extreme case when $\lambda_+ \gg l$, the problem is reduced to the field of a plane condenser generated by the charges of the atmospheric electrode layer, which is pressed by the SW, and the opposite charges induced in the ground. The solution of this simplified problem is given by

$$E_{\max} \approx \rho_0 l V_+ / (\varepsilon_0 U). \qquad (11.35)$$

It should be noted that the exact solution of the problem derived by Soloviev and Surkov (1994) on the basis of Maxwell equations and Eq. (11.31) can be reduced to Eqs. (11.34) and (11.35) in the above extreme cases.

It follows from the observations that the amplitude of the electric field variations is approximately proportional to the parameter $\lambda_+ V_+$ and this tendency keeps for different normalized distances and masses of explosives. It appears that the case of short aerial wave ($\lambda_+ \ll l$) occurs in practice. Substituting the following numerical values $\rho_0 = 8$–$80\,\text{nC/m}^3$, $\lambda_+ = 10\,\text{m}$, $V_+ = 35\,\text{m/s}$, $U = 350\,\text{m/s}$ and $l = 10^2\,\text{m}$ into Eq. (11.34) we obtain $E_{\max} = 0.2$–$2\,\text{kV/m}$. Both this estimate and experimental data shown in Fig. 11.12 are found to be of the same order. It should be noted that the shock compression may result in the release of ions which are in bound state. This leads to the enhancement of the charge density in the SW which results in the increase of the electric field amplitude.

11.2.5 Ionospheric and Magnetospheric Effects

As noted in Chap. 10, the aerial waves is the most efficient way to transfer the energy from the EQs and strong explosions to the ionosphere.

When the aerial SW enters the ionosphere, the entrainment of the ionospheric plasma with neutral flow results in the generation of ionospheric currents and local perturbations of the Earth magnetic field. One of the pioneering studies possibly related to the excitation of the ionosphere by SWs was the variations of the geomagnetic field observed at Irkutsk magnetic observatory after the detonation of Tunguska meteorite in 1908 (e.g., see Ivanov 1961). The abnormal behavior of the geomagnetic field has been observed 2.3 min after the detonation and lasted for several hours.

Considerable attention has been paid in the past to the study of man-made excitation of the ionosphere by the SWs produced by atmospheric nuclear explosions (Daniels et al. 1960; Lawrie et al. 1961; Stoffregen 1962, 1972; Gassmann 1963; Kotadia 1967; Breitling et al. 1967; Baker and Davies 1968; Baker and Cotten 1971; Kanellakos and Nelson 1972; Lomax and Nielson 1972), surface (Barry et al. 1966; Najita et al. 1975) and underground nuclear explosions (Blanc 1984, 1985; Pokhotelov et al. 1995). SW-induced oscillations in the lower thermosphere followed by the ionospheric perturbations have been observed after a 5 kt chemical explosion (Jacobson et al. 1988). Among the other sources of the strong acoustic waves in the atmosphere and ionosphere are volcano eruptions, spacing flights of supersonic airplanes (Marcos 1966) and rocket launch (Rao 1972; Karlov et al. 1980).

In standard geophysical practice the techniques of vertical, oblique, and Doppler sounding are used in order to examine the ionospheric response to natural and anthropogenic forcing on the Earth's ionosphere. In addition, the technique of continuous VLF electromagnetic transmission probing of the Earth-ionosphere waveguide is in routine use (e.g., see Surkov 2000; Molchanov and Hayakawa 2008). The experimental evidences on the impact of surface and underground explosions on the ionospheric F, E, and D layers have been demonstrated on the basis of these techniques (e.g., see Barry et al. 1966; Broche 1977; Blanc 1984, 1985). In the epicentral region the effect of explosion on the ionosphere is predominantly due to the upward-propagating atmospheric acoustic wave. This wave is generated when the underground SW reflects from the ground surface. The amplitude of mass velocity in the aerial wave increases with altitude due to the exponential fall off of the atmospheric density. Even a weak upward propagating wave can be converted into an SW because of nonlinear properties of the atmosphere. This nonlinear transformation of the wave shape and wave-front breaking occurs at the altitude $H_* \approx 2P_0/\left(\rho_g g\right) \sim 20\,\mathrm{km}$ where P_0 and ρ_g are the pressure and air density at the sea level (e.g., see Whitham 1974). As the altitude is larger than H_*, the wave profile becomes universal. The wave front has a triangular shape. It is imperative that the compression phase is followed by the rarefaction phase in such a way that the wave profile resembles a letter N, as shown in Fig. 11.13. It follows from the principle of conservation of momentum that the positive and negative phases of the N-wave are equal in square (e.g., Landau and Lifshitz 1959). In the bottom of the ionosphere; that is, in the altitude range of 90–100 km, the amplitude of mass velocity and length of the N-wave can reach several tens m/s and several km, respectively. The exponential increase of the SW amplitude ceases at the altitudes over 100 km. The cause of this effect is the enhancement of the gas viscosity at these altitudes due to the increase of mean free path of molecules (Enstrom et al. 1972). Then the dissipative processes result in both the decrease in the pressure and mass velocity gradients at the SW front and the increase in the wavelength which can reach a few tens km in F-region of the ionosphere.

The numerical stimulations have shown that the effect of SW impact on the lower ionosphere is maximal in the circular area with radius of the order of 100 km (Orlov and Uralov 1984), and the influence of shock upon the ionosphere diminishes essentially outside this area. This effect is due to the refraction of sound in the

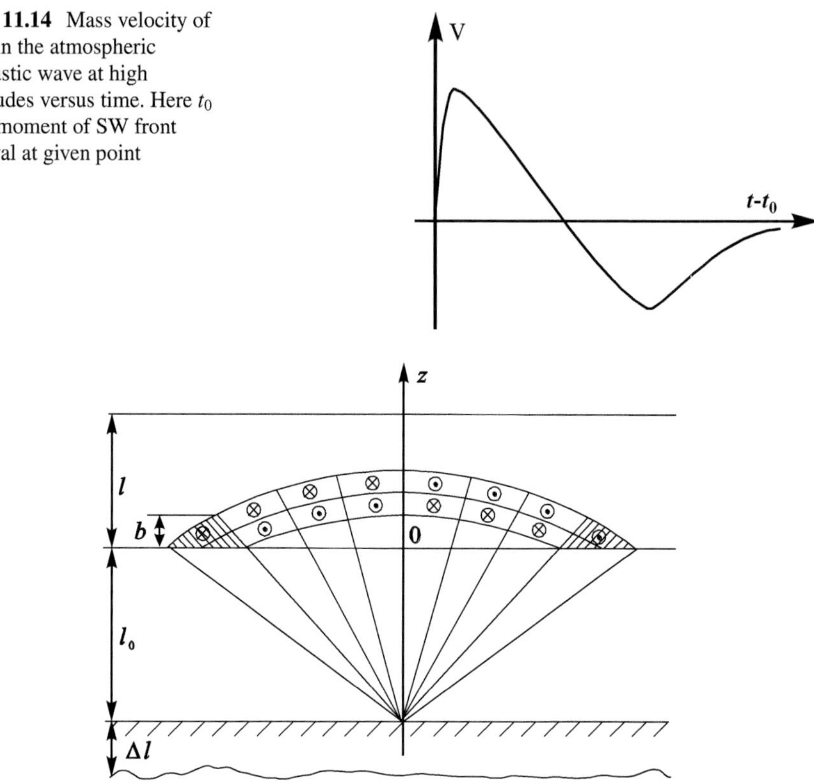

Fig. 11.14 Mass velocity of gas in the atmospheric acoustic wave at high altitudes versus time. Here t_0 is a moment of SW front arrival at given point

Fig. 11.15 A model of the atmosphere ($-l_0 < z < 0$) and the ionospheric E-layer ($0 < z < l$). An aerial SW is originated from the source/explosion located on the ground surface $z = -l_0$. A bundle of direct lines represents the directions of SW propagation. The circles with crosses and points in the middle show the opposite directions for wind-driven/extrinsic currents. These currents were not compensated in the shaded region of the ionosphere. A polygonal line at the bottom of this figure shows a depth Δl of the layer of sedimentary rocks which have a high conductivity

stratosphere and thermosphere which results in the sound ray curvature and rotation down to the earth.

In what follows we study the ionospheric perturbations caused by the axially symmetrical SW propagating from the explosion site situated at the ground surface towards the ionosphere. The mass velocity has a contour of "N-wave" which is shown in Fig. 11.14. We use the plane-stratified model of the medium which consists of the conducting ground ($z < -l_0$), the nonconducting atmosphere ($-l_0 < z < 0$) and gyrotropic E-layer of the ionosphere ($0 < z < l$). The origin of coordinate system is placed on the symmetry axis of the wave at the lower boundary of the ionosphere. In Fig. 11.15 the bundle of direct lines sketches the directions of SW propagation from the explosion site. In this region the acoustic wave refraction is ignored.

To study the ionospheric perturbations in a little more detail we consider first a polar ionosphere. The geomagnetic field $\mathbf{B}_0$ is therefore vertically parallel to z axis. In this case Maxwell equations in the ionospheric E-layer can be written in cylindrical coordinates

$$-\partial_z B_\varphi = \mu_0 \left(\sigma_P E_r - \sigma_H E_\varphi + \sigma_H V_r B_0 \right), \tag{11.36}$$

$$\partial_z B_r - \partial_r B_z = \mu_0 \left(\sigma_P E_\varphi + \sigma_H E_r - \sigma_P V_r B_0 \right), \tag{11.37}$$

$$r^{-1}\partial_r \left(r B_r \right) + \partial_z B_z = 0, \tag{11.38}$$

where B_z, B_r, and B_φ are the components of the geomagnetic field perturbations, E_r and E_φ are the components of the electric perturbations. Pedersen and Hall conductivities of the ionospheric plasma, i.e. σ_P and σ_H, are considered as constant values. The field-aligned ionospheric conductivity is assumed to be infinite and hence $E_z = 0$. At given orientation of the vector $\mathbf{B}_0$, only radial component V_r of the gas velocity enters the set of Eqs. (11.36)–(11.38).

The area of gas flow due to the acoustic wave is shown in Fig. 11.15 with three arcs. This area is of a form of narrow band since the longitudinal size of the acoustic wave is much smaller than a width l of the ionospheric E region. The upper arc corresponds to the N-wave front. Because of the bipolar form of the acoustic N-wave, at first the gas moves forward in radial directions and then it moves back. The first area is bounded by the upper and middle arcs, while the next area is restricted by the middle and lower arcs. The currents generated by the motion of conductive media are oppositely directed in these areas. The changes in the current direction occur in the middle portion of the wave where the gas velocity vanishes. In Fig. 11.15 the opposite directions of the extrinsic currents are represented by the circles with cross and point.

The total extrinsic current is proportional to the integral of the gas velocity over the area covered by the wave. Now we divide this area into narrow sectors in the z, r plane formed by rays diverging from the explosion point. Notice that the currents flowing through the upper and lower portions of these sectors are oppositely directed. The total current of each sector is proportional to the integral of the gas velocity along the corresponding ray. For N-wave this integral is equal to zero, so that the corresponding extrinsic current vanishes as well. As is seen from Fig. 11.15, only the last shaded cells on the right and on the left contain the unbalanced currents. Certainly, this approach is valid for the short N-waves. Usually the longitudinal size of the acoustic waves $b \approx 1-3$ km while the width of the E-layer is about 20–30 km so that the condition $b \ll l$ is true.

Thus, the uncompensated extrinsic current arises only at the lower boundary of the ionosphere. The cross section of this ring current is on the order of the longitudinal wave size. This implies that the extrinsic current density is nonzero only inside the narrow layer $b \ll l$ in the vicinity of the ionosphere bottom, that is at $z = 0$. To simplify the problem, we formally assume that the velocity altitude distribution is described by a delta-function, that is $V_r = b\delta\left(z \right) V_r \left(r, t \right)$. Performing integration of Eqs. (11.36) and (11.37) over z from zero to ε and then assuming $\varepsilon \to 0$ we come to the following boundary conditions at $z = 0$

$$B_{\varphi a} - B_{\varphi i} = \mu_0 b J_H, \quad B_{ra} - B_{ri} = \mu_0 b J_P, \quad B_{za} = B_{zi}, \tag{11.39}$$

where the subscripts a and i are related to the atmosphere and ionosphere, respectively. The extrinsic current densities, $J_H = \sigma_H V_r B_0$ and $J_P = \sigma_P V_r B_0$, are assumed to be given functions.

Equations (11.36) and (11.37) can be solved for E_r and E_φ. Substituting these components of electric field into Maxwell equation $\nabla \times \mathbf{E} = -\partial_t \mathbf{B}$ and rearranging yields

$$\partial_t B_z = D_P \nabla^2 B_z - \frac{D_H}{r} \partial_r \left(r \partial_z B_\varphi \right), \tag{11.40}$$

$$\partial_t B_r = D_P \left(\nabla^2 B_r - \frac{B_r}{r^2} \right) + D_H \partial_{zz}^2 B_\varphi, \tag{11.41}$$

$$\partial_t B_\varphi = D_P \nabla^2 B_\varphi - D_H \left(\nabla^2 B_r - \frac{B_r}{r^2} \right), \tag{11.42}$$

where

$$D_P = \frac{\sigma_P}{\mu_0 \left(\sigma_P^2 + \sigma_H^2 \right)}, \quad D_H = \frac{\sigma_H}{\mu_0 \left(\sigma_P^2 + \sigma_H^2 \right)}, \tag{11.43}$$

are the coefficients of diffusion in a gyrotropic medium and the operator ∇^2 is given by

$$\nabla^2 = r^{-1} \partial_r + \partial_{rr}^2 + \partial_{zz}^2. \tag{11.44}$$

The set of Eqs. (11.40)–(11.42) can be solved with respect to each component of the magnetic perturbation. For example, excluding B_φ from Eq. (11.40) gives (Surkov 1996)

$$\partial_t \left(\partial_t B_z - D_P \nabla^2 B_z \right) = \partial_{zz}^2 \left\{ D_P \partial_t B_z - \left(D_P^2 + D_H^2 \right) \left(\nabla^2 B_z - B_z / r^2 \right) \right\}. \tag{11.45}$$

Applying Eq. (11.36) to the atmosphere and taking into account that the components of the conductivity tensor are equal to zero, we obtain that $B_\varphi = 0$ everywhere in the atmosphere. Other components of the GMP can be determined from the Laplace equation:

$$\nabla^2 B_r = B_r / r^2, \quad \nabla^2 B_z = 0. \tag{11.46}$$

Suppose that the acoustic wave reaches the lower boundary of the ionosphere at the moment $t = 0$. The region of interaction between the wave and ionosphere increases in time, but we shall assume the simultaneous arrival of waves at the ionospheric boundary. Owing to the wave refraction in the atmosphere, the ionospheric region interacting with the wave is limited by the radius $a \sim l_0 \approx 100$ km. As alluded to earlier in Chap. 7, the GMP will propagate into the ionosphere in accordance with the diffusion law. For the time t the diffusion front climbs to the altitude $z_d \sim 2 \left\{ t / \left(\mu_0 \sigma_p \right) \right\}^{1/2}$ and thus the front will reach the upper boundary of the ionospheric E-layer at the moment $t_d \sim \mu_0 \sigma_p l^2 / 4$. In what follows we restrict our study to the short interval $0 < t < t_d$ which corresponds to the initial stage of signal. During this interval the solution of Eqs. (11.40)–(11.42) weakly depends on the boundary conditions of the problem at $z = l$. Hence, we replace this condition by the requirement that the solution must be finite when $z \to \infty$.

Considering the GMP diffusion in a horizontal direction, we note that for the time $t < t_d$ the diffusion front propagates at the distance much smaller than a. So, we will neglect the lateral expansion of the diffusion region in the ionosphere and focus on the vertical propagation of the GMP. This implies that in the region $r < a$ the terms $\partial_{rr}^2 B_z$, $r^{-1} \partial_r B_z$, and B_z / r^2 in Eq. (11.45) are much smaller than $\partial_{zz}^2 B_z$. As a first approximation, we assume that B_z is not a function of r. Notice that the components B_r and B_φ are equal to zero at $r = 0$ and their dependence on r should already be taken into account in the first approximation.

Laplace transformation with respect to time can be applied to all the equations with boundary conditions. Let b_z, b_r, b_φ, and j be Laplace transforms of the magnetic perturbations and extrinsic current density, respectively. Taking the notice of the above approximations, one can reduce Eq. (11.45) to the following:

$$\frac{d^4 b_z}{d z^4} - 2 p \mu_0 \sigma_P \frac{d^2 b_z}{d z^2} + p^2 \mu_0^2 \left(\sigma_P^2 + \sigma_H^2 \right) b_z = 0, \tag{11.47}$$

where p denotes the parameter of Laplace transformation. When the finiteness of b_z at $z \to \infty$ is taken into account, the solution of (11.47) is given by

$$b_{zi} = C_1 \exp\left(-\lambda_+ z \right) + C_2 \exp\left(-\lambda_- z \right); \quad \lambda_\pm = \left\{ \mu_0 p \left(\sigma_P \pm i \sigma_H \right) \right\}^{1/2}, \tag{11.48}$$

where C_1 and C_2 are the arbitrary constants, and $\mathrm{Re}\lambda_\pm > 0$. Other components of the magnetic perturbations in the ionosphere can be expressed through b_{zi} via

$$b_{ri} = -\frac{1}{r} \int_0^r r' \frac{d b_{zi}}{d z} d r', \tag{11.49}$$

$$b_{\varphi i} = -\frac{1}{r} \int_0^r r' \left\{ \frac{1}{p \mu_0 \sigma_H} \frac{d^3 b_{zi}}{d z^3} - \frac{\sigma_P}{\sigma_H} \frac{d b_{zi}}{d z} \right\} d r', \tag{11.50}$$

In the same approximation the equation for the atmosphere $(-l_0 < z < 0)$ reads: $d^2 b_z / dz^2 = 0$. If the ground is considered as a perfect conductor, the solution of this equation is given by

$$b_{za} = C_0 (z + l_0). \qquad (11.51)$$

The arbitrary constants C_0, C_1, and C_2 can be found from Eqs. (11.48)–(11.51) and the boundary conditions given by Eq. (11.39).

Now consider the low frequency perturbations when the corresponding skin-depth in the ground is greater than the depth $\Delta l \approx 1$–$2\,$km of upper layer of sedimentary rocks which possess a high conductivity. The lower boundary of the sedimentary rock layer is shown in Fig. 11.15 with wavy line. In this extreme case we can consider this layer as if it were transparent for the GMP. To simplify the problem, we assume that formally l_0 goes to infinity. Then the inverse Laplace transformation of the solution can be reduced to the simple quadratures. For example, in the ionosphere $(z \geq 0)$ the result can be written as follows (Surkov 1996)

$$B_{zi}(z, r, t) = -b \left(\frac{\mu_0}{2\pi\sigma_H} \right)^{1/2} r^{-1} \partial_r r \int_0^t G_z(z, t') J_H(r, t - t') dt', \qquad (11.52)$$

$$B_{\varphi i}(z, r, t) = \frac{\mu_0 z b}{2} \left(\frac{\mu_0 \sigma_H}{2\pi} \right)^{1/2} \int_0^t G_\varphi(z, t') J_H(r, t - t') dt', \qquad (11.53)$$

where the functions G_z and G_φ are given by

$$G_z = \frac{1}{t^{1/2}} \left(\gamma_+ \cos \frac{\alpha z^2}{t} + \gamma_- \sin \frac{\alpha z^2}{t} \right) \exp \left(-\frac{\beta z^2}{t} \right), \qquad (11.54)$$

$$G_\varphi = \frac{1}{t^{3/2}} \left(\chi_+ \sin \frac{\alpha z^2}{t} - \chi_- \cos \frac{\alpha z^2}{t} \right) \exp \left(-\frac{\beta z^2}{t} \right). \qquad (11.55)$$

Here the following abbreviations are introduced

$$\chi_\pm = m\gamma_\pm - \gamma_\mp, \qquad \gamma_\pm = \left\{ (1 + m^2)^{1/2} \pm m \right\}^{1/2},$$

$$\alpha = \frac{\mu_0 \sigma_H}{4}, \qquad \beta = \frac{\mu_0 \sigma_P}{4}, \qquad m = \frac{\sigma_P}{\sigma_H}. \qquad (11.56)$$

In order to analyze the features of this solution, we choose the pulsed source as a simple model of extrinsic current, that is $J_H(r, t) = \sigma_H B_0 V_* r \eta(a - r) T \delta(t) / a$, where $\delta(t)$ denotes δ-function, $\eta(a - r)$ is the step-function, V_* is the amplitude of mass velocity at the lower boundary of the ionosphere, and T is the typical time scale. As is seen from this expression, the extrinsic current flows in the circle with radius a and it vanishes outside this area. In this case the integrals in Eqs. (11.52) and (11.53) can be easily calculated (Surkov 1996)

$$B_{zi} = -\frac{2B_0 V_* bT}{a} \left(\frac{\mu_0 \sigma_H}{2\pi}\right)^{1/2} G_z(z,t), \quad B_{ri} = -\frac{r}{2}\partial_z B_{zi}, \tag{11.57}$$

$$B_{\varphi i} = \frac{B_0 V_* bT zr}{a(2\pi)^{1/2}} (\mu_0 \sigma_H)^{3/2} G_\varphi(z,t). \tag{11.58}$$

where the functions G_z and G_φ are determined by Eqs. (11.54) and (11.55).

We recall that Eqs. (11.57) and (11.58) are valid in the interval $t < t_d$, which can be applied to the front of electromagnetic perturbations. Formally this solution describes the case of infinite gyrotropic conductive half-space bordering the atmosphere.

The factor $\exp\{-\mu_0 \sigma_P z^2/(4t)\}$ is indicative of the diffusion character of the GMP propagation across the ionospheric E-layer. Danilov and Dovzhenko (1987) have noted that this factor determines the length of an electromagnetic precursor for acoustic wave. This effect is similar to the electromagnetic forerunner of seismic wave that we have examined in more detail in Chap. 7. Substituting σ_p for σ in Eq. (7.20) we obtain the estimate of the precursor length $\lambda \sim (\mu_0 \sigma_P C_a)^{-1}$, where C_a is the acoustic wave velocity.

The damping factor in Eqs. (11.54) and (11.55) is analogous to the skin effect in conductive media. However, the oscillating factors in these equations lead to a new property of this effect because the diffusion perturbations propagate in a form of damped oscillation. The phase of the oscillations $\mu_0 \sigma_H z^2/(4t)$ depends merely on the Hall conductivity, which means that the effect essentially depends on the presence of magnetized electrons in the ionospheric plasma of the E layer. The oscillation period increases in time and the oscillations cease at $t > \mu_0 \sigma_H z^2/(4\pi)$. By analogy with the above line of reasoning, one can estimate the "oscillatory" length of the electromagnetic precursor as $\lambda_o \sim (\mu_0 \sigma_H C_a)^{-1}$.

The same regime of diffusion has been demonstrated to be excited in the ionosphere for the case of horizontal geomagnetic field (Surkov 1990a,b). In the Hall medium, the analogous type of micropulsations propagating along the geomagnetic field has been termed the Schrödinger mode (Greifinger and Greifinger 1965). Another way to explain the oscillatory structure of the electromagnetic forerunner in the magnetoactive plasma is to take into account the radiations of helicon waves which are known as whistler mode in the geophysical practice. As the electrons are magnetized whereas the ions are not yet, the dispersion relation for the field-aligned helicons reads (e.g., Lifshitz and Pitaevskii 1981)

$$\omega = \frac{k^2 V_A^2}{\Omega_H} \tag{11.59}$$

where k is the wave number and Ω_H is the ion gyrofrequency. In the coordinate system which moves at the acoustic wave velocity, the dispersion relation is given by (Danilov and Teselkin 1984)

$$\omega' - kC_a = \frac{k^2 V_A^2}{\Omega_H}. \tag{11.60}$$

In the stationary problem under consideration, the frequency $\omega' = 0$ whence it follows the typical size of the precursor is about $\lambda_o \sim k^{-1} = V_A^2 / (C_a \Omega_H)$. We recall that these equations are valid if the electrons are magnetized, that is, $\omega_H \gg \nu_e$, and the ions are not yet, that is $\Omega_H \ll \nu_{in}$. Taking into account that in this case $\sigma_H \approx e^2 n / (m_e \omega_H) = en / B_0$ and substituting $\Omega_H = eB_0 / \overline{m_i}$ and $V_A^2 = B_0^2 / (\mu_0 n \overline{m_i})$ into the above estimate, we obtain the length of "oscillatory" portion of the electromagnetic precursor $\lambda_o \sim (\mu_0 \sigma_H C_a)^{-1}$ which coincides with the above estimate.

It appears that a high-power surface detonation can have influence not only on the ionosphere but also on the magnetosphere. For example, the magnetic pulses with amplitude 100 nT have been detected onboard AUREOL-3 satellite at the altitude about 750 km several minutes after the surface detonation known as experiment MASSA-1 (Galperin and Hayakawa 1996). The detonation of HE with TNT equivalent 288 t has been performed in a sandy desert 60 km to the north of Alma-Ata (former USSR) on November 28, 1981. In order to discuss the plausibility of the coupling mechanisms operating between the surface detonation and the magnetosphere, it is necessary at this point to estimate the magnitude of the signals produced by the surface detonation at the magnetospheric altitudes. To do this, we suppose that Eqs. (11.40)–(11.42) are justified in the altitude range $0 < z < l$ while above this layer; that is at $z > l$, there take place the equations for a cold collisionless plasma. Assuming that the field-aligned plasma conductivity is infinite, we come to the standard equations describing Alfvén and FMS plasma waves in this region

$$\partial_{tt}^2 B_z = V_A^2 \nabla^2 B_z, \qquad \partial_{tt}^2 B_\varphi = V_A^2 \partial_{zz}^2 B_\varphi. \tag{11.61}$$

where V_A is the Alfvén wave speed. Disregarding, as before, the derivatives of r in the operator ∇^2, we choose the solution of Eq. (11.61) in a form of upgoing waves. The proper boundary condition at $z = l$ is that the solution would transform continuously into that of Eqs. (11.40)–(11.42).

It is easy to show that the solution of Eq. (11.61) appears as Eqs. (11.57)–(11.58) where z and t should be replaced by l and $\tau = t - (z - l) / V_A$, respectively. This means that the same temporal dependence holds if we use the coordinate system which moves at the Alfvén wave velocity. The components B_z, B_r, and E_φ describe Alfvén wave, while the components E_r, E_z, and B_φ correspond to FMS wave. The numerical modeling of the GMP in the magnetosphere versus τ is displayed in Fig. 11.16. Here we made use of Eqs. (11.57)–(11.58) and the typical parameters

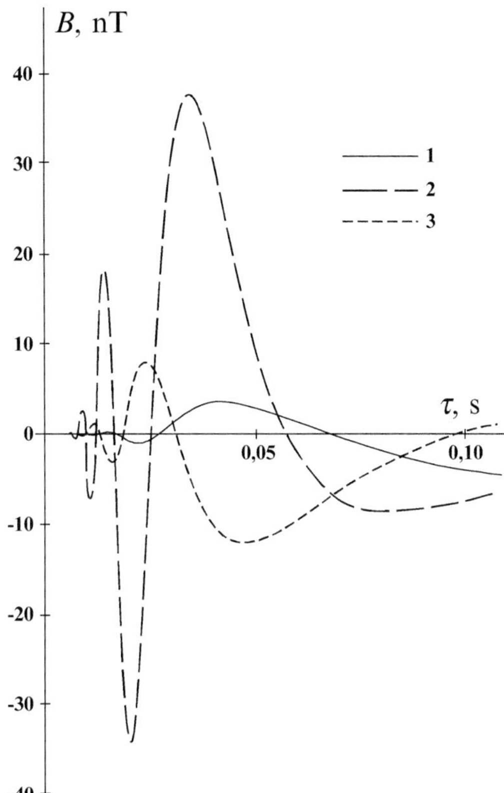

Fig. 11.16 Numerical simulation of magnetic perturbations in the polar magnetosphere. The extrinsic current at the lower boundary of the ionosphere is modeled by the delta-function of time. The components B_z, B_r, and B_φ versus $\tau = t - (z - l)/V_A$ are shown with lines 1, 2, and 3, respectively (Surkov 1996)

of the dayside ionospheric E-layer; that is, $\sigma_P = 0.5\sigma_H = 2.5 \cdot 10^{-4}\,\mathrm{S/m}$, $V_A = 300\,\mathrm{km/s}$ as well as the following numerical values $b = 3\,\mathrm{km}$, $a = 100\,\mathrm{km}$, $r = 90\,\mathrm{km}$, $T = 0.1\,\mathrm{s}$.

Next consider the case when the source function/extrinsic current is modeled by a step function of time, that is $J_H(r, t) = \sigma_H B_0 V_* r \eta (a - r) T \eta (t)/a$. In this case the integrals in Eqs. (11.52) and (11.53) are not expressed by the elementary functions though the component B_r can be written as

$$B_{ri} = -\frac{B_0 V_* b r}{a V_A} \left(\frac{\mu_0 \sigma_H}{2\pi}\right)^{1/2} G_z(l, \tau). \tag{11.62}$$

For this extreme case the results of numerical calculations are shown in Fig. 11.17.

As is seen from Figs. 11.16 and 11.17, the risetime of the signals is approximately coincident with that of the exponential factor $\exp\left\{-\mu_0 \sigma_P l^2/(4\tau)\right\}$. This time is of the order of diffusion time through the conducting E-layer $t_d \sim \mu_0 \sigma_P l^2/4 \approx 0.07\,\mathrm{s}$. The oscillations with the phase $\mu_0 \sigma_H l^2/(4\tau)$ are due to the Hall conductivity of the ionospheric plasma. The period of oscillations increases in time until they disappear at the moment $t > t_o = \mu_0 \sigma_H l^2/4 \approx 0.05\,\mathrm{s}$. The substitution of the step function

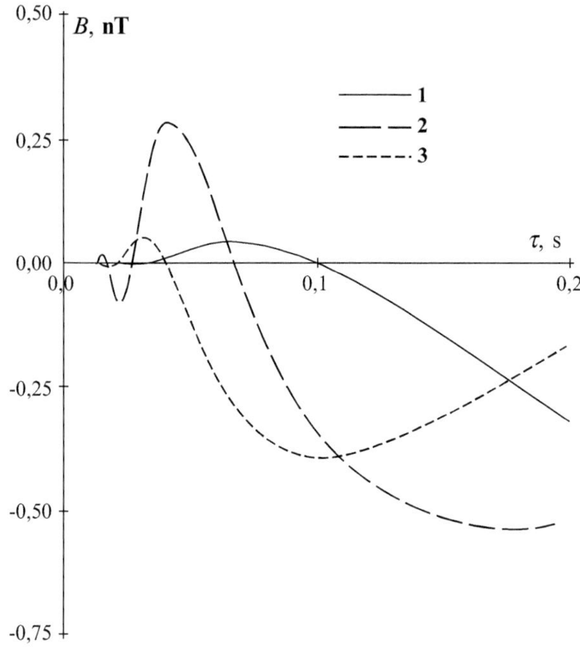

Fig. 11.17 Numerical simulation of magnetic perturbations in the polar magnetosphere. The extrinsic current at the lower boundary of the ionosphere is modeled by the step function of time. The components B_z, B_r, and B_φ versus $\tau = t - (z - l)/V_A$ are shown with lines 1, 2, and 3, respectively (Surkov 1996)

of time in Eqs. (11.52) and (11.53) results in smoothing these oscillations though a few oscillations remain in the initial part of the signal. As we noted above, such a structure is typical of the nonstationary diffusion process in gyrotropic media. In the model of the magnetosphere the wave profile is steady in the reference frame moving at the Alfvén wave velocity. It is not surprising, then, that this structure of perturbations is saved in the Alfvén and FMS waves and hence it is transferred upward at the Alfvén velocity.

It should be noted that the solution at $z > l$ is not entirely correct because the approximation of collisionless plasma is inapplicable to the typical wave frequencies $\nu = 0.1$–$1\,\text{Hz}$ at the altitudes of F layer. The decrease of the wave amplitude due to energy dissipation in the F layer can be roughly estimated by means of the attenuation factor $\exp\left(-z^2\mu_0\sigma_P\nu/4\right)$. Substituting $\sigma_P = 10^{-5}\,\text{S/m}$, $z = 200\,\text{km}$ as a mean altitude of the F layer and the above frequencies into this factor gives a decrease of the amplitude at 1.1–3.5 times. Certainly, a strong fall off of the spectrum should be expected in the frequency range above 1 Hz.

The effect of opposite polarity can arise approximately 1–2 min the after acoustic wave arrival at the bottom of the ionosphere when the wave will cross the upper boundary of the E layer. In such a case the area of uncompensated extrinsic current appears at this boundary. This current flows oppositely to direction of the extrinsic current at lower surface of the ionosphere that results in the generation of GMP of the opposite polarity. The amplitude of these GMPs can significantly exceed the original perturbations since the velocity amplitude and the length of the acoustic wave increase with altitude.

As one might expect, the front structure of the Alfvén and FMS waves in the magnetosphere is correlated with the processes in the ionospheric E layer. The front duration is of the order of t_d or t_0 while the typical front length is $V_A t_d$ or $V_A t_0$, that is about 15–20 km at the altitudes of a few hundreds km. The total duration of perturbations is determined by the time of acoustic wave passage through E layer and thus it can be far beyond the duration of the wave front (Surkov 1992a,b). The area of perturbations is extended along the geomagnetic field lines. The lateral size of this area is about 100 km which appear to be much less than the field-aligned scale. The sharp front and gradual drop of the signals shown in Figs. 11.16 and 11.17 are in qualitative agreement with the onboard observations though the predicted amplitudes of the signals ($\sim$0.1–10 nT) are much smaller than that measured by the satellite AUREOL-3 ($\sim$ 100 nT). It should be noted that on a basis of Maxwell equations one can find the following simple estimate of the GMP amplitude (e.g., see Danilov and Dovzhenko 1987)

$$\Delta B \sim \mathrm{Rm} B_0 \Delta p / p, \tag{11.63}$$

where Δp is excess pressure and $\mathrm{Rm} = \mu_0 \sigma_P \lambda V_a$ is magnetic Reynolds number. Taking the numerical values of the wavelength $\lambda = 1$ km and $\Delta p / p = 0.1$ we obtain the value $\Delta B \sim 1$ nT which is in agreement with the satellite observations (Pokhotelov et al. 1995). Certainly these rough estimates essentially depend on the parameters of the ionosphere, diurnal variations, and so on.

The splitting of the perturbations into two types in the lower ionosphere has occasionally been observed under nuclear explosions (Daniels et al. 1960). In the upper ionosphere these vertically traveling perturbations had different velocities. It appears that the slow perturbation corresponded to the conventional sound wave, whereas the velocity of the fast perturbation was increased roughly 2 times. It was hypothesized by Wickersham (1970) that this effect can be due to the excitation of ion-sound mode at the altitude range of 160–200 km. The velocities of the sound C_a and ion-sound V_i waves are given by

$$V_a = \left(\frac{\gamma k_B T}{\overline{m}}\right)^{1/2}, \quad V_i = \left(\frac{k_B \{Z \gamma_e T_e + \gamma_i T_i\}}{\overline{m_i}}\right)^{1/2}, \tag{11.64}$$

where Z denotes the ion charge; $\overline{m}$ and $\overline{m_i}$ are the average masses of neutrals and ions; γ_e, γ_i and γ stand for adiabatic exponents of the electrons, ions, and neutral gases; and T_e, T_i, and T are their temperatures, respectively. In the theory the ion sound can be generated in a strongly anisothermic plasma ($T_e \gg T_i$) when the frequency of inelastic collisions between the charged and neutral particles exceeds the frequency of their elastic collisions. However in the media of neutral particles the ion sound mode undergoes a strong attenuation because of the collisions between ions and neutrals. In order to overcome this difficulty, one should assume a possibility for some wave-induced exothermic reactions which results in the enhancement of the particle temperature (Wickersham 1970).

The powerful explosions can generate not only acoustic but also internal gravity waves (IGWs) in the atmosphere. It is well known that IGWs develop in media whose density varies with altitude and, in particular, in the stratified media. Basically, these waves propagate horizontally along the Earth surface at the velocities up to 400–500 m/s. At the epicentral distance of 1,200 km the period of IGW is about 7 min while the period of the acoustic wave is approximately equal to 1–2 min (Broche 1977).

The satellite observations have shown the increase of the electric noises in the frequency range of 0.1–1 kHz 6–7 min after the surface detonation MASSA-1 (Galperin and Hayakawa 1996). The enhanced noises were detected within ± 200 km around the magnetoconjugate tube with $L \approx 1.5$. The field-aligned electric components exhibited the greatest noise amplitude while the most spectral intensity is related to the frequency region below 100 Hz. Taking the notice of weak magnetic perturbations in this region, the observed effect is assumed to be the result of electrostatic turbulence induced by Alfvén waves propagating along the magneto-conjugate paths (Pokhotelov et al. 1994). The similar effect has been observed in the vicinity of the magnetoconjugate tube during the experiment MASSA-2 (Galperin and Hayakawa 1996). The region of the electric noise expanded at the velocity about 0.6 km/s up to the altitudes about 10^3 km.

Of interest in the analysis of satellite observations is the strong Alfvén pulses (with amplitudes 117 and 50 nT) measured by AUREOL-3 with onboard magnetometers and electric field sensors several minutes after HE detonations MASSA-1 and MASSA-3. The above estimates have shown that the acoustic channel of the explosion energy transfer to the ionosphere cannot be so effective in order to excite the pulses with so high amplitudes. It has been speculated that this effect can be attributed to the electric discharge generated at the SW front (Galperin and Hayakawa 1996, 1998; Surkov and Galperin 2000). The thermal ionization and changes in constants of chemical and ionization equilibrium can lead to an increase of the conductivity at the SW. In this notation, the SW surface with the enhanced conductivity and the bottom of the ionosphere form a peculiar kind of capacitor which can be charged by chance. For example, as the aerial SW propagates through the thundercloud or dust cloud or the wave flank crosses them, then a portion of the charge can flow from the cloud to the wave surface. Assuming for the moment that the total charge captured from the cloud is about 20 C and considering the SW as a hemisphere with radius of 60 km, the average surface charge density has to be about $0.9\,nC/m^2$ which corresponds to the electric field 10^2 V/m. This value is close to the air breakdown threshold, 130–250 V/m, at the altitudes 60–70 km. So one might expect the generation of the electric discharges between the SW and the ionosphere such as BJs or so on. It was hypothesized that this kind of discharge can be initiated by a meteor-burst channel of ionization. Certainly, this is only the maximal estimate of the effect because the charge decreases continuously due to the atmospheric conductivity. In addition, the favorable circumstances such as appropriate meteor path are desirable to explain this exotic phenomenon.

Appendix I: Magnetic Perturbations Caused by Underground Detonation

High-Heated Plasma Ball Expanding in Ambient Magnetic Field

Let us consider an expanding homogeneous plasma ball immersed into the uniform magnetic field with induction $\mathbf{B}_0$ (Ablyazov et al. 1988). At the time $t = 0$ the ball radius begins to increase in accordance with the dependence $R(t) = R_0 \beta(t)$ where R_0 is the initial ball radius and $\beta(t)$ is a given function, which is equal to unity at $t = 0$. The plasma conductivity obeys the known law $\sigma_p = \sigma_p(t)$ as well. We assume that the plasma ball is surrounded by the non-magnetic rock ($\mu = 1$) whose conductivity is everywhere negligible compared with the plasma one. As the ball is situated at the depth which is much greater than the ball radius, one can neglect the influence of the atmosphere in calculating the field in the vicinity of the ball. In this approach the magnetic induction $\mathbf{B}$ in the plasma ball is described by the quasi-stationary Maxwell equations ($0 < r < R$)

$$\partial_t \mathbf{B} = \nabla \times (\mathbf{V} \times \mathbf{B}) + \frac{1}{\mu_0 \sigma_p} \nabla^2 \mathbf{B}, \qquad \nabla \cdot \mathbf{B} = 0, \qquad (11.65)$$

where $\mathbf{V}$ is the plasma velocity.

Since the rock conductivity is ignored, Maxwell equations outside the ball are given by $\nabla \times \mathbf{B} = 0$ and $\nabla \cdot \mathbf{B} = 0$. In this region we seek for the solution of these equations as a sum of the uniform field, $\mathbf{B}_0$, and of the field of effective magnetic dipole whose moment is proportional to $\mathbf{B}_0$. If the origin of spherical coordinate system is placed in the ball center, the solution of the problem can be represented as follows ($r > R$)

$$\mathbf{B} = B_0 \left[\cos\theta \left(\frac{2\chi R_0^3}{r^3} + 1 \right) \hat{\mathbf{r}} + \sin\theta \left(\frac{\chi R_0^3}{r^3} - 1 \right) \hat{\boldsymbol{\theta}} \right], \qquad (11.66)$$

where the angle θ is measured from the direction of $\mathbf{B}_0$ and $\hat{\mathbf{r}}$ and $\hat{\boldsymbol{\theta}}$ denote the unit vectors of spherical coordinate system. Here the dimensionless function $\chi(t)$ is related to the magnetic moment of the plasma ball through the following relationship $\mathbf{M}(t) = 4\pi \chi(t) R_0^3 \mathbf{B}_0 / \mu_0$.

The plasma ball is assumed to expand uniformly so that the plasma moves in radial directions. Consequently, the radius-vector of an elementary plasma volume can be written as $\mathbf{r} = \mathbf{r}_0 \beta(t)$, where $\mathbf{r}_0$ denotes the initial coordinate of the elementary volume. Whence it follows that the plasma velocity is given by: $\mathbf{V} = \mathbf{r}_0 d\beta/dt = \mathbf{r}(d\beta/dt)/\beta$. Substituting this expression into Eq. (11.65) and then transforming Euler's variables r, t to Lagrange's ones; that is to r_0, t, we come to

$$\partial_t \mathbf{B} + 2\mathbf{B}\frac{\dot{\beta}}{\beta} = \frac{1}{\mu_0 \sigma_p \beta^2} \nabla_{\mathbf{r}_0}^2 \mathbf{B}, \qquad \nabla_{\mathbf{r}_0} \cdot \mathbf{B} = 0, \tag{11.67}$$

where $\dot{\beta}$ denotes the time-derivative and the subscript $\mathbf{r}_0$ stands for derivatives with respect to Lagrange's variables. For simplicity, we will omit the subscript 0 having in mind that now $\mathbf{r}$ is the Lagrange's variable. We seek for the solution of Eq. (11.67) in the form

$$\mathbf{B} = \cos \theta B_1 (r,t) \hat{\mathbf{r}} - \sin \theta B_2 (r,t) \hat{\boldsymbol{\theta}}. \tag{11.68}$$

Substituting the Eq. (11.68) into Eq. (11.67) we come to the following equations for the unknown functions B_1 and B_2:

$$\begin{aligned}
\partial_t B_1 + 2B_1 \frac{\dot{\beta}}{\beta} &= \frac{1}{\mu_0 \sigma_p \beta^2} \left[B_1'' + \frac{2B_1'}{r} - \frac{4(B_1 - B_2)}{r^2} \right], \\
\partial_t B_2 + 2B_2 \frac{\dot{\beta}}{\beta} &= \frac{1}{\mu_0 \sigma_p \beta^2} \left[B_2'' + \frac{2B_2'}{r} + \frac{2(B_1 - B_2)}{r^2} \right], \\
B_1' + \frac{2}{r} (B_1 - B_2) &= 0.
\end{aligned} \tag{11.69}$$

Here the primes denote derivatives with respect to r.

Now we turn to new unknown dimensionless functions $f = (B_1 - B_2)/B_0$ and $g = (B_1 + 2B_2)/B_0 - 3$. These functions satisfy the new set of equations

$$\partial_t f + 2f \frac{\dot{\beta}}{\beta} = \frac{1}{\mu_0 \sigma_p \beta^2} \left(f'' + \frac{2f'}{r} - \frac{6f}{r^2} \right); \tag{11.70}$$

$$\partial_t g + 2 (g + 3) \frac{\dot{\beta}}{\beta} = \frac{1}{\mu_0 \sigma_p \beta^2} \left(g'' + \frac{2g'}{r} \right); \tag{11.71}$$

$$2f' + g' + 6f/r = 0. \tag{11.72}$$

Suppose at the initial time there is a uniform magnetic field $\mathbf{B}_0$ everywhere. The initial conditions for the functions f and g are as follows then

$$f (r,0) = g (r,0) = 0. \tag{11.73}$$

The normal and tangential components of $\mathbf{B}$ have to be continuous at the ball surface. Considering this requirement and combining Eqs. (11.66) and (11.68) we come to the following boundary conditions:

$$f (R_0,t) = 3\chi/\beta^3, \qquad g (R_0,t) = 0. \tag{11.74}$$

The set of Eqs. (11.70)–(11.72) with initial and boundary conditions (11.73) and (11.74) contains only two unknown function. However, we shall demonstrate that this set has a single solution. First of all we note that Eq. (11.72) can be solved for f under a requirement that f is finite when $r \to 0$

$$f = -\frac{1}{2r^3} \int_0^r r_1^3 \partial_{r_1} g\,(r_1, t)\, dr_1. \tag{11.75}$$

Now we prove that if g satisfies Eq. (11.71) then the function f given by Eq. (11.75) must satisfy Eq. (11.70). For this purpose we take the operator $\partial_t + 2\dot{\beta}/\beta$ in order to act on Eq. (11.75). Using Eq. (11.71) we find

$$\partial_t f + 2f \frac{\dot{\beta}}{\beta} = -\frac{1}{2\mu_0 \sigma_p \beta^2 r^3} \int_0^r r_1^3 \partial_{r_1} \left(g'' + \frac{2g'}{r_1} \right) dr_1. \tag{11.76}$$

Taking the integral several times by parts and applying Eq. (11.75) we can reduce Eq. (11.76) to the form which is identical with Eq. (11.70). If the solution of Eq. (11.71) under requirements by Eqs. (11.73) and (11.74) is found, then substituting this solution; that is the function g, in Eq. (11.75) gives the function f. Thus one can obtain the unique solution of the problem.

We expand this solution into a series

$$g\,(r, t) = \frac{1}{r} \sum_{n=1}^{\infty} \gamma_n\,(t) \sin \frac{\pi n r}{R_0}, \tag{11.77}$$

where the eigenfunctions $\sin(\pi n r / R_0)$ form a complete orthogonal system which satisfies the boundary conditions given by Eq. (11.74). The undetermined functions $\gamma_n\,(t)$ can be found by substituting Eq. (11.77) for g into Eq. (11.71). Taking into account the initial conditions given by Eq. (11.73), we obtain

$$\gamma_n\,(t) = \frac{(-1)^n\, 6R_0}{\pi n \beta^2\,(t)} \int_0^t \frac{d\beta^2}{dt'} \exp\left(-\int_{t'}^t \frac{\pi^2 n^2}{\mu_0 \sigma_p R_0^2 \beta^2} dt'' \right) dt'. \tag{11.78}$$

Substituting Eqs. (11.77) and (11.78) for g and γ_n into Eq. (11.75), and performing integration with respect to r, yield

$$f\,(r, t) = -\frac{R_0^2}{2\pi^2 r^3} \sum_{n=1}^{\infty} \frac{\gamma_n\,(t)}{n^2} \left\{ \frac{3\pi n r}{R_0} \cos \frac{\pi n r}{R_0} \right.$$
$$\left. + \left[\left(\frac{\pi n r}{R_0} \right)^2 - 3 \right] \sin \frac{\pi n r}{R_0} \right\}. \tag{11.79}$$

Combining Eqs. (11.74) and (11.79), we finally obtain the dimensionless magnetic moment of the plasma ball

$$\chi = -\frac{3\beta(t)}{\pi^2} \sum_{n=1}^{\infty} \frac{1}{n^2} \int_0^t \frac{d\beta^2}{dt'} \exp\left(-\int_{t'}^{t} \frac{\pi^2 n^2}{\mu_0 \sigma_p R_0^2 \beta^2} dt''\right) dt'. \tag{11.80}$$

Residual Magnetic Field

First of all consider Eq. (11.23) in the region $R_c < r < R_e$. All the values in Eq. (11.23) are independent of azimuthal angle due to the cylindrical symmetry of the problem. This implies that only azimuthal component of the curl is nonzero. Therefore, substituting of Eq. (9.29) for $\Delta \mathbf{J}$ into Eq. (11.23) gives

$$\frac{1}{r}\left[\partial_r (rB_\theta) - \partial_\theta B_r\right] = -\mu_0 C_m J d_r s_{rr} \sin\theta, \tag{11.81}$$

where B_r and B_θ are the radial and tangential components of magnetic field and d_r denotes derivative with respect to r, that is $d_r s_{rr} = ds_{rr}/dr$. Maxwell equation $\nabla \cdot \mathbf{B} = 0$ can be written in the form

$$\frac{1}{r^2}\partial_r \left(r^2 B_r\right) + \frac{1}{r\sin\theta}\partial_\theta (\sin\theta B_\theta) = 0. \tag{11.82}$$

We seek for the solution of Eqs. (11.81) and (11.82) in the form $B_r = B_1(r)\cos\theta$ and $B_\theta = B_2(r)\sin\theta$, where $B_1(r)$ and $B_2(r)$ are unknown functions. This yields

$$d_r (rB_2) + B_1 = -r\mu_0 C_m J d_r s_{rr}, \tag{11.83}$$

and

$$d_r \left(r^2 B_1\right) + 2r B_2 = 0. \tag{11.84}$$

For the inner and outside areas, i.e. at $r < R_c$ and $r > R_e$, the right-hand side of Eq. (11.83) is equal to zero whereas Eq. (11.84) is valid in the whole space. Integrating of Eq. (11.83) over short intervals $(R_c - \varepsilon, R_c + \varepsilon)$ and $(R_e - \varepsilon, R_e + \varepsilon)$, where $\varepsilon \longrightarrow 0$, gives the boundary conditions for tangential component B_2

$$B_2(R_c + 0) - B_2(R_c - 0) = -\mu_0 C_m J s_{rr}(R_c), \tag{11.85}$$

$$B_2(R_e + 0) - B_2(R_e - 0) = \mu_0 C_m J s_{rr}(R_e). \tag{11.86}$$

The continuity of normal component of the magnetic induction results in

$$B_1\left(R_c - 0\right) = B_1\left(R_c + 0\right), \quad B_1\left(R_e - 0\right) = B_1\left(R_e + 0\right), \tag{11.87}$$

Eliminating the function B_2 from the set of Eqs. (11.83) and (11.84), we obtain

$$r d_r^2 B_1 + 4 d_r B_1 = 2\mu_0 C_m J d_r s_{rr} \tag{11.88}$$

The general solution of Eq.(11.88) is given by

$$B_1 = \frac{2\mu_0 C_m J}{r^3} \int_{R_c}^{r} \left(r'\right)^2 s_{rr}\left(r'\right) dr' + \frac{c_1}{r^3} + c_2, \tag{11.89}$$

where c_1 and c_2 are arbitrary constants.

At the regions $r < R_c$ and $r > R_e$ the solution of problem should be limited as $r \longrightarrow 0$ and $r \longrightarrow \infty$. So, we obtain that $B_1 = c_3$ if $r < R_c$ and $B_1 = c_4/r^3$ if $r > R_e$ where c_3 and c_4 are the arbitrary constants. These solutions should fit Eq. (11.89) at the boundaries $r = R_c$ and $r = R_e$. Taking into account the boundary conditions given by Eqs. (11.85)–(11.87) one can find the constants $c_1 - c_4$. Whence it follows that $c_1 = c_2 = c_3 = 0$. So the magnetic field is equal to zero in the inner area at $r < R_c$. For the region $R_c < r < R_e$ one can find

$$B_r = \frac{2\mu_0 C_m J \cos\theta}{r^3} \int_{R_c}^{r} r'^2 s_{rr}\left(r'\right) dr', \tag{11.90}$$

$$B_\theta = \mu_0 C_m J \sin\theta \left(\frac{1}{r^3} \int_{R_c}^{r} r'^2 s_{rr}\left(r'\right) dr' - s_{rr}\left(r\right)\right). \tag{11.91}$$

Note that these formulas are more correct than that obtained by Surkov (1989) in the framework of simplified approach which leaves out of account the boundary conditions and thereby the contribution of the surface magnetization currents at $r = R_c$ and $r = R_e$.

References

Ablyazov MK, Surkov VV, Chernov AS (1988) Distortion of an external magnetic field by an expanding plasma sphere located in a slightly conductive semispace. J Appl Mech Tech Phys 29(6):778–784

Adushkin VV, Soloviev SP (1988) Low-frequency electric fields in the atmospheric surface layer during underground explosion. Rep USSR Acad Sci (Dokl Akad Nauk SSSR) 299(4):840–844 (in Russian)

Adushkin VV, Soloviev SP, Surkov VV (1990) Electrical field arising during ejection explosion. Combust Explosion Shock Waves 26(4):478–482

Adushkin VV, Soloviev SP (1996) Generation of low-frequency electric fields by explosion crater formation. J Geophys Res 101B:20165–20173

Allison FE (1965) Shock induced polarization in plastics. 1. Theory. J Appl Phys 36:2111–2113

Baker DM, Davies K (1968) Waves in the ionosphere produced by nuclear explosions. J Geophys Res 73:448–450

Baker DM, Cotten DE (1971) Interpretation of high-frequency Doppler observations of waves from nuclear and natural sources. J Geophys Res 76:1803–1810

Barry GH, Griffiths LJ, JC Taenzer (1966) HF radio measurements of high-altitude acoustic wave from a ground-level explosion. J. Geophys. Res 71:4173–4182

Barsukov OM, Skovorodkin YuP (1969) Magnetic observations in the region of Medeo detonation. Proc USSR Acad Sci Phys Earth (Izvestiya Akad Nauk SSSR Fizika Zemli) 5:68–69 (in Russian)

Baum FA, Orlenko LP, Stanyukovich KP, Chelyshev VP, Shekhter BI (1975) Physics of explosion. In: Stanyukovich KP. Nauka, Moscow, 704 pp (in Russian)

Blanc E (1984) Interaction of an acoustic wave of artificial origin with the ionosphere as observed by vertical HF sounding at total reflection levels. Radio Sci 19:653–664

Blanc E (1985) Observations in the upper atmosphere of infrasonic waves from natural or artificial sources: A summary. Ann Geophys. 3:673–688

Breitling WJ, Kupferman RA, Gassmann GJ (1967) Traveling ionospheric disturbances associated with nuclear detonation. J Geophys Res 72:307–315

Broche P (1977) Propagation des ondes acoustico-gravitationnells excitees par des explosions. Ann Geophys 33:281–288

Chadwick P, Cox AD, Hopkins HG (1964) Mechanics of deep underground explosion. Phil Trans Roy Soc Lond A256:235–300

Chalmers JA (1967) Atmospheric electricity, 2nd edn. Pergamon Press, New York

Colvin JD, Mitchell CK, Greig JR, Murphy DP, Pechacek RE, Raleigh M (1987) An empirical study of the nuclear explosion-induced lightning seen on IVY-MIKE. J Geophys Res 92D:5696–5712

Daniels FB, Bauer SJ, Harris AK (1960) Vertically traveling shock waves in the ionosphere. J Geophys Res 65:1848–1850

Danilov AV, Teselkin SF (1984) A structure of shock wave precursor in weakly ionized anisothermic magnetoactive plasma. Phys Plasma (Fizika Plazmy) 10(4):735–740 (in Russian)

Danilov AV, Dovzhenko VA (1987) On the excitation of electromagnetic fields by acoustic pulses entering the ionosphere. Geomagn Aeron (Geomagnetizm i Aeronomiya) 27(5):772–777 (in Russian)

Enstrom JE, Brode HL, Gilmore FR (1972) Shock propagation in exponential atmospheres. R & D Associates, RDATR042DNA, Santa Monica, California

Erzhanov ZhS, Kurskeev AK, Nysanbaev TE, Bushuev AV (1985) Geomagnetic observations during experiment MASSA. Proc USSR Acad Sci Phys Earth (Izvestiya Akad Nauk SSSR Fizika Zemli) 11:80–82 (in Russian)

Galperin YuI, Hayakawa M (1996) On the magnetospheric effects of experimental ground explosions observed from AUREOL-3. J Geomagn Geoelectr 48:1241–1263

Galperin YuI, Hayakawa M (1998) On a possibility of parametric amplifier in the stratosphere-mesosphere suggested by active MASSA experiments with the AUREOL-3 satellite. Earth Planets Space 1:253–258

Gassmann GJ (1963) Electron density profiles of wave-motions in the ionosphere caused by nuclear detonations, AFCRL Research report. Report US Department of Commerce AD626-694, pp 63–440

Gilinsky V (1965) Kompaneet's model for radio emission from a nuclear explosion. Phys Rev 137A(1):50–55

Gilinsky V, Peebls G (1968) The development of a radio signal from a nuclear explosion in the atmosphere. J Geophys Res 73(1):405–414

Gorbachev LP, Glushkov AI, Kotov YuB, Semenova TA, Skryl'nik AA (1999) Geomagn Aeron (Geomagnetizm i Aeronomiya), Estimate of camouflet explosion power using initial stage of geomagnetic perturbations 39(5):77–82 (in Russian)

Gorbachev LP, Kotov YuB, Semenova TA (1999) Generation of geomagnetic perturbations at late stage of camouflet explosion. Appl Mech Tech Phys (Prikladnaya Mekhanika i Tekhnicheskaya Fizika) 40(4):16–24 (in Russian)

Gorbachev LP, Semenova TA (2000a) Generation of geomagnetic perturbations caused by nuclear underground explosions under decoupling. Pt. 1. Mechanisms of generation. Eng Phys (Inzhenernaya Fizika) 2:66–72 (in Russian)

Gorbachev LP, Semenova TA (2000b) Generation of geomagnetic perturbations caused by nuclear underground explosions under decoupling. Pt. 2. Explosions in evaporated chambers filled by water vapors. Eng Phys (Inzhenernaya Fizika) 4:84–88 (in Russian)

Greifinger C, Greifinger P (1965) Transmission of micropulsations through the lower ionosphere. J Geophys Res 70:2217–2231

Grigor'iev VG, Nemirov AS, Sirotkin VK (1979) Structure of shock waves in elastic-plastic relaxation media. Appl Mech Tech Phys (Prikladnaya Mekhanika i Tekhnitseskaya Fizika) 1:153–160 (in Russian)

Grover MK (1981) Some analytical model for quasi-static source region EMP: Application to nuclear lightning. IEEE Trans Nucl Sci NS-28(1):990–994

Hasbrouk WP, Allen JH (1972) Quasi-static magnetic field changes associated with CANNIKIN nuclear explosions. Bull Seism Soc Am 62(6):1479–1480

Herbst RF, Werth GC, Springer DL (1961) Use of large cavities to reduce seismic waves from underground explosions. J Geophys Res 66(3):959–978

Hill RD (1973) Lightning induced by nuclear bursts. J Geophys Res 78:6355–6358

Holzer RE (1972) Atmospheric electrical effects of nuclear explosions. J Geophys Res 77:5845–5855

Hoppel WA (1967) Theory of electrode effect. J Atmosph Terr Phys 29:709–721

Ivanov KG (1961) Geomagnetic phenomena observed at Irkutsk magnetic observatory after detonation of Tunguska meteorite. Meteoritics 21:46–48 (in Russian)

Jacobson AR, Carlos RC, Blanc E (1988) Observation of ionospheric disturbances following a 5 kt chemical explosion. 1. Persistent oscillation in the lower thermosphere after shock passage. Radio Sci 23:820–830

Johnson MH, Lippman BA (1960) Electromagnetic signals from nuclear explosions in outer space. Phys Rev Ser II 119(3):827–828

Kanellakos DP, Nelson RA (1972) Comparison of computed and observed shock behavior from multikiloton near surface nuclear explosions. Effects of atmospheric acoustic gravity waves on electromagnetic wave propagation. AGARD Confer Proc CP 115(19):240–251

Karlov VD, Kozlov SI, Tkachev GN (1980) Large-scale perturbations in the ionosphere generated by rocket with engine-on flight. A review. Cosmic Res (Kosmicheskie issledovaniya) 18(2):266–277 (in Russian)

Karzas WJ, Latter R (1962a) The electromagnetic signal due to the interaction of nuclear explosions with Earth's magnetic field. J Geophys Res 67(12):4635–4640

Karzas WJ, Latter R (1962b) Electromagnetic radiation from nuclear explosions in space. Phys Rev 126(6):1919–1926

Karzas WJ, Latter R (1965) Detection of the electromagnetic radiation from nuclear explosions in space. Phys Rev 137B(5):1369–1378

Kompaneets AS (1958) Radioemission of atomic explosion. J Exp Theor Phys (Zhurnal eksperimental'noy i teoreticheskoy fiziki) 35(6)(12):1538–1544 (in Russian)

Kompaneets AS (1977) Radio-emission of atomic detonation. II. Physicochemical and relativistic gas dynamics, Selected papers, Nauka, Moscow, pp 83–91 (in Russian)

Kotadia KN (1967) Ionospheric effects of nuclear explosions. Ann Geophys 23:1–11

Landau LD, Lifshitz EM (1959) Fluid mechanics, Vol 6 of A course of theoretical physics. Pergamon Press, New York

Latter AL, LeLevier RE, Martinelly EA, McMillan WG (1961a) A method of concealing underground nuclear explosion. J Geophys Res 66(3):943–946

Latter R, Herbst RF, Watson KM (1961b) Detection of nuclear explosions. Ann Rev Nucl Sci 11:371–375

Lawrie JA, Gerard VB, Gill PJ (1961) Magnetic effects resulting from the Johnston Island high altitude nuclear explosions. N Z J Geology Geophys 4(2):109–124

Leypunskiy OI (1960) On possible magnetic effect under high-altitude detonations of atomic bombs. J Exp Theor Phys (Zhurnal eksperimental'noy i teoreticheskoy fiziki) 38(1):302–304 (in Russian)

Lifshitz EM, Pitaevskii LP (1981) Physical kinetics, course of theoretical physics, Vol 10, Tranls. by J. B. Sykes and R. N. Franklin, Oxford

Lomax JB, Nielson DL (1972) Nuclear weapon effects on the ionosphere (F-region disturbances). Effects of atmospheric acoustic gravity waves on electromagnetic wave propagation. AGARD Confer Proc CP 115(38):495–506

Malik J, Fitzhugh R, Hormuth F (1985) Electromagnetic signals from underground nuclear explosions. Report LA-10545-MS. Los Alamos Natl. Lab., Los Alamos, New Mexico

Marcos FA (1966) Aircraft induced ionospheric disturbances, Sci. Rep. Airforce Surveys in Geophys, vol 175, AFCRL-66-229, Bedford Massachusetts

Medvedev YuA, Stepanov BM, Fedorovich GV (1970) Electric field excited by gamma-quantum pulse in air. J Appl Mech Tech Phys (Zhurnal Prikladnoi Mekhaniki i Tekhnicheskoi Fiziki) 4:3–8 (in Russian)

Medvedev YuA, Fedorovich GV (1975) On the question about radiation of electromagnetic pulses. J Tech Phys (Zhurnal Tekhnicheskoi fiziki) 45(4):697–704 (in Russian)

Medvedev YuA, Stepanov BM, Fedorovich GV (1980) Physics of radiative excitation of electro-magnetic fields. Atomizdat, Moscow, 104 pp (in Russian)

Molchanov OA, Hayakawa M (2008) Seismo-electromagnetics and related phenomena: history and latest results. TERRAPUB, Tokyo, 189 pp

Najita K, Huang YN, Fang GTN (1975) Ionospheric disturbances detected over Hawaii after the 1968 French thermonuclear explosion. Ann Geophys 31:301–310

Orlov VV, Uralov AM (1984) The response of the atmosphere to weak surface explosion. Proc USSR Acad Sci Phys Atmos Ocean (Izvestiya Akademii Nauk SSSR, Fizika Atmosfery i Okeana) 6:476–483 (in Russian)

Patterson D (1966) Nuclear decoupling, full and partial. J Geophys Res 71:3427–3436

Pokhotelov OA, Pilipenko VA, Fedorov EN, Stenflo L, Shukla PK (1994) Induced electromagnetic turbulence in the ionosphere and the magnetosphere. Phys Scripta 50:600–605

Pokhotelov O, Parrot M, Fedorov E, Pilipenko V, Surkov V, Gladychev V (1995) Acoustic response of the ionosphere to natural and man-made sources. In: Console R, Nikolaev A (eds) Earthquakes induced by underground nuclear explosions (environmental and ecological problems). NATO ASI series 2/4, Springer, New York, pp 267–279

Rao GL (1972) Ionospheric disturbances caused by long period sound waves generated by Saturn-Apollo launches. Effects of atmospheric acoustic gravity waves on electromagnetic wave propagation. AGARD Confer Proc CP 115(26):329–337

Rodionov VN, Adushkin VV, Kostyuchenko VN, Nikolaevskiy VN, Romashev AN, Tsvetkov VM (1971) In: Sadovsky MA (ed) Mechanical effect of underground explosion. Nedra, Moscow (in Russian)

Sandmeier HA, Dupree SA, Hansen GE (1972) Electromagnetic pulse and time-dependent escape of neutrons and gamma rays from a nuclear explosion. Nucl Sci Eng 48(3):343–352

Semenov AS (1974) Electrical survey based on natural electric field. Nedra, Moscow, 200 pp (in Russian)

Soloviev SP, Surkov VV (1994) Electric perturbations in the atmospheric surface layer caused by an aerial shock wave. Combust Explosion Shock Waves 30(1):117–121

Soloviev SP, Surkov VV (2000) Electrostatic field and lightning generated in the gaseous dust cloud of explosive products. Geomagn Aeron 40(1):61–69

Soloviev SP, Surkov VV, Sweeney JJ (2002) Quadrupolar electromagnetic field from detonation of high explosive charges on the ground surface. J Geophys Res 107(B6):2119

Sorokin VM (2007) Plasma and electromagnetic effects in the ionosphere related to the dynamics of charged aerosols in the lower atmosphere. Russ J Phys Chem B 1(2):138–170

Stacey FD (1964) The seismomagnetic effect. Pure Appl Geophys 58:5–23

Stoffregen W (1962) Ionospheric effects observed in connection with nuclear explosions at Novaja Semlja on 23 and 30 October 1961, FOA3-Rept. A517. Res. Inst. Natl. defense electron. Dept., Stockholm, Sweden

Stoffregen W (1972) Traveling ionospheric disturbances initiated by low-altitude nuclear explosions. Effects of atmospheric acoustic gravity waves on electromagnetic wave propagation. AGARD Confer Proc CP 115(35):471–479

Straker EA (1971) The effect of the ground on the steady-state and time-dependent transport of neutrons and secondary gamma rays in the atmosphere. Nucl Sci Eng 46(3):334–355

Surkov VV (1986) Electromagnetic field produced by a shock wave propagating in a condensed medium. J Appl Mech Tech Phys 27(1):25–31

Surkov VV (1989) Local variations of geomagnetic and geoelectric fields caused by near-surface deformations. Izvestiya Acad Sci USSR Phys Solid Earth 25(5):421–424

Surkov VV (1990a) The propagation of geomagnetic pulsations in the E layer of the ionosphere. Geomagn Aeron 30(1):94–98

Surkov VV (1990b) Propagation characteristics of low-frequency electromagnetic signals due to disturbances in the E layer. Geomagn Aeron 30(5):682–686

Surkov VV (1992a) Excitation of low-frequency geomagnetic oscillations upon the propagation of a vertical acoustic wave in the E layer of the ionosphere. Geomagn Aeron 32(3):332–336

Surkov VV (1992b) Dispersion equations of low-frequency geomagnetic disturbances in the E-layer. Radiophys Quantum Electron 35(9–10):477–486

Surkov VV (1996) Front structure of the Alfvén wave radiated into the magnetosphere due to excitation of the ionospheric E layer. J Geophys Res 101(A7):15403–15409

Surkov VV (2000) Electromagnetic effects caused by earthquakes and explosions, MEPhI, Moscow, 448 pp (in Russian)

Surkov VV, Galperin YuI (2000) Electromagnetic impulse in magnetosphere resulted from impulse of electric current near lower boundary of ionosphere. Cosmic Investig 38(6):602–613

Sweeney JJ (1989) An investigation of the usefulness of extremely low-frequency electromagnetic measurements for treaty verification. Report UCRL-53899, Lawrence Livermore Natl. Lab., Livermore, CA

Sweeney JJ (1995) Low-frequency electromagnetic signals from underground explosions on site inspection: research progress report. Report UCRL-ID-122067. Lawrence Livermore Natl. Lab., Livermore, CA

Sweeney JJ (1996) Low-frequency electromagnetic measurements as a zero-time discriminant of nuclear and chemical explosions – OSI research final report. Report UCRL-ID-126780. Lawrence Livermore Natl. Lab., Livermore, CA

Troitskaya VA (1960) Effects of terrestrial currents caused by high-altitude atomic explosions. Proc USSR Acad Sci Ser Geophys (Izvestiya Akad Nauk SSSR Seriya Geofiziki) 9:1321–1327 (in Russian)

Uman MA, Seacord DF, Price GH, Pierce ET (1972) Lightning induced by thermonuclear detonations. J Geophys Res 77:1591–1596

Undzenkov BA, Shapiro VA (1967) Seismomagnetic effect at magnetite deposit. Proc USSR Acad Sci Phys Earth (Izvestiya Akad Nauk SSSR Fizika Zemli) 1:121–126 (in Russian)

Wåhlin L (1986) Atmospheric electrostatics. Research Studies Press, Wiley, New York, Chichester, Toronto, Brishbane, Singapore, 130 pp

Wait JR (1961) The electromagnetic fields of a horizontal dipole in the presence of a conducting half-space. Can J Phys 39(7):1017–1028

Wait JR (1970) Electromagnetic waves in stratified media, 2nd edn. Pergamon Press, Oxford

Wickersham AF (1970) Detection and analysis of propagating ion-acoustic waves in ionosphere. In: Davies K (ed) Phase and frequency instabilities in electromagnetic wave propagation, AGARD Confer Proc, vol 33. The Advisory Group for Aerospace Research and Development of NATO, pp 421–437

Williams ER, Cooke CM, Wright KA (1988) The role of electric space charge in nuclear lightning. J Geophys Res 93D:1679–1688

Whitham GB (1974) Linear and nonlinear waves. Wiley, New York

Zablocki CJ (1966) Electrical transient observed during underground nuclear explosions. J Geophys Res 71:3523–3542

Zeldovich YaB, Raizer YuP (1963) Physics of shock waves and high-temperature hydrodynamical phenomena. Fizmatgiz, Moscow, 632 pp (in Russian)

Index

A

Abnormal VHF wave propagation, 402
Antenna, 65, 122, 131, 279, 336, 353, 361,
  430, 444
Atmosphere, 17, 45, 57, 109, 148, 228, 277,
  325, 355, 396, 435, 436
Atmospheric electricity, 57–97
Auroral electrojet (AE), 33–36
Azimuthal harmonics, 216, 218

B

Blue jet (BJ), 73–97, 411
Blue starter, 73, 74
Boltzmann
  constant, 186, 343, 344
  transport equation, 88
Bow shock, 21–26, 238, 240
Breakdown
  air, 62–64, 82, 83, 88, 470
  runaway, 63, 64, 82, 85, 86, 88, 89

C

Cavity
  earth-ionosphere, 109–142
  mode, 36, 215, 223–224, 236, 238, 241
Charge
  density, 6, 61, 63, 77, 261, 342, 344–349,
    352, 358, 359, 361, 366, 378, 379, 404,
    457, 458, 470
  moment change, 79, 80, 94, 97, 128
Chemical high explosive (HE), 447
Co-seismic signal, 280, 287, 288, 331
Compression, 9, 12, 13, 16, 20, 30, 41, 49,
  54–59, 145, 151,152, 159, 161, 164,
  189–190, 192, 202, 211, 215, 216, 223,
  225, 292, 293, 330, 335, 338, 340, 341,
  345, 348–350, 354, 355, 362–366, 458,
  459
  uniaxial, 354
Conditions
  boundary, 22, 23, 83, 112–114, 118, 125,
    136, 154–156, 158, 162, 172, 173,
    195–197, 199, 202, 203, 213, 214,
    216, 217, 219, 220, 223–226, 245–247,
    264–267, 283, 291, 300, 308, 311, 315,
    325, 358, 359, 385, 398, 438, 449,
    462–464, 466, 472–475
  initial, 291, 292, 308, 312, 341, 363, 472,
    473
Conductivity
  atmospheric, 58, 59, 83, 111, 118, 157, 259
  field-aligned, 155
  ground, 65, 81, 83, 84, 157, 172, 278–301,
    308–315, 327, 376, 380, 381, 388, 398,
    445, 454, 456, 457, 460
  hall, 51, 53, 54, 112, 146, 155, 156,
    159–164, 171, 172, 202, 203, 219, 228,
    245, 465, 467
  ionospheric, 112, 165, 416
  parallel, 21, 51–54, 155, 173, 203, 210,
    219, 244
  pedersen, 53, 54, 155, 156, 160, 161, 216,
    219–221, 223, 228, 244
  plasma, 20, 21, 32, 47, 50–55, 77, 118, 146,
    148, 154, 155, 173, 189, 210, 216, 221,
    243, 244, 249, 251, 411, 435, 439, 440,
    466, 471
Conservation
  of energy, 7
  of momentum, 7, 459

V. Surkov and M. Hayakawa, *Ultra and Extremely Low Frequency Electromagnetic*
Fields, Springer Geophysics, DOI 10.1007/978-4-431-54367-1,
© Springer Japan 2014